OXIDATIVE STRESS

OXIDATIVE STRESS

ITS MECHANISMS AND IMPACTS ON HUMAN HEALTH AND DISEASE ONSET

HAROLD I. ZELIGER
Zeliger Research, LLC Cape Elizabeth, ME, United States

Academic Press is an imprint of Elsevier
125 London Wall, London EC2Y 5AS, United Kingdom
525 B Street, Suite 1650, San Diego, CA 92101, United States
50 Hampshire Street, 5th Floor, Cambridge, MA 02139, United States
The Boulevard, Langford Lane, Kidlington, Oxford OX5 1GB, United Kingdom

Notices
Knowledge and best practice in this field are constantly changing. As new research and experience broaden our understanding, changes in research methods, professional practices, or medical treatment may become necessary.

Practitioners and researchers must always rely on their own experience and knowledge in evaluating and using any information, methods, compounds, or experiments described herein. In using such information or methods they should be mindful of their own safety and the safety of others, including parties for whom they have a professional responsibility.

To the fullest extent of the law, neither the Publisher nor the authors, contributors, or editors, assume any liability for any injury and/or damage to persons or property as a matter of products liability, negligence or otherwise, or from any use or operation of any methods, products, instructions, or ideas contained in the material herein.

ISBN: 978-0-323-91890-9

For information on all Academic Press publications visit our website at https://www.elsevier.com/books-and-journals

Publisher: Stacy Masucci
Acquisitions Editor: Kattie Washington
Editorial Project Manager: Maria Elaine D. Desamero
Production Project Manager: Niranjan Bhaskaran
Cover Designer: Victoria Pearson

Typeset by TNQ Technologies

Dedication

To my wife Gail. Thank you for your encouragement and for asking the difficult questions whose answers made this book possible.

Contents

Preface xiii

I
Oxidative stress and disease

1. Introduction

1.1 Oxidative stress 3
1.2 Oxidative stress measuremnt 5
1.3 Major book topics 5
References 5

2. Chemicals and chemical mixtures

2.1 Historical perspective 7
2.2 Chemical characteristics—octanol: water partition coefficients 10
2.3 Lipophilic organic chemicals 11
2.4 Hydrophilic organic compounds 11
2.5 Metals 13
2.6 Nonmetallic inorganic chemicals 14
2.7 Chemical mixtures 15
2.8 Traditional toxicology 16
2.9 Chemical mixtures 17
2.10 Unanticipated effects of mixtures 19
2.11 Lipophile—hydrophile mixtures 21
2.12 Lipophile—lipophile mixtures 21
2.13 Sequential absorption 21
2.14 Diseases caused by chemicals 24
2.15 Chemically caused disease and oxidative stress 26
References 26

3. Particles and fibers

3.1 Introduction 29
3.2 Particle and fiber sources and content 29
3.3 Oxidative stress (OS) 29
3.4 Dust 33
References 34

4. Air pollution and oxidative stress

4.1 Introduction 37
4.2 Ambient air pollutants 37
4.3 Confined air pollutants 44
4.4 Air pollution and oxidative stress 44
References 46

5. Water and soil pollution

5.1 Introduction 47
5.2 Water polluting chemicals 47
5.3 Chemical reactions in groundwater 56
5.4 Disinfection bye-products 64
5.5 Soil pollution 65
5.6 Plant absorption of soil toxins and bioacculation 65
5.7 Effects of mixtures 66
5.8 Oxidative stress from polluted water and soil 67
References 68

6. Alcohol

6.1 Introduction 71
6.2 Effect of alcohol on nutrients and pharmaceuticals 72
6.3 Effect of alcohol on toxicities of other chemicals 72
6.4 Alcohol metabolism 73
6.5 Alcohol and oxidative stress 74
References 75

7. Tobacco

7.1 Tobacco toxicity: introduction 77
7.2 Tobacco and cancer 78
7.3 Tobacco smoke and disease rate synergism 78
7.4 Tobacco and oxidative stress 84
7.5 Electronic cigarettes 84
References 85

8. Electromagnetic radiation

8.1 Introduction 87
8.2 The electromagnetic spectrum 87
8.3 Ionizing radiation 88
8.4 Ionizing radiation and chemical mixtures 89
8.5 Ultraviolet radiation 89
8.6 Ultraviolet radiation and toxic chemical mixtures 90
8.7 Nonionizing radiation 92
8.8 Visible light radiation 95
8.9 Radiation and oxidative stress 95
References 96

9. Inflammation

9.1 Introduction 101
9.2 Acute inflammation 101
9.3 Chronic inflammation 102
9.4 Inflammation and oxidative stress 106
References 107

10. Food

10.1 Introduction 111
10.2 Uptake from soil and plant surfaces 111
10.3 Animal ingestion 112
10.4 Persistent organic pollutants (POPs) 112
10.5 Mercury in food 113
10.6 Food preparation. Deliberate incorporation of additive chemicals into food 114
10.7 Chemical impurities in food - allowable xenobiotics 115
10.8 Allowable chemical impurities in food 115
10.9 Artificial colors 116
10.10 Flavor enhancers 120
10.11 Esthetic and storage additives 121
10.12 Volatile organic compounds (VOCs) in food 121
10.13 Chemicals in food packaging 121
10.14 Irradiated food 127
10.15 Chemical preservatives in food 128
10.16 Food chemicals and oxidative stress 128
10.17 Dietary choices and oxidative stress 128
References 131

11. Sleep deprivation

11.1 Introduction 137
11.2 Temperature extremes 137
11.3 Sleep apnea 138
11.4 Circadian cycle interruption 139
11.5 Prevalent health conditions and lifestyle choices 139
11.6 Sleep deprivation and oxidative stress 139
References 140

12. Pharmaceuticals

12.1 Introduction 143
12.2 Pharmaceutical use 143
12.3 Pharmaceutical adverse drug reactions 144
12.4 Excipients in pharmaceuticals 146
12.5 Pharmaceuticals and oxidative stress 150
References 150

13. Psychological stress

13.1 Introduction 153
13.2 Chronic stress and disease 154
13.3 Psychological stress and oxidative stress 157
References 157

14. Genetics and epigenetics

14.1 Introduction 159
14.2 Genetics 159
14.3 Epigenetics 161
14.4 Chemical environmental and other factors in epigenetic effects 162
14.5 Role of oxidative stress in genetic and epigenetic effects 163
References 164

15. Aging

15.1 Introduction 167
15.2 Clinical conditions associated with aging 167
15.3 Lowering age of disease onset 168
15.4 Age-related disease and oxidative stress 168
15.5 Hallmarks of aging 170
15.6 Summary 173
References 173

16. Diseases and comorbidities

16.1 Introduction 177
16.2 Systems, organs, oxidative stress and disease 184
16.3 Diseases associated with oxidative stress 184
16.4 Multimorbidity 188
16.5 Conclusion 190
References 190

17. Total oxidative stress and disease

17.1 Introduction 195
17.2 Diseases and oxidative stress 195
17.3 Malondialdehyde as an indicator of oxidative stress 195
17.4 Oxidative stress additivity - infectious disease 202
17.5 Multimorbidity 202
17.6 Disease prevention strategies 203
17.7 Obesity 205
17.8 Limits to oxidative stress-mediated disease prevention 205
17.9 Disease onset prediction 206
17.10 Conclusions 206
References 207

18. Free radicals

18.1 Introduction 211
18.2 Free radical stability 212
18.3 The Fenton reaction 215
18.4 Free radical reactions with DNA 217
18.5 Free radical reactions with proteins 217
18.6 Free radical reactions with lipids 219
18.7 Free radical signaling 220
18.8 Diseases associated with free radicals 220
18.9 Antioxidants 220
18.10 Immune system and free radicals 220
18.11 Hypothalamus-pituitary-adrenal (HPA) axis 223
18.12 Summary 225
References 225

II

Mechanisms of oxidative stress driven disease

19. Aging mechanism

19.1 Introduction 229
19.2 Hallmarks of aging 229
19.3 Summary 234
References 235

20. Obesity

20.1 Introduction 239
20.2 Statistics 239
20.3 Causes of obesity 240
20.4 Biomarkers of obesity 240
20.5 Comorbidities 240
20.6 Obesity, adipose tissue and chemical toxicity 241
20.7 Obesity and oxidative stress 244
20.8 Psychological impact of obesity 245
20.9 Summary 245
References 245

21. Cancer

21.1 Introduction 249
21.2 Hallmarks of cancer—Hanahan and Weinberg 252
21.3 Other hallmarks 254
21.4 Cancer initiation promotion and progression 256
21.5 Exogenous carcinogens 258
21.6 Metals, metalloids and cancer 260
21.7 Food and cancer 260

21.8 Metastasis 261
21.9 Mechanisms associated with specific cancers 262
21.10 Summary 278
References 278

22. Atherosclerosis

22.1 Introduction 285
22.2 Hallmark of atherosclerosis 285
22.3 Atherosclerosis risk factors 285
22.4 Atherosclerosis progression 285
22.5 Mechanism of onset 286
22.6 Oxidative stress and Artherosclerosis 287
References 288

23. Alzheimer's disease

23.1 Introduction 291
23.2 Hallmarks of Alzheimer's disease 291
23.3 Risk factors for Alzheimer's disease 292
23.4 Alzheimer's disease co-morbidity data 294
23.5 Mechanisms of Alzheimer's disease onset 294
References 295

24. Type 2 diabetes (T2D)

24.1 Introduction 299
24.2 Risk factors for type 2 diabetes 300
24.3 Chemical exposure as a cause of type 2 diabetes 301
24.4 Mechanisms of type 2 diabetes onset 307
24.5 Effects of ROS 308
24.6 Type 2 diabetes and oxidative stress 308
24.7 Type 2 diabetes complications 308
24.8 Diabetes outlook 309
References 311

25. Rheumatoid arthritis

25.1 Introduction 317
25.2 Complications of rheumatoid arthritis 318
25.3 Biomarkers of rheumatoid arthritis 318
25.4 Risk factors for rheumatoid arthritis 318
25.5 Mechanisms of rheumatoid arthritis onset 322
25.6 Oxidative stress in rheumatoid arthritis 323
References 324

26. Asthma

26.1 Introduction 329
26.2 Phenotypes of asthma 329
26.3 Biomarkers of asthma 334
26.4 Risk factors for asthma 335
26.5 Cross sensitization 339
26.6 Mechanisms of asthma and oxidative stress 339
26.7 Summary 340
References 340

27. Liver cirrhosis

27.1 Introduction 345
27.2 Liver fibroproliferative diseases 345
27.3 Oxidative stress and liver disease 351
References 351

28. Chronic kidney disease (CKD)

28.1 Introduction 353
28.2 Causes of chronic kidney disease 353
28.3 Hallmarks of chronic kidney disease 353
28.4 Risk factors for chronic kidney disease 354
28.5 Mechanisms of chronic kidney disease and oxidative stress 354
References 357

29. Disease comorbidities

29.1 Introduction 359
29.2 Comorbidity 359
29.3 Oxidative stress, the common thread 362
References 364

III

Disease onset prediction

30. Oxidative stress index

30.1 Introduction 371
30.2 Oxidative Stress Index (OSI) 371
30.3 OSI basis 373
30.4 Oxidative Stress Index questionnaire 373

30.5 OSI score and disease 377
30.6 Additional applications 389
References 389

31. OSI condensed questionnaire

31.1 Introduction 391
31.2 Age 392
31.3 Weight 392
31.4 Preexisting chronic conditions 392
31.5 Medications regularly taken 393
31.6 Genetics 393
31.7 Education level 393
31.8 Place of residence 394
31.9 Chronic psychological stress 396
31.10 Condensed OSI form 397
31.11 Condensed OSI applications 399
31.12 Conclusions 399
References 399

32. OSI and Alzheimer's disease

32.1 Introduction 401
32.2 Alzheimer's disease and oxidative stress 401
32.3 Alzheimer's disease and dose response relationship (DRR) 402
32.4 Parameters known to increase likelihood of AD onset 402
32.5 Late onset AD 402
32.6 The OSI and AD 406
32.7 AD prevention 411
32.8 Questionnaire use 413
32.9 Strengths and limitations 413
32.10 Conclusions 414
References 414

33. OSI public health surveys

33.1 Introduction 419
33.2 Toxicity as a function of emission distance 420
33.3 Disease impact as a function of exposure time 422
33.4 Summary 428
References 428

34. Predicting COVID-19 severity

34.1 Introduction 431
34.2 Oxidative stress and COVID-19 severity 431
34.3 Oxidative stress index and COVID-19 severity 433
34.4 Validation of OSI use in predicting COVID-19 severity 435
34.5 Covid-19 as a cause of oxidative stress 437
34.6 Summary 438
References 438

IV

Prevention

35. Disease prevention: oxidative stress control, antioxidants, and social factors

35.1 Introduction 443
35.2 Oxidative stress overproduction control 443
35.3 Antioxidants 444
35.4 Antioxidants and disease prevention 445
35.5 Diet 445
35.6 Treating prevalent illness 449
35.7 Psychological stress 449
35.8 Exercise 452
35.9 Treating illness 453
35.10 Inequalities 453
35.11 Summary 454
References 454

36. Global warming, oxidative stress and disease

36.1 Introduction 457
36.2 Global warming, oxidative stress 458
36.3 Stress 467
36.4 Human extinction? 467
References 470

Index 473

Preface

Oxidative stress is well established as both the cause and consequence of all diseases. Accordingly, one's level of oxidative stress can serve as a predictor of the likelihood of disease onset. Factors that raise oxidative stress include genetics, lifestyles, preexisting conditions, and environmental exposures. It is total elevated oxidative stress, be it caused by a single parameter or a combination of parameters that is indicative of the likelihood of disease onset.

Numerous biomarkers can serve as indicators of oxidative stress levels, but all involve invasive procedures and each can vary widely from day to day and even hourly depending upon environmental factors. Recently, the Oxidative Stress Index (OSI), a more reliable method of measuring oxidative stress, has been developed. The OSI is based upon a noninvasive questionnaire that takes all oxidative stress raising factors into account, is not influenced by short-term varying effects, and can be rapidly completed.

This book examines the known causes of oxidative stress, disease-causing mechanisms, its remediation, and the use of OSI scores in predicting disease onset likelihood and severity.

Harold I. Zeliger
Cape Elizabeth, Maine
September, 2022

PART I

Oxidative stress and disease

CHAPTER

1

Introduction

1.1 Oxidative stress

Reactive oxygen species (ROS) and reactive nitrogen species (RNS) are naturally produced endogenously in the body as by-products of cellular metabolic activities. These are highly reactive, free radical intermediates whose production can occur by molecules losing or gaining single electrons. ROS and RON have beneficial effects that include protection against invading pathogens, wound healing and tissue repair and also act as essential signaling molecules Overproduction of these, however, results in an imbalance known as oxidative stress (OS) (Bhattacharyya et al., 2014). In homeostasis, excess ROS/RON quantities are neutralized by natural antioxidant defenses and OS is prevented from obtaining (Bhattacharyya et al., 2014; Liguori et al., 2018).

The term oxidative stress was coined about 35 years ago and OS continues to be extensively studied Breitenbach and Eckl (2015); Seis (2015). To date, approximately a quarter of a million entries for it appear in PubMed.

The imbalance of free radicals in OS leads to accelerated aging and the onset of numerous diseases. OS is both a cause and a consequence of disease and is associated with all disease (Liguori et al., 2018). Absorption of all exogenous chemicals results in the OS elevation. Though chemical exposures are a primary cause of OS (Zeliger, 2011), there are, however, many other causes. These include: age, lifestyle choices (diet, smoking and drug use), environmental exposures, pre-existing chronic disease and chronic psychological stress (Zeliger, 2016).

OS is a direct cause of environmental (non-communicative) diseases as well as an indirect cause of communicative diseases via its impact on the immune system (Akaike, 2001). For disease to occur, critical levels of OS, which interrupt body homeostasis, must be reached. Such levels can result from single sources or combinations of two or more sources, as it is total oxidative stress that is the determining factor for disease onset (Zeliger and Lipinski 2015; Zeliger, 2016).

Causes of oxidative stress include both endogenous and exogenous parameters. These are listed in Table 1.1.

Oxidative Stress
https://doi.org/10.1016/B978-0-323-91890-9.00031-3

TABLE 1.1 Exogenous and endogenous causes of oxidative stress.

EXOGENOUS
Chemicals
Fibers and particles
Air pollution
Water and soil pollution
Radiation
Tobacco
Alcohol
Pharmaceutical drugs
Dietary choices
Inflammation
Trauma
Extreme temperature exposure
Pathogens
Endogenous
Birth defects and abnormalities
Genetics and epigenetics
Pre-existing diseases and disease symptoms
Obesity
Psychological stress
Sleep depravation
Immune system response to chronic insult
Metabolic and elimination system responses to exogenous agents
Aging

Disease onset is dose-response related and may occur following acute exposure to a high dose of a single OS elevating factor, such as a chemical, or after long term chronic exposure to single or multiple OS raising factors, such as the presence of multiple diseases. This book examines these known causes of OS, the mechanisms of its onset and disease induction, OS measurement, prediction of the likelihood of OS-caused disease onset and OS reducing strategies that lower disease onset likelihood and severity.

1.2 Oxidative stress measuremnt

OS levels in the body have traditionally been determined via biomarkers in serum (Nielsen et al., 1997). Recently, however, the Oxidative Stress Index (OSI), a non-invasive protocol, based on a questionnaire, has been developed that produces similar results. The OSI has been shown to predict the likelihood of disease onset, identify primary causes of specific diseases, demonstrate community health effects attributable to environmental chemical releases and predict the severity of infectious disease (Zeliger, 2019, 2020).

1.3 Major book topics

Major book topics include:

- Chemical causes of OS and disease
- Non-chemical causes of OS and disease
- Dose response relationship between OS and disease
- Mechanisms of disease induction
- Free radical formation and stability
- Antioxidant action in homeostasis
- Immune system and gut microbiome roles in oxidative stress and disease
- Spiraling disease
- Disease co-morbidities
- Oxidative stress measurement
- The Oxidative Stress Index as a predictor of disease likelihood, disease severity, disease clusters and other applications
- Disease prevention

References

Akaike, T., 2001. Role of free radicals in viral pathogenesis and mutation. Rev. Med. Virol. 11 (2), 87–101.

Bhattacharyya, A., Chattopadhyay, R., Mitra, S., Crowe, S.E., 2014. Physiol. Rev. 94 (2), 329–354.

Breitenbach, M., Eckl, P., 2015. Introduction to oxidative stress in biomedical and biological research. Biomolecules 5, 1169–1177.

Liguori, I., Russo, G., Curcio, F., Bulli, G., Aran, L., Della-Morte, D., et al., 2018. Oxidative stress, aging and diseases. Clin. Interv. Aging 13, 757–772.

Nielsen, F., Mikkelsen, B.B., Niesen, J.B., Andersen, H.R., Grandjean, P., 1997. Plasma malondialdehyde as biomarker for oxidative stress: reference interval and effects of lifestyle factors. Clin. Chem. 43 (7), 1209–1214.

Seis, H., 2015. Oxidative stress: a concept in redox biology and medicine. Redox Biol. 4, 180–183.

Zeliger, H.I., 2011. Human Toxicology of Chemical Mixtures, second ed. Elsevier, Oxford.

Zeliger, H.I., Lipinski, B., 2015. Physiochemical basis of human degenerative disease. Interdiscipl. Toxicol. 8 (1), 15–21.

Zeliger, H.I., 2016. Predicting disease onset in clinically healthy people. Interdiscipl. Toxicol. 9 (2), 39–54.

Zeliger, H.I., 2019. Oxidative Stress Index: disease onset prediction and prevention. EC Pharm. Toxicol. 7 (9), 1022–1036.

Zeliger, H.I., 2020. Oxidative Stress Index (OSI) condensed questionnaire. Euro. J. Med. Health Sci. 2 (1). https://doi.org/10.24018/ejmed.2020.2.1.163. (Accessed 5 February 2021).

CHAPTER

2

Chemicals and chemical mixtures

2.1 Historical perspective

Chemicals and chemical mixtures are the primary exogenous causes of oxidative stress and hence, noncommunicative disease. The following provides an historical perspective on the ever-growing effect of chemical exposures on human health.

Before 1828, it was believed that organic chemicals could only be formed under the influence of the Vital Force in the bodies of animals and plants. Vital Force, also referred to as vital spark, energy and soul, is a tradition in all cultures, including Eastern as well as Western ones. Until 1828, this vitalism, and only it, was believed to be responsible for all factors affecting life, including the synthesis of all organic molecules. It was inconceivable that man could create such material. In 1828, Friedrich Wohler accomplished the first synthesis of urea, a naturally occurring component of human urine. Once it was demonstrated that such synthesis was possible, chemists were freed to pursue other such work, and since then, numerous other naturally occurring compounds have been synthetically prepared. Organic synthesis, however has not limited itself to duplicating nature. Hundreds of thousands of new, previously unknown to nature, chemicals have been synthesized.

The Industrial revolution led to exponential growth in the production, synthesis and use of both natural and man-made chemicals. The following timeline is an indicator of the growth of chemical use that resulted from the Industrial Revolution (Zeliger, 2019).

- 1746. Introduction of the Chamber Process for the large-scale production of sulfuric acid, which until this day is the chemical produced in greatest volume with the exception of water. This achievement in many ways triggered the chemical revolution.
- 1824. The first man made carbon containing chemical, oxalic acid, a compound still in use to this day for rust removal, as a rat poison and other applications. This was a huge advance in chemistry, for it introduced the notion that carbon-containing compounds, which are the back bones of the chemical revolution could be made by man.
- 1828. The first synthesis of urea, a component of urine. This represented a severe blow to vitalism, a belief at the time that organic chemicals (chemicals from

Oxidative Stress
https://doi.org/10.1016/B978-0-323-91890-9.00001-5

living things) had a vital force and could only be made by biological sources. This discovery opened the door to the chemical revolution which ensued.

- 1856. Production of the first synthetic dyes to replace those derived from plants. This ultimately led to the large-scale introduction of cancer causing azo dyes that have been also been associated with hyperactivity in children. Though initially used in textiles, these compounds are still used today as colorants in foods.
- 1856. The synthesis of Parkesine, the world's first man-made plastic.
- 1864. First production of chlorine. The Chloralkali process was ultimately introduced in the in 1890s for the large-scale production of chlorine. This, in turn, led to wide scale water disinfection, and the production of pesticides, but also resulted in the release of toxic mercury into rivers, streams, lakes and the ocean.
- 1872. The first synthesis of PVC (polyvinylchloride) now widely used in countless plastic applications, including piping, plastic pails and shower curtains.
- 1873. The synthesis of Acetaminophen. First used medically in 1893 as a replacement for aspirin. It went on sale in the United States in 1955 as "Tylenol."
- 1874. The first synthesis of the insecticide DDT. Its pesticide properties were discovered in 1939 and its discoverer given a Nobel Prize in 1948. DDT was banned in the United States following the publication of Rachel Carson's book, "Silent Spring" in 1962.
- 1907. Bakelite plastic introduced, making possible the manufacture of strong structural components.
- 1909. The introduction of the Haber Process to convert atmospheric nitrogen into ammonia ultimately used for the production of synthetic fertilizers. It was also used to produce nitric acid, a precursor to the manufacture of munitions.
- 1911. Arsphenamine, the first man made antibiotic was developed.
- 1927. The introduction of nylon, a synthetic replacement for silk.
- 1930. Polystyrene invented. Widely known as the styrofoam used in packaging material and as insulated food containers.
- 1931. PCBs (polychlorinated biphenyls) introduced into the marketplace for applications in electrical insulators, adhesives, paints, resins and numerous other applications. These known cancer-causing chemicals were banned in 1979, but still are ubiquitous almost everywhere in the world as they are carried by winds, ocean currents and through the food chains of plants and animals.
- 1933. Polyethylene introduced. Widely used as plastic sheeting and in packaging containers.
- 1941. PET (polyethylene terephthalate) plastic introduced. Now the world's most widely used plastic beverage container material.

Each new chemical added to our environment potentially creates a vast number of new chemical mixtures with unknown health consequences. The number of compounds is multiplied by chemical reactions of newly released compounds with existing released compounds as well as with naturally occurring species to create yet more toxic molecules. An example of this is photochemical smog formation from the reaction of nitrogen oxides with volatile organic compounds in the presence of sunlight. The Earth's flora and fauna, including humans, are guinea pigs who are afflicted by the myriad number of mixtures that are thus produced with the results of these multiple exposures often only becoming evident after people are stricken, all too often with unknown precise causes.

Research into the toxic effects of single chemicals often produce conflicting results when investigators fail to consider the presence of species other than the ones being studied. For example, different effects have been reported following the inhalation of formaldehyde when it was admixed with other chemicals (Alexandersson et al., 1982).

There are numerous sources of both indoor and outdoor environmental chemicals as listed in Table 2.1.

After chemicals, the second major contributor to increasing disease prevalence brought about by the Industrial Revolution is the greatly increased energy production required to power it. The first large scale energy production fuel was coal. With the discovery of oil and natural gas reserves in multiple locations around the world, petrochemicals soon eclipsed coal as the primary energy fuels. The amount of energy produced by burning of these fossil fuels increased by more than 10,000-fold from 1900 to 2020, accompanied by a corresponding increase in natural and synthetic chemicals released into our environment.

Human population growth has also been a major contributor to chemical releases into the environment. The quantities of fertilizers and pesticides needed for the farming necessary to feed the ever-growing human population are major contributors to the toxic chemical load. These insecticides, herbicides and fungicides, often used in combination, result in massive releases of persistent organic pollutants into the air water and soil, from which they are absorbed and passed up the food chains to affect most life forms. At the same time, global warming has

TABLE 2.1 Sources of environmental chemicals.

Organic solvents
Industrial chemicals
Mining dusts and leachates
Manufacturing processes
Electrical and transportation power production
Paints
Adhesives
Fertilizers
Pesticides
Food additives
Cosmetics and personal care products
Cleaners, deodorants and disinfectants
Building materials
Home furnishings

resulted in warming oceans, rising sea levels and sharp increases in violent storms that distribute these chemicals to all parts of the globe. Increased levels of the prevalence of numerous diseases have accompanied the increased production of synthetic chemicals, pesticide use, air and water pollution and energy production. Indeed, a plot of these parameters versus time from 1945 to 2015 produces the same hyperbolic curve shown in Fig. 2.1.

2.2 Chemical characteristics—octanol: water partition coefficients

Environmental exposures to all chemicals elevate OS as these attack body organs and cause disease. These attacks, in turn, trigger the body's metabolic, elimination and immune system responses to their presence, thus compounding the increase in OS.

OS raising chemicals fall into five categories. These include:

- Lipophilic organic chemicals
- Hydrophilic organic chemicals
- Transition metal ions
- Nonmetallic inorganic compounds
- Mixtures of these

Organic chemicals are characterized as being lipophilic or hydrophilic. Lipophilic chemicals are species with low polarity, and are more soluble in nonpolar solvents than in hydrophilic ones. Hydrophilic chemicals are those with higher polarity that are more soluble in water than in lipophilic solvents.

Octanol:Water partition coefficients of chemicals (K_{ow}) are indicative of the polarity characteristics of molecules. K_{ow} is the logarithm of the ratio of a chemical that dissolves in the

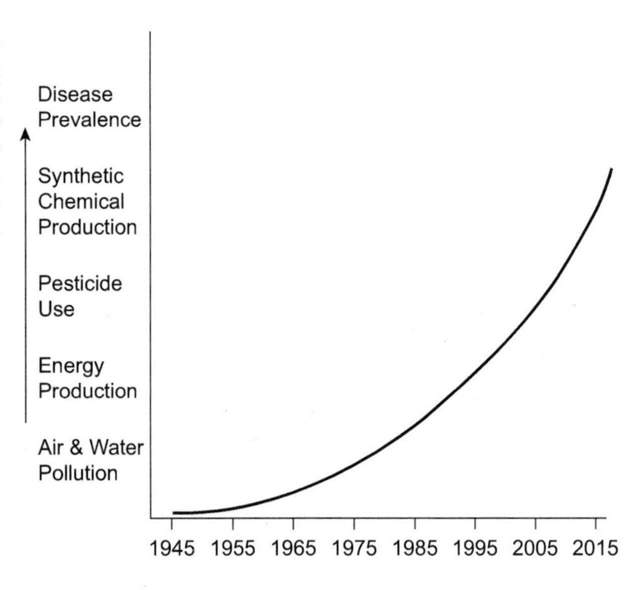

FIG. 2.1 Relationship between disease prevalence, synthetic chemical production, pesticide use, worldwide energy production from fossil fuel production and increased air and water pollution from 1945–2015. Values for all parameters follow the same hyperbolic curve. *Reproduced with permission from Zeliger HI. Predicting disease onset in clinically healthy people. Interdiscp Toxicol 2016;9(2):39–54.*

octanol phase of a octanol:water mixture. K_{ow} numbers have no units and range from −1.0 to greater than 6.0 for most organic compounds (Sangster, 1989; Zeliger, 2003, 2011).

2.3 Lipophilic organic chemicals

Lipophilic organic chemicals are those with K_{ow} values of equal to or greater than 2.0. As a rule, K_{ow} values increase with molecular weight in homologous series of chemicals. The K_{ow} values for n-alcohols in Table 2.2 are illustrative of this.

Lipophilic compounds are readily absorbed through the lipophilic cell membranes that enclose body cells. These include low molecular weight aliphatic and aromatic hydrocarbons (LMWHCs), polynuclear aromatic hydrocarbons (PAHs) and persistent organic pollutants (POPs). POPs, which absorb into white adipose tissue (WAT), are long-lasting in the body and responsible for multiple illnesses (Neuberger et al., 1998; Perez-Maldonado et al., 2005; Ha et al., 2007; Michalowicz et al., 2013; Harada et al., 2016; Maheshwari et al., 2019). These include organo-chlorine pesticides (OCs), polychlorinated biphenyls (PCBs), polybrominated diphenyl ethers (PBDEs), dioxins furans. Table 2.3 lists representative examples of LMWHCs, PAHs and POPs. Table 2.4 lists K_{ow} values for representative lipophilic compounds (Zeliger, 2011).

2.4 Hydrophilic organic compounds

Hydrophilic organic chemicals are those with K_{ow} values of less than 2.0. These water-soluble compounds are not readily absorbed through cell membranes. Hydrophilic compounds include low molecular weight alcohols, aldehydes, ketones, esters, amines, sulfides

TABLE 2.2 K_{ow} values for a homologous series of n-alcohols.

Alcohol	K_{ow}
Methanol	−0.77
Ethanol	−1.31
n-propanol	0.25
n-butanol	0.88
n-pentanol	1.51
n-hexanol	2.03
n-heptanol	2.62
n-octanol	3.00

TABLE 2.3 Representative lipophilic organic compounds.

Low molecular weight hydrocarbons
C-1 to C-8 hydrocarbons
Benzene
Ethyl benzene
Styrene
Toluene
Xylenes
Cumene
Polynuclear aromatic hydrocarbons
Naphthalene
Acenaphthene
Fluorine
Phenanthrene
Anthracene
Pyrene
Benzo[a]anthracene
Chrysene
Persistent or organic pollutants
DDT
PCBs
Dioxins
Furans
PBDEs
Dieldrin
Aldrin
Endosulfan

and inorganic compounds Table 2.5 lists representative hydrophilic chemicals and their K_{ow} values (Zeliger, 2011).

TABLE 2.4 Representative lipophilic organic compounds and their K_{ow} values.

Chemical	K_{ow}
Amyl acetate	2.26
Atrazine	2.61
Benzene	2.13
Benzophenone	3.18
Butylated hydroxytoluene	5.10
Cyclohexane	3.44
DDT	6.79
n-decane	5.01
Endosulfan	2.81
Ethyl benzene	3.15
n-heptane	4.66
n-hexane	3.90
Isobutene	2.76
Naphthalene	3.30
Pentachlorophenol	4.89
Styrene	2.95
2,3,7,8-Tetrachlorodibenzo-p-dioxin (TCDD)	6.64
Tetrachloroethylene	3.40
Toluene	2.73
1,1,1-Trichloroethane	2.49
Xylene	3.15

2.5 Metals

All transition metals are toxic. Some, however, are essential for homeostasis in small quantities. Several heavy metals, which have no essential biological function, but are stored in the body, are extremely toxic (Jaishankar et al., 2014; Hunter, 2015; Maret, 2016). Table 2.6 lists these two groups.

TABLE 2.5 Hydrophilic organic compounds and their K_{ow} values.

Chemical	K_{ow}
Acetaldehyde	−0.34
Acetic acid	−0.17
Acetone	−0.24
Aldicarb	1.13
n-butanol	0.88
2-Butoxyethanol	0.83
Carbon disulfide	1.94
Chloroform	1.97
Di-chloroaniline	0.94
2,4-Dichlorophenoxyacetic acid (2,4-D)	0.65
Dimethylformamide	−1.01
Epichlorohydrin	0.45
Ethanol	−0.31
2-Ethoxyethylacetate	0.59
Ethyl acetate	0.73
Ethylene glycol	−1.36
Ethylene oxide	−0.30
Methanol	−0.77
Methylene chloride	1.25
Methylethyl ketone	0.29
Methylisobutyl ketone	1.19
i-propanol	0.05
n-propyl acetate	1.24
Vinyl acetate	0.73
Vinyl chloride	1.62

2.6 Nonmetallic inorganic chemicals

All nonmetallic inorganic compounds are hydrophilic. Examples of these are shown in Table 2.7.

TABLE 2.6 Essential and highly toxic metals.

Essential	Highly toxic
Iron	Cadmium
Copper	Chromium
Cobalt	Lead
Molybdenum	Mercury
Selenium[a]	Arsenic[a]

[a]*Selenium and arsenic are metalloid elements which exhibit both metallic and nonmetallic properties.*

TABLE 2.7 Nonmetallic inorganic compounds and their K_{ow} values.

Chemical	K_{ow}
Ammonia	−1.38
Dimethyl sulfoxide	−1.35
Formaldehyde	0.35
Hydrofluoric acid	0.23
Hydrogen cyanide	−0.25
Nitric acid	0.21

2.7 Chemical mixtures

For single chemical exposures, it is well known that most individuals are affected by very high concentrations. Individuals who are genetically predisposed and/or have been previously sensitized react to lower concentrations of a chemical. Effects at different concentration levels for single chemical exposures are, for the most part, known and predictable, enabling proper precautions to be taken.

Exposures to mixtures of chemicals produce effects that are, largely, unknown and unpredictable (Zeliger, 2003). These are:

1. Enhanced effects.
2. Low level reactions.
3. Unpredicted points of attack.

An enhanced effect is defined as one where exposure to a chemical mixture produces a reaction at a target organ that is anticipated for one of the chemicals in the mixture, but is a reaction that is far in excess of that anticipated from the toxicology of the individual chemical species.

A low-level reaction is one where exposure to a mixture of chemicals, in which each all are present at concentrations far below those known to produce reactions, does indeed impact a target organ that is known to be affected by for one of the chemicals.

An unpredicted point of attack reaction occurs when exposure to a mixture of chemicals results in attack on an organ not known to be impacted by any of the individual chemicals in the mixture.

2.8 Traditional toxicology

Toxicology is the science devoted to the study of the effects of toxins on living organisms. Such toxins can be biological; poisons, bacteria, viruses and fungi; or chemical.

The toxic effects of chemical poisons are dose related. At high doses, severe injury or death can quickly occur. At low enough doses, organisms can be exposed to even the most toxic of substances without suffering a deleterious impact. Traditional Toxicology attempts to define dose levels that produce the various degrees of responses for individual chemicals. These dose response relationships (DRRs) are used to numerically assign exposure values to levels that range from no adverse health responses (and are considered safe) though levels that cause maximum health effects.

2.8.1 Toxicological data

Toxicological data are presented in a number of different ways. These and their commonly used abbreviations are given here.

NOEL: No observed effect level. This is the highest level at which no toxicological effects are noted. This level is often presented as.

NOAEL: No observed adverse effect level.

NOEC: No observed effect concentration. This is datum is identical to NOEL.

MOEL: Minimum observed effect level. This is the lowest concentration at which adverse effects are note. This level is often presented as MOAEL – minimum observed adverse effect level.

PEL: Permissible exposure level. PEL data are those established by the U. S. Occupational Safety and Health Administration (OSHA) for inhalation exposures in the workplace.

TWA: Time weighted average. TWA data are for exposures in the workplace. These are set by the National Institute of Occupational Safety and Health (NIOSH) for inhalation of airborne contaminants.

TLV: Threshold limit values. These are similar to PEL data, but are set by the American Conference of Governmental Industrial Hygienists (AGCIH). TLV data tend to be more conservative, i.e., lower levels than PEL data.

STEL: Short term exposure limit. Recommended inhalation exposure level for exposures up to 20 minutes.

IDLH: Immediately dangerous to life or health. Airborne concentrations at which even momentary exposure can kill or seriously injure.

MCL: Maximum contaminant level. This value is generally given for contaminants dissolved in drinking water.

Inhalation data, PEL, TWA, TLV, STEL and IDLH data are generally presented in units of parts per million (PPM), parts per billion (PPB) or milligrams per cubic meter of air (MPCM). Ingestion data MCL data are generally presented in PPM or PPB, or in moles, millimoles or micromoles per liter or milliliter of water.

The exposure limits listed for individual chemicals are arrived at via a combination of scientific and political considerations, with different groups looking at the same data arriving at different exposure limit recommendations. As an example of this let us consider methyl isobutyl ketone (MIBK). MIBK targets the eyes, skin, respiratory system, central nervous system, liver and kidneys. The OSHA TWA for MIBK is 100 ppm while NIOSH and ACGIH recommend a TWA of 50 ppm. Such differences can arise from a difference of scientific opinion and/or the vested interests of those who manufacture and sell a particular chemical. The data nevertheless are a reflection of the body's ability to protect itself against the hazards posed by a particular xenobiotic. A higher exposure level value indicates a reduced danger. In the MIBK example, TWA of 50 ppm indicates a greater hazard for this chemical than a value of 100 ppm.

2.8.2 Chemical impact

Chemicals can impact all body systems and organs. Several processes are involved when such impact occurs. These include:

Exposure
Absorption
Distribution
ROS/RON generation
Impact on target organs
Metabolism
Immune system response
Endocrine system response
Excretion.

Toxicology assesses these impacts and these are incorporated into establishing the data points just discussed.

2.9 Chemical mixtures

Exposures to chemical mixtures present problems in trying to assess the various toxicological data points. What happens when one is exposed to more than one toxic chemical at a time? This question is a difficult one to answer. The enormous increase of chemical releases to the environment accompanied by the wide-spread distribution of these chemicals has made it impossible for anyone to be impacted solely by single chemicals outside of controlled laboratory settings. Though large single chemical exposure impacts can be studied for their impacts, mixtures present a challenge to toxicologists.

Traditionally, toxicologists have addressed the effects of chemical mixtures as being additive, antagonistic, potentiated or synergistic (Ballantyne, 1985). The effects of sequential absorption of different chemicals have been added to these in recent years (Zeliger, 2011; Conde, 1985; Djordjevic et al., 1998; Maellaro et al., 1990).

2.9.1 Additivity

Additive effects occur when two or more substances with the same toxicity (i.e., attack the same organ) are present together. The total or additive effect is the sum of the individual effects. Additive effects are observed when mixtures consist of species that are similar, i.e., act identically on a target organ. Additive effects may be observed, e.g., when a mixture of two compounds, each below the no observed exposure level (NOEL) produce a predicted toxic effect when the sum of their concentrations is greater than the threshold level for toxic action.

Examples of chemical mixtures that produce additive effects are:

1. n-hexane and methyl-n-butyl ketone (peripheral neuropathy) (Baselt, 2000).
2. trichloroethylene and tetrachloroethylene (liver and kidney toxins) (Stacey, 1989).
3. toluene and xylene (brain function loss) (Dennison et al., 2005).

2.9.2 Antagonism

Antagonism occurs when two chemicals interfere with each other's effect. The result is a reduction in the effect predicted for the individual species. Antagonistic mixtures need not be structurally similar. One species may stimulate the metabolism of a second one or somehow interfere with its sorption. Antagonism can be considered the antithesis of synergism (see below).

Examples of chemical mixtures that produce antagonisms and their effects are:

1. DDT and parathion (DDT induces and parathion inhibits enzymatic activity) (Chapman and Leibman, 1971).
2. Oxygen and carbon monoxide (oxygen competes with CO for receptor sites) (Thom, 1990).
3. Toluene and benzene (toluene inhibits benzene metabolism and reduces its toxicity) (Hseih et al., 1990).

2.9.3 Potentiation

A potentiated effect is observed when the effect of a chemical is enhanced by the presence of one or more other compounds that are nontoxic or only slightly toxic. One compound can potentiate a second one toxicologically, e.g., by producing the same metabolites in the body.

Examples of potentiated effects chemical mixtures are:

1. Organophosphorothiolate esters potentiate malathion (CNS) (Fukuto, 1983).
2. Isopropanol potentiates carbon tetrachloride (liver) (Folland et al., 1976).

3. Methyl ethyl ketone potentiates n-hexane (CNS and peripheral nervous system) (Shibuta et al., 1990).

2.9.4 Synergism

Synergism is observed when the effect of exposure to a mixture is much greater than or different from that expected from an additive effect. In such instances, exposures to mixtures of chemicals that are substantially different from each other induce responses not predicted by the known toxicology of the individual chemical species. When synergistic effects are observed, one of the chemicals of the mixture changes the body's response in a qualitative or quantitative (additive) way. A qualitative response results in much greater response than would be observed for an additive effect, by resulting in attack at a different target organ than one that is predicted.

Examples of chemical mixtures that produce synergism and their effects are:

1. Nitrate and aldicarb (immune, endocrine and nervous system) (Porter and Jaeger, 1999).
2. Carbon disulfide and carbon tetrachloride, (nervous system) (Peters et al., 1986).
3. Cigarette tar and nitric oxide (carcinogenic).

2.10 Unanticipated effects of mixtures

As noted in the introduction, exposures to chemical mixtures can produce enhanced effects, low level reactions and unpredicted points of attack. The toxicological literature reported these, but until recently was at a loss to offer an explanation. The following published studies are illustrative of how toxicologists viewed the unexpected effects of exposures to mixtures prior to 2003.

Alessio reviewed the literature and reported on the exposure of workers to multiple solvents in the workplace. His study showed that exposures to some solvent mixtures resulted in the inhibition of the metabolism of the solvents, while exposures to other solvent mixtures enhanced the metabolism of the solvents (Alessio, 1996). No explanations of the effects noted were offered.

Feron studied the effects of mixtures administered at the "No-Observed-Adverse-Effect-Level" (NOAEL) and the "Minimum-Observed-Adverse-Effect-Level" (MOAEL). Evidence of an increased hazard was found when combinations of chemicals when administered at the NOAEL of each of the components, despite the fact that exposures to the individual chemicals had no adverse effects. When mixtures were administered at the MOAEL levels of the individual components some severe adverse effects noted (Feron et al., 1995).

Alexandersson studied the effects of exposure of carpenters to formaldehyde, terpenes and dust particles. The mean formaldehyde levels were far below the threshold value. The terpenes levels were very low and frequently undetectable and dust levels were about one tenth of the threshold levels. At the concentration levels recorded, no respiratory effects would be expected, yet dyspnea (shortness of breath), nose and throat irritation, chest tightness and productive cough were observed (Alexandersson et al., 1982). These results were reported without explanation.

Formaldehyde exposure is not known to cause neurobehavioral symptoms or disturbed mental of neurologic function. Kilburn et al., however, found that exposure by hospital histology technicians to formaldehyde, xylene and toluene produced such effects (Kilburn et al., 1985). No attempt was made to explain these results.

A study of rubber workers exposed to a mixture of resorcinol, formaldehyde and ammonia revealed that these workers suffered acute drops in lung function and other respiratory symptoms over a work shift. The levels of exposure of the chemicals were low. The researchers concluded that the cause for the observed effects was unknown (Gamble et al., 1976).

Brooks reported several instances of reactive airways dysfunction syndrome (RADS) following exposures to mixtures of chemicals each of which contained no compounds known to cause respiratory sensitization. In the first instance, a store clerk was stricken with RADS following application of a floor sealant containing a mixture of aliphatic and aromatic hydrocarbons and epichlorohydrin. In the second instance, two painters were stricken after spray painting primer in an apartment. The primer contained a mixture consisting of ammonia, aluminum chlorohydrin and other unidentified additives. In another case, a woman was stricken within 15 minutes of the application of a fumigant containing polyoxyethylated vegetable oil, dipropylene glycol, a turpine hydrocarbon, sodium nitrate, an unsaturated aldehyde and isobornyl acetate (Brooks et al., 1985). No attempt was made to account for the observed effects.

Lee reported on the prevalence of pulmonary and upper respiratory tract symptoms experienced by pressmen exposed to low levels of aliphatic hydrocarbons, limonene, glycol ethers, isopropyl alcohol and mineral oil. The airborne levels of these solvents were below the permissible exposure limits (Lee et al., 1997). No explanation was offered to account for the observed results.

Waterborne paints are generally low in volatile organic compounds and not thought of as being particularly dangerous. Typical formulations include glycol ethers, esters, glycols, formaldehyde and amines. Hansen et al. investigated the waterborne paints used in Denmark and their effects on painters. They reported that mucous membrane irritation was observed in these painters even though the airborne concentrations of the volatiles were, for the most part, below the known irritation levels for the single chemicals. The researchers concluded that irritation due to the combined action of the chemicals cannot be excluded but offered no explanation for this conclusion (Hansen et al., 1987).

Liver damage was reported in chemical workers exposed to low levels of carbon disulfide, isopropanol, toluene and other chemicals in trace quantities (Dossig and Ranek, 1984). The researchers suggested that the liver injury was caused by the combined action of organic solvents and compared it to the known synergistic effect of isopropanol on the hepatotoxicity of carbon tetrachloride (Folland et al., 1976).

The above case studies demonstrate that chemicals can cause disease via a large-scale acute exposure (Brooks et al., 1985) or via long term chronic exposure (Dossing and Ranek, 1976; Lee et al., 1977; Kilburn et al., 1985). High concentration acute exposures overwhelm the body's ability to remove attacking chemicals and cause immediate tissue damage. Low concentration chronic exposures result in a constant state of oxidative stress that lead to disease onset. The mechanisms that apply to disease onset due to chronic OS are discussed in detail in Part II of this book.

2.11 Lipophile–hydrophile mixtures

In 2003, Zeliger reported for the first time that in all cases cited in the above section and, indeed, in virtually all literature reports of chemical mixture of unusual mixture effects, each of such mixtures contained at least one lipophilic and one hydrophilic chemical. Numerous additional examples of these were subsequently reported (Zeliger, 2003, 2011). Lipophiles promote the permeation of hydrophiles through mucous membranes resulting in the absorption of greater quantities of hydrophilic species than would be absorbed if the lipophile were not present. As discussed above, once absorbed, such mixtures of chemicals may affect the body in ways not anticipated from the actions of single chemicals alone. The unanticipated mixture effects, enhanced effects, low level reactions or unpredicted points of attack can follow either acute exposures of high chemical concentrations or chronic exposures to low chemical concentrations (Zeliger, 2003, 2011).

It is well known that increases in oxidative stress, and the diseases caused by such increases due to chemical exposures occur in a response relationship (DRR) (Zhu et al., 2015; Alfanie, 2015; Zeliger, 2016). As exposures to mixtures of lipophiles and hydrophiles increase the amounts of hydrophiles absorbed, such mixtures increase the dose of hydrophiles and thus are more toxic than hydrophiles alone.

2.12 Lipophile–lipophile mixtures

It has been observed that mixtures of lipophiles also exhibit synergistic exposure effects. An example of this phenomenon is that observed with exposure to a mixture of the ubiquitous and highly toxic 2,3,7,8-tetrachlorodibenzo-p-dioxin (TCDD) ($K_{ow} = 6.64$) and endosulfan ($K_{ow} = 2.81$). In such instances the two lipophiles have K_{ow} values that differ by 3.0 units or more (3.83 here). It is theorized that the species with the greater K_{ow}, which preferentially absorbs through cell membranes, carries the less lipophilic species with it (Rainey et al., 2017). Lipophilic chemical mixtures of species with similar K_{ow} values are observed to demonstrate additive effects.

When Kow values of the individual components of lipophilic chemical mixtures are of similar values, it is total lipophilic load that influences disease onset. This has been demonstrated for type 2 diabetes, cardiovascular diseases and neurological diseases (Zeliger, 2013, 2013a, 2013b). Cardiovascular diseases triggered by total lipophilic load are shown in Table 2.8.

Neurological diseases associated with total lipophilic load include neurological impairments, neurodevelopmental disorders and neuro degenerative diseases. These are shown in Table 2.9.

2.13 Sequential absorption

Sequential effects arise when one chemical is absorbed first and a second chemical is absorbed at a finite, but important time, thereafter resulting in an effect not observed from

TABLE 2.8 Cardiovascular diseases associated with total lipophilic load.

Myocardial infarction
Atherosclerosis
Hypertension
Coronary heart disease
Peripheral heart disease
Ischemic heart disease
Cardiac autonomic dysfunction

TABLE 2.9 Neurological diseases associated with total lipophilic load.

Neurological impairments
Cognitive effects
Motor deficits
Sensory deficits
Peripheral neurological system effects
Neurodevelopmental disorders
Autism and autism spectrum disorders (ASD)
Attention deficit hyperactivity disorder (ADHD)
Mental retardation
Cerebral palsy
Neural tube defects
Hearing loss
Neurodegenerative diseases
Alzheimer's disease
Parkinson's disease
Amyotrophic lateral sclerosis (ALS)

exposure to either one of the single chemicals (Ashauer et al., 2017). Examples that demonstrate this effect are:

1. The administering of ethanol or acetone prior to administering acetaminophen results in a marked increase in the hepatotoxity of acetaminophen (due to enzyme induction by ethanol) (Djordjevic et al., 1998).
2. Pretreatment with diethyl maleate increases toxicity of bromobenzene (due to depletion of glutathione by diethyl maleate) (Maellaro et al., 1990).

Sequential absorption of lipophilic and hydrophilic chemicals has been associated the toxic effects of lipophile:hydrophile mixture effects. Many lipophiles are stored in body fat and re retained in the body for years, and, in some instances, decades (Zeliger, 2011). Examples of these are so called "Stockholm Dirty Dozen," a group of highly toxic pesticides, industrial chemicals and chemical by-products. The Stockholm Convention on Persistent Organic Pollutants, which was signed in 2001 is aimed at the elimination or restricted production of these chemicals (Stockholm Convention, 2019). The original compounds are listed Table 2.10. Since its adoption, an additional group of nine polynuclear aromatic hydrocarbons and brominated flame retardants, known as the "Nasty Nine," have been added to the original list and yet others have been proposed.

Hydrophiles, being water soluble, are more rapidly eliminated than lipophiles. Accordingly, retained lipophiles can facilitate the absorption of hydrophiles to which one is exposed even if hydrophilic exposure occurs at a sequential time. Thus, lipophilic exposure can occur

TABLE 2.10 Stockholm Dirty Dozen of persistent organic pollutants.

Pesticides
Aldrin
Chlordane
DDT
Dieldrin
Endrin
Heptachlor
Hexachlorobenzene
Mirex
Toxaphene
Industrial chemicals
PCBs
By-products
Polychlorinated dibenzo-p-dioxins (PCDD)
Polychlorinated dibenzofurans (PCDF)

from breathing polluted air and hydrophilic exposure can sequentially follow ingestion of a hydrophile-containing food, for example.

Retained aromatic lipophiles can also facilitate the absorption of subsequently obtained metal ions (Zeliger and Lipinski, 2015). The mechanism of this phenomenon, as well as the other examples of sequential absorption just presented, are presented in detail in part II of this book.

Heavy metals, some of which are also retained in the body for years or even decades, are also participants in sequential absorption toxicity. These metals include (Jaishankar et al., 2014):

Arsenic
Cadmium
Chromium
Lead
Mercury

In this instant, it is the hydrophiles are first absorbed, followed by the lipophiles

2.14 Diseases caused by chemicals

Exposure to exogenous chemicals is the primary cause of the accelerated levels of disease that have accompanied the Industrial Revolution. Though, as will be seen in the succeeding chapters, other OS raising parameters also contribute to increased disease, chemicals are primarily responsible. Unequivocal evidence for this is the numerous clusters of disease that have affected groups of people for whom other contributing factors can be ruled out. Two examples of such clusters are:

1. A childhood leukemia cluster in Woburn, Massachusetts from 1969 to 1979 in which the affected children were exposed to chlorinated hydrocarbons, PAHs, arsenic and chromium in their drinking water (Cutler et al., 1986).
2. Clusters of chloracne shortly following and non-Hodgkin's lymphoma long tern in residents of Seveso, Italy from dioxin exposure following an explosion at a trichlorophenol manufacturing plant in 1976 (Bertzzi et al., 1998).

Exogenous chemical induced disease can be caused via acute high-level exposure, or chronic low-level exposure. RADS onset following single high concentration chemical exposures is an example of this (Brooks et al., 1985). Late onset of type 2 diabetes following chronic exposure to DDE, the metabolite of DDT, exemplifies this response (Turyk et al., 2009).

Some chemicals attack primarily single systems or organs. Toluene diisocyanate or trimellitic anhydride, for example, primarily attack the respiratory system and can cause respiratory sensitization (Zeliger, 2011). Other chemicals have multiple system and organ impacts. An example of this is the large number of health effects attributable to exposures to PCBs (Carpenter, 2006; ATSDR, 2016). These include:

Cancer
Ischemic heart disease

Asthma
Hypertension
Diabetes
Neurological impacts including decline in cognitive function and altered behavioral effects
Liver disease
Arthritis
Hypothyroidism
Infertility

Some diseases are known to be caused by a wide spectrum of different chemicals. Two examples are:

1. Diabetes mellitus (type 2 diabetes) has been shown to ensue following chronic exposures to the following chemicals (Zeliger, 2013b):
   DDT and other pesticides
   Bisphenol A
   PCBs
2. Multiple parameters have been attributed to the onset of Alzheimer's disease (Zeliger, 2019a). Included in these are exposures to the following:
   DDT and other pesticides
   Air pollutant chemicals
   Heavy metals
   Tobacco smoke chemicals
   Nitrosamines

Mixtures of chemicals generally attack multiple systems. It has been shown that people who exhibit respiratory sensitization, and thus extreme reactions to airborne chemicals, also have diseases of multiple other systems (Zeliger et al., 2012). Table 2.11 shows the systems affected and the percent of individuals with such co-morbidities.

TABLE 2.11 Systemic disease co-morbid with respiratory sensitization from chemicals.

System	% Affected
Nervous	100
Immune	100
Cardiovascular	77
Gastrointestinal	77
Musculoskeletal	33
Hormonal	10

2.15 Chemically caused disease and oxidative stress

Organic and inorganic chemicals cause oxidative stress leading to disease when absorbed by the human body in a dose response relationship (Zeliger, 2016). The mechanisms for OS induced disease onset are explored in the chapters devoted to specific diseases.

In our complex environment, all living entities are constantly exposed to numerous mixtures of chemicals of far-ranging concentrations, as well as to other sources of OS. The immediately following chapters discuss these other sources.

References

Alexandersson, R., Kolmodin-Hedman, B., 1982. Exposure to formaldehyde: effects on pulmonary function. Arch. Environ. Health 17 (5), 279–284.

Alfanie, I., 2015. Effect of heavy metal on malondialdehyde and advanced oxidation protein products concentration: a focus on arsenic, cadmium and mercury. J. Med. Bioengen 4 (4), 332–337.

Alessio, I., 1996. Multiple exposures to solvents in the workplace. Int. Arch. Occup. Environ. Health 69, 1–4.

Ashauer, R., O'Connor, I., Escher, B.I., 2017. Toxic mixtures in time – the sequence makes the poison. Environ. Sci. Technol. 51, 3084–3092.

ATSDR, 2016. Environmental Health and Medicine Education. Polychlorinated Biphenyls (PCBs) Toxicity Clinical Assessment. https://atsdr.cdc.gov/csem/csem.asp?csem=30&po=11 (Accessed 3 January 2021).

Ballantyne, B., 1985. Evaluation of hazards from mixtures of chemicals in the occupational environment. J. Occup. Med. 27, 85–94.

Baselt, R.C., 2000. Disposition of Toxic Drugs and Chemicals in Man, fifth ed. Chemical Toxicology Institute, Fresno City, CA, pp. 416–417.

Bertzzi, P.A., Bernucci, I., Brambilla, G., et al., 1998. The Seveso studies on early and long-term effects of dioxin exposure: a review. Environ. Health Perspect. 106 (Suppl. 2), 625–633.

Brooks, S.M., Weiss, M.A., Bernstein, I.L., 1985. Reactive airways dysfunction syndrome (RADS): persistent asthma syndrome after high level exposures. Chest 88 (3), 376–384.

Carpenter, D.O., 2006. Polychlorinated biphenyls (PCBs): routes of exposure and effects on human health. Rev. Environ. Health 21 (1), 1–23.

Chapman, S.K., Liebman, K.C., 1971. The effects of chlordane, DDT and 3-methylchloroanthene upon the metabolism and toxicity of diethyl-4-nitrophenyl phosphorothionate (parathion). Toxicol. Appl. Pharmacol. 18 (4), 977–987.

Condie, L.W., 1985. Target organ toxicology of halocarbons commonly found in drinking water. Sci. Total Environ. 47, 433–442.

Cutler, J.J., Parker, G.S., Rosen, S., et al., 1986. Childhood leukemia in Woburn, Massachusetts. Public Health Rep. 101 (2), 201–205.

Dennison, J.E., Bigolow, P.L., Mumtaz, M.M., et al., 2005. Evaluation of potential toxicity from co-exposure to three CNS depressants (toluene, ethylbenzene and xylene) under resting and working conditions using PBPK modeling. J. Occup. Environ. Hyg. 2 (3), 127–135.

Djordjevic, D., Nikolic, J., Stefanovic, V., 1998. Ethanol interactions with other cytochrome P450 substrates including drugs, xenobiotics and carcinogens. Pathol. Biol. 46 (10), 760–770.

Dossing, M., Ranek, I., 1984. Isolated liver damage in chemical workers. Br. J. Ind. Med41, 142–144.

Feron, V.J., Groten, J.P., Jonker, D., Cassee, F.R., van Bladeren, P.J., 1995. Toxicology of chemical mixtures: challenges for today and the future. Toxicology 105 (2–3), 415–427.

Folland, T.R., Schaffner, W., Ginn, H.E., 1976. Carbon tetrachloride toxicity potentiated by isopropyl alcohol. J. Am. Med. Assn. 236 (18), 1853–1856.

Fukuto, T.R., 1983. Toxicological properties of trialkyl phosophorothionate and dialkyl alkyl- andaryl phosphonothionate esters. J. Environ. Sci. Health B 18 (1), 898–117.

Gamble, J.F., McMichael, A.J., Williams, T., Battigelli, M., 1976. Respiratory function and symptoms: an environmental-epidemiological study of rubber workers exposed to a phenol-formaldehyde type resin. Am. Ind. Hyg. Assoc. J. 1976, 499–513.

Ha, M.H., Lee, D.H., Jacobs, D.R., 2007. Association between serum concentrations of persistent organic pollutants and self-reported cardiovascular disease prevalence: results from the National Health and Nutrition Examination Survey, 1999–2002. Environ. Health Perspect. 115 (8), 1204–1209.

Hansen, M.K., Larsen, M., Cohr, K.H., 1987. Waterborne paints: a review of their chemistry and toxicology and the results of determinations made during their use. Scand. J. Work. Environ. Health 13, 473–485.

Harada, T., Takeda, M., Kojima, S., et al., 2016. Toxicity and carcinogenicity of dichlodiphenyltrichloroethane (DDT). Toxicol. Res. 32 (1), 21–33.

Hseih, G.C., Parker, R.D., Sharma, R.P., Hughes, B.J., 1990. Subclinical effects of groundwater contaminants, III. Effects of repeated oral exposure to combinations of benzene and toluene on immunologic responses in mice. Arch. Toxicol. 64 (4), 320–328.

Hunter, P., 2015. Essentially deadly: living with toxic elements. EMBO Rep. 16 (12), 1605–1608.

Jaishankar, M., Tseten, T., Anbalagan, N., et al., 2014. Toxicity, mechanism and health effects of some heavy metals. Interdiscipl. Toxicol. 7 (2), 60–72.

Kilburn, K.H., Seidman, B.C., Warshaw, R., 1985. Neurobehavioral and respiratory symptoms of formaldehyde and xylene exposure in histology technicians. Arch. Environ. Health 40 (4), 229–233.

Lee, B.W., Kelsey, K.T., Hashimoto, D., et al., 1997. The prevalence of pulmonary and upper respiratory tract infections and spirometric test finding among newspaper pressroom workers exposed to solvents. J. Occup. Environ. Med. 39 (10), 960–969.

Maellaro, E., Casini, A.F., Del Bello, B., Comporti, M., 1990. Lipid peroxidation and antioxidant systems in the liver injury produced by glutathione depleting agents. Biochem. Pharmacol. 39 (10), 1513–1521.

Maheshwari, N., Khan, F.H., Mahmood, R., 2019. Pentahclorophenol-induced cytotoxicity in human erythrocytes: enhanced generation or ROS and RNS, lowered antioxidant power, inhibition of glucose metabolism, and morphological changes. Environ. Sci. Poll. Res. Int. 26 (13), 12985–13001. https://doi.org/10.1007/s11356-019-04736-8.

Maret, W., 2016. The metals in the biological periodic system of the elements: concepts and conjectures. Int. J. Mol. Sci. 17, 66. https://doi.org/10.3390/ijms17010066 (Accessed 3 January 2021).

Michalowicz, J., Mokra, K., Rosiak, K., et al., 2013. Chlorobenzenes, lindane and dieldrin induce apoptotic alterations inn human peripheral blood lymphocytes (in vitro study). Environ. Toxicol. Pharmacol. 36 (3), 979–988.

Neuberger, M., Kundi, M., Jager, R., 1998. Chloracne and morbidity after dioxin exposure. Toxicol. Lett. 96–97, 347–350.

Perez-Maldonado, I.N., Herrera, C., Batres, L.E., et al., 2005. DDT-induced oxidative damage in human blood mononuclear cells. Environ. Res. 98 (2), 177–184.

Peters, H.A., Levine, R.L., Matthews, C.G., et al., 1986. Synergistic toxicity of carbon tetrachloride/carbon disulfide (80/20 fumigants) and other pesticides in grain storage workers. Acta Pharmacol. Toxicol. 59 (Suppl. 7), 535–546.

Porter, W.P., Jaeger, J.W., Carlson, I.H., 1999. Endocrine, immune and behavior effects of aldicarb (carbamate), atrazine (triazine) and nitrate (fertilizer) mixtures at groundwater concentrations. Toxicol. Ind. Health 15 (1–20), 133–150.

Rainey, N., Saric, A., Leberre, E., Dewailly, D., Slomianny, C., Vial, G., Zeliger, H.I., Petit, P., 2017. Synergistic cellular effects including mitochondrial destabilization, autophagy and apoptosis following low-level exposure to a mixture of lipophilic persistent organic pollutants. Sci. Rep. 7, 4728–4748.

Sangster, J., 1989. Octanol-water partition coefficients of simple organic compounds. J. Phys. Chem. Ref. Data 18 (3), 1111–1127.

Shibata, E., Huang, J., Hisanaga, N., et al., 1990. Effects of MEK on kinetics of n-hexane metabolites in serum. Arch. Toxicol. 64 (30), 247–250.

Stacey, N.H., 1989. Toxicity of mixtures of trichloroethylene, tetrachloroethylene and 1,1,1-trichloroethane: similarity of in vitro and in vivo responses. Toxicol. Ind. Health 5 (3), 441–450.

Stockholm Convention, 2019. The 12 Initial POPs under the Stockholm Convention. http://chm.pops.int/TheConvention/ThePOPs/The12InitialPOPs/tabid/296/Default.aspx.

Thom, S.R., 1990. Antagonism of carbon monoxide- mediated brain lipid peroxidation by hyperbaric oxygen. Toxicol. Appl. Pharmacol. 105 (2), 340–344.

Turyk, M., Anderson, M.A., Knobeloch, L., Pim, P., Persky, V.W., 2009. Organochlorine exposure and incidence of diabetes in a cohort of Great Lakes sport fish consumers. Environ. Health Perspect. 117, 1076–1082.

Zeliger, H.I., 2003. Toxic effects of chemical mixtures. Arch. Environ. Health 58 (1), 23–29.

Zeliger, H.I., 2011. Human Toxicology of Chemical Mixtures, second ed. Elsevier, Oxford.

Zeliger, H.I., Pan, Y., Rea, W.J., 2012. Predicting co-morbidities in chemically sensitive individuals from exhaled breath analysis. Interdiscipl. Toxicol. 5 (3), 123–126.

Zeliger, H.I., 2013. Lipophilic chemical exposure as a cause of cardiovascular disease. Interdiscipl. Toxicol. 6 (2), 101–108.

Zeliger, H.I., 2013a. Exposure to lipophilic chemicals as a cause of neurological impairments, neurodevelopmental disorders and neurodegenerative diseases. Interdiscipl. Toxicol. 6 (3), 101–108.

Zeliger, H.I., 2013b. Lipophilic chemical exposure as a cause of type 2 diabetes (T2D). Rev. Environ. Health 28 (1), 9–20.

Zeliger, H.I., Lipinski, B., 2015. Physiochemical basis of human degenerative disease. Interdiscipl. Toxicol. 8 (1), 15–21.

Zeliger, H.I., 2016. Predicting disease onset in clinically healthy people. Interdiscipl. Toxicol. 9 (2), 39–54.

Zeliger, H.I., 2019. A Pound of Prevention. Universal Publications, Boca Raton, FL.

Zeliger, H.I., 2019a. Predicting Alzheimer's disease onset. Euro. J. Med. Health Sci. 1 (No 1). https://doi.org/10.24018/ejmed.2019.1.1.16.

Zhu, Q.X., Shen, T., Ding, R., Liang, Z.Z., Zhang, X.J., 2015. Cytotoxicity of trichloroethylene and perchloroethylene on normal human epidermal keratinocytes and protective role of vitamin E. Toxicology 209 (1), 55–67.

CHAPTER

3

Particles and fibers

3.1 Introduction

Though particles and fibers are composed of chemicals, they are not soluble in water, organic media or body fluids. Most particles and fibers are rejected by the body and not absorbed. Some, however, become imbedded in body tissues while others are small enough to be absorbed through the skin, respiratory system and the digestive tract into the blood stream and carried to the far reaches of the body, including the brain. All retained particles and fibers elevate oxidative stress and lead to disease. This chapter addresses this relationship.

It is generally accepted that particles above 10.0 microns (PM_{10}) in size are not absorbed into the body, nor do they penetrate deeply into the lungs. Particles 2.5 microns ($PM_{2.5}$) or smaller however, deeply penetrate the lungs and elevate oxidative stress (Donaldson et al., 2005). Ultrafine particles (nanoparticles) are small enough to pass through body membranes and enter the blood and are considered responsible for many detrimental body effects (Houdy et al., 2011).

3.2 Particle and fiber sources and content

There are numerous indoor and outdoor sources of fine particles in the environment. Some of these are listed in Table 3.1.

Examples of particles and fibers that are absorbed by the body include both airborne and water carried species. The most prominent of these are listed in Table 3.2.

3.3 Oxidative stress (OS)

All particles and fibers that are absorbed into the body elevate oxidative stress and are associated with multiple illnesses. The most studied of these are crystalline silica, asbestos, tobacco smoke, coal fly ash, heavy metals and plastics. The impact of these to oxidative stress and disease is addressed here.

https://doi.org/10.1016/B978-0-323-91890-9.00007-6

TABLE 3.1 Particle and fiber sources.

Fossil fuel combustion
Wood combustion
Vehicle exhausts
Tobacco smoke
Mining
Manufacturing
Construction
Farming
Natural occurrences - forest fires, volcanos, severe storms
Suspended Silt
Rain runoff
Cooking and indoor heating
House Dust
Plastics use

3.3.1 Crystalline silica (CS)

Respirable CS is a component of coal fly ash (Hicks and Yager, 2006), as well as a product of mining, glass working and ceramic use (NIOSH, 1988) and a result of municipal waste incineration (Shih et al., 2008). It has long been associated with silicosis of the lungs as well as production of lung and lymphatic tumors (NISOH, 1988).

Evidence for CS induced OS comes from lipid peroxidation and DNA damage to cells contacted by CS. This induced OS is ascribed to surface-related free radical activity on CS (Saffiotti et al., 1994; Deshpande et al., 2002).

3.3.2 Asbestos

The name asbestos covers a range of naturally occurring fibrous silicates that differ from each other chemically with varying amounts of sodium, magnesium and iron (Houdy et al., 2011).

Both acute high-level exposure and chronic low-level exposure to asbestos fibers cause inflammation, a cause of OS, and produce ROS/RON and cytokine factors leading to asbestosis, mesothelioma and numerous other cancers (Houdy et al., 2011). Though the precise mechanisms for the formation of these diseases remain unknown, ROS production is believed

TABLE 3.2 Airborne and water carried particles and fibers.

Carbon black
Coal dust
Fly ash
Heavy metals
Radioactive nucleotides
Vehicle exhausts
Polynuclear aromatic hydrocarbons
Crystalline silica
Asbestos
Glass Insulation
Fertilizers
Pesticides
Soil
Plastics – Including microfibers

to be the primary driver via three pathways; fiber surface reactivity that produces hydroxyl radicals, release of inflammatory cells and mitochondrial activity impact (Liu et al., 2013; Walter et al., 2019).

Further evidence for the association of asbestos exposure and the onset of OS comes from a murine peritoneal macrophage exposure study which showed asbestos-induced cytotoxicity, ROS generation and release of lipid peroxidation markers malondialdehyde and 8-iso Prostaglandin F2α, which are indicative of OS elevation (Pietrofesa, et al., 2017).

3.3.3 Heavy metals

Combustion of coal and petroleum produces particulates containing heavy metals, the precise composition of which depends upon the coal or petroleum burned (Sushil and Batra, 2006; Dahl et al., 2008; Smith et al., 2009; Sijakova-Ivanova et al., 2011). Heavy metals contained in fuel combustion particulates are listed in Table 3.3.

Several of the metals in Table 3.3, such as iron and zinc, are essential for good health. When absorbed in solid particulate form, however, they produce free radicals in the lungs and are leading to cancer, cardiovascular disease, diabetes, atherosclerosis, neurodegenerative diseases (Alzheimer's disease and Parkinson's disease), chronic inflammation and other diseases. The mechanism for the onset of these diseases is metal-mediated free radical formation via redox cycling that generates reactive oxygen and reactive nitrogen species (ROS/RON) such as hydroxyl radical, superoxide radical, hydrogen peroxide and nitric oxide

TABLE 3.3 Heavy metals contained in particles produced by coal and petroleum combustion.

Lead
Cadmium
Mercury
Arsenic
Zinc
Iron
Copper
Chromium
Vanadium
Cobalt
Nickel
Manganese

and leads to oxidative stress and disease onset. This mechanism for responsible for such metal activity is the Fenton reaction, shown here (Valko et al., 2005; Chen and Lippmann, 2009; Jomova and Valko, 2011).

$$Fe^{2+} + H_2O_2 + H^+ \rightarrow Fe^{3+} + OH + H_2O$$

3.3.4 Radionuclides

The burning of coal releases radioactive isotopes of natural elements. These include uranium and thorium as well as their numerous decay products, including isotopes of radium, radon and potassium (Gabbard, 1993; USGS, 1997).

It is well known that radioactive isotopes cause the release of alpha and beta particles as well as gamma rays, which break chemical bonds, producing free radicals and thereby cause oxidative stress (Azzam et al., 2012; Volkova et al., 2017). Though the quantity of radiation produced from the burning of coal is modest (Gabbard, 1993), it does add to the levels of natural radiation that people are constantly exposed to and thus raises the radiation impact of oxidative stress. It has been found, however, that Americans living proximate to coal-fired electrical power plants are exposed to higher doses of radiation than those residing near nuclear power plants (McBride et al., 1978).

3.3.5 Fly ash

Fly ash produced from combustion of coal and petroleum, as well as from waste incineration contains nano sized particle of silica, heavy metals and radioactive isotopes, all of which, as just discussed, elevate oxidative stress and induce lung diseases, including cancer (USGS, 1997; Donaldson et al., 2005).

3.3.6 Plastics

Plastics are ubiquitous, making life easier for us in uncountable ways, but are also insidious, causing illness in unimagined ways. Microscopic plastic particles and fibers are now found worldwide in the air, in water, in medications, personal care products, detergents, fertilizers, and in clothes (nylon, spandex, polyester and other fibers) (ECHA Europa, 2020).

Plastics are soft and abrade easily from clothes, upholstery, carpets, food packaging and countless other forms to produce both micro-sized (smaller than 5 mm) and nano-sized (less than 1 μm) particles and species which are hazardous to health, as the these can penetrate through biological molecules (Sarasamma et al., 2020).

Microbeads are tiny plastic particles that are deliberately added to cosmetics and other products for their abrasive properties. For example, they are put into toothpastes to help scour teeth and in shower gels to abrade away dry cells from skin surfaces. Microbeads also enter foods via absorption through plant roots and are passed up food chains. Microbeads easily pass through water filtration plants, and flow out into rivers and streams that lead to the oceans. Ocean current movement has now spread microbeads to all seven seas (ECHA Europa, 2020; Kapp and Miller, 2020; Singh et al., 2020).

Micro- and nano-plastics affect oxidative stress by directly causing OS via surface activity initiation, with toxicity inversely associated dependent upon particle size (Jeong et al., 2017; Hu and Palic, 2020; Sarasamma et al., 2020).

A second way in which microplastics affect oxidative stress is by acting as carriers of both heavy metals (mercury, copper and zinc are examples) and persistent organic pollutants (PCBs are an example) and desorbing these OS elevating species when taken up by the body (Brennecke et al., 2016; Barboza et al., 2018; Bayo et al., 2018; Gerdes et al., 2019). Microbeads are not chemically reactive, don't readily biodegrade and persist in the environment for decades or longer from where they are absorbed by marine species that pass them up the food chains.

Marine species are not the only ones impacted by OS raising microplastics. A recent study has been shown that mosquito larvae ingest micro plastics and retain them into adulthood. Thus, birds, bats and any other animals that feed on mosquitoes are also helping pass microplastics up the food chain (Al-Jaibachi et al., 2018).

3.4 Dust

Dusts that people are exposed to are complex particle mixtures that vary widely, depending upon geographic location and material choices made by individuals. Dust sources include, but are not limited to, mining, agriculture, soil erosion, manufacturing operations, smoke from fires, volcanoes, tobacco smoking and biological residues (including animal fur and dander, molds and mites, as well as materials, furnishings and chemical products used in the home (Paustenback et al., 1997; Zeliger, 2011; EPA, 2020).

The toxicity of solid dust particles is magnified by the propensity of dust to absorb toxic lipophilic and hydrophilic chemicals and transport these into the body (Lv et al., 2016; Pelley, 2017; Shin et al., 2019).

Numerous studies have demonstrated the association dust inhalation and increased OS. The following references are illustrative of this phenomenon (Chan et al., 2016; Lv et al., 2016; Xiang et al., 2016; Shin et al., 2019; Sly et al., 2019). Inhalation of combinations of dust and other chemicals has been reported to magnify oxidative stress levels compared with the inhalation of dust and a given chemical alone, an example being simultaneous inhalation of house dust and ozone (Jantzen et al., 2018).

References

Al-Jaibachi, R., Cuthbert, R.N., Callaghan, A., 2018. Up and away: ontogenic transference as a pathway for aerial dispersal of microplastics. Biol. Lett. 14. https://doi.org/10.1098/rsbl.2018.0479.

Azzam, E.I., Jay-Gerin, J.P., Pain, D., 2012. Ionizing radiation-induced metabolic oxidative stress and prolonged cell injury. Cancer Lett. 327 (1–2), 48–60.

Barbosa, G.A.B., Vieira, L.R., Branco, V., Figueriedi, N., Carvalho, F., Carvalho, C., Gueilhermino, L., 2018. Microplastics cause neurotoxicity, oxidative damage and energy-related changes and interact with the bioaccumulation of mercury in European seabass, *Dicentrarchus labrax* (Linneaeus, 1758). Aquat. Toxicol. (Amst.) 195, 49–57.

Bayo, J., Guillen, M., Olmos, S., Jimenez, P., Sanchez, E., Roca, M.J., 2018. Microplastics as vector for persistent organic pollutants in urban areas: the role of polychlorinated biphenyls. Int. J. Dev. Plann. 13 (4), 671–682.

Bennecke, D., Duarte, B., Paiva, F., Cacador, D., Canning-Clode, J., 2016. Microplastics as vector for heavy metal contamination from the marine environment. Estuar. Coast Shelf Sci. 178, 189–195.

Chan, T.K., Loh, X.Y., Peh, H.Y., Tan, W.N.F., Tan, W.S.D., Li, N., et al., 2016. House dust mite- induced asthma causes oxidative damage and DNA soluble-strand breaks in the lungs. J. Allergy Clin. Immunol. 138 (1), 84–96.

Chen, L.C., Lippmann, M., 2009. Effects of metals within ambient air particulate matter (PM) on human health. Inhal. Toxicol. 21 (1), 1–31.

Dahl, O., Poykio, R., Nurmesniemi, H., 2008. Concentrations of heavy metals in fly ask from coal-fired power plant with respect to Finnish limit values. J. Mater. Cycles Waste Manag. 10, 87–92.

Deshpande, A., Narayanan, P.K., Lehnett, B.E., 2002. Silica-induced generation of extracellular factor(s) increases reactive oxygen species in human bronchial epithelial cells. Toxicol. Sci. 67, 275–283.

Donaldson, K., Tran, L., Jimenez, L.A., Duffin, R., Newby, D.E., Mills, N., et al., 2005. Combustion-derived nanoparticles: a review of their toxicology following inhalation exposure. Fiber Particle Toxicol. 2, 10. https://doi.org/10.1186/1743-8977-2-10. (Accessed 1 February 2021).

ECHA Europa, 2020. European union. Microplastics. https://echa.europa.eu/hot-topics/microplastics. (Accessed 15 February 2021).

EPA, 2020. Exposure Assessment Tools by Media – Soil and Dust. https://www.epa.gov/node/81729/view. (Accessed 15 February 2021).

Gabbard, A., 1993. Coal combustion: nuclear resource or danger. ORNL Rev. 26 (3–4), 1–9. http://www.ornl.gov/info/ornlreview/rev26-34/text/colmain.html.

Gerdes, Z., Ogonowski, M., Nyboom, I., Ek, C., Adoslfsson-Erici, M., Gorokhava, E., 2019. Microplastic-mediated transport of PCBs? A depuration study with Daphnia magna. PLoS One. https://doi.org/10.1371/journal.pone.0205379. (Accessed 19 February 2021).

Hicks, J., Yager, J., 2006. Airborne crystalline silica concentrations at coal-fired power plants associated with coal fly ash. J. Occup. Environ. Hyg. 3 (8), 448–455.

Houdy, P., et al. (Eds.), 2011. Nanoethics and Nanotechnology. Springer-Verlag, Berlin. https://doi.org/10.1007/978-3-642-20177-6_1. (Accessed 25 January 2021).

Hu, M., Palic, D., 2020. Micro- and nano-plastics activation of oxidative and inflammatory adverse outcome pathways. Redox Biol. https://doi.org/10.1016/j.redox.2020.101620. (Accessed 16 February 2021).

Jantzen, K., Jansen, A., Kermanizadeh, A., Elmholm, G., Sigsgaard, T., Moller, P., Roursgaard, M., Loft, S., 2018. Inhalation of house dust and ozone alters systemic levels of endothelial progenitor cells, oxidative stress, and inflammation in elderly subjects. Toxicol. Sci. https://doi.org/10.1093/toxsci/kfy027. (Accessed 18 February 2021).

Jeong, C.B., Kang, H.M., Lee, M.C., Kim, D.H., Han, J., Hwang, D.S., et al., 2017. Adverse effects of microplastics and oxidative stress-induced MAPK/Nrf2 pathway-mediated defense mechanisms in the marine copepod Parachclopina nana. Sci. Rep. https://doi.org/10.1038/srep41324. (Accessed 19 February 2021).

Jomova, K., Valko, M., 2011. Advances in metal-induced oxidative stress and human disease. Toxicology 283 (2–3), 65–87.

Kapp, K.J., Miller, R.Z., 2020. Electric clothes dryers: an underestimated source of microfiber pollution. PLoS One. https://doi.org/10.1371/journal.pone.0239165. (Accessed 10 February 2021).

Liu, G., Cheresh, P., Kamp, D.W., 2013. Molecular basis of asbestos-induced lung diseases. Ann. Rev. Path. 8, 161–187. https://doi.org/10.1146/annrev-pathol-020712-163942. (Accessed 18 February 2021).

Lv, Y., Rui, C., Dai, Y., Pang, O., Li, Y., Fan, R., Lu, S., 2016. Exposure of children to BPA through dust and the association of urinary BPA and triclosan with oxidative stress in Guangzho, China. Environ. Sci. Processes Imports. https://pubs.rsc.org/en/content/articlelanding/2016/EM/c6em00472e. (Accessed 19 February 2021).

McBride, J.P., Moore, R.E., Witherspoon, J.P., Blanco, R.E., 1978. Radiological impact of airborne effluents of coal and nuclear plants. Science 202 (4372), 1045–1050.

NIOSH, 1988. Center for disease control. Silica. Crystal. https://www.cdc.gov/niosh/pel88/14808-60.html. (Accessed 20 February 2021).

Paustenbach, D.J., Finley, B.L., Long, T.F., 1997. The critical role of house dust in understanding the hazards posed by contaminated soils. Int. J. Toxicol. 16, 339–362.

Pelley, J., 2017. Tracing the chemistry of household dust. Chem. Eng. News 95 (7). https://cen.acs.org/articles/95/17/Tracing-chemistry-household-dust.html. (Accessed 5 February 2021).

Pietrofesa, R.A., Woodruff, P., Hwang, W.T., Patel, P., Chatterjee, S., Albelda, S., 2017. The synthetic lignan secoisolariciresinal diglucoside prevents asbestos-induced NLRP3 Inflammasome activation in murine macrophages. Ox. Med. Cellular Long. https://doi.org/10.1155/2017/7395238 (Accessed 22 August 2022).

Saffiotti, U., Daniel, L.N., Mao, Y., Shi, X., Williams, A.O., Kaighn, M.E., 1994. Mechanisms of carcinogenesis by crystalline silica in relaton ot oxygen radicals. Environ. Health Perspect. 102 (Suppl. 10), 159–163.

Sarasamma, S., Audira, G., Siregar, P., Malhotra, N., Lai, Y.H., Liang, S.T., et al., 2020. Nanoplastics cause neurobehavioral impairments, reproductive and oxidative damages, and biomarker responses in zebrafish: throwing up alarms of wide spread health risk of exposure. Int. J. Mol. Sci. https://doi.org/10.3390/ijms21041410. (Accessed 1 February 2021).

Shih, T.S., Lu, P.Y., Chen, C.H., Soo, J.C., Tsai, C.L., Tsai, P.J., 2008. Exposure profiles and source identification for workers exposed to crystalline silica during a municipal waste incinerator relining period. J. Hazard Mater. 154 (1–3), 469–475.

Shin, H.M., Moschet, C., Young, T.M., Bennett, D.H., 2019. Measured concentrations of consumer product chemicals in California house dust: implications for sources, exposure, and toxicity potential. Int. J. Indoor Environ. Health. https://doi.org/10.1111/ina.12607. (Accessed 20 February 2021).

Sijakova-Ivanova, T., Panov, Z., Blazev, K., Zajkova-Paneva, V., 2011. Investigation of fly ash heavy metals content and physico chemical properties from thermal power plant, Republic of Macedonia. Int. J. Eng. Sci. Technol. 3 (12), 8219–8226.

Singh, R.P., Mishra, S., Das, A.P., 2020. Synthetic microfibers: pollution toxicity and remediation. Chemosphere. https://doi.org/10.1018/j.chemosphere.2020.127199.

Sly, P.D., Cormier, S.A., Lomnicki, S., Harding, J.N., Grimwood, K., 2019. Environmentally persistent free radicals: linking air pollution and poor respiratory health? Am. J. Respir. Crit. Care Med. 200 (8). https://doi.org/10.1164/rccm.201903-0675LE. (Accessed 17 February 2021).

Smith, A.H., Ercumen, A., Yuan, Y., Steinmaus, C.M., 2009. Increased ling cancer risks whether arsenic is ingested on inhaled. J. Expo. Sci. Environ. Epidemiol. 19 (4), 343–348.

Sushil, S., Batra, V.S., 2006. Analysis of fly ash heavy metal content and disposal in three thermal power plants in India. Fuel 85 (17–18), 2676–2679.

USGS, 1997. Fact Sheet FS-163-97. Elements in Coal and Fly Ash: Abundance, Forms and Environmental Significance. https://pubs.usgs.gov/fs/1997/fs163-97/FS-163-97.html. (Accessed 19 February 2021).
Valko, M., Morris, H., Cronic, M.T., 2005. Metals, toxicity and oxidative stress. Curr. Med. Chem. 12 (10), 1161–1208.
Volkova, P.U., Geras'kin, S.A., Kazakova, E.A., 2017. Radiation exposure in the remote period after the Chernobyl accident caused oxidative stress and genetic effects in Scot pine populations. Sci. Rep. https://doi.org/10.1038/srep43009. (Accessed 18 February 2021).
Walter, M., Scheinkeveld, W.D.C., Reissner, M., Gille, L., Kraemer, S.M., 2019. The effect of pH and biogenic ligands on the weathering of Chrysotile asbestos: the pivotal role of tetrahedral Fe in dissolution kinetics and radical formation. Chemistry 25 (13), 3286–3300. https://doi.org/10.1002/chem.201804319. (Accessed 19 February 2021).
Xiang, P., He, R.W., Han, Y.H., Sun, H.J., Cui, X.Y., Ma, L.Q., 2016. Mechanisms of housedust-induced toxicity in primary human corneal epithelial cells: oxidative stress, proinflammatory response and mitochondrial dysfunction. Environ. Int. 89–90, 30–37.
Zeliger, H.I., 2011. Human Toxicology of Chemical Mixtures, second ed. Elsevier, London.

CHAPTER

4

Air pollution and oxidative stress

4.1 Introduction

World Health Organization (WHO) data show that nine out of 10 people in the world breath heavily polluted air and that polluted air causes some eight million deaths each year – 4.2 million caused by breathing polluted ambient air and 3.8 million from breathing indoor air polluted with chemicals in polluted fuels (WHO 2020).

4.2 Ambient air pollutants

Ambient air pollutants are those present in the general outdoor atmosphere to which all individuals will be exposed. These result from releases of toxicants into the air and from chemical reactions that take place in the atmosphere. Such pollutants can be vapors, aerosols or particulates and primarily arise from industrial manufacturing, mining, agricultural activity, volcanoes, fires and fossil fuel combustion.

The United States Environmental Protection Agency (EPA) produces a regularly updated list of hazardous air pollutants (EPA, 2020). Worldwide, other nations and groups of nations produce similar lists and numerous literature references identify still other known air pollutants. Table 4.1, toxic chemicals in the air, which was compiled from these sources, shows the range of pervasive toxic chemicals that are found in ambient air. All species listed in this table cause an increase in oxidative stress when absorbed by the body.

The EPA National Air Toxics Assessment, which is based upon analysis of 33 air pollutants, is used to produce results that are representative of air quality and its effects on human health (EPA 2020a). The chemicals on Table 4.1 that are on this list are starred (*) in the table.

The World Health Organization's IARC Monograph-109 maintains a list of air pollutants that are established or probable carcinogens [IARC Mongraph-109. The chemicals on Table 4.1 that IARC so identifies are double starred (**) in the table.

Not all the species listed in Table 4.1 are found in all locations and the concentrations of these xenobiotics vary with proximity to source points and prevailing meteorological conditions. On the other hand, numerous other air pollutants are found in different locations.

Ambient air is also contaminated with microscopic particles and fibers. As discussed in chapter 3, these are toxic by themselves and also act as adsorbents of many of the species

Oxidative Stress
https://doi.org/10.1016/B978-0-323-91890-9.00028-3

TABLE 4.1 Toxic chemicals in the air.

Acetaldehyde**
Acetamide
Acetonitrile
Acetophenone
2-acylaminofluorene
Acrolein**
Acrylamide
Acrylic acid
Acrylonitrile**
Allyl chloride
4-aminobiphenyl
Ammonia
Aniline
o-anisidine
Antimony compounds
Arsenic compounds (including arsine)* **
Asbestos*
Benzene* **
Benzidine
Benzo[a]pyrene*
Benzotrichloride
Benzyl chloride
Beryllium compounds* **
Biphenyl
Bis(2-ethylhexyl)phthalate
Bis(chloromethyl)ether
Bromoform
1,3-butadiene* **
n-butanol
Cadmium compounds* **
Calcium cyanamide

TABLE 4.1 Toxic chemicals in the air.—cont'd

Caprolactum
Captan
Carbaryl
Carbon disulfide
Carbon monoxide
Carbon tetrachloride**
Catechol
Chloramben
Chlordane
Chlorine
Chloroacetic acid
2-chloroacetophenone
Chlorobenzene
Chlorobenzilate
1-chloro-1,1-difluoromethane
Chloroform**
Chloromethyl methyl ether
Chloroprene
Chromium compounds* **
Cresols
Cresylic acid
Cumene
Cobalt compounds
Coke oven emissions (including carbon black)
Copper compounds
Cyanides (sodium and potassium)
Cylcopenta[cd]pyrene
2,4-D, salts and esters
DDE
Diazomethane
Dibenzofurans

(Continued)

TABLE 4.1 Toxic chemicals in the air.—cont'd

1,2-dibromo-3-chloropropane
Dibutylphthalate
p-1,4-dichlorobenzene
3,3-dichlorobenzidine
1,1-chloro-1-fluoroethane
Dichloroethyl ether
1,3-dichloropropene**
Dichlorvos
Diesel particulate matter**
Diethanolamine
N,N-diethyl aniline
Diethyl sulfate
3,3-dimethoxybenzindine
Dimethyl aminoazobenzene
3,3′-dimethyl benzidine
Dimethyl carbamoyl chloride
Dimethyl formamide
1,1-dimethyl hydrazine
Dimethyl phthalate
Dimethyl sulfate
4,6-dinitro-o-cresol and salts
2,4-dinitrophenol
2,4-dinitrotoluene
1,4-dioxane
1,2-diphenylhydrazine
Epichlorohydrin
1,2-epoxybutane
Ethyl acetate
Ethyl acrylate
Ethyl benzene
Ethyl carbamate
Ethyl chloride
Ethylene dibromide* **

TABLE 4.1 Toxic chemicals in the air.—cont'd

Ethylene dichloride
Ethylene glycol
Ethylene imine
Ethylene oxide* **
Ethylene thiourea
Ethylidene dichloride
Fine mineral fibers
Formaldehyde* **
Glycol ethers
Heptachlor
Hexachlorobenzene**
Hexachlorobutadiene
Hexachlorocyclopentadiene
Hexachloroethane
Hexamethylene-1,6-diisocyanate
Hexamethylphosphoramide
Hexane
Hydrazine**
Hydrochloric acid
Hydrogen fluoride
Hydrogen sulfide
Hydroquinone
Isophorone
Lead compounds* **
Lindane
Maleic anhydride
Manganese compounds**
Mercury compounds**
Methanol
Methoxychlor
Methyl bromide
Methyl chloride
Methyl ethyl ketone

(*Continued*)

TABLE 4.1 Toxic chemicals in the air.—cont'd

Methyl hydrazine
Methyl iodide
Methyl isobutyl ketone
Methyl isocyanate
Methyl methacrylate
Methyl t-butyl ether
4,4-methylene bis(2-chloroaniline)
Methylene chloride**
Methylene diphenyl diisocyanate
4,4-methylenedianiline
Naphthalene
Nickel compounds* **
Nitrates and nitric acid
Nitrogen oxides (NOx)
4-nitrobiphenyl
4-nitrophenol
2-nitropropane
N-nitroso-N-methylurea
N-nitrosodimethyleneamine
N-nitrosomorpholine
N-methyl-2-pyrrolidone
Ozone
Parathion
Pentachloronitrobenzene
Pentachlorophenol
Perchloroethylene**
Phenol
p-phenylenediamine
Phosgene
Phosphine
Phosphorous
Phthalic anhydride

TABLE 4.1 Toxic chemicals in the air.—cont'd

Polybrominated biphenyls*
**Polychlorinated biphenyls (PCBs)*
Polynuclear aromatic hydrocarbons (PAHs)**
1,3-propane sulfone
Beta-propiolactone
Propionaldehyde
Propoxur
Propylene
Propylene dichloride**
Propylene oxide
1,2-propyleneimine
Quinoline**
Quinone
Radionuclides (including Radon)
Selenium compounds
Silica*
Sulfuric acid (including sulfur dioxide and trioxide)
Styrene
Styrene oxide
2,3,7,8-tetrachlorodibenzo-p-dioxin*
1,1,2,2-tertachloroethane**
Tetrachloroethylene*
Titanium tetrachloride
2,4-toluene diamine
2,4-toluene diisocyanate
o-toluidine
Toxaphene
1,2,4-trichlorobenzene
1,1,1-trichloroethane
1,1,2-trichloroethane
Trichloroethylene*

(Continued)

TABLE 4.1 Toxic chemicals in the air.—cont'd

Trichlorophenols
1,1,2-trichloropropane*
Triethylamine
Trifluralin
2,2,4-trimethylpentane
Vinyl acetate
Vinyl bromide*
Vinyl chloride* **
Vinyl fluoride*
Xylenes

listed in Table 4.1, facilitating their uptake by the body. An example of this phenomenon is the absorption of polychlorinated dibenzo-p-dioxins and dibenzofurans onto municipal incinerator fly ash (Buser et al., 1978).

With solar radiation acting as a catalyst, many of the chemicals in Table 4.1 react with themselves as well as with oxygen and nitrogen compounds in the air to generate additional toxicants. The formation of ozone is an example of such reactivity. Reactive organic molecules and nitrogen oxides react with atmospheric oxygen in the presence of sunlight via free radical mechanisms to produce ozone and a large number of new toxicants. Some of these reactions produce aerosols, small droplets of liquids suspended in air, which present dangers that far exceed those of vapors, as these droplets introduce very large quantities of toxicants relative to the inhalation of vapors of the same chemicals (Zeliger, 2011).

4.3 Confined air pollutants

Confined air pollutants are those pollutants present in a building, a room, a confined area or an outdoor area with limited circulation to which all individuals present will be exposed. These result from limited releases of toxicants into well-defined areas. Confined air pollutants include those in ambient air as well as others emitted indoors (Chang, 1994; Zeliger, 2011). Table 4.2 lists sources of these.

4.4 Air pollution and oxidative stress

The chemicals listed in Table 4.1 contain numerous hydrophiles and lipophiles. In combination with particulates present in polluted air, the numbers of possible mixtures possible are endless. Accordingly, health effects are not easily ascribable to specific pollutants or mixtures, but are categorized by the effects rather than the composition of the polluted air.

TABLE 4.2 Sources of confined air pollutants.

Cooking fuels
Tobacco smoke
Upholstery
Carpets
Adhesives
Paints
Varnishes
Deodorant air sprays
Disposable diapers
Candles
Particle board
Plywood
Cleaning products
Pesticides
Mold and mildew growth
Animal fur and dander

Respiratory and cardiovascular diseases are well known to result from chronic exposure to polluted air (Huynh et al., 2010; Lodovici and Bigagli, 2011; Lee et al., 2014). Air pollution has also been shown to accelerate aging by erosion of telomeres (Grahame and Schlessinger, 2012). Recent findings have associated air pollution with other health effects, including: lowering of bone mineral content (Ranzani et al., 2020), macular degeneration (Chua, et al., 2020), dementia (Zeliger, 2019; Grande et al., 2020) higher rates of infection and mortality of viruses, including COVID-19 (Wu et al., 2020), and chronic inflammatory skin disease (Araviiskaia et al., 2019).

Though the effects just noted are exhibited by different and unrelated body organs, all are a result of oxidative stress. As discussed in chapter 3, air pollution is primarily composed of fine particulates ($PM_{2.5}$ of smaller silicates and heavy metals) atmospheric gases (ozone, oxides of nitrogen and sulfur) and organic compounds pollutants (such as those listed in Table 4.1). Each of these produces oxidative stress, albeit via different mechanisms. Silicates induce inflammation, heavy metals act via oxidative cycling as shown by the Fenton reaction, atmospheric gases undergo chemical reactions in the atmosphere to produce persistent free radicals and organic chemicals produce free radicals via metabolic pathways (Lodovici and Bigagli, 2011; Xu et al., 2020). These mechanisms are discussed in detail in section II of this book.

References

Araviiskiai, E., Berardesca, E., Birber, T., Gontijo, G., Viera, M.S., Marrot, L., Chuberre, B., 2019. The impact of airborne pollution on skin. J. Euro. Acad. Dermatol. Venereal 33, 1496–1505.

Buser, H.R., Bosshardt, H.P., Rappe C Lindahl, R., 1978. Identification of polychlorinated dibenzofuran isomers in fly ash and PCB pyrolysis. Chemosphere 5, 419–429.

Chang, C.C., Ruhl, R.A., Halpern, G.M., 1994. Building components contributors of the sick building syndrome. J. Asthma 31 (2), 127–137.

Chua, S.Y.L., Warwick, A., Peto, T., Balaskas, K., Moore, A.T., Reisman, C., et al., 2020. Association of air pollution with aige-related macular degeneration and retinal thickness in UK biobank. Br J Ophthamol. https://doi.org/10.1136/bjophthomol-2020-316218. (Accessed 1 February 2021).

EPA, 2020. United States Environmental Protection Agency. The original list of hazardous air pollutants. www.epa.gov/cgi-bin/epaprintonly.cgi. (Accessed 20 January 2021).

EPA, 2020a. United States Environmental Protection Agency. National-Scale air toxic assessment overview: the 33 pollutants. www.epa.gov/ttn/atw/nata/34poll.html. (Accessed 20 January 2021).

Grahame, T.J., Schlesinger, R.B., 2012. Oxidative stress-induced telomeric erosion as a mechanism underlying airborne particulate matter-related cardiovascular disease. Part. Fibre Toxicol. http://www.particleandfibretoxicology.com/content/9/1/21. (Accessed 10 February 2021).

Grande, G., Ljungman, P.L.S., Eneroth, K., Bellander, T., Rizzuto, D., 2020. Association between cardiovascular disease and long-term exposure to air pollution with risk of dementia. JAMA Meurol. 77 (7), 801–809. https://doi.org/10.1001/jamaneurol.2019.4914. (Accessed 10 November 2020).

Huynh, P., Salam, M.T., Morphew, T., Kwong, K.Y.C., Scott, L., 2010. Residential proximity to freeways is associated with uncontrolled asthma in inner-city Hispanic children and adolescents. J. Allergy. https://doi.org/10.1155/2010/157249. (Accessed 15 January 2021).

IARC Monograph – 109. World Health Organization. International Agency for Research on Cancer. Outdoor Air pollution. https://monographs.iarc.who.int/. Accessed February 22, 2021.

Lee, B.J., Kim, B., Lee, K., 2014. Air pollution exposure and cardiovascular disease. Toxicol. Res. 30 (2), 71–75.

Lodovici, Bigagli, 2011. Oxidative stress and air pollution exposure. J. Toxicol. https://doi.org/10.1155/2011/487074. (Accessed 15 January 2021).

Ranzani, O.T., Mila, C., Kulkarni, B., Kinra, S., Tonne, C., 2020. Association of ambient and household air pollution with bone mineral content among adults in Peri-urban south India. JAMA Netw. Open. https://doi.org/10.1001/jamanetworkopen.2019.18504. (Accessed 15 January 2021).

WHO, 2020. World Health Organization. World health statistics 2020. https://www.who.int/data/gho/whs-2020-visual-summary. (Accessed 22 February 2021).

Wu, X., Nethery, R.C., Sabath, M.B., Braun, D., Dominici, F., 2020. Air pollution and COVID-19 mortality in the United States: strengths and limitations of an ecological regression analysis. Sci. Adv. 6, eabd4049, 2020.

Xu, Y., Yang, L., Wang, X., Zheng, M., Li, C., Zhang, A., et al., 2020. Ecotoxicol Environ Safety; 196. https://www.sciencedirest.com/science/artricle/abs/pii/S0147651320304103. (Accessed 23 February 2021).

Zeliger, 2011. Toxic Effects of Chemical Mixtures, second ed. Elsevier, London.

Zeliger, H.I., 2019. Predicting Alzheimer's disease onset. European J. Med. Health Sci. 1 (1). https://doi.org/10.24018/ejmed.2019.1.1.16. (Accessed 23 February 2021).

CHAPTER

5

Water and soil pollution

5.1 Introduction

There are virtually no sources of drinking water on earth that are not contaminated with xenobiotics. Rain water cleanses the atmosphere as it forms and falls. Surface collection basins from which potable water is drawn; rivers, streams and lakes, accumulate ground level pollutants from runoff in addition to those carried in rain water. Underground water, which is somewhat filtered and generally contains lesser quantities of pollutants than surface water, may itself be contaminated by ground releases of toxicants and by contaminants produced by chemical reactions in the soil and water.

Biologically contaminated water is treated to remove bacteria, viruses, fungi and parasites, via aeration, chlorination, ozone, ultra violet radiation or a combination of these methods to make it safe for human consumption.

Major sources of water pollutants include mining, manufacturing, farming, power production, transportation, urban runoff, chemical products used in the home and personal care products. Virtually all oxidative stress raising species are present in both water and soil and readily transfer between both (Zeliger, 2011).

5.2 Water polluting chemicals

Water polluting chemicals come from multiple sources, which include natural as well as anthropogenic activities.

Many naturally occurring chemicals that contribute to OS elevation are present in drinking water, the concentrations of which vary widely. In some areas, the quantities of these chemicals are high enough to require removal prior to be safe for human consumption (Zeliger, 2011). Table 5.1 lists these chemicals for U.S. potable water.

5.2.1 Priority pollutants

1979, the United States EPA published a list of 129 priority water pollutants from human activities most responsible for the contamination of water (EPA, 1979). Table 5.2 lists these chemicals as well as their sources.

Oxidative Stress
https://doi.org/10.1016/B978-0-323-91890-9.00009-X

TABLE 5.1 Naturally occurring OS elevating chemicals in U.S. drinking water.

Aluminium
Ammonia
Arsenic
Bromine
Chromium
Copper hydrogen sulfide
Lead
Manganese
Mercury
Nitrate
Nitrite
Phosphate
Selenium
Silver
Sulfate
Hexavalent chromium
Lithium
Phosphorus
Chloromethane
Alpha particle activity (excluding radon and uranium)
Alpha particle activity (including radon and uranium)
Radon
Total uranium
Uranium-234
Uranium-235
Uranium-238
Total radium
Radium-226
Radium-228
Alpha particle activity (suspended)
Gross beta particle activity (dissolved)
Gross beta particle activity (suspended)
Potassium-40

TABLE 5.2 EPA 129 priority pollutants and their product sources.

Priority pollutant	Source
1. Acenaphthene	Manufacturing of insecticides, fungicides, dyes plastics
2. Acrolein	Chemical manufacturing intermediate
3. Acrylonitrile	Chemical manufacturing
4. Benzene	Organic chemicals, solvents, petroleum products
5. Benzidine	Manufacturing of chemicals, rubber and dyes
6. Carbon tetrachloride	Manufacture of chlorinated hydrocarbons, chemical intermediate
7. Chlorobenzene	Chemical manufacturing, degreaser
8. 1,2,4-trichlorobenzene	Manufacturing of chemicals, heat transfer lubricant
9. Hexachlorobenzene	Fungicide for wood preservation
10. 1,2-Dichloroethane	Cleaners and wax removers
11. 1,1,1-Trichloroethane	Degreasers, cleaners
12. Hexachloroethane	Insecticides
13. 1,1-Dichloroethane	Degreasers
14. 1,1,2-Trichloroethane	Waxes, cleaners, photographic products
15. 1,1,2,2-Tetrachloroethane	Fumigant, garden sprays
16. Chloroethane	Waxes, cleaners
17. Bis (chloromethyl) ether	Chemical manufacturing
18 Bis (2-chloroethyl) ether	Paints, varnishes, callus removers
19. 2-Chloroethyl vinyl ether	Waterproofing compounds
20. 2-Chloronaphthalene	Engine oil additive
21. 2,4,6-Trichlorophenol	Adhesives, cleaners, disinfectants
22. p-chloro-m-cresol	Glue, paint and shampoo preservatives
23. Chloroform	Manufacturing solvents, disinfection byproduct
24. 2-Chlorophenol	Disinfectants, cleaners, paints
25. 1,2-Dichlorobenzene	Waxes, polishes, cleaners, deodorizers, preservatives
26. 1,3-Dichlorobenzene	Waxes, polishes, cleaners, deodorizers, preservatives
27. 1,4-Dichlorobenzene	Fruit spray, household cleaners, dyes, disinfectants
28. 3,3-Dichlorobenzidine	Dye manufacture
29. 1,1-Dichloroethylene	Plasticizer, environmental decomposition of trichloroethylene
30. 1,2-Dichloroethylene	Solvents, cleaners

(*Continued*)

TABLE 5.2 EPA 129 priority pollutants and their product sources.—cont'd

Priority pollutant	Source
31. 2,4-Dichlorophenol	Wood preservatives, insect repellants cosmetics
32. 1,2-Dichloropropane	Tar removers, waxes, degreasers
33. 1,3-Dichloropropylene	Tar removers, waxes, degreasers
34. 2,4-Dimethylphenol	Asphalt products, shampoos, skin treatments
35. 2,4-Dinitrotoluene	Manufacture of TNT
36. 2,6-Dinitrotoluene	Manufacture of TNT
37. 1,2-Diphenylhydrazine	Manufacture of chemicals
38. Ethylbenzene	Solvents, manufacture of plastics, petroleum fuels
39. Fluoranthene	Coal tars, antibiotic creams, shampoos, skin treatments
40. 4-Chlorophenyl phenyl ether	Dielectric fluid
41. 4-Bromophenyl phenyl ether	Dielectric fluid
42. Bis (2-chloroisopropyl) ether	Waxes, paint removers, degreasers
43. Bis (2-chloroethoxy) methane	Manufacture of adhesives, sealants
44. Methylene chloride	Solvents, degreasers, cleaners
45. Methyl chloride	Manufacture of chemicals and, herbicides
46. Methyl bromide	Manufacture of crop fumigants
47. Bromoform	Solvents, manufacture of pharmaceuticals, disinfection byproduct
48. Dichlorobromomethane	Drinking water treatments, waxes, greases
49. Trichlorofluoromethane	Aerosol propellants, perfumes, deodorants
50. Dichlorodifluoromethane	Aerosol propellants, perfumes deodorants
51. Chlorodibromomethane	Aerosol propellants, perfumes, deodorants, fire extinguishers
52. Hexachlorobutadiene	By-product of trichloroethylene manufacturing
53. Hexachlorocyclopentadiene	Pesticide manufacture
54. Isophorone	Solvents, pesticide manufacture, degreasers
55. Naphthalene	Deodorants, detergents, moth repellant, skin treatments
56. Nitrobenzene	Textiles, dyes
57. Nitrophenol	Manufacture of dyes and chemicals
58. 4-Nitrophenol	Pesticide manufacture
59. 2,4-Dinitrophenol	Manufacture of pesticides, photographic products
60. 4,6-Dinitro-o-cresol	Pesticides

TABLE 5.2 EPA 129 priority pollutants and their product sources.—cont'd

Priority pollutant	Source
61. N-nitrosodimethylamine	Manufacture of dyes
62. N-nitrosodiphenylamine	Manufacture of rubber
63. N-nitrosodi-N-propylamine	Manufacture of organic chemicals
64. Petachlorophenol	Wood preservatives
65. Phenol	Adhesives, preservatives, disinfectants, callus removers
66. Bis (2-ethylhexyl) phthalate	Plasticizer
67. Butylbenzyl phthalate	Plasticizer
68. Di-n-butyl phthalate	Plasticizer
69. Dioctyl phthalate	Plasticizer
70. Diethyl phthalate	Plasticizer
71. Dimethyl phthalate	Plasticizer
72. Benzo[a]anthracene	Cigarette smoke, asphalt products, petroleum combustion
73. Benzo[a]pyrene	Cigarette smoke, asphalt products, petroleum combustion
74. 3,4-Benzofluoranthene	Cigarette smoke, asphalt products, petroleum combustion
75. 11,12-Benzofluoanthene	Cigarette smoke, asphalt products, petroleum combustion
76. Chrysene	Cigarette smoke, asphalt products, petroleum combustion
77. Acenaphthylene	Dye manufacturing, petroleum combustion
78. Anthracene	Dye manufacturing, petroleum combustion
79. Benzo[c]perylene	Dye manufacturing, petroleum combustion
80. Phenanthrene	Dye, manufacturing, petroleum combustion
81. Indeno[1,2,3-cd]pyrene	Dye manufacturing, petroleum combustion
82. Dizenzo[a,h]anthracene	Dye manufacturing, petroleum combustion
83. Indeno[1,2,3-cd]pyrene	Dye manufacturing, petroleum combustion
84. Pyrene	Dye manufacturing, petroleum combustion
85. Tetrachloroethylene	dry cleaning solvent, degreasers
86. Toluene	Solvents, adhesives, paints, varnishes
87. Trichloroethylene	Degreasers, solvents
88. Vinyl chlorid	Manufacture of PVC resins, adhesives, environmental decomposition of trichloroethylene
89. Aldrin	Insecticides

(Continued)

TABLE 5.2 EPA 129 priority pollutants and their product sources.—cont'd

Priority pollutant	Source
90. Dieldrin	Insecticides
91. Chlordane	Insecticides
92. 4,4′-DDT	Insecticides
93. 4,4′-DDE	Environmental decomposition of 4,4′-DDT
94. 4,4′-DDD	Impurity in 4,4′-DDT
95. Endosulfan-alpha	Acaricides
96. Endosulfan-beta	Acaricides
97. Endosulfan sulfate	Acaricides
98. Endrin	Insecticides, rodenticides
99. Endrin aldehyde	Insecticides, rodenticides
100. Heptachlor	Insecticides, rodenticides
101. Heptachlor epoxide	Insecticides, rodenticides
102. BHC-alpha	Insecticide, fungicide
103. BHC-beta	Insecticide, fungicide
104. BHC-gamma	Insecticide, fungicide
105. BHC-delta	Insecticide, fungicide
106. TCDD	Manufacture of chlorophenols, herbicide contaminant, incineration of chemical wastes
107. Toxaphene	Insecticides
108. Arochlor 1016 (PCB)	Electrical transformers, carbonless copy papers
109. Arochlor 1221 (PCB)	Electrical transformers, carbonless copy papers
110. Arochlor 1232 (PCB)	Electrical transformers, carbonless copy papers
111. Arochlor 1242 (PCB)	Electrical transformers, carbonless copy papers
112. Arochlor 1248 (PCB)	Electrical transformers, carbonless copy paper
113. Arochlor 1254 (PCB)	Electrical transformers, carbonless copy paper
114. Arochlor 1260 (PCB)	Electrical transformers, carbonless copy paper
115. Antimony	Mining, smelting, paints, fireproofing compounds
116. Arsenic	Insecticides, herbicides, pressure treated woods
117. Asbestos	Construction materials, insulation materials

TABLE 5.2 EPA 129 priority pollutants and their product sources.—cont'd

Priority pollutant	Source
118. Beryllium	Electrical components, manufacturing of chemicals
119. Cadmium	Welding rods, alkaline storage batteries, fluorescent lamps
120. Chromium	Chrome plating, paints, chemical manufacturing
121. Copper	Corrosion in plumbing, algicides, bronze plating
122. Cyanide	Ore refining, combustion of plastics, electroplating
123. Lead	Corrosion in plumbing, paints applied pre-1973, storage batteries
124. Mercury	Electrical switches, pharmaceutical preservatives, mining of gold and silver
125. Nickel	Mining, smelting, alloy production
126. Selenium	Mining, coal combustion, glass manufacture
127. Silver	Dentistry, photographic film, jewelry
128. Thallium	Smelting and refining of lead and zinc, pesticides
129. Zinc	Textile finishing, skin treatments, electrical batteries

Since this priority list was first drawn up in 1979, many other hazardous contaminants in water have been identified.

5.2.2 Agricultural chemicals

Agricultural activities are a major producer of water pollutants. Although pesticides are the primary pollutants, other water-contaminating chemicals are also attributable to agriculture (Zeliger, 2011). Table 5.3 lists agricultural pollutants in U.S. drinking water.

5.2.3 Urban runoff

Urban runoff from roads, lawns and human waste contributes the water polluting chemicals in Table 5.4 (Zeliger, 2011 and references therein).

5.2.4 Industrial chemicals

Industrial activities result in the discharge of a large number and wide variety of chemicals into drinking water (Zeliger, 2011 and references therein). Table 5.5 lists these water pollutants.

TABLE 5.3 Agricultural pollutants in U.S. drinking water.

Ammonia
Chlorate
Nitrate and nitrite mix
Nitrate (alone)
Nitrite (alone)
Phosphate
Sulfate
Thallium
MBAS (surfactants)
Phosphorus
Endrin
Desethylatrazine
Desisopropyatrazine
Lindane
Methoxychlor
Toxaphene
Carbaryl
Methomyl
Baygon (propoxur)
Methiocarb
Acetochlor
Paraquat
Prometon
2,4-Bis-6-(isopropylamino)
Dalapon
Diquat
Endothall
Glyphosate
Oxamyl (Vydate)
Simazine
Pichloram

TABLE 5.3 Agricultural pollutants in U.S. drinking water.—cont'd

Dinoseb
Aldicarb sulfoxide
Aldicarb sufone
Metolachlor
Carbofuran
Aldicarb
Atrazine
Alaclor
EPTC (Eptam)
Butylate (Sutan)
Cyanazine (Bladex)
Trifluralin
Ethion
Hepatochlor
3-Hydroxycarbofuran
Hepatachlor epoxide
Endosulfan I
Dieldrin
DDT
Butachlor
Propachlor
Bromacil
Dacthal
Diuron
2,4-D
2,4-DB
2,4,5-TP (Silvex)
2,4,5-T
Chloramben
Dichloroprop
Bromomethane
Isophorone
Alpha-lindane

(Continued)

TABLE 5.3 Agricultural pollutants in U.S. drinking water.—cont'd

Beta-lindane
Aldrin
1,3-Dichloropropene
Dicamba
Iodomethane
Chloropicrin
Metribuzin
Bentazon (Basagran)
Molinate (Ordram)
Thiobencarb (Bolero)
Foaming agents
Phenols
1,2-Dribomo-3-chloropropane
Ethylene dibromide
Chlordane
m-dichlorobenzene
Ethylbenzene
Perchlorate
Aldicarb
Alpha chlordane

5.3 Chemical reactions in groundwater

In addition to those chemicals that are released directly, chemical reactions in water and soil generate still more pollutants. After being discharged onto soil and into groundwater, some pollutants are degraded into chemicals that are more toxic than the parent compounds. Such degradation occurs via chemical oxidation, reduction, hydration and bacterial action.

The transformations of the chlorinated hydrocarbons perchloroethylene (PCE), trichloroethylene (TCE) and 1,1,1-trichloroethane (TCA) are illustrative of these types of reactions. Cis- and trans-1,2-dichloroethylene (1,2-DCE), 1,1-dichloro ethylene (1,1-DCE) and vinyl chloride (VC), a known liver carcinogen, are often found in well water which has been contaminated only with tetrachloroethylene and/or trichloroethylene (Parsons et al., 1984). Some of the transformations of PCE and TCE are biologically catalyzed and occur

TABLE 5.4 Urban runoff water pollutants in U.S. drinking water.

Ammonia
Arsenic
Cadmium
Copper
Hydrogen sulfide
Lead
Mercury
Nitrate and nitrite mix
Nitrate
Nitrite
Phosphate
Antimony
Lithium
Molybdenum
Oil and grease total
Phosphorus
Lindane
Baygon (propoxur)
Paraquat
Glyphosate
Trifluralin
Isopropyl alcohol
Trichlorofluoromethane
Acetone
Naphthalene
Methyl tertiary butyl ether
Fluorine
Phenanthrene
Anthracene
Dimethylphthalate
Diethylphthalate
Fluoranthene

(Continued)

TABLE 5.4 Urban runoff water pollutants in U.S. drinking water.—cont'd

Pyrene
di-n-butyl phthalate
Butyl benzyl phthalate
Benzo[a]anthracene
Benzo[b]fluoanthene
Benzo[k]fluoanthene
Benzo[a]pyrene
Indeno[1,2,3-cd]pyrene
Dibenz[a,h]anthracene
Benzo[g,h,i]perylene
Alpha-lindane
Beta-lindane
Tert-butylbenzene
Sec-butylbenzene
Chloropicrin
Trichlorofluoroethane
Phenols
Xylenes (total)
p-xylene
o-xylene
m-xylene
Tetrachloroethylene
Benzene
Bromobenzene
n-propylbenzene
Ethyl-t-butyl ether

in soil and sediment, while others are believed to occur in groundwater (Parsons et al., 1985; Barrio-Lage et al., 1987). TCA hydrolyses to 1,1-DCE, a human carcinogen, in ground water (Cline et al., 1984).

Such transformations help shed light on how some of the "mysterious" human illnesses and cancers are triggered by the discharge of pollutants into water that are not by themselves

TABLE 5.5 Industrial pollutants in U.S. drinking water.

Aluminum
Ammonia
Bromide
Arsenic
Chlorate
Barium
Cadmium
Chromium
Cyanide
Hydrogen sulfide
Lead
Manganese
Mercury
Nitrate and nitrite (mix)
Nitrate
Nitrite
Phosphate
Selenium
Silver
Strontium
Sulfate
Antimony
Beryllium
Chromium (hexavalent)
Lithium
Molybdenum
Thallium
Vanadium
MBAS (surfactants)
Oil and grease (total)
Phosphorus

(*Continued*)

TABLE 5.5 Industrial pollutants in U.S. drinking water.—cont'd

Carbon disulfide
Lindane
p-isopropyltoluene
di(2-ethylhexyl)adipate
di(2-ethylhexyl)phthalate
Hexachlorocylcopentadiene
1,4-Dioxane
Endosulfan I
Butyl acetate
Ethyl ether
Isopropyl alcohol
Chloromethane
Dichlorodifluoromethane
Bromomethane
Chloroethane
Trichlorofluoromethane
n-nitrosodiphenylamine
Aniline
1,2-Dibromoethylene
Acrylonitrile
Acetone
Isopropyl ether
Hexachlorobutadiene
Methyl ethyl ketone
Naphthalene
Methyl isobutyl ketone
Methyl tertiary butyl ether
Nitrobenzene
Acenaphthylene
Acenaphthene
Dimethylphthalate
Diethylphthalate
Fluoranthene

TABLE 5.5 Industrial pollutants in U.S. drinking water.—cont'd

Pyrene
di-n-butyl phthalate
Butyl benzyl phthalate
Methyl methacrylate
Chrysene
Indeno[1,2,3-cd]pyrene
Dibenz[a,h]anthracene
Pentachlorophenol
n-hexane
1,2,4-Trichlorobenzene
cis-1,2-Dichloroethylene
Total PCBs
Arochlor 1016
Arochlor 1221
Arochlor 1232
Arochlor 11242
Arochlor 1248
Arochlor 1252
Arochlor 1260
1,1-Dichloropropene
1,3-Dichloropropane
1,2,3-Trichloropropane
2,2-Dichloropropane
1,2,4-Trimethylbenzene
1,2,3-Trichlorobenzene
n-Butylbenzene
sec-Butylbenzene
tert-Butylbenzene
1,3,5-Trimethylbenzene
Bromochloromethane
Chloropicrin
2-Nitropropane

(Continued)

TABLE 5.5 Industrial pollutants in U.S. drinking water.—cont'd

Glyoxal
Trichlorotrifluoroethane
Foaming agents
Phenols
Ethylene dibromide
Xylenes (total)
o-xylene
m-xylene
p-xylene
Meta and para xylene (mix)
Formaldehyde
Methylene chloride
o-chlorotoluene
p-chlorotoluene
o-dichlorobenzene
m-dichlorobenzene
p-dichlorobenzene
Vinyl chloride
1,1-Dichloroethylene
1,1-Dichloroethane
1,2-Dichloroethane
trans-1,2-Dichloroethylene
1,1,1-Trichloroethane
Carbon tetrachloride
1,2-Dichloropropane
Trichloroethylene
Tetrachloroethylene
1,1,2-Trichloroethane
1,1,1,2-Tetrachloroethane
1,1,2,2-Tetrachloroethane
Chlorobenzene
Benzene
Toluene

TABLE 5.5 Industrial pollutants in U.S. drinking water.—cont'd

Ethylbenzene
Bromobenzene
Isopropylbenzene
Styrene
n-propylbenzene
Perchlorate
Ethyl-t-butyl ether
Dichlorofluoromethane
Alpha particle activity (including radon and uranium)
Alpha particle activity (excluding radon and uranium)
Uranium (total)
Uranium-234
Uranium-235
Uranium-238
Radium (total)
Radium-226
Radium-228
Alpha particle activity (suspended)
Gross beta activity (dissolved)
Gross beta activity (suspended)
Potassium-40
Tritium
Gross beta particles & photon emitters (man-made)
Manganese-54
Strontium-90

highly toxic. The degradation sequence of tetrachloroethylene and trichloroethylene are schematically presented in Fig. 5.1.

The schematic just presented demonstrates a sequence that accounts for the presence of OS producing chemicals in water and soil. Most OS, related to ingestion of contaminated water, however, water can only be attributed to unknown chemicals, due to the endless possibilities of chemical mixtures (Zeliger, 2011).

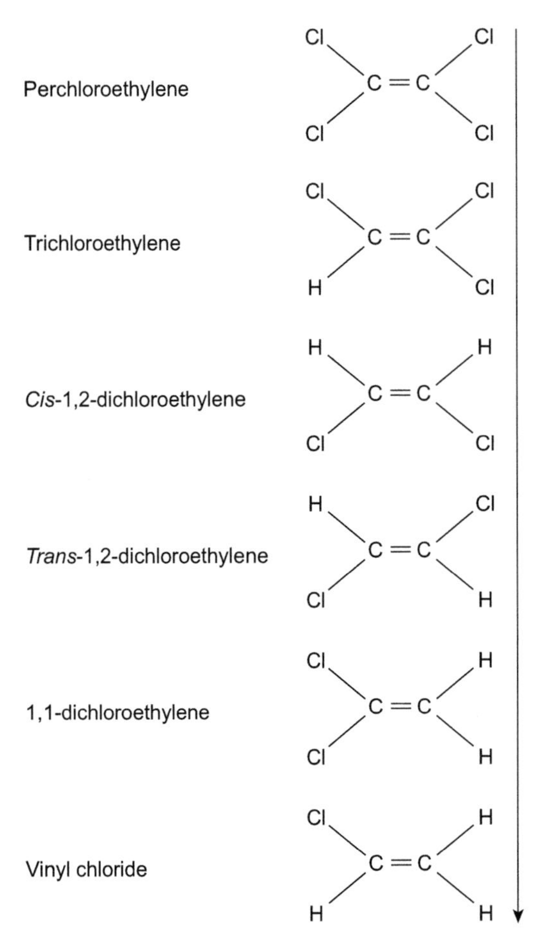

FIG. 5.1 Degradation sequence of tetrachloroethylene. *Reproduced with permission from Zeliger, H.I., 2011. Human toxicology of chemical mixtures. Second Ed. Elsevier, Oxford.*

5.4 Disinfection bye-products

Drinking water has been disinfected with chlorine for approximately 100 years to protect against waterborne infectious diseases. In addition to chlorination, other methods of drinking water disinfection include the use of chlorine dioxide (either alone or in combination with chlorine), the addition of ammonia to chlorine to form chloramines, ozone treatment, oxidation with potassium permanganate and ultra violet radiation. Of these, chlorination, however, is by far the most widely used method. Treatment with chlorine has virtually eliminated cholera, typhoid, dysentery, hepatitis A and other waterborne diseases (Water Quality and Health Counsel, 2001).

Chlorination of raw water high in organic content and/or infused with sea water results not only in the disinfection of water, but also in the formation of disinfection by-products

(DBPs). These include; trihalomethanes (THMs), haloacetic acids (HAAs) and haloacetonitriles (HANs), all of which are individually toxic and have been associated with cancer, liver disease, kidney disease, birth defects and reproductive failures (Kolvusalo and Vertiainen, 1997; Lilly et al., 1997; Reif et al., 1996). Consumption of water containing mixtures of DBPs has produced spontaneous abortion rates greater than normal in areas drinking water was high in DBPs (Waller et al., 1998; Savitz et al., 2006).

5.5 Soil pollution

Soil is defined as: "(i) The unconsolidated mineral or organic material on the immediate surface of the earth that serves as a natural medium for the growth of land plants. (ii) The unconsolidated mineral or organic matter on the surface of the earth that has been subjected to and shows the effects of genetic and environmental factors of: climate (including water and temperature effects), and macro- and microorganisms, conditioned by relief, acting on parent material (material in which soils form) over a period of time. A product-soil differs from the material from which it is derived in many physical, chemical, biological and morphological properties and characteristics" (USDA, 2021).

This broad definition accommodates numerous variations in soil composition, with more 20,000 different kinds of soil identified in the United States alone (Soil Science Society of America, 2021). Soil formation is a dynamic, ongoing process dependent upon climate (wind, water availability temperature), topography, time and biological activity of plants, animals and microorganisms.

The chemical composition of soil is complex, naturally containing numerous inorganic and organic compounds. Pollution of soil may be accomplished via the addition of chemical species that are alien to soil, such as organochlorine pesticides, or through the addition of quantities of naturally occurring chemicals which, at elevated concentrations, are toxic. Examples of the latter being chlorides and heavy metals such as copper and selenium.

The toxicological impact of polluted soil on humans is indirect. Chemical pollutants in soil can dissolve in ground water and be carried to aquifers and surface sources of water used for human consumption. Pollutants in soil also affect its ability to support plant life (fertility) by depressing microorganistic and soil dwelling animal activities. Plants growing in polluted soil absorb toxic chemicals through their root systems and induce toxic effects in humans when those plants are ingested. Polluted soil also adversely affects humans via bioaccumulation of toxic chemicals in animals when plants grown in it are ingested or dermally absorbed by animals that come in contact with it (Zeliger, 2011).

5.6 Plant absorption of soil toxins and bioacculation

Plants that grow in contaminated soils can absorb those contaminants. Once absorbed, such toxic chemicals are taken up by animals that feed on the contaminated plants and passed up the food chain to humans. Examples of this effect follow.

Prairie grass absorbs pesticides present in the soil in which it grows, including Atrazine, Arachlor, Metolachlor and Pendimetnalin (Belden et al., 2004). Cattle grazing on grass so contaminated bioaccumulate these pesticides and pass them along to humans who eat the affected cattle. PCBs, polychlorinated dibenzo-p-dioxins and polychlorinated dibenzofurans, all of which are well known to induce OS, are transferred to biota that grow in contaminated soils.

Foraging animals (chickens, turkeys and ducks) that eat biota contaminated with pesticides have been shown to be highly contaminated with those toxins. People eating such foul are thus exposed to extremely toxic chemicals (Zhao et al., 2006).

Other lipophilic chemicals, including polynuclear aromatic hydrocarbons (PAHs), are also readily absorbed from soil and passed up the food chain to humans (Roos et al., 2004).

Heavy metals absorbed by crops from contaminated soils are also passed up the food chain. Cadmium, zinc and lead are absorbed by wheat and rice plants (Muramoto et al., 1990). Cultivated ryegrass, fed to cattle, absorbs cadmium from polluted soil and passes it up the food chain to humans (Moreno et al., 2006).

5.7 Effects of mixtures

Some of the effects of toxic chemical mixtures on soil pollution are predictable. Acidic soils dissolve otherwise insoluble metal oxides and salts, thereby increasing available metal concentrations and toxicity to flora and fauna. Available copper content, e.g., is inversely proportional to increased pH of soil (De La Iglesia et al., 2006). Earthworm mortality in soil polluted by lead increases as pH decreases (Bradham et al., 2006). The addition of ethylenediaminetetraacetic acid (EDTA) and its disodium salt to soil contaminated with cadmium, lead and zinc increases the availability of these metals to plants and results in significant increases of uptake of these in plants (Lai et al., 2004).

Other effects of toxic chemical mixtures on soil may not be predictable. Mixtures of fertilizers and pesticides produce enhanced toxic effects. The additions of urea, superphosphate and potash enhance the toxicities of carbaryl and carbofuran insecticides to nitrogen fixing bacteria in soil (Padhy, 2001). Soil cocontaminated with arsenic and DDT does not break down DDT as rapidly as soil contaminated with DDT alone. This results in a persistence of DDT in the environment (Van Zwieten et al., 2003).

Chlorinated hydrocarbons are persistent volatile organic compound (VOC) pollutants that infiltrate soil from disposal of dry-cleaning fluids, degreasing solvents, food extraction solvents and paint strippers. Trichloroethylene (TCE) is illustrative of these compounds.

TCE, a widely used solvent and degreaser, is believed to be a carcinogen and mutagen. It is a dense nonaqueous phase liquid (DNAPL) which displaces water, sinks into the soil subsurface and permeates through the soil into groundwater. Some of the TCE, however, is bonded to lipophilic soil molecules which help retain it in the soil. TCE also accumulates in soil voids where it persists for long periods of time (Howard, 1990). Bacteria slowly biotransform TCE to the dichloroethanes, dichloroethylenes and vinyl chloride, as is shown in Fig. 5.1 above. This process is slow, however, resulting in the persistence of TCE in soil for years (Parons et al., 1984). TCE in soil acts as a solvent for other organic molecules and pesticides and contributes to the retention of these toxic compounds in soil for long periods of time and forms toxic mixtures with unknown consequences.

Contaminated sites, such as industrial chemical disposal locations, contain mixtures of numerous toxicants. These include multiple lipophiles and hydrophiles that can undergo chemical reactions, migrate, contaminate water and soil and be absorbed by plants and animals (Fent, 2004). Such sites are often acutely and chronically toxic, environmentally persistent and lead to bioaccumulation of toxicants in food webs. An excellent example of such a site is Love Canal in New York State.

From 1942–52, Love Canal was used as a disposal site for over 21,000 tons of chemical wastes, including halogenated organics, pesticides, chlorobenzenes and dioxins. In 1953, the landfill was covered and deeded to the Niagara Falls Board of Education. Subsequently, the area near the landfill was extensively developed with an elementary school and numerous homes constructed. In the 1960s and 1970s, ground water levels under the landfill rose and more than 100 lipophilic and hydrophilic organic compounds leached out, contaminating the air and drinking water supplies of thousands of people. The health effects of exposure to these chemicals include high rates of birth defects, immune system suppression, various cancers and chromosome damage. Children exposed were diagnosed with elevated numbers of seizures, learning problems, hyperactivity, skin rashes, eye irritation, abdominal pain and incontinence (Brown and Clapp, 2002). Though some of these effects could be attributed to individual compounds, such as the immunosuppression effects of TCDD (Silkworth et al., 1989), most of the problems reported could not be ascribed to the individual chemicals found. It is hypothesized that the unexpected effects were caused by exposures to mixtures of lipophilic and hydrophilic chemicals (Zeliger, 2003).

5.8 Oxidative stress from polluted water and soil

As is shown in the previous sections of this chapter, the toxic chemicals contained in polluted water and soil are essentially identical. This is not surprising, given that polluting OS increasing chemicals readily transfer between water and soil in the environment. A study on the environmental movement of perfluorinated substances (PFAS), which are used in numerous applications that include: use as clothing water repellants and incorporaton into leather, paper, cookware and cosmetics. PSAS are found in drinking water and are persistent throughout the globe.

PFAS are toxic to the thyroid, liver, kidneys and metabolic function and their OS producing effects are demonstrated by peroxidation of cell membrane lipids (via malondialdehyde increase), alteration of redox balance, enzyme inactivation and DNA damage (Bonato et al., 2020). PFAS are components sewage treatment plant sludge (which is often used as fertilizer) from which they readily transfer between water and soil, as is schematically shown in Fig. 5.2.

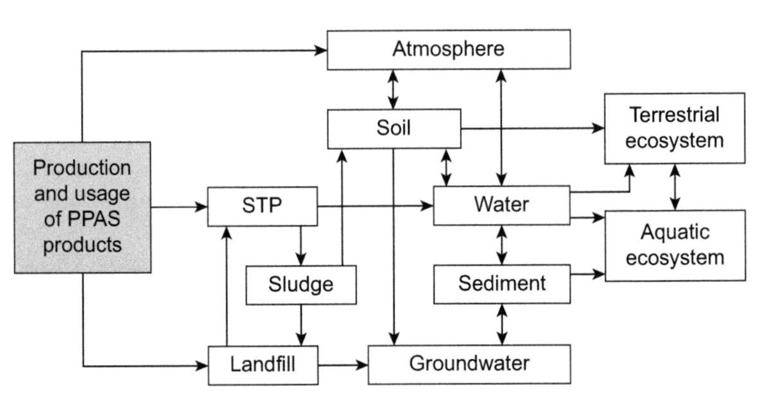

FIG. 5.2 Pathways of PFAS into the environment and their fate. *STP*, sewage treatment plant. *Reproduced with permission from Bonato, M., Corra, F., Bellio, N., Guidolin, L., Tallandini, L., Irato, P., Santovino, G. 2020. PFAS environmental pollution and antioxidant responses: an overview of the impact on human field. Int J Environ Res Public Health 2020 (17), 8020. https://dx.doi.org/10.3390/ijerph.17218020.*

References

Barrio-Lage, G.A., Parsons, F.Z., Nassar, R.S., Lorenzo, P.A., 1987. Biotransformation of trichloroethene in a variety of subsurface materials. Environ. Toxicol. Chem. 6, 571–578.

Belden, J.B., Phillips, T.A., Coats, J.R., 2004. 2004. Effect of prairie grass on the dissipation, movement and bioavailability of selected herbicides in prepared soil columns. Environ. Toxicol. Chem. 23 (1), 125–132.

Bonato, M., Corra, F., Bellio, M., Guidolin, L., Tallandini, L., Irato, I., Santovito, G., 2020. PFAS environmental pollution and antioxidant responses: an overview of the impact on human field. Int. J. Environ. Res. Publ. Health. https://doi.org/10.3390/ijerph17218020. (Accessed 26 February 2021).

Bradham, K.D., Cayton, E.A., Basta, N.T., 2006. Effect of soil properties on lead bioavailability and toxicity earthworms. Environ. Toxicol. Chem. 25 (3), 769–775.

Brown, P., Clapp, R., 2002. Looking back at Love canal. Publ. Health Rep. 117, 95–98.

Cline, P.V., Delfino, J.J., Cooper, W.J., 1984. Hydrolysis of 1,1,1-trichloroethane: Formation of 1,1-dichloroethane. Proc National Well Water Assoc Conf on petroleum hydrocarbons and organic chemicals in groundwater: prevention and detection, Houston, TX.

De La Iglesia, R., Castro, R., Ginocchio, R., 2006. Factors influencing the composition of bacterial communities found at abandoned copper-tailing dumps. Appl. Microbiol. 100 (3), 537–544.

Fent, K., 2004. Ecotoxicological effects at contaminated sites. Toxicology 205 (3), 223–240.

Howard, P.H., 1990. Handbook of Environmental Fate and Exposure Data for Organic Chemicals. Lewis Publishers, Chelsea MI, pp. 467–474.

Koivusalo, M., Vartiainen, T., 1997. Drinking water chlorination by-products and cancer. Rev. Environ. Health 12 (2), 81–90.

Lai, H.Y., Chen, Z.S., 2004. Effects of EDTA solubility of cadmium, zinc and lead and their uptake by rainbow pink and vetiver grass. Chemosphere 55 (3), 421–430.

Lilly, P.D., Ross, T.M., Pegram, R.A., 1997. Trihalomethane comparative toxicity: acute renal and hepatic toxicity of chloroform and bromodichloromethane following aqueous gavage. Fund. Appl. Toxicol. 40 (1), 101–110.

Moreno, J.L., Sanchez-Marin, A., Hernandez, T., Garcia, C., 2006. Effect of cadmium on microbial activity and a ryegrass crop in two semiarid soils. Environ. Man 37 (5), 626–633.

Muramoto, S., Nishizaki, H., Aoyama, I., 1990. The critical levels and the maximum metal uptake for wheat and rice plants when applying metal oxides to soil. J. Environ. Sci. Health B 25 (2), 273–280, 1273-280990.

Padhy, R.N., 2001. Monitoring of chemical fertilizers on toxicity of two carbamate insecticides to the cyanobacterium *Anabaena* PCCC 7120. Microbios 106 (415), 165–175.

Parons, F., Wood, P.R., DeMarco, J., 1984. Transformations of tetrachloroethene and trichloroethene in microcosms and groundwater. J. Am. Water Works Assc. 76, 56–59.

Parsons, F., Barrio-Lage, G., Rice, R., 1985. Biotransformation of chlorinated organic solvents in static microcosms. Environ. Toxicol. Chem. 4, 739–742.

Reif, J.S., Hatch, M.C., Bracken, M., et al., 1996. Reproductive and developmental effects of disinfection by-products in drinking water. Environ. Health Perspect. 104, 1056–1061.
Roos, P.H., Tschirbs, S., Pfeifer, F., et al., 2004. Risk potentials for humans or original and remediated PAH-contaminated soils: application of biomarkers of effect. Toxicology 205 (3), 181–194.
Savitz, D.A., Singer, P.C., Herring, A.H., Hartman, K.E., Weinberg, H.S., Makarushka, C., 2006. Exposure to drinking water disinfection by-products and pregnancy loss. Am. J. Epidemiol. 164 (11), 1043–1951.
Silkworth, J.B., Cutler, D.S., Sack, G., 1989. Immunotoxicity of 2,3,7,8-tetrachlorodibenzo-p-dioxin in a complex environmental mixture from the Love Canal. Fund. Appl. Toxicol. 12 (2), 303–312.
Soil Science Society of America, 2021. Glossary of Soil Science Terms 2006. https://www.soils.org/about-soils. (Accessed 25 February 2021).
USDA, United States Department of Agriculture, 2021. What is Soil?. https://www.nrcs.usda.gov/wps/portal/nrcs/detail/soils/edu/?cid=nrcs142p2_054280. (Accessed 25 February 2021).
U. S. Environmental Protection Agency, 1979. Environmental Fate of 129 Priority Pollutants. EPA-440/4-79-029.
Van Zwieten, L., Ayres, M.R., Morris, S.G., 2003. Influence of arsenic con-contamination on DDT breakdown and microbial activity. Environ. Pollut. 124 (2), 331–339.
Waller, K., Swan, S.H., DeLoorenze, G., Hopkins, B., 1998. Trihalomethanes in drinking water and spontaneous abortion. Epidemiology 9 (2), 134–140.
Water Quality and Health Council, 2001. Drinking Water Chlorination. A Review of Disinfection Practices and Issues. www.waterandhealth.org/drinkingwater/wp.html. (Accessed 25 February 2021).
Zeliger, H.I., 2003. Toxic effects of chemical mixtures. Arch. Environ. Health 58 (1), 23–29.
Zeliger, H.I., 2011. Human Toxicology of Chemical Mixtures, second ed. Elsevier, Oxford.
Zhao, X., Zheng, M., Zhang, B., et al., 2006. Evidence for the transfer of polychlorinated biphenyls, polychlorinated dibenzo-p-dioxins and poly chlorinated dibenzofurans from soils to biota. Sci. Total Environ. 368 (2–3), 744–752.

CHAPTER

6

Alcohol

6.1 Introduction

Humans have been imbibing ethanol (alcohol) since the Stone Age (circa 10,000 B.C.). Alcoholic beverages were consumed by Chinese about 7000 B.C, by the Egyptians around 4000 B.C., by the Babylonians around 2700 B.C and by the Greeks about 2000 B.C. The Hebrews were introduced to wine during their captivity in Egypt and passed it down to Christianity. Throughout history, alcoholic beverages have been part of the religious rites of man (Hanson, 1997). As a result, the use of alcohol is deeply ingrained into most of the cultures of the world and It is consumed by vast numbers of people for religious and recreational purposes.

Alcohol readily diffuses through cell membranes and is toxic in a dose-response relationship to the following parts of the human body (Rusyn and Bataller, 2013):

- Digestive tract
- Liver
- Cardiovascular system
- Brain
- Reproductive system
- Fetal development
- Excretory system
- Respiratory system

Cancers associated with alcohol abuse include those of the oral cavity, pharynx, larynx, esophagus, colorectum and breast (Baan et al., 2007; Klein et al., 2019).

It is not implied here that all consumption of alcohol is hazardous to one's health. Indeed, positive effects of moderate alcohol consumption, including reduced coronary disease, have been reported in the literature (Di Castelnuovo et al., 2006). With notable exceptions (e.g., use by alcoholics and diabetics), it is the excessive drinking of alcohol that is responsible for its toxic effects.

Oxidative Stress
https://doi.org/10.1016/B978-0-323-91890-9.00011-8

6.2 Effect of alcohol on nutrients and pharmaceuticals

Alcohol has a negative impact on the efficacy of nutrients and pharmaceutical drugs. Following are examples of these effects.

1. Retinol is metabolized by the same enzymes that oxidize alcohol. As a result, prolonged use of alcohol, which induces those degradative enzymes, results in the breakdown of retinol to toxic metabolites. Alcohol also interferes with the conversion of beta-carotene, a precursor of vitamin A, to retinol. Thus, alcohol both promotes the deficiency of vitamin A and enhances its toxicity and that of beta-carotene (Leo et al., 1999).
2. Acetaminophen (APAP) is a widely used analgesic. Pretreatment with APAP followed by alcohol intake increases the metabolism of APAP and thereby its toxicity. Binge drinking of alcohol increases the hepatotoxicity of APAP (McCuskey et al., 2005).
3. Caffeine also activates the metabolism of APAP and it, too, increases the hepatotoxicity of APAP and the combination of alcohol and caffeine significantly increases the liver toxicity of APAP (DiPetrillo et al., 2002). Caffeine also causes intracellular calcium levels to increase and thereby increases apoptosis induction. Alcohol promotes higher calcium levels and higher apoptotic rates than caffeine. When caffeine and alcohol are combined, however, increased calcium levels and apoptosis are markedly elevated, indicating that the apoptotic effect of alcohol is potentiated by a when it is mixed with caffeine (Hirata et al., 2006).

6.3 Effect of alcohol on toxicities of other chemicals

Combined exposure to alcohol and numerous other chemicals produces enhanced toxic effects. The following are examples of such interactive effects.

1. Benzene's toxicity is attributed to its metabolites. These, which can be measured in urinary output, are significantly lowered when exposures are to mixtures of benzene and alcohol compared with exposure to benzene alone in test animals. Reduction in benzene metabolism due to chronic alcohol ingestion exacerbates benzene myelotoxity, by prolonging benzene residence time in the body (Maruhini et al., 2003).
2. The human central nervous system (CNS) toxicity of inhaled m-xylene is enhanced by alcohol ingestion. This effect is attributed to the fact that both m-xylene and ethanol are metabolized by the CYP2E1 enzyme and that alcohol is preferentially metabolized, resulting in higher residual levels of m-xylene (MacDonald et al., 2002).
3. Alcohol potentiates the toxicity of carbon tetrachloride. In a case study, acute liver and kidney poisoning ensued in two workers following their exposure to carbon tetrachloride vapors from a discharged fire extinguisher. Other workers exposed to the same vapors for the same period of time showed no toxic signs or symptoms. Upon investigation, it was determined that the two injured individuals were chronic alcohol users, with daily consumptions of 120 and 250 g, respectively. Each of their non-affected co-workers consumed less than 50 g of alcohol per day. The authors do not offer a mechanistic reason for the observed effect (Manno et al., 1996). It is hypothesized that, here too, preferential metabolism of alcohol resulted from the greater retention time of carbon tetrachloride.

4. Exposure to 2,2-dichloro-1,1,1-trifluoroethane (HCFC-123), a widely used substitute for ozone-depleting chlorofluorocarbons, produces reversible liver lesions in laboratory animals. When exposure to HCFC-123 is coupled with alcohol ingestion, liver toxicity is markedly increased. The increase, due to CPY2E1 induction, greatly enhances the metabolism of HCFC-123 to compounds that are toxic to the liver (Hoet, at al., 2002).
5. Heavy metals target the central nervous system (CNS) and the toxic CNS effects of heavy metals are enhanced by consumption of alcohol (Papp et al., 2005). Laboratory animals treated with a combination of lead and alcohol showed enhanced CNS toxicity compared with animals treated with lead alone. It was shown that co-exposure to lead and alcohol caused a significant decline in the rate of mitochondrial respiration compared to that induced by lead alone (Verma et al., 2005).
6. N,N-dimethyl-m-toluamide (DEET) is the active ingredient in numerous commercial mosquito repellants. DEET by itself, or as formulated into commercial products permeates through human skin and dermal exposures to DEET and to mixtures containing DEET have been associated with toxic encephalopathy in children (Briassoulis et al., 2001). It has been shown that alcohol, the solvent for DEET, significantly enhances the permeation of DEET through the skin (Stinecipher and Shah, 1997). Toxic effects attributed to DEET absorption may, therefore, be enhanced by the incorporation of alcohol into commercial product formulations.
7. Alcohol is frequently consumed with other recreational drugs. Cocaine abuse is well known to increase of catastrophic cardiovascular events such as myocardial infarction, ventricular arrhythmias, angina pectoris and sudden death. Animal experiments as well as empirical human observations show that the combination of imbibing alcohol first followed by cocaine use generates synergistic cardiovascular effects in both animals and humans (Mehta et al., 2002; Farre et al., 1997). The findings are believed to be due to the inhibition effect of alcohol on cocaine metabolism (Roberts et al., 1993). The combination of alcohol consumption with cocaine use also increases the toxicity of cocaine to human hepatocytes. This effect is the result of a depletion of hepatocyte GSH by alcohol, thereby increasing the sensitivity of human hepatocytes to cocaine-induced oxidative damage (Ponsoda et al., 1999).
8. Alcohol also enhances the effects of marijuana smoking. The blood levels of delta(9)-tetrahydrocannabinol (THC) and subjective mood states of human volunteers are increased by the combination of marijuana smoking and alcohol consumption relative to marijuana smoking alone. The volunteers reported that the marijuana effects were noted more rapidly when co-consumption with alcohol took place. The accelerated subjective effects corresponded to high plasma THC levels following co-consumption of the two drugs with the enhanced effects attributed to alcohol increasing the absorption of THC (Lucas et al., 2001).

6.4 Alcohol metabolism

Alcohol metabolism occurs primarily in the liver with Cytochrome P450 2E1 (CYP2E1) serving as the key enzyme. CYP2E1 is inducible by chronic alcohol consumption. Its activity in the liver is three to five-fold greater in chronic abusers and accounts for alcohol tolerance in such individuals. This metabolic tolerance persists for several days following cessation of alcohol consumption. CYP2E1 also oxidizes other toxic chemicals including benzene, trichloroethylene, carbon tetrachloride, other organic solvents and nitrosamines (present in food and tobacco smoke) to toxic metabolites and carcinogens. Accordingly, heavy consumption of alcohol which induces CYP2E1 increases the individual susceptibility of those who are exposed to these other xenobiotics even when adsorption of these occurs subsequent to alcohol ingestion via on-the-job and/or in the home exposures to chemicals, polluted air, contaminated water, dietary intake and other sources (Meskar et al., 2001; Dennison et al., 2004; Djordjevic et al., 1998).

6.5 Alcohol and oxidative stress

Alcohol contributes to OS in the body by promoting the generation of ROS via stimulating the activity of cytochrome P450 enzymes, and reduces antioxidant levels that eliminate ROS. Several processes and factors contribute to alcohol-induced OS (Wu and Cederbaum, 2020). These are:

- Changes in NAD^+/NADH cellular ratio
- Production of acetaldehyde during alcohol metablolism
- Mitochondrial damage resulting in decreased ATP production
- Damage to cell membranes
- Immune system impact leading to cytokine production
- Increase in ability of endotoxin to enter the bloodstream
- Increase in the activity of cytochrome P450 2E1
- Increase in free iron level
- Reduction in glutathione (GSH) antioxidant
- Generation of free radicals derived from alcohol
- Conversion of xanthine dehydrogenase into xanthine oxidase

Alcohol in combination with other OS producing species produce enhanced toxic effects. Two studies exemplify this phenomenon.

1. The greatly enhanced oxidative stress resulting from the combined effects of alcohol and iron is believed to be responsible for increased incidences of hepatocellular carcinomas in individuals who have high liver iron levels and consume alcohol (Petersen, 2005). This phenomenon demonstrates the dose-response relationship of total OS and disease arising from more than one OS source.
2. Hydrogen peroxide (H_2O_2) is cytotoxic. Treatment of PC12 cells with H_2O_2 results in nuclear damage, decrease in the mitochondrial transmembrane potential, increase in reactive oxygen species (ROS), and formation and depletion of glutathione (GHS). Combining H_2O_2 treatment with alcohol, produces synergistic effects. Relative to H_2O_2 alone, the combination of H_2O_2 and alcohol results in increased cell death as a function of exposure time. Nuclear damage, change in mitochondrial membrane permeability and ROS are all increased and GSH levels in cells are decreased. It is concluded that alcohol enhances H_2O_2 viability loss by promoting mitochondrial membrane permeability change, which is associated with increased ROS formation and glutathione (GHS) depletion (Lee et al., 2005).

References

Baan, R., Straif, K., Grosse, Y., El Ghissassi, F., Altieri, A., Cogliano, V., 2007. Carcinogenicity of alcoholic beverages. Lancel. Oncol. 8 (4), 292–293.

Briassoulis, G., Narlioglou, M., Hatzis, T., 2001. Toxic encephalopathy associated with the use of DEET insect repellants. A case analysis of its toxicity in children. Hum. Exp. Toxicol. 20 (1), 8–14.

Dennison, J.E., Bigelow, P.L., Andersen, M.E., 2004. Occupational exposure limits in the context of solvent mixtures, consumption of ethanol and target tissue dose. Toxicol. Ind. Health 20 (6–10), 165–175.

Di Castelnuovo, A., Costanzo, S., Bagnardi, V., et al., 2006. Alcohol dosing and total mortality in men and women: an updated meta-analysis of 34 prospective studies. Arch. Intern. Med. 166 (22), 2437–2445.

DiPetrillo, K., Wood, S., Kostrubsky, V., et al., 2002. Effect of caffeine on acetaminophen hepatotoxicity cultured hepatocytes treated with ethanol and isopentanol. Toxicol. Appl. Pharrmacol. 185 (2), 91–97.

Djordjevic, D., Nikolic, J., Stefanovic, V., 1998. Ethanol interactions with other cytochrome P450 substrates including drugs, xenobiotics and carcinogens. Pathol. Biol. 46 (10), 760–770.

Farre, M., de la Torre, R., Gonzalez, M.L., et al., 1997. Cocaine and alcohol interactions in humans: neuroendocrine effects and cocaethylene metabolism. J. Pharmacol. Exp. Therapeut. 283 (1), 164–176.

Hanson, D.J., 1997. History of Alcohol and Drinking Around the World. www2.potsdam.edu/hansondj/Controversies/1114796842.html. (Accessed 1 March 2021).

Hirata, H., Machado, L.S., Okuno, C.S., et al., 2006. Apoptotic effect of ethanol is potentiated by caffeine-induced calcium release in rat astrocytes. Neurosci. Lett. 393 (2–23), 136–140.

Hoet, P., Buchet, J.P., Sempoux, C., et al., 2002. Potentiation of 2,2-dichloro-1,1,1-trifluoro (HCFC-123)-induced liver toxicity by ethanol in Guinea-pigs. Arch. Toxcol. 76 (12), 707–714.

Klein, W.M.P., Jacobson, P.B., Jelzlsouer, K.J., 2019. Alcohol and cancer risk. Clinical and research implications. JAMA Online. https://doi.org/10.1001/jama.2019.19133. (Accessed 14 December 2020).

Lee, C.S., Kim, Y.J., Ko, H.H., Han, E.S., 2005. Synergistic effects of hydrogen peroxide and ethanol on cell viability loss in PC12 cells by increase in mitochondrial permeability transition. Biochem. Pharmacol. 70 (2), 317–325.

Leo, M.A., Lieber, C.S., 1999. Alcohol, vitamin A, and beta-carotene: adverse interactions, including hepatotoxicity and carcinogenicity. Am. J. Clin. Nutr. 69 (6), 1071–1085.

Lukas, S.E., Orozco, S., 2001. Ethanol increases plasma delta(9)-tetrahydrocannabinol (THC) levels and subjective effects after marihuana smoking in human volunteers. Drug Alcohol Depend. 64 (2), 143–149.

MacDonald, A.J., Rostami-Hodjegan, A., Tucker, G.T., Linkens, D.A., 2002. Analysis of solvent central nervous system toxicity and ethanol interactions using a human population physiologically based kinetic and dynamic model. Regul. Toxicol. Pharmacol. 35 (2 Pt 1), 165–176.

Manno, M., Rezzadore, M., Grossi, M., Sbrana, C., 1996. Potentiation of occupational carbon tetrachloride toxicity by ethanol abuse. Hum. Exp. Toxicol. 15 (4), 294–300.

Marrubini, G., Castoldi, A.F., Coccini, T., Manzo, L., 2003. Prolonged ethanol ingestion enhances benzene myelotoxicity and lowers urinary concentrations of benzene metabolite levels in CD-1 mice. Toxicol. Sci. 75 (1), 16–24.

McCuskey, R.S., Bethea, N.W., Wong, J., et al., 2005. Ethanol binging exacerbates sinusoidal endothelial and parenchymal injury elicited by acetaminophen. J. Hepatol. 42 (3), 371–377.

Mehta, M.C., Jain, A.C., Billie, M., 2002. Effects of cocaine and alcohol alone and in combination on cardiovascular performance in dogs. Am. J. Med. Sci. 324 (2), 76–83.

Meskar, A., Plee-Gautier, E., Amet, Y., et al., 2001. Alcohol-xenobiotic interactions. Rove of cytochrome P4502E1. Pathol. Biol. 49 (9), 696–702, 2004.

Papp, A., Pecze, L., Vezer, T., 2005. Acute effects of lead mercury and manganese on the central and peripheral nervous system in rats in combination with alcohol exposure. Arch. Hig. Rada. Tokiskol. 56 (3), 241–248. ([abstract only - PubMed]).

Petersen, D.R., 2005. Alcohol, iron assisted oxidative stress and cancer. Alcohol 35 (3), 243–249.

Ponsoda, X., Bort, R., Jover, R., et al., 1999. Increased toxicity of cocaine on human hepatocytes induced by ethanol: role of GSH. Biochem. Pharmacol. 58 (10), 1579–1585.

Roberts, S.M., Harbison, R.D., James, R.C., 1993. Inhibition by ethanol of the metabolism of cocaine to benzoylecgonine and ecgonine methyl ester in mouse and human liver. Drug Metab. Dispos. 21 (3), 537–541.

Rusyn, I., Bataller, R., 2013. Alcohol and toxicity. J. Hepatol. 59 (2), 387–388.

Stinecipher, J., Shah, J., 1997. Percutaneous permeation of N,N-diethyl-m-toluamide (DEET) from commercial mosquito repellants and the effect of solvent. J. Toxicol. Environ. Health 52 (2), 119–135.
Verma, S.K., Dua, R., Gill, K.D., 2005. Impaired energy metabolism after co-exposure to lead and ethanol. Basic Clin. Pharmacol. Toxicol. 96 (6), 475–479, 2005.
Wu, D., Cederbaum, A.J., 2020. Alcohol, Oxidative Stress, and Free Radical Damage. National Institute on Alcohol Abuse and Alcoholism, 2020. https://pubs.niaaa.nih.gpv/publications/arh27-4/277-284.htm. (Accessed 27 November 2020).

CHAPTER

7

Tobacco

7.1 Tobacco toxicity: introduction

Cigarettes, cigars, pipe tobacco and smokeless tobacco are made up of dried tobacco leaves and hundreds of components added for flavor and other properties. More than 7000 individual toxic lipophilic and hydrophilic chemical compounds, of which greater than 70 are carcinogens, have been identified in tobacco and tobacco smoke (FDA, 2012). Accordingly, all tobacco and tobacco smoke exposure is of necessity, to mixtures of toxic chemicals.

Tobacco and tobacco smoke exposure has been identified with numerous health effects. Each year, almost 450,000 people in the United States and more than seven million worldwide die from tobacco use (CDC, 2014; WHO, 2020). Cigarette smoking is associated with numerous cancers with smoking being the cause of at least 30% of all cancers and 87% of lung cancers (CDC, 2002; American Cancer Society, 2020).

Cancers only account for about half of the smoking related deaths. Smoking is a major cause of bronchitis, emphysema, chronic obstructive pulmonary disease (COPD) heart disease and stroke. Tobacco smoking is also is associated with female reproductive health, increases in miscarriage rates, early delivery, stillbirth, infant death and low birth rate. It is estimated that nine million Americans suffer from tobacco related illnesses at any given time (CDC, 2003).

Passive smoking (also known as exposure to secondhand smoke, or environmental tobacco smoke (ETS)), has been shown to produce the same health effects on those exposed to it as to actual smokers and is responsible for 1.2 million deaths worldwide annually (WHO, 2020), and approximately 3000 lung cancer deaths in the United States annually (EPA, 1993), increases the risks of stroke (Bonita et al., 1999) and myocardial infarction (Teo et al., 2006) and also impacts respiratory health in those exposed (EPA, 1993).

It should be noted that deleterious health effects resulting from tobacco use are almost independent of the form of the tobacco. Thus, cigarette smoking, cigar smoking, pipe smoking, chewing tobacco and snuff inhalation produce many of the same adverse health outcomes in people.

The complexity of tobacco smoke makes it difficult to ascribe a particular health effect to a single component, though some inroads have been made in this regard. Complicating matters

https://doi.org/10.1016/B978-0-323-91890-9.00035-0

further is the fact that human exposure to tobacco smoke is almost never without coexposure to other toxic chemicals. It is known, however, that when individuals are co-exposed to tobacco smoke and other chemicals not contained in the smoke, the health effects are observed are different from those seen from tobacco smoke exposure alone.

In the United States, the Food and Drug Administration has established a list of harmful and potentially harmful constituents in tobacco products and tobacco smoke (FDA, 2012). This list is shown in Table 7.1.

7.2 Tobacco and cancer

The question of whether the induction of cancer in humans by cigarette smoke results from the action of a single carcinogen or the action of more than one carcinogen is yet to be resolved. It is known, however, that tobacco smoking has been shown to be associated with numerous cancers (CDC, 2021). These are listed in Table 7.2.

7.3 Tobacco smoke and disease rate synergism

Smokers have increased rates of disease when coexposed to other disease-causing factors.

7.3.1 Lung cancer

Tobacco smoking is the major cause of lung cancer, accounting for almost 90% of all lung cancers. Lung cancer risks for smokers, however, increase dramatically when smokers are exposed to other lung carcinogens. The following examples are illustrative.

1. The cancer risk for uranium miners who smoke is much higher than that for uranium miners who don't smoke. This epidemiological finding is supported by laboratory experiments showing an increased incidence of pulmonary tumors in hamsters simultaneously exposed to benzo[a]pyrene and alpha radiation relative to exposure to the PAH or the radiation alone (Little, et al., 1978).
2. Tobacco smoking alone (ACS, 2020) and exposure to asbestos alone ((Vainio and Boffetta, 1994; Bofetta and Nyberg, 2003) are risk factors for developing lung cancer. The combination of smoking and working with asbestos carries a multiplicative risk for developing lung cancer (Selikoff and Hammond, 1979). The data in Table 7.3 clearly illustrate this.

   The mechanism for the synergism between asbestos and cigarette smoke is thought to be related to the formation of hydroxyl radicals catalyzed by the iron contained in asbestos (Kovacic and Jacintho, 2001; Jackson et al., 1987). It has also been found, however, that ceramic fibers as well as asbestos fibers have a synergistic effect on the formation of tumor necrosis factor (an indicator of carcinogenesis) by alveolar macrophages in rats (Moriomoto et al., 1993). A mechanistic explanation is that the solid, insoluble fibers serve as active surfaces for the adsorption of carcinogenic molecules contained in tobacco smoke.

TABLE 7.1 Established list of the chemicals and chemical compounds identified by FDA as harmful and potentially harmful constituents in tobacco products and smoke. Categories include: carcinogen (CA), toxicant (CT), reproductive or developmental toxicant (RDT), addictive (AD).

Constituent	Category
Acetaldehyde	CA, RT, AD
Acetamide	CA
Acetone	RT
Acrolein	RT, CT
Acrylamide	CA
Aflatoxin B1	CA
4-Aminobiphenyl	CA
2-Aminonaphthalene	CA
Ammonia	RT
Anabasine	AD
o-Anisidine	CA
Arsenic	CA, CT, RDT
A-α-C(2-amino-9H-pyrido[2,3-b]indole)	CA
Benz[a]anthrancene	CA, CT
Benz-[j]aceanthrylene	CA
Benzene	CA, CT, RDT
Benzo[b]fluoroanthene	CA, CT
Benzo[b]furan	CA
Benzo[a]pyrene	CA
Benzo[c]phenanthrene	CA
Beryllium	CA
1,3-Butadiene	CA, RT, RDT
Cadmium	CA, RT, RDT
Caffeic acid	CA
Carbon monoxide	RDT
Catechol	CA
Chlorinated dioxins/furans	CA, RDT
Chromium	CA, RT, RDT
Chrysene	CA, CT
Cobalt	CA, CT
Coumarin	Banned in food

(Continued)

TABLE 7.1 Established list of the chemicals and chemical compounds identified by FDA as harmful and potentially harmful constituents in tobacco products and smoke. Categories include: carcinogen (CA), toxicant (CT), reproductive or developmental toxicant (RDT), addictive (AD).—cont'd

Constituent	Category
Cresols (o-, m-, and p-)	CA, RT
Crotonaldehyde	CA
Cyclopenta[c,d]pyrene	CA
Dibenz[ah]anthracene	CA
Dibenzo pyrenes (ae, ah, aj, al)	CA
2,6-Dimethylanaline	CA
Ethyl carbamate (urethane)	CA, RDT
Ethylbenzene	CA
Ethylene oxide	CA, RT, RDT
Formaldehyde	CA, RT
Furan	CA
Glu-P-1 (2-amino-6-methyldipyrido [1,2-a:3′,2′-d]imidazole	CA
Glu = P-2 (2-aminodipyrido[1,2-a:3′,2′-d] Imidazole)	CA
Hydrazine	CA, RT
Hydrogen cyanide	RT, CT
Indeno[1,2,3-cd]pyrene	CA
IQ (2-amino-3-methylimidazo[4,5-f] Quinoline)	CA
Isoprene	CA
Lead	CA, CT, RDT
MeA-α-C (2-amino-3-methyl)-9H-pyrido [2,3-b]indole	CA
Mercury	CA, RDT
Methyl ethyl ketone	RT
5-Methylchrysene	CA
4-(methylnitrosoamino)-1-(3-pyridyl-1-Butanone	CA
Naphthalene	CA, RT
Nickel	CA, RT
Nicotine	RDT, AD
Nitrobenzene	CA, RT, RDT
9 N-nitroso amines	CA
Nornicotine	AD
Phenol	RT, CT

TABLE 7.1 Established list of the chemicals and chemical compounds identified by FDA as harmful and potentially harmful constituents in tobacco products and smoke. Categories include: carcinogen (CA), toxicant (CT), reproductive or developmental toxicant (RDT), addictive (AD).—cont'd

Constituent	Category
PhIP (2-amino-1-methyl-6 Phenylimidazo[4,5-b]pyridine)	CA
Polonium-210	CA
Propionaldehyde	RT, CT
Propylene oxide	CA, RT
Quinoline	CA
Selenium	RT
Styrene	CA
o-toluidine	CA
Trp-P-1 (3-amino-1,4-dimethyl 5H-pyrido[4,3-b]indole)	CA
TrP-2 (1-methyl-3-amino-5H-pyrido [4,3b]indole	CA
Uranium-235	CA
Uranium-238	CA
Vinyl acetate	CA, RT
Vinyl chloride	CA

TABLE 7.2 Cancers associated with smoking tobacco.

- Blood (acute myeloid leukemia)
- Bladder
- Cervix
- Colon
- Rectum
- Esophagus
- Kidney
- Larynx
- Liver
- Lung
- Trachea
- Bronchia
- Mouth
- Stomach
- Pancreas

TABLE 7.3 Lung cancer risks associated with tobacco smoking and occupational asbestos exposure.

Exposures	Relative risk for lung cancer
Did not work with asbestos	1.00
Did not smoke	
Worked with asbestos	5.17
Did not smoke	
Did not work with asbestos	10.85
Did smoke	
Did work with asbestos	53.24
Did smoke	

3. Chromium (VI) is a well known human carcinogen and occupational exposure to it is strongly associated with lung cancer. Polynuclear aromatic hydrocarbons (PAHs) are considered the major lung carcinogens in tobacco smoke. Together, chromium (VI) and PAHs act synergistically and account for the high incidence of lung cancer in those exposed to both. It has been shown that chromium (VI) exposure greatly enhances the mutagenicity and cytotoxicity of PAHs by inhibiting the cellular nucleotide excision repair (Hu et al., 2004).
4. Radon present in indoor air is a known lung carcinogen. It is estimated that between one and 5% of all lung cancers can be attributed to radon inhalation in a dose-response relationship (Bofetta, 2006). Radon and cigarette smoking, however, have a multiplicative synergistic effect on lung cancer rate (Lee et al., 1999). No mechanism has been proposed for the observed combined effect of the two carcinogens, though increase in ROS caused by radon radioactive decay coupled with that induced by tobacco smoke is thought be a factor.
5. Smoking enhances the immunotoxicity of aromatic solvents. Tobacco smoke and organic solvents acting alone are immunotoxins that reduce antibody levels in blood serum. The reduction of serum IgA, IgG and IgM levels are enhanced (relative to solvent or smoking effects alone) when smokers are occupationally exposed to benzene and its homologs (Moszczynski et al., 1989).
6. Alcohol consumption and tobacco use are corelated. Alcohol consumers smoke more than nonsmokers and smokers are more likely than nonsmokers to consume alcohol. Cardiovascular, immunological, as well as carcinogenic effects are enhanced by the coconsumption of alcohol and tobacco (Nair et al., 1990).
7. Human volunteers pretreated with nicotine via a transdermal patch and who then consumed ethanol reported that alcohol's effects (feeling drunk and euphoria) were enhanced relative to those not pretreated with nicotine. It was found that heart rates were increased by nicotine and that ethanol-induced heart rates were further increased

by nicotine. The authors conclude that the results of this study may help explain the high prevalence of elevated heart rates in those using the combination of alcohol and tobacco (Kouri et al., 2004).

8. Mice preexposed to cigarette smoke show increased levels of cardiac lipid peroxidation. These levels are further increased in animals coexposed to alcohol. Decreases in glutathione levels are noted in animals exposed to cigarette smoke and greater glutathione decreases are observed in those coexposed to cigarette smoke and alcohol. It was concluded that alcohol potentiates the cigarette smoke-induced peroxidative damage to the heart and thereby lowers the cardiac antioxidant defense system (Sandhir et al., 2003).
9. In an in vitro study involving human lymphocytes, it was demonstrated that alcohol and nicotine at noninhibitory levels when added alone, show significant suppression of natural killer cell activity when added combined (Nair et al., 1990).
10. Heavy smoking and excessive ethanol consumption are the primary risk factors for upper digestive tract cancers. The cancer risk is dose-dependent and the combination of alcohol consumption and tobacco smoking acts synergistically and multiplicatively to increase the risk of cancer in abusers (Salaspuro, 2003).

7.3.2 Noncarcinogenic synergism

Synergistic effects of mixtures of tobacco smoke and other chemicals are not limited to carcinogenesis. The following effects are illustrative.

1. Tobacco smoking alone greatly enhances the risk for coronary heart disease (CHD) (Teo et al., 2006). Smoking and elevated serum cholesterol level synergistically increase the levels of CHD observed (Perkins, 1985). No mechanism for the synergism was proffered by the study.
2. Coke oven workers are at risk for developing chronic obstructive pulmonary disease (COPD) with a dose-dependent relationship between exposure to the benzene soluble fraction of coke oven emissions and COPD observed. Smoking in coke oven workers synergistically increases the incidence of COPD (Hu et al., 2006). Though many of the compounds in cigarette smoke are identical to those in coke oven emissions, this alone cannot account for the observed synergism.
3. Dairy farmers have high incidences of cough, phlegm production and chronic bronchitis due to their exposure to numerous irritants, allergens, including grain dust and other particulates, airborne bacteria and chemicals that target the respiratory system. Though smoking alone is associated with the same symptoms, the combination of dairy farming and smoking synergistically elevates chronic cough (Dalphin et al., 1998).
4. Mucin is the primary component of mucous. Factors associated with COPD, such as bacterial infections and cigarette smoke, individually induce respiratory mucin production in vivo and in vitro. Cigarette smoke, in combination with bacterial infection, synergistically induces the hyperproduction of mucin in those with COPD (Baginski et al., 2006).

7.4 Tobacco and oxidative stress

The examples noted in Section 7.3 serve to demonstrate the association of oxidative stress with tobacco-induced illness. Tobacco contains numerous PAHs and polyphenols which form ROS/RON when metabolized. Radioactive isotopes which emit ionizing radiation, and heavy metals which catalyze free radical formation are also causes of ROS/RON production. The synergistic effects observed when tobacco smoking is coupled with exposures to other OS-raising factors demonstrates that it is total OS, from all sources, which is responsible for the onset of numerous diseases.

7.5 Electronic cigarettes

Electronic cigarettes, also called e-cigarettes, vapes and other names, are devices used to create nicotine-containing aerosols for inhalation. Aerosols thus produced contain a large number of flavor-enhancing chemicals in addition to nicotine, all of which induce oxidative stress (Muthumalage et al., 2018; ALA, 2020). A partial list of chemicals contained in e-cigarettes is contained in Table 7.4. The starred species, are those also chemicals contained in tobacco smoke. The double starred chemicals are carcinogens.

Though vaping is considered less toxic than tobacco smoking, there is reason to question such an assumption. Vaping triggers oxidative stress and is associated with the onset of

TABLE 7.4 Chemicals contained in electronic cigarettes.

Propylene glycol
Nicotine*
Acetaldehyde* **
Formaldehyde* **
Acrolein*
Diacetyl
Nickel* **
Lead* **
Tin **
Cadmium* **
Benzene* **
Cinnamaldehyde **
Acetoin
Maltol
2,3-Pentanedione**

diseases of the respiratory system (Muthumalage et al., 2018). Indeed, more than half of the chemicals listed in Table 7.4 are carcinogens.

References

ACS, 2020. American Cancer Society Cancer Facts and Figures. http://www.cancer.org (Accessed 20 February 2021).

ALA, American Lung Association, 2020. What's in an E-Cigarette? https://www.lung.org/quit-smoking/e-cigarettes-vaping/whats-in-an-e-cigarette (Accessed 27 February 2021).

Baginski, T.K., Dabbagh, K., Satjawatchcharaphong, C., Swinney, D.C., 2006. Cigarette smoke synergistically enhances respiratory mucin induction by proinflammatory stimuli. Am. J. Respir. Cell Mol. Biol. 35 (2), 165–174.

Boffetta, P., Nyberg, F., 2003. Contribution of environmental factors to cancer risk. Br. Med. Bull. 68, 71–94.

Boffetta, P., 2006. Human cancer from environmental pollutants: the epidemiological evidence. Mutat. Res. 608 (2), 157–162.

Bonita, R., Duncan, J., Truelsen, T., et al., 1999. Passive smoking as well as active smoking increases the risk of acute stroke. Tobac. Control 8 (2), 156–160.

CDC, Centers for Disease Control and Prevention, 2002. Annual smoking-attributable mortality, years of potential life lost and economic costs – United States, 1995-1999. Morb. Mortal. Wkly. Rep. 51, 300–303. www.cdc.gov/mmwr//preview/mmwrhtml/mm5114a2.htm (Accessed 2 March, 2021).

CDC, Centers for Disease Control and Prevention, 2003. Cigarette smoking-attributable morbidity – United States 2000. MMWR (Morb. Mortal. Wkly. Rep.) 52 (35), 842–844. www.cdc.gov/mmwr/preview/mmwrhtml/mm5235a4.htm (Accessed 2 March 2021).

CDC, Centers for Disease Control and Prevention, 2014. Smoking & Tobacco Use. Health Effects. https://www.cdc.gov/tobacco/basic_information/health_effects/index.htm (Accessed 2 March 2021).

Dalphin, J.C., Dubiez, A., Monnet, E., et al., 1998. Prevalence of respiratory symptoms in dairy farmers in the French province of the Doubs. Am. J. Respir. Crit. Care Med. 158 (5 Pt 1), 1493–1498.

EPA, 2002. United States Environmental Protection Agency. 2002. Respiratory Health Effects of Passive Smoking (Also Known as Exposure to Secondhand Smoke or Environmental Tobacco Smoke ETS). EPA Document Number EPA/600/6-90/006F, 1992. http://cfpub.epa.gov/ncea/cfm/recordisplay.cfm?deid=2835 (Accessed 23 February 2021).

EPA, United States Environmental Protection Agency, 2003. Fact Sheet: Respiratory Health Effects of Passive Smoking. EPA Document Number 43-F-93-003, 1993. www.epa.gov/smokefree/pubs/etsfs.html (Accessed 22 February 2021).

FDA, 2012. Harmful and Potentially Harmful Constituents in Tobacco Products and Tobacco Smoke: Established List. https://www.fda.gov/tabacco-products/public-health-education (Accessed 2 March 2021).

Hu, W., Feng, Z., Tang, M.S., 2004. Chromium(VI) enhances (+/–)-anti7beta,8alpha-dihydroxy-9alpha,10alpha-epoxy-7,8,9,10-tetrahydrobenzo[a]pyrene-induced cytotoxicity and mutagenicity in mammalian cells through its inhibitory effect on nucleotide excision repair. Biochemistry 43 (44), 14282–14289.

Hu, Y., Chen, B., Yin, Z., et al., 2006. Increased risk of chronic obstructive pulmonary disease in coke oven workers: interaction between occupational exposure and smoking. Thorax 61 (4), 290–295.

Jackson, J.H., Schraufstatter, I.U., Hyslop, P.A., et al., 1987. Role of oxidants in DNA damage. Hydroxyl radical mediates the synergistic DNA damaging effects of asbestos and cigarette smoke. J. Clin. Invest. 80 (4), 1090–1095.

Kouri, E.M., McCarthy, E.M., Faust, A.H., Lukas, S.E., 2004. Pretreatment with transdermal nicotine enhances some of ethanol's acute effects in men. Drug Alcohol Depend. 75 (1), 55–65.

Kovacic, P., Jacintho, J.D., 2001. Mechanisms of carcinogenesis: focus on oxidative stress and electron transfer. Curr. Med. Chem. 8, 773–796.

Lee, M.E., Lichtenstein, E., Andrews, J.A., et al., 1999. Radon-smoking synergy: a population-based behavioral risk reduction approach. Prev. Med. 29 (3), 222–227.

Little, J.B., McGandy, R.B., Kennedy, A.R., 1978. Interactions beltween polonium-210 alpha radiaiton, benzo(a)pyrene, and 0.9% NaCl solution instillations in the induction of experimental lung cancer. Cancer Res. 38 (7), 1929–1935.

Morimoto, Y., Kido, M., Tanaka, I., et al., 1993. Synergistic effects of mineral fibers and cigarette smoke on the production of tumor necrosis factor by alveolar macrophages in rats. Fr. J. Ind. Med. 50 (10), 955–960.

Moszczynski, P., Lisiewicz, J., Slowinski, S., 1989. Synergistic effect of organic solvents and tobacco smoke on serum immunoglobin levels in humans. Med. Pr. 40 (6), 337–341 (abstract only).

Muthumalage, T., Prinz, M., Ansah, K.D., Gerloff, J., Sundar, I.K., 2018. Inflammatory and oxidative responses induced by exposure to commonly used e-cigarette flavoring chemicals and flavored e-liquids without nicotine. Front. Physiol. https://doi.org/10.3389/fphys.2017.01130 (Accessed 22 February 2021).

Nair, M.P., Kronfol, S.A., Schwartz, S.A., 1990. Effects of alcohol and nicotine on cytotoxic functions of human lymphocytes. Clin. Immunol. Immunopathol. 54 (3), 395–409.

Perkins, K.A., 1985. The synergistic effect of smoking and serum cholesterol on coronary heart disease. Health Psychol. 4 (4), 337–360.

Salaspuro, M.P., 2003. Alcohol consumption and cancer of the gastrointestinal tract. Best Pract. Res. Clin. Gastroenterol. 17 (4), 679–694, 2003.

Sandhir, R., Subramanian, S., Koul, A., 2003. Long-term smoking and ethanol consumption accentuates stress in hearts of mice. Cardiovasc. Toxicol. 3 (2), 135–140.

Selikoff, I.J., Hammond, E.C., 1979. Asbestos and smoking. J. Am. Med. Assoc. 242 (5), 458–459.

Teo, K.K., Ounpuu, S., Hawken, S., et al., 2006. Tobacco use and risk of myocardial infraction in 52 countries in the INTERHEART study: a case-control study. Lancet 368, 647–658.

Vainio, H., Boffetta, P., 1994. Mechanisms of the combined effect of asbestos and smoking in the etiology of lung cancer. Scand. J. Work Environ. Work 20 (4), 235–242.

WHO, 2020. No smoking. World Health Organization. https://www.who.int/health-topics/tobacco (Accessed 2 March 2021).

CHAPTER

8

Electromagnetic radiation

8.1 Introduction

The effects of ionizing radiation have been well known since the time of Marie Curie's death following her Nobel Prize winning experiments. The effects of nonionizing radiation have also been extensively studied but the literature, until recently, has been replete with conflicting findings of the effects of such exposure. Both ionizing and nonionizing radiation have been shown to produce the three hallmarks of oxidative stress; lipid peroxidation, DNA damage and protein damage (Akbari et al., 2019).

The effects of combined coexposure to radiation and toxic chemicals that have been studied show synergism compared to radiation or chemical exposure alone. This chapter examines the effects of electromagnetic radiation (EMR) at all frequencies and the combined effects of radiation and toxic chemical exposure which demonstrate that total oxidative stress, from all sources, is responsible for disease onset.

The interaction between toxic chemicals and EMR can occur in three different ways. These are (Zeliger, 2011):

1. EMR activates xenobiotics to forms that react with endogenous molecules.
2. EMR activates endogenous molecules to forms that react with the xenobiotics.
3. Both the endogenous molecule and the xenobiotic are activated by EMR and reaction between the activated species occurs.

Many of the interactions between xenobiotics and EMR occur via as yet unknown mechanisms. It is known, however, that free radicals and ROS play large roles in the toxicities of EMR/chemical mixtures.

8.2 The electromagnetic spectrum

The electromagnetic spectrum, and applications associated with different frequencies is shown in Fig. 8.1.

Biological effects of EMR are best ascribed to five regions of the spectrum. All occur naturally and all can also be man-made.

Oxidative Stress
https://doi.org/10.1016/B978-0-323-91890-9.00005-2

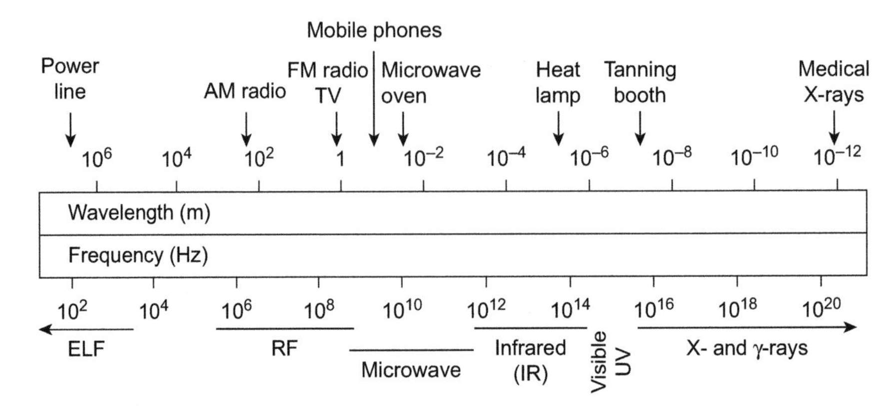

FIG. 8.1 The electromagnetic spectrum, and applications associated with different frequencies.

1. Ionizing radiation is that in the high frequency area of the spectrum where chemical bonds are broken, free electrons generated and cellular destruction occurs. Gamma radiation, cosmic radiation and X-rays are included in these frequencies. Ionizing radiation is commonly used for therapeutic purposes and when used in this manner, targeted and as well as nontargeted neighboring cells are often attacked via redox-modulated intercellular communication mechanisms (Azzam et al., 2012).
2. Ultraviolet radiation, exposure to which occurs naturally via sunlight and artificially in tanning booths, is that with wavelengths of 290–420 Hz, which correspond to the spectral region where multiple bonds are cleaved and free radicals are formed. Cellular damage may occur on prolonged exposure or via excitation of phototoxic and photoallergic species (De Jager et al., 2017).
3. Microwave and radiofrequency wavelengths are referred to as extremely low frequency (ELF) and comprise the spectral region where cellular heating is induced. Examples are microwaves and radio transmissions (Cahill and Elder, 1984).
4. Power frequency. The spectral region where energies are sufficiently low so that cellular heating is not readily induced. The most common and most studied example is that of electrical power line generated waves (Savitz and Zuckerman, 1987).
5. Visible frequencies. The narrow band of 400–700 nm that enables vision, it too may also lead to oxidative stress (Liebel et al., 2012).

8.3 Ionizing radiation

Ionizing radiation (IR), which includes emissions from radioisotopes and X-rays, is an effective killer of microorganisms and cancerous tissue. Accordingly, it is used to sterilize medical devices, to irradiate food and to treat malignant tumors. IR breaks chemical bonds, producing single electrons that cause ROS formation and induce oxidative stress (Dong et al., 2020). It is well known that exposure to excessive quantities of ionizing radiation can induce mutations, carcinogenesis and trigger epigenetic effects (Tharmalingam et al., 2017; Wei et al., 2019).

Ionizing radiation, which is extensively used in cancer treatments has many side effects which are attributable to the oxidative stress IR produces. Cardiotoxicity, manifest by increases in myocardial infarction, angina and pericarditis is an example of this effect (Ping et al., 2020).

8.4 Ionizing radiation and chemical mixtures

Simultaneous exposure to ionizing radiation and chemicals have been reported to induce unanticipated effects. Following are examples of such interactions.

1. Simultaneous exposure to 1,2-dibromoethane and low doses of X-rays induce somatic mutations. This synergistic effect has been ascribed as due to single strand legions in DNA (Leenhouts et al., 1986).
2. Temozolomide, a chemotherapeutic drug used to treat cancerous human brain tumors, enhances radiation response in human glioblastoma cells when concurrently administered with radiation. This effect has been shown to increase the degree of radiation-induced double strand DNA damage (Chakravarti et al., 2006).
3. Interleukin-13 receptor-targeted cytotoxin (IL-13) is highly toxic to human glioblastoma cells (GMB). Prior irradiation of GMB cell lines followed by the administering of IL-13 does not enhance cytotoxicity. Concomitant treatment of radiation with IL-13, however, produces greatly enhanced cytotoxic effects (Kawakami et al., 2005). No mechanism for the observed synergism was offered by the study's authors.
4. A synergistic antitumor effect is observed in human chronic lymphocytic leukemia cells when treated with radiation and a combination of 2′-deoxycoformycin and deoxyadenosine. The effects noted are greater than the predicted additive ones and synergism is enhanced by increasing radiation or by raising the concentration of deoxyadenosine. The authors of the study conclude that the combination of 2′deoxycoformycin and deoxyadenosine act as a radiosensitizer (Begleiter et al., 1988).

The studies just noted, as well as others that are similar in nature, point out the synergistic effects observed when ionizing radiation is coadministered with toxic chemicals. Though most of the studies found in the literature were designed to address the issue of combating carcinogenesis, it should be noted that combinations of ionizing radiation and toxic chemicals can produce adverse effects in healthy tissues and that intentional exposure to ionizing radiation (taking X-rays, for example) can prove to be more toxic than anticipated when one is simultaneously exposed to toxic chemicals (Leenhouts et al., 1986).

8.5 Ultraviolet radiation

Ultraviolet radiation (UV) from sun that impinges upon the earth's surface is in wavelengths between 290 and 400 nm (nm). UV that reaches the earth arrives is classified as UVA (long length waves of 320–400 nm) and UVB (short length waves of 290–320 nm).

UV is well known to excite organic molecules, generating free radicals which are responsible for substitution, elimination and polymerization reactions. UV is also well established as a generator of ROS, the overabundance of which causes oxidative stress (Hernandez et al., 2019). The relationship between UV exposure and OS is elegantly demonstrated by an experiment in which pretreatment with melatonin, a potent antioxidant, is shown to significantly protect rodents from oxidative damage caused by UV radiation (Goswami et al., 2013).

UVA plays an essential role in the formation of Vitamin D by the human skin, but is harmful by causing sunburn and cataract formation in the eyes (de Jager et al., 2017). UVB causes damage at the molecular level via absorbed by DNA that alters its structure.

8.6 Ultraviolet radiation and toxic chemical mixtures

Exposures to both UVA and UVB concurrent with exposures to xenobiotic chemicals produce unanticipated toxic effects. Following are examples of such mixture effects.

1. Both hydrogen peroxide and UV kill bacteria spores. Applied together, however, they act synergistically via free radical formation (Reidmiller et al., 2003). A similar synergism is observed between ozone and UV. While ozone is a stronger disinfectant than UV, coupling ozone with UV is more effective than ozone or UV alone. The observed synergy is attributed to generation of hydroxyl radicals by ozone photolysis (Magbanua et al., 2006).
2. Inorganic arsenic is an established human carcinogen. The combined action of arsenic and UV increases the cancer risk. Pretreatment of human lymphoblastoid cells in vitro with arsenic (III) followed by UV exposure results in the inhibition of the repair of UV-induced pyrimidine dimer-related DNA damage and thereby enhanced mutagenesis (Danaee et al., 2004).
3. Retinol (an isomer of vitamin A) is regularly used as a supplement to combat various illnesses. UV exposure results in the generation of free radicals, oxidative damage to biomolecules and decreased cellular viability in cultured mammalian cells. Cotreatment of cells with retinol and UV results in significant increases in UV-mediated free radical formation, lipid peroxidation, DNA fragmentation and mitochondrial oxidative damage relative to UV treatment alone. Retinol, rather than protecting against free radical generation, enhances UV-mediated oxidative damage. The authors of the study suggest that retinol-enhanced uptake of iron increases ROS generated via the Fenton reaction and acts synergistically to cause oxidative damage to cells (Klamt et al., 2003).
4. UV induces nonmelanoma skin cancer. Ethanol and aloe emodin alone do not induce skin tumors in the absence of UV. An ethanol solution of aloe emodin painted onto the skin of mice in conjunction with UVB exposure results in mice developing melanin-containing skin tumors. The mechanism for the observed carcinogenesis induction by the mixture is unknown (Dubakiene and Kupiene, 2006).
5. UV exposure induces photosensitive responses in humans. Photosensitivity is an adverse reaction that occurs when a chemical or drug is applied to the skin or taken internally at the same time one is exposed to UV. Not all people are photosensitive to the same agents. Two distinct responses are observed; phototoxic and photoallergic reactions.

Phototoxicity is independent of immunological responses and affects almost anyone, causing direct cellular damage when sufficient dosage is applied or when taken concurrent with UV exposure. UV absorption produces either excited state chemicals or metabolites of these chemicals. These, in turn, are converted either to free radicals or singlet oxygen, either of which results in biomolecular oxidation (Svensson et al., 2001).

Photosensitizers are chemicals that induce photoallergic responses. These can be drugs, industrial chemicals, agricultural chemicals or cosmetics and include artificial sweeteners, petroleum products, deodorants, hair sprays, makeup, antibiotics, antihistamines, antifungals, cardiovasculars, diuretics, nonsteroidal antiinflammatory drugs (ibuprofen, e.g.), sunscreens (p-aminobenzoic acid, PABA, e.g.) and various fragrances (Dubakiene and Kupiene, 2006).

Photoallergy, which is immunologically mediated and develops in sensitized individuals is independent of dose, though higher doses of photoallergins induce stronger reactions. Cross-sensitivity is often observed, where one's photosensitivity to one chemical increases the likelihood of reacting to a second chemical. Photoallergic reactions may not be predictable from a consideration of the chemical(s) to which exposure occurs. Reactions to first exposures of photoallergins and UV are not generally observed, since hypersensitivity responses require immune system activation and hence an incubation period (Tokura, 2003). Once sensitized, subsequent exposures of an individual to the same or another photoallergen, can induce a more rapid response (Vassileva et al., 1998). Though photoallergic responses primarily occur on skin areas exposed to UV, they can spread to other areas and produce systemic responses which are often difficult to characterize (Epstein, 1999).

Phototoxicity and photoallerginicity reactions to many antibiotics have been reported. The fluoroquinolone group of antibiotics is known to induce both types of reactions when administered to patients who are coexposed to UV. Administering of a single oral dose of each of seven fluoroquinolones (nalidixic acid, norfloxacin, ofloxacin, enoxacin, ciprofloxacin lomefloxacin and tosufloxacin) followed by UVA exposure induces phototoxicity in guinea pigs. Photoallergic reactions are also induced for two of these antibiotics (nalidixic acid and lomefloxacin) by an aminoadjuvent (cyclophosphamide) pretreatment followed by UVA exposure (Horio et al., 1994).

A number of the tetracycline derivatives and phototoxic. Chlortetracycline, doxycycline and dimethylchlortetracycline treatment of normal human skin fibroblasts results in total cell death within 14 days when coadministered with UVA. Dimethylchlortetracycline and UVA cotreatment also shows a strong photosensitizing effect when administered for 7 days. These results, too, are consistent with clinically reported reactions in humans (Horio et al., 1994).

Testing human volunteers has shown that doxycycline is a potent photosensitizer. Subjects were given the drug for 3 days. On the third day, they were exposed to UVA and UVB and evaluated 24 h after to the UVA treatment. Half the subjects developed strong sensitizing symptoms (Bjellerup and Ljunggren, 1987).

Mixtures of pesticides and UV are also phototoxic. Hairless dogs treated with maneb (a fungicide) and UVA showed epidermal degradation, vasodilation and intradermal infiltration of inflammatory cells. Animals treated with zineb (a general use pesticide) and UVA produced comedones with well developed pilosebaceous glands (Kimura et al., 1998).

Cosmetic and personal care products are widely used for esthetic purposes. Many cosmetic products are complex mixtures of many chemical species, including some known to be toxic. Several studies have identified phototoxic effects that are attributable to cosmetic ingredients. The following are illustrative.

1. Methylparaben (MP), a widely used preservative in cosmetics, significantly increases oxidative stress, nitric oxide production and lipid peroxidation when cells were exposed to UVB in vitro (Handa et al., 2006).
2. The photomutagenic sunscreen Padimate-O (octyl dimethyl PABA) generates free radicals and attacks DNA, producing strand breaks and lesions when illuminated with simulated sunlight on cells in vitro. While preventing sunburn, this ingredient contributes to sunlight related cancers (McHugh and Knowland, 1997; Knowland et al., 1993).
3. Titanium dioxide, another ingredient of sunscreens, absorbs about 70% of incident UV. In aqueous solutions, this UV absorption leads to the generation of free radicals and damages human cells both in vitro and in vivo (Dunford et al., 1997).
4. Petrolatum and basis cream are commonly used skin emollients. In a human volunteer test it was shown that both increase the minimal erythema dose upon exposure to UV (Fetil et al., 2006).

8.7 Nonionizing radiation

Nonionizing radiation consists of parts of the electromagnetic spectrum that correspond to microwaves, also identified as radiofrequency waves (RF) and extremely low frequency waves (ELF). RF waves range from 300 MHz to 30 GHz and correspond to AM radio, FM radio, TV, mobile telephone and microwave oven transmissions. ELF waves are in the 50–60 Hz range and correspond to electrical transmission power line emissions.

Exposure to nonionizing radiation is associated with the onset of cancer (Havas, 2017; Kocaman et al., 2018) and neurodegeneration (Consales et al., 2012), as well as adversely impacting both female and male fertility (Santini et al., 2018). These studies and numerous others have cited the ubiquitous use of modern electronic devices including; mobile phones, computers, wi-fi, microwave ovens, cellular phone base stations, broadcast antennas, overhead power lines and radar installations as the sources of ELF exposure.

Nonionizing radiation has been shown to be directly attributable to oxidative stress. ELF and RF radiation exposure have been documented in numerous studies as sources of free radical damage in humans, animals, plants and microorganisms (Havas, 2017 and references therein; Singh et al., 2020).

Nonionizing radiation consists of parts of the electromagnetic spectrum that correspond to microwaves, also identified as radiofrequency waves (RF) and extremely low frequency waves (ELF). RF waves range from 300 MHz to 30 GHz and correspond to AM radio, FM radio, TV, mobile telephone and microwave oven transmissions. ELF waves are in the 50–60 Hz range and correspond to electrical transmission power line emissions.

Most studies to date have been aimed at addressing nonionizing radiation exposure alone, often resulting in conflicting conclusions regarding human toxicity. Fewer studies have addressed the effects of mixtures of toxic chemicals and nonionizing radiation. These studies, discussed in the following sections, offer valuable insights and strongly suggest that the spurious results found in the radiation-only research may have overlooked coexposure to chemicals.

8.7.1 RF radiation and chemical mixtures

RF radiation is harmful to human tissue when converted into heat following absorption. As a result, the amount of RF energy absorbed is critical to ascribing damage. Whole body average specific absorption rate (SAR) is used to quantize the amount of RF that is absorbed with SAR expressed in watts per kilogram (W/kg). Biological effects occur at SAR of 1 W/kg or greater in adults and children. RF radiation exposures of less than 1.0 SAR are generally considered safe (Cahill and Elder, 1984).

Human exposures to RF radiation arise from military use, industrial use, broadcasting and cellular phone use. These exposures have been linked to increased numbers of spontaneous abortion, neurological effects, altered red and white blood cell counts, increased somatic mutation rates in lymphocytes, cardiovascular effects, increased cancer risk and increased childhood cancers (Goldsmith, 1987; Lin, 2004; Huber et al., 2004; Kundi et al., 2004). Other studies, however, have refuted these findings (Cahill and Elder, 1984; Rapacholi, 1997, 1998; Knave, 2001).

The mechanism associated with thermal health effects of RF radiation is induction of ROS species altering gene expression alteration and DNA damage through both genetic and epigenetic processes (Belpomme et al., 2018).

As stated above, relatively few studies have addressed the combined effects of toxic chemical and RF exposure. A thorough search of the literature shows that studies which do address such combined effects have not been refuted. Following are illustrative examples of these mixture studies.

1. The combination of chromium trioxide and RF radiation has synergistic mutagenic effects upon *Vicia faba* root tip cells upon exposure. Mutagenesis is sharply elevated relative to treatment with chromium trioxide alone (Qian et al., 2006).
2. Combined exposure to RF and the glycol ether 2-methoxyethanol (2-ME) produces increased teratogenicity in rats. Combined exposures enhance the adverse effects produced by either RF of 2-ME alone (Nelson et al., 1991, 1999).
3. Mitomycin C (MMC) and 4-nitroquinoline-1-oxide (4NQO) are known to be mutagenic to human lymphocytes. When cells were exposed to each of these and RF, synergistic effects were observed and DNA damage was greater than that observed for MMC or 4NQO alone (Baohong et al., 2005).
4. Tumor formation is generally regarded as involving an initial damage to DNA by mutagenic chemicals or ionizing radiation followed by a second step in which nonmutagenic (promoting) agents promote tumor formation. Tumor promoters include DDT, PCBs, saccharin and phorbol esters (plant lectins). Nonionizing RF radiation alone is not known to promote carcinogenesis. In combination with phorbol esters, however, RF does promote cancer (Byrus et al., 1988; Adey, 1986). It has also been shown that phorbol ester treatment of embryonic fibroblasts previously irradiated with X-rays (ionizing) and microwaves (nonionizing) increases transformation frequencies to rates greater than those observed when preirradiation is with X-rays only under the conditions tested (Adey, 1990). These findings strongly implicate the carcinogenicity of RF/toxic chemical mixtures.

8.7.2 ELF radiation and chemical mixtures

All electrical devices generate extremely low frequency electromagnetic waves (ELF) in the range from 50 to 60 Hz. Epidemiological studies have associated environmental exposure to ELF waves with human malignancies, including brain cancer and leukemia in children living close to high tension power lines (Wertheimer and Leeper, 1979; Savitz and Zuckerman, 1987; Ahlbom, 1988; Shelikh, 1986). Other studies have refuted these findings (Fulton et al., 1980; Meyers et al., 1990; London et al., 1991; Repacholi and Greenbaum, 1990). As best as can be determined, none of these studies considered coexposures to chemical agents.

It has been theorized that the connection between ELF exposure and cancer suggests that ELF alters certain cellular processes that subsequently lead to strand breaks in DNA and other chromosomal aberrations. This theory is based upon free radical production caused by acute and chronic ELF exposure. It opines that acute exposure can lead to phagocytosis and consequently free radical production, macrophage activation and direct stimulation of free radical production, as well as an increase in free radical lifetimes. It further opines that chronic exposure can lead to chronically increased free radical levels (Simko and Mattsson, 2004).

It is generally accepted that magnetic fields at flux densities below 2 T (T) do not induce adverse health effects (Repacholi and Greenbaum, 1999). This, however, is refuted by a study that reported that a 50 Hz ELF of 1 mT strength is genotoxic to cells in vitro (Fatigoni et al., 2005). This study did not consider coexposures to chemicals.

Several studies have been carried out on mixtures of ELF and toxic chemicals that demonstrate the combined toxicities. Illustrative examples follow.

1. Environmental magnetic fields (1.2 micro T, 60 Hz) significantly reduce the inhibitory action of physiological levels of the hormone melatonin on the growth on human breast cancer cells in vitro. A similar inhibitory effect is also found from exposure of these cells to a pharmacological level of the cancer inhibitor tamoxifen and the same ELF (Harland and Liburdy, 1997). This result was reproduced by a second team of researchers with identical results (Blackman et al., 2001). Both sets of authors conclude that environmental level magnetic fields can act to modify the action of a hormone or a drug on the regulation of cell proliferation.
2. Benzene is a known leukemogen and it is widely accepted that the benzene's free radical metabolites are the ultimate leukemogens. An in vitro study shows that coexposures to the benzene metabolite hydroquinone and ELF (50 Hz, 1 mT) produced a clear genotoxic effect. Coexposure to the same ELF and 1,2,4-benzenetriol (BTL) led to a marked increase in the genotoxicity of BTL (Moretti et al., 2005). In a second in vitro study, ELF (50 Hz, 5 mT) plus N-methyl-N′-nitro-N-nitrosoguanidine (MNNG) or 1,4-benzenediol (BD), both known carcinogens, were shown to increase the extent of DNA damage relative to MNNG and BD alone. Under the conditions of the testing, ELF alone did not cause primary DNA damage (Scasellati-Sforzolini et al., 2004).
3. A study in which human peripheral blood leukocytes from four different donors were exposed to ELF (at 3 mT) plus the genotoxic xenobiotics N-methyl-N′-nitro-N-nitrosoguanidine (MNNG) and 4-nitroquinoline N-oxide (4NQO) found that ELF increased primary DNA damage of both MNNG and 4NQO. In this study, ELF alone did not cause primary DNA damage (Villarini et al., 2006).

4. Metal mediated formation of free radicals causes modifications in DNA bases, enhanced lipid peroxidation and altered calcium and sulfhydryl homeostasis that is attributed to increases in ROS (Valko et al., 2005). Rat lymphocytes incubated with low concentrations of ferrous chloride (10 μg/mL) alone do not produce detectible DNA damage. Exposure of these lymphocytes to ELF (50 Hz, 7 mT) alone also does not increase the number of cells with DNA damage. When cells are simultaneously exposed to both the iron and the ELF, however, the number of damaged cells is significantly increased. It is hypothesized that the reason for the increased damage is that the combination of ELF and iron acts to substantially increase the number of reactive oxygen species generated relative to iron itself (Zmylslony et al., 2000).

8.8 Visible light radiation

Daily dermal exposure to solar radiation is essential for vision and Vitamin D production as well as detrimental by inducing skin cells to produce ROS. The effect of the visible light part of the spectrum on oxidative stress elevation has not been as extensively studied as have the effects of other parts of the EMR spectrum, but, it too, raises OS. Two examples follow.

1. Skin irradiation with visible light produces ROS, proinflammatory cytokines and matrix metalloproteinase (MMP)-1 expression, all indicative of OS elevation (Liebel et al., 2012).
2. Light exposure at night is associated with increased prevalence of breast cancer in women (Breastcancer.org, 2020; James et al., 2017; Spivey, 2010). These studies have shown that women who work at night (factory workers, doctors, nurses and police officers, e.g., have higher risks of developing breast cancer compared with women who work in the day. These studies have also found that women living in areas with high levels of external night light (street lights, e.g.) are also at higher risk for breast cancer than their cohorts living in lower night-lit areas. Increased night-light induced cancer has been linked to melatonin levels. Production of melatonin, a natural antioxidant, peaks at night and is lower during daylight hours, when light exposure, as registered by the eyes, is greater. Routine exposure to light at night lowers melatonin production, leading to increases in oxidative stress (Breastcancer.org, 2020; James et al., 2017; Spivey, 2010).

The melatonin example just noted is just one of many of the impacts of EMR on antioxidant systems in the body. EMR affects both exogenous and endogenous antioxidant functions, thereby increasing oxidative stress (Kivrak et al., 2017). This subject is discussed in more detail in Chapter 21.

8.9 Radiation and oxidative stress

The studies discussed above demonstrate that all types of radiation increase oxidative stress and when combined with chemicals that also increase OS and produce synergistic effects not attributable to either the radiation alone or the chemicals alone. Similar results have been found with radiation combined with other OS raising parameters. An example of this is the enhanced OS associated with radiation and Ketogenic diets, which are high in OS elevating fats, and lead to increased risk of breast cancer (Allen et al., 2013; Rockenback et al., 2011) (discussed in Chapter 21). These studies show the need to consider total OS from all sources as the primary cause of disease onset. The chapters that follow address how other OS sources and combinations of these produce similar effects.

References

Adey, W.R., 1986. The sequence and energies of cell membrane transductive coupling to intracellular enzyme systems. Bioeng 15, 447–456.

Adey, W.R., 1990. Joint actions of environmental nonionizing electromagnetic fields and chemical pollution in cancer promotion. Environ. Health Perspect. 86, 297–305.

Ahlbom, A., 1988. A review of the epidemiological literature on magnetic fields and cancer. Scand. J. Work. Environ. Health 14 (6), 337–343.

Akbari, A., Jelodar, C., Nazifi, S., Afsar, T., Nasiri, K., 2019. Oxidative stress as the underlying biomechanism of detrimental outcomes of ionizing and non-ionizing radiation on human health: antioxidant protective strategies. Zahedan J. Res. Med. Sci. 21 (4), e85655. https://doi.org/10.5812/zjrms.85655. (Accessed 24 February 2021).

Allen, B.G., Bhatia, S.K., Buatti, J.M., Brandt, K.E., Lindholm, K.E., Button, A., et al., 2013. Ketogenic diets enhance oxidative stress and radio-chemo-therapy responses in ling cancer xenografts. Clin. Cancer Res. 19 (14), 3905–3913.

Azzam, E.I., Jay-Gerin, J.P., Pain, D., 2012. Ionizing radiation-induced metabolic oxidative stress and prolonged cell injury. Cancer Lett. 327 (1), 48–60.

Baohong, W., Jiliang, H., Lifen, J., et al., 2005. Studying the synergistic damage effects induced by 1.8 GHz radiofrequency field radiation (RFR) with four chemical mutagens on human lymphocyte DNA using comet assay in vitro. Mutat. Res. 578 (1–2), 149–157.

Begleiter, A., Pugh, L., Israels, L.G., Johnston, J.B., 1988. Enhanced cytotoxicity and inhibition of DNA damage repair in irradiated Murine L5178Y lymphoblasts and human chronic lymphocytic leukemia cells treaded with 2′deoxycoformycin and deoxyadenosine in vitro. Cancer Res. 48, 3981–3986.

Bjellerup, M., Ljunggren, B., 1987. Double blind cross-over studies on phototoxicity to three tetracycline derivatives in human volunteers. Photo Dermatol. 4 (6), 281–287.

Byrus, C.V., Lundak, K., Pieper, S., et al., 1988. Increased ornithine decarboxylase activity in cultured cells exposed to low energy modulated microwave fields and phorbol ester tumor promoters. Cancer Res. 48, 4222–4226.

Belpomme, D., Hardell, L., Belyaev, I., Burgio, E., Carpenter, D.O., 2018. Therma and non-thermal health effects of low intensity non-ionizing radiation: an international perspective. Environ. Pollut. 242, 643–658. https://doi.org/10.1016/j.envpol.2018.07.0129. (Accessed 10 March 2021).

Blackman, C.F., Benane, S.G., House, D.E., 2001. The influence of 1.2 micro T, 60 Hz magnetic fields on melatonin- and tamoxifen-induced inhibition of MCF-7 cell growth. Bioelectromagnetics 22 (2), 122–128.

Breastcancer.org, 2020. Light Exposure at Night. https://www.breastcancer.org/risk/factors/light_exp. (Accessed 10 March 2021).

Cahill, D.F., Elder, J.A. (Eds.), 1984. Biological Effects of Radiofrequency Radiation. U.S. Environmental Protection Agency, Research Triangle, NC. EPA Report No. EPA-600/8-83-026F.

Chakravarti, A., Erkkinen, M.G., Nestler, U., et al., 2006. Temozolomide-mediated radiation enhancement in glioblastoma: a report on underlying mechanisms. Clin. Cancer Res. 12 (15), 4738–4746.

Consales, C., Merla, C., Marino, C., Benassi, B., 2012. Electromagnetic fields, oxidative stress, and neurodegeneration. Int. J. Cell Biol. https://doi.org/10.1155/2012/683897. Article ID 683897. (Accessed 9 March 2021).

Danaee, H., Nelson, H.H., Liber, H., Little, J.B., Kelsey, R.T., 2004. Low dose exposure to sodium arsenite synergistically interacts with UV radiation to induce mutations and alter DNA repair in human cells. Mutagensis 19 (2), 143–148.

De Jager, T.I., Cockrell, A.E., Du Plessis, S.S., 2017. Ultraviolet light induced generation of reactive oxygen species. Adv. Exp. Med. Biol. 996, 15–23, 1.1007/978-3-319-56017-5_2. (Accessed 8 March 2021).

Dong, S., Lyu, X., Yuan, S., Wang, S., Wei, L., Chen, Z., et al., 2020. Oxidative stress: A critical hint in ionizing radiation induced pyroptosis. Rad. Med. Protect. 1. https://doi.org/10.1016/j.radmp.2020.10.001. (Accessed 24 February 2021).

Dubakiene, R., Kupriene, M., 2006. Scientific problems of photosensitivity. Medicina (Kaunas) 42 (8), 619–624.

Dunford, R., Salinaro, A., Cai, L., et al., 1997. Chemical oxidation and DNA damage catalysed by inorganic sunscreen ingredients. FEBA Lett. 418 (1–1), 87–90.

Epstein, J.H., 1999. Phototoxicite and photoallergy. Semin. Cutan. Med. Surg. 18 (4), 274–284.

Fatigoni, C., Dominici, L., Moretti, M., 2005. Genotoxic effects of extremely low frequency (ELF) magnetic fields (MF) evaluated by Tradescantia-micronucleus. Environ. Toxicol. 20 (6), 585–591.

Fetil, E., Akarsu, S., Ilknur, T., et al., 2006. Effects of some emollients on the transmission of ultraviolet. Photodermatol. Photoimmunol. Photomed. 22 (3), 37–40.

Fulton, J.P., Cobb, S., Preble, L., et al., 1980. Electrical wiring configurations and childhood leukemia in Rhode Island. Am. J. Epidemiol. 111 (3), 292–296.

Goldsmith, J.R., 1987. Epidemic evidence of radio frequency radiation (microwave) effects and health in military, broadcasting and occupational studies. Int. J. Occup. Environ. Med. 1 (1), 47–57.

Goswami, S., Sharma, S., Haldar, C., 2013. The oxidative damages cased by ultraviolet radiation type C (UVC) to a tropical rodent Funambulus pennanti: role of melatonin. J. Photochem. Photobiol. B 125, 19–25.

Handa, O., Kokura, S., Adachi, S., et al., 2006. Methylbaraben potentiates UV-induced damage of skin keratinocytes. Toxicology 227 (1–2), 62–72.

Harland, J.D., Liburdy, R.P., 1997. Environmental magnetic fields inhibit the antiproliferative action of tamoxifen and melatonin in a human breast cancer cell line. Bioelectromagnetics 18 (8), 555–562.

Havas, M., 2017. When theory and observation collide: can non-ionizing radiation cause cancer? Environ. Pollut. 221, 501–505. https://doi.org/10.1016/j.envpol.2016.10.008. (Accessed 9 March 2021).

Horio, Miyauchi, H., Asada, Y., et al., 1994. Phototoxicity and photoallergenicity of quinolones in Guinea pigs. J. Dermatol. Sci. 7 (2), 130–135.

Hernandez, A.R., Vallejo, B., Ruzgas, T., Bjorklund, S., 2018. The effect of UVB irradiation and oxidative stress on the skin barrier – a new method fot evaluate sun protective factor based on electrical impedance spectroscopy. Sensors 2019, 2376. https://doi.org/10.3390/s19102376. (Accessed 8 March 2021).

Huber, R., Schruder, J., Graf, T., et al., 2004. Radio frequency electromagnetic field exposure in humans: estimation of SAR distribution in the brain, effects on sleep and heart rate. Bioelectromagnetics 24 (4), 262–276.

James, P., Bertrand, K.A., Hart, J.E., Schernhammer, E., Tamimi, R.M., Laden, F., 2017. Outdoor light at night and breast cancer incidence in the nurses' health study II. Environ. Health Perspect. https://doi.org/10.1289/ehp935. (Accessed 10 March 2021).

Kawakami, K., Kawakami, M., Liu, Q., Puri, R.K., 2005. Combined effects of radiation and interleukin-13 receptor-targeted cytotoxin on glioblastoma cell lines. Int. J. Radiat. Oncol. Biol. Phys. 63 (1), 230–237.

Kimura, R., Kuroki, K., Doi, K., 1998. Dermatotoxicity of agricultural chemicals in the dorsal skin of hairless dogs. Toxicol. Pathol. 26 (3), 442–447.

Kivrak, E.G., Yurt, K.K., Kaplan, A.A., Alkan, I., Altun, G., 2017. Effects of electromagnetic fields exposure on the antioxidant defense system. J. Microscopy Ultrastructure 5, 167–176.

Klamt, R., Dal-Pizzol, F., Bernard, E.A., Moreira, J.C., Enhanced, 2003. UV-mediated free radical generation; DNA and mitochondrial damage caused by retinol supplementation. Photochem. Photobiol. Sci. 2 (8), 856–860.

Knowland, J., McKenzie, E.A., McHugh, P.J., Cridland, N.A., 1993. Sunlight-induced mutagenicity of a common sunscreen ingredient. FEBS Lett. 324 (3), 309–313.

Knave, B., 2001. Electromagnetic fields and health outcomes. Ann. Acad. Med. Singapore 30 (5), 489–493.

Kocaman, A., Altun, G., Kaplan, A.A., Deniz, O.G., Yurt, L.L., Kaplan, S., 2018. Genotoxic and carcinogenic effects of non-ionizing electromagnetic fields. Environ. Res. https://doi.org/10.1016/j.envres.2018.01.034. (Accessed 9 March 2021).

Kundi, M., Mild, K., Hardell, L., Mattsson, M.O., 2004. Mobile telephones and cancer—a review of epidemiological evidence. J. Toxicol. Environ. Health B Crit. Rev. 7 (5), 351–384.

Leenhouts, H.P., Sijsma, M.J., Cebulska-Wasilewska, A., Chadwick, K.H., 1986. The combined effect of DBDE and x-rays on the induction of somatic mutations in tradescantia. Int. J. Radiat. Biol. Relat. Stud. Phys. Chem. Med. 49 (1), 109–119.

Liebel, F., Kaur, S., Ruvolo, E., Kollias, N., Southall, M.D., 2012. Irradiation of skin with visible light induces reactive oxygen species and matrix-degrading enzymes. J. Invest. Dermatol. 132, 1901–1907.

London, S.J., Thomas, D.C., Bowman, J.D., et al., 1991. Exposure to residential electric and magnetic fields and risk of childhood leukemia. Am. J. Epidemiol. 134 (9), 923–937.

Magbanua Jr., B.S., Savant, G., Truax, D.D., 2006. Combined ozone and ultraviolet inactivation of *Escherichia coli*. J. Environ. Sci. Health A Tox. Hazard. Subst. Environ. Eng. 41 (6), 1043–1055.

McHugh, P.J., Knowland, J., 1997. Characterization of DNA damage inflicted by free radicals from a mutagenic sunscreen ingredient and its location using an in vitro genetic reversion assay. Photochem. Photobiol. 66 (2), 276–281.

Meyers, A., Clayden, A.D., Cartwright, R.A., Cartwright, S.C., 1990. Childhood cancer and overhead power lines: a case-control study. Br. J. Cancer 62, 1008–1014.

Moretti, M., Villarini, M., Simonucci, S., et al., 2005. Effects of co-exposure to extremely low frequency (ELF) magnetic fields and benzene of benzene metabolites determined in vitro by the alkaline comet assay. Toxicol. Lett. 157 (2), 119–128.

Nelson, B.K., Conover, D.L., Brightwell, W.S., et al., 1991. Marked increase in the teratogenicity of the combined administration of the industrial solvent 2-methoxyethanol and radiofrequency radiation in rats. Teratology 43 (6), 621–634.

Nelson, B.K., Snyder, D.L., Shaw, P.B., 1999. Developmental toxicity interactions of methanol and radiofrequency radiation of 2-methoxyethanol in rats. Reprod. Toxicol. 13 (20), 137–145.

Ping, Z., Peng, Y., Lang, H., Xinyong, C., Wu, X., Hong, Z., Liang, S., 2020. Oxidative stress in radiation-induced cardiotoxicity. Oxidat. Med. Cell Long. https://doi.org/10.1155/3579143. Article ID 3579143. (Accessed 10 March 2021).

Qian, X.W., Luo, W.H., Zheng, O.X., 2006. Joint effects of microwave and chromium trioxide on root tip of *Vicia faba*. J. Zhejiang Univ. Sci. B 7 (3), 221–227 (abstract only).

Reidmiller, J.S., Baldeck, J.D., Rutherford, G.C., Marquis, R.E., 2003. Characterization of UV-peroxide killing of bacterial spores. J. Food Prod. 66 (7), 1233–1240.

Repacholi, M.H., Greenbaum, B., 1990. Interaction of static and extremely low frequency electric and magnetic fields with living systems: health effects and research needs. Bioelectromagnetics 20 (3), 133060.

Repacholi, M.H., 1997. Radiofrequency field exposure and cancer: what do the laboratory studies suggest? Environ. Health Perspect. 105 (Suppl. 6), 1565–1568.

Repacholi, M.H., 1998. Low-level exposure to radiofrequency electromagnetic fields: health effects and research needs. Bioelectromagnetics 19 (1), 1–19.

Rockenbach, G., Di Pietro, P.F., Ambrosi, C., Vieira, F.G.K., Cripps, C.G., Da Silva, E.L., Fausto, M.A., 2011. Dietary intake and oxidative stress in breast cancer: before and after treatments. Nutr. Hosp. 26, 737–744.

Santini, S.J., Cordone, V., Falone, S., Mijit, M., Tatone, C., Amicarelli, F., Di Emidio, G., 2018. Role of mitochondria in the oxidative stress induced by Electromagnetic fields: focus on reproductive systems. Ox. Med. Cell Long. https://doi.org/10.1155/2018/5076271. Article ID 5076271. (Accessed 9 March 2021).

Savitz, D.A., Zuckerman, D.L., 1987. Childhood cancer in the Denver metropolitan area 1976-1983. Cancer 59 (8), 1539–1542.

Scasellati-Sforzolini, G., Moretti, M., Villarini, M., et al., 2004. Evaluation of genotoxic and/or co-genotoxic effects in cells exposed in vitro to extremely low frequency electromagnetic fields. Ann. Ig 16 (1–2), 321–340 (abstract only).

Shelikh, K., 1986. Exposure to electromagnetic fields and the risk of leukemia. Arch. Environ. Health 41 (1), 56–63.

Simko, M., Mattsson, M.O., 2004. Extremely low frequency electromagnetic fields as effectors of cellular responses in vitro: possible immune cell activation. J. Cell. Biochem. 93 (1), 82–93.

Singh, K.V., Gautam, R., Meena, R., Nirala, J.P., Jha, S.K., Rajamani, P., 2020. Effect of mobile phone radiation on oxidative stress, inflammatory response and contextual fear in Wistar rat. Environ. Sci. Pollut. Res. 27, 19340–19351. https://link.springer.com/article/10.1007/s11356-020-07916-z. (Accessed 2 March 2021).

Spivey, A., 2010. Light at night and breast cancer risk worldwide. Environ. Health Perspect. 118 (12), A525.

Svensson, C.K., Cowen, E.W., Gaspari, A.A., 2001. Cutaneous drug reactions. Pharmacol. Rev. 53 (3), 357–379.

Tharmalingam, S., Sreetharan, S., Kulesza, A.V., Boreham, D.R., Tai, T.C., 2017. Low-dose ionizing radiation exposure, oxidative stress and epigenetic programming of health and disease. Radiat. Res. 188 (4.2), 525–538.

Tokura, Y., 2003. Immunological and molecular mechanisms of photoallergic contact dermatitis. J. UOEH 25 (4), 387–395.

Valko, M., Morris, H., Cronin, M.T., 2005. Metals, toxicity and oxidative stress. Curr. Med. Chem. 12 (10), 1161–1208.

Vassileva, S.G., Mateev, G., Parish, L.C., 1998. Antimicrobial photosensitive reactions. Arch. Intern. Med. 158, 1993–2000.

Villarini, M., Moretti, M., Scassellati-Sforzolini, G., et al., 2006. Effects of co-exposure to extremely low frequency (50 Hz) magnetic fields and xenobiotics determined in vitro by the alkaline comet assay. Sci. Total Environ. 361 (1–3), 208–219.

Wei, J., Wang, B., Wang, H., Meng, L., Zhao, Q., Li, X., et al., 2019. Radiation-induced normal tissue damage: oxidative stress and epigenetic mechanisms. Ox. Med. Cell Long. https://doi.org/10.1155/2019/3010342. Article ID 3010342. (Accessed 9 March 2021).

Wetheimer, N., Leeper, E., 1979. Electrical wiring configurations and childhood cancer. Am. J. Epidemiol. 109 (3), 273–284.

Zeliger, H.I., 2011. Human Toxicology of Chemical Mixtures, second ed. Elsevier, London.

Zmylslony, M., Palus, J., Jajte, J., et al., 2000. DNA damage in rat lymphocytes treated in vitro with iron cations and exposed to 7 mT magnetic fields (static or 50 Hz). Mutat. Res. 453 (1), 89–96.

CHAPTER

9

Inflammation

9.1 Introduction

Inflammation is the body's protective response to microbial invasion, injury and other challenges to homeostasis. Acute inflammation rapidly occurs anywhere needed in the body to accomplish its protective mission and quickly "fix" a problem and subside. When inflammation fails to subside, acute inflammation becomes chronic, a condition that not only fails to resolve the "problem" at hand, but also becomes a cause of further disease.

Oxidative stress, caused by an imbalance in the production of ROS and reduced availability of antioxidants leads to the onset of inflammation. Inflammation causes an immune system response, as well as a reduction in cellular antioxidant capacity, which, together, lead to further OS. Thus, OS is both a cause and a consequence of chronic inflammation in a self-perpetuating cycle which, if unabated, can lead to further disease, organ failure and death (Varizi and Rodriguez-Iturbe, 2006; Khansari et al., 2009; Chatterjee, 2016).

Diseases of multiple systems and organs are associated with chronic inflammation include those of the autoimmune, metabolic, respiratory, cardiovascular, skeletal and neurological, kidney diseases and cancer. The mechanisms for chronic inflammation-induced diseases are all tied to OS.

Symptoms of inflammation include:

Heat
Redness
Pain
Swelling

All four of these symptoms are signs of elevated oxidative stress (Zeliger, 2016).

9.2 Acute inflammation

Acute inflammation is the body's response to psychological stress, injury, major surgery, burns, attack by pathogens, sepsis and trauma via mechanisms that elevate oxidative stress in the short term, but generally subside when the crisis is abated (Neher et al., 2011).

Oxidative Stress
https://doi.org/10.1016/B978-0-323-91890-9.00037-4

9.3 Chronic inflammation

Acute inflammation normally subsides as danger to body is eliminated or removed and a return to homeostatic status occurs. When the triggering inflammatory event fails to resolve, chronic inflammation and oxidative stress persist (Ahmed, 2011).

Chronic inflammation can be caused by persistent infection (chronic disease), continual exposure to exogenous toxins, a regular regimen of some pharmaceuticals, smoking tobacco, living in an area of high air pollution, diet, chronic trauma, non-healing wounds, aging and cancer. Oxidative stress induces inflammation and inflammation leads to OS, setting up a vicious cycle of chronic inflammation (Reuter et al., 2010).

Air pollution, chemical exposures, tobacco use, diet and pharmaceutical use as sources of oxidative stress and associated inflammation have been or will be covered in other chapters. What follows are other parameters associated with chronic inflammation, including infectious and non-infectious disease, chronic trauma, aging and cancer, all of which chronically elevate oxidative stress.

9.3.1 Persistent infection

Bacterial and viral and agents cause a wide variety of chronic infections that result in continual inflammation and chronic oxidative stress (Beck and Levender, 2000; Ivanov et al., 2017; CDC, 2021; Grant and Hung, 2013; Gotkin and Smith, 2016; Lee, 2018, Schwartz, 1996; Peterhans, 1997; Beck, 2000). Examples of such pathogens and the diseases they cause are listed in Table 9.1.

9.3.2 Chronic non-infectious diseases

The inflammation-oxidative stress cycle is responsible for chronic non-infectious diseases (Smallwood et al., 2018; Ramani et al., 2020; Zeliger, 2016; Vaziri et al., 2006; Aimo et al., 2019). These illnesses, which include autoimmune diseases, are those that affect almost all body systems and organs, including the metabolic, respiratory, cardiovascular, musculoskeletal and nervous systems, liver, kidney, pancreatic and intestinal tract diseases, skin, mitochondrial disfunction, obesity and cancer. Representative examples of these are listed in Table 9.2 (Ahmed, 2011; Vaziri et al., 2006; Aimo et al., 2019; Miller et al., 2018; Solleiro-Villavicencio and Rivas-Arancibia, 2018; Collins et al., 2018; McGarry et al., 2018; Chen et al., 2018; Hotamisligil, 2006; Luc et al., 2019; Bartsch and Nair, 2006; Khanna et al., 2014; Reuter et al., 2010; Khansari et al., 2009; Muller et al., 2019; Bickers and Athar, 2006; Baek and Lee, 2016; Bertino et al., 2020).

9.3.3 Cancer

Inflammation is critical to the promotion of tumorigenesis, cancer cell growth and the increase of mutation rates via the production of ROS and RON intermediates which damage DNA. The longer that inflammation exists, the greater the risk of cancer (Grivennikof et al., 2010; Federico et al., 2007).

TABLE 9.1 Bacterial and viral agents that cause chronic inflammation, chronic oxidative stress and chronic disease.

Pathogen	Disease
Bacterial	
Mycobacterium tuberculosis	Tuberculosis
Helicobacter pylori	Gastritis
Salmonella typhi	Salmonella
Escherichia coli	Recurrent urinary tract infections
Hemophilus influenza	Recurrent Otitis media
Mycobacterium leprae	Leprosy
Borrelia burgdorferi	Lyme
Viral	
Human Immunodeficiency	HIV
Human Papilloma	Cervical cancer
Epstein–Barr v	Herpes
SARS-CoV-2	COVID-19

Chronic inflammation increases cancer risk by impacting every step of tumorigenesis from initiation to tumor promotion and ultimately to metastatic progression (Ikemura et al., 2013; Hanahan and Weinberg, 2011; Milara and Cortijo, 2012) and is considered a hallmark of cancer (Diakos et al., 2014; Melnik, 2015; Liu et al., 2014). This connection underscores the importance of limiting the factors that induce oxidative stress and inflammation in preventing the onset of further disease.

The mechanisms associated with cancer and its metastasis are discussed in Chapter 21.

9.3.4 Aging

Aging, a natural consequence of living, is the progressive decline in physiological function following the peak of physical prowess. It is accelerated by excessive OS via accumulation of free radical-induced damage to biomolecules, including DNA, lipids and protein. Chronic inflammation accelerates the aging process via a continued production of ROS that cause this damage (Khansari et al., 2009). A widely held theory is that OS within mitochondria damages the mitochondria, which in turn leads to the production of increased quantities of ROS that cause further damage. Aging also severely impacts immune system function, increasing susceptibility to infectious disease and contributing to the oxidative stress-inflammation-caused diseases listed in Table 9.2 (Muller et al., 2019). Once this aging cycle starts, it leads to further damage and correspondingly more aging. OS has been shown to shorten telomere length (Kawanishi and Oikawa, 2004; von Zglinicki, 2002).

TABLE 9.2 Diseases associated with chronic inflammation (AARDA, 2021).

Metabolic
Type 2 diabetes
Metabolic syndrome
Obesity
Hyperlipidemia
Respiratory
Asthma
Chronic obstructive pulmonary disease
Chemical sensitivity
Cardiovascular
Hypertension
Heart failure
Arteriosclerosis
Atherosclerosis
Myocardial infarction
Stroke
Musculoskeletal
Rheumatoid arthritis*
Osteoarthritis
Osteoporosis
Tendinopathy
Nervous
Chronic posttraumatic stress disorder
Alzheimer's disease
Parkinson's disease
Multiple sclerosis*
Amyotrophic lateral sclerosis
Liver
Hepatitis
Cirrhosis
Non-alcoholic fatty liver disease
Kidney
Chronic kidney disease

TABLE 9.2 Diseases associated with chronic inflammation (AARDA, 2021).—cont'd

End stage renal disease
Renal vascular disease
Glomerulosclerosis
Pancreas
Pancreatitis
Intestinal tract
Inflammatory bowel disease*
Crohn disease
Ulcerative colitis*
Peptic ulcer
Mitochondial disfunction
Hypoxia
Obesity
Hyperglycemia
Skin
Acne
Atopic dermatitis
Eczema
Psoriasis*
Rosacea
Urticaria

Autoimmune diseases are starred (*).

Telomeres, the repetitive DNA sequences at the ends of eukaryotic chromosomes are shortened in each somatic cell division. Reduced telomere length is associated with aging, the onset of cancer and other age-related diseases (Epel et al., 2004; Hou et al., 2015). Lowering of OS, however, has been shown to delay the shortening of telomere length, thus prolonging life and reducing cancer incidence (Crous-Bou et al., 2014).

Nine hallmarks of aging have been identified (Lopez-Otin et al., 2013; Meiners et al., 2015). These are:

- Genomic instability
- Telomere attrition
- Epigenetic alterations
- Loss of proteostasis
- Deregulated nutrient-sensing

- Mitochondrial dysfunction
- Cellular senescence
- Stem cell exhaustion
- Altered cellular communication

All of these hallmarks have been shown to be negatively impacted by OS and associated with chronic inflammation (Kawanishi and Oikawa, 2004; von Zglinicki, 2002; Lopez-Otin et al., 2013; Meiners et al., 2015). These results suggest that chronic inflammation and oxidative stress be added to the list of hallmarks of aging.

9.3.5 Non-healing wounds

Non-healing (chronic) wounds occur when wound-healing processes are impaired due to the presence of chronic disease, vascular insufficiency, diabetes, malnutrition, tissue damage or aging that keep the wound in a prolonged and heightened oxidative stress state due to an inflammatory state in which ROS are continually produced. Examples of such wounds are diabetic ulcers and infected wounds (Frykberg and Banks, 2015; Zhao et al., 2016).

9.3.6 Chronic trauma

Chronic trauma, the continual injury, even in a mild form, to a given body site results in chronic oxidative stress at that site. Chronic traumatic encephalopathy (CTE) is an example of such trauma. CTE, also known as dementia pugilistica (punch-drunk syndrome) was first associated with boxers and more recently with athletes playing American football, rugby, soccer, or ice hockey, as well as in military combatants. CTE has been shown to be caused by mild traumatic brain injury (TBI) from repetitive concussions associated with contact sports and exposure to repeated explosions (Kulbe and Hall, 2017; Albayram et al., 2020).

TBI, which in most cases, comes from a single blow to the head produces inflammation in the brain which generally resolves with time. Chronic TBI, which ensues from taking repeated hits to the head, as for example in playing football, results in chronic inflammation and steady-state oxidative stress which leads to permanent brain injury (CTE), neurodegenerative symptoms and psychiatric disorders. Indeed, those with CTE have higher incidences of Alzheimer's disease, Parkinson's disease and Amyotrophic lateral sclerosis, as well as increased losses in motor and cognitive abilities (Cruz-Haces et al., 2017; McKee et al., 2018; Gunther et al., 2018; Jarrahi et al., 2020).

9.4 Inflammation and oxidative stress

Oxidative stress plays an essential role in the development of inflammation and in maintaining it in all stages of inflammatory response (Lugrin et al., 2014). The interdependent relationship between inflammation and oxidative stress is well established as a cause of chronic disease and aging (Millal et al., 2014). Immune system responses to inflammation include release of oxidative stress-raising cytokines, inflammatory proteins and inflammatory enzymes. These are listed in Table 9.3.

In the inflammatory process, activated phagocytes such as neutrophils and macrophage generate large numbers of ROS, RON and chlorinated species that include superoxide,

TABLE 9.3 Immune system responses to inflammation.

Cytokines
IL-1β
IL-6
TNA-α
Inflammatory proteins
C-reactive protein
Haptoglobin
Serum amyloid A
Fibrinogen
Alpha 1-acid glycoprotein
Inflammatory enzymes
Superoxide dismutase
Glutathione peroxidase
NADPH oxidase
Inducible nitric oxide synthase
Cyclooxygenase

hydrogen peroxide, hydroxyl free radicals, nitric oxide, peroxynitrite and hypochlorous acid to eliminate invading pathogens (Fialkow et al., 2007).

Given the inflammation-oxidative stress interrelationship, steps to lower inflammation must also include to lower OS (Reuter et al., 2010; Aimo et al., 2019). This subject is addressed in Part IV of this book.

References

AARDA, 2021. Autoimmune Disease List. American Autoimmune and Related Diseases Association. https://aarda.org/diseaselist/ (Accessed 19 March 2021).

Aimo, A., Castiglioni, V., Borrelli, C., Saccaro, L.F., Franzini, M., Masi, S., et al., 2019. Oxidative stress and inflammation in the evolution of heart failure: from pathophysiology to therapeutic strategies. Euro J. Prevent. Cardiol. https://doi.org/10.1177/2047487319870344 (Accessed 19 March 2021).

Ahmed, A.U., 2011. An overview of inflammation: mechanism and consequences. Front. Biol. 6 (4), 274–281.

Albayram, O., Albayram, S., Mannix, R., 2020. Chronic traumatic encephalopathy – a blueprint fot he bridge between neurological and psychiatric disorders. Transl. Psychiatry 10, 424. https://doi.org/10.1038/s41398-020-01111 (Accessed 17 March 2021).

Baek, J., Lee, M.G., 2016. Oxidative stress and antioxidant strategies in dermatology. Redox Rep. 21 (4), 164–169.

Beck, M.A., 2000. The role of oxidative stress in viral infections. Ann. N. Y. Acad. Sci. 917, 906–912.

Bartsch, H., Nair, J., 2006. Chronic inflammation and oxidative stress in the genesis and perpetuation of cancer: role of lipid peroxidation, DNA damage and repair. Langerbecks Arch. Surg. 391 (5), 499–510.

Bertino, L., Guarneri, F., Cannavo, S.P., Casciaro, M., Pioggia, G., GAngemi, S., 2020. Oxidative stress and atopic dermatitis. Antioxidants. https://doi.org/10.3390/antiox9030196, 2020. (Accessed 18 March 2021).

Bickers, D.R., Athar, M., 2006. Oxidative stress in the pathogenesis of skin disease. J. Invest. Dermatol. 126 (12), 2565–2575.

CDC, 2021. Centers for disease control and prevention. Lyme Dis. https://www.cdc.gov/lyme/index.html (Accessed 18 March 2021).

Chatterjee, S., 2016. Oxidative stress, inflammation and disease. Sci. Direct. https://doi.org/10.1016/B978-0-803269-5.00002-4 (Accessed 14 March 2021).

Chen, L., Deng, H., Cui, H., Fang, J., Zuo, Z., Deng, J., et al., 2018. Inflammatory responses of inflammation-associated diseases in organs. Oncotarget 9 (6), 7204–7218.

Collins, K.H., Herzog, W., MacDonald, G.Z., Reimer, R.A., Rios, J.L., Smith, I.C., et al., 2018. Obesity, metabolic syndrome and musculoskeletal disease: common inflammatory pathways suggest a central role for loss of muscle integrity. Front. Physiol. 9. https://doi.org/10.3389/fphys.2018.00112 article 112. (Accessed 15 March 2021).

Cruz-Haces, M., Tang, J., Acosta, G., Fernandez, J., Shiu, R., 2017. Pathalogical correlations between traumatic brain injury and chronic neurodegenerative diseases. Transl. Neurodegener. 6, 20. https://doi.org/10.1186/s40035-017-0088-2, 2017. (Accessed 17 March 2021).

Crous-Bou, M., Fung, T.T., Prescott, J., Julin, B., Du, M., Sun, Q., et al., 2014. Mediterannean diet and telomer length in Nurses' Health Study: population based cohort study. BMJ 349, g6674. https://doi.org/10.1136/bmj.g6674 (Accessed 18 March 2021).

Diakos, C.I., Charles, K.A., McMillan, D.C., Clarke, S.J., 2014. Cancer-related inflammation and treatment effectiveness. Lancet Oncol. 15 (11), e493–503.

Epel, E.S., Blackburn, E.H., Lin, J., Dhabhar, F.S., Adler, N.E., MOrrow, J.D., Cawthon, R.M., 2004. Accelerated telomer shortening in response to life stress. Proc. Nat. Acad. Sci. USA 101 (49), 17312–17315.

Federico, A., Morgillo, F., Tuccillo, C., Ciardiello, F., Loguercio, C., 2007. Chronic inflammation and oxidative stress in human carcinogenesis. Int. J. Cancer 121 (11), 2381–2386.

Fialkow, L., Wang, Y., Downey, G.P., 2007. Reactive oxygen and nitrogen species as signaling molecules regulating neutrophil function. Free Rad. Biol. Med. 42 (2), 153–164.

Frykberg, R.G., Banks, J., 2015. Challenges in the treatment of chronic wounds. Adv. Wound Care 4 (9), 560–582.

Goodkin, A., Smith, K.A., 2016. Chronic infections with viruses or parasites: breaking bad to make good. Immunology 150, 389–396.

Grant, S.S., Hung, D.T., 2013. Persistent bacterial infections, antibiotic tolerance, and the oxidative stress response. Virulence 4 (4), 273–283.

Grivennikof, S.I., Greten, F.R., Karin, M., 2010. Immunity, inflammation and cancer. Cell 140 (6), 883–899.

Gunther, M., Al, Nimer, Piehl, F., Risling, M., Mathiesen, T., 2018. Susceptibility to oxidative stress is determined by genetic background in neuronal cell cultures. eNeuro. https://doi.org/10.1523/ENEURO.0335-17.2018 (Accessed 17 March 2021).

Hanahan, D., Weinberg, R.A., 2011. Hallmarks of cancer: the next generation. Cell 144, 646–674.

Hotamisligil, G.S., 2006. Inflammation and metabolic disorders. Nature 444 (7121), 860–867.

Hou, L., Joyce, B.T., Gao, T., Liu, L., Zheng, Y., Penedo, F.J., et al., 2015. Blood telomere length attrition and cancer development in the normative aging study cohort. EBioMedicine 13;2 (6), 591–596.

Ikemura, M., Nishikawa, M., Kusamori, K., Fukuoka, M., Yamashita, F., Hashida, M., 2013. Pivotal role of oxidative stress in tumor metastasis under diabetic conditions in mice. J. Contr. Release 170 (2), 191–197.

Ivanov, A.V., Bartosch, B., Isaguliants, M.G., 2017. Oxidative stress in infection and consequent disease. Oxid. Med. Cell. Longev. https://doi.org/10.1155/2017/3496043, 2017. (Accessed 18 March 2021).

Jarrahi, A., Braun, M., Ahluwalia, M., Gupta, R.V., Wilson, M., Munie, S., et al., 2020. Revisiting traumatic brain injury: from molecular mechanisms to therapeutic interventions. Biomedicines 8, 389. https://doi.org/10.3390/biomedicines8100389, 2020. (Accessed 17 March 2021).

Kawashini, S., Oikawa, S., 2004. Mechanism of telomere shortening by oxidative stress. Ann. NY Acad. Sci. 1019, 278–284.

Khanna, H.D., Karki, J., Pande, D., Nagi, R., Khanna, R.S., 2014. Inflammation, free radical damage, oxidative stress and cancer. Interdiscip. J. Microinflamm. 1, 1. https://doi.org/10.4172/ijm.1000109 (Accessed 16 March 2021).

Khansari, N., Shakiba, Y., Mahmoudi, M., 2009. Chronic inflammation and oxidative stress as a major cause of age-related diseases and cancer. Recent Pat. Inflamm. Allergy Drug Discov. 3 (1), 73–80.
Kulbe, J.R., Hall, E.D., 2017. Chronic traumatic encephalopathy of canonical traumatic brain injury secondary injury mechanisms with tau pathology. Prog. Nuerobiol. 158, 15–44.
Lee, C., 2018. Therapeutic modulation of virus-induced oxidative stress via the Nrf2-dependent antioxidative pathway. Oxidat. Stress Microb. Dis. Pathogen Host Ther. https://doi.org/10.1155/2018/6208067 (Accessed 18 March 2021).
Liu, Y.C., Zou, X.B., Chai, Y.F., Yao, Y.M., 2014. Macrophage polarization in inflammatory diseases. Int. J. Biol. Sci. 10 (5), 520–529.
Lopez-Otin, C., Blasco, M.A., Partridge, L., Serrano, M., Kroemer, G., 2013. The hallmarks of aging. Cell 153 (6), 1194–1217.
Luc, L., Schramm-Luc, A., Guzik, T.J., Mikolajczyk, T.P., 2019. Oxidative stress and inflammatory markers in prediabetes and diabetes. J. Physiol. Pharmacol. 70 (6), 809–824.
Lugrin, J., Rosenthal-Velin, N., Parapanov, R., 2014. The role of oxidative stress during inflammatory processes. Biol. Chem. https://doi.org/10.1515/hsz-2013-0241, 2014. (Accessed 15 March 2021).
McGarry, T., Biniecka, M., Veale, D.J., Fearon, U., 2018. Hypoxia, oxidative stress and inflammation. Fred. Radic. Biol. Med. 125, 15–24.
Mckee, A.C., Abdolmohammadi, B., Stein, T.D., 2018. The neuropathology of chronic traumatic encephalopathy. Handb. Clin. Neurol. 158, 297–307, 2018.
Meiners, S., Eickelberg, O., Konigshoff, M., 2015. Hallmarks of ageing lung. Eur. Respir. H 45 (3), 807–827.
Melnik, B.C., 2015. MiR21: an environmental driver of malignant melanoma? J. Transl. Med. 13, 202. https://doi.org/10.1186/s12967-015-0570-5.
Milara, J., Cortijo, J., 2012. Tobacco, inflammation, and respiratory tract cancer. Curr. Pharmaceut. Des. 18, 3901–3938.
Millal, M., Siddiqui, M.R., Tran, K., Reddy, S.P., Malik, A.B., 2014. Reactive oxygen species in inflammation and tissue injury. Antioxidants Redox Signal. 20 (7), 1126–1167.
Miller, M.W., Lin, A.P., Wolf, E.J., Miller, D.R., 2018. Oxidative stress, inflammation and neuroprogression in chronic PTSD. Harv. Rev. Psychiatr. 26 (2), 57–69.
Muller, L., Di Benedetto, S., Pawelec, G., 2019. The immune system and its dysregulation with aging. Subcell. Biochem. 91, 21–43.
Neher, M.D., Weckbach, S., Hierl, M.A., Huber-Lang, M.S., Stahel, P.F., 2011. Molecular mechanisms of inflammation and tissue injury after major trauma-is complement the "bad guy". J. Biomed. Sci. 18, 90, 2011. https://www.jbiomedsci.com/content/18/1/90 (Accessed 15 March 2021).
Peterhans, E., 1997. Oxidants and antioxidants in viral diseases: disease mechnaisms and metabolic regulation. J. Nutr. 127 (5 Suppl. l), 962S–965S.
Ramani, S., Pathak, A., Dalal, V., Paul, A., Biswas, A., 2020. Oxidative stress in autoimmune diseases: an under dealt malice. Curr. Protein Pept. Sci. https://doi.org/10.2174/1389203721666200214111816, 2020. (Accessed 18 March 2021).
Reuter, S., Gupta, S.C., Chaturvedi, M.M., Aggarwal, B.B., 2010. Oxidative stress, inflammation and cancer: how are they linked? Free Rad. Biol. Med. 49 (11), 1603–1616.
Schwartz, K.B., 1996. Oxidative stress during viral infection: a review. Free Rad. Biol. Med. 21 (5), 641–649.
Smallwood, M.J., Nissim, A., Knight, A.R., Whiteman, M., Haigh, R., Winyard, P.G., 2018. Oxidative stress in autoimmune rheumatic diseases. Free Radic. Biol. Med. 125, 3–14.
Solleiro-Villavicencio, H., Rivas-Arancibia, S., 2018. Effect of chronic oxidative stress on neuroinflammatory response mediated by $CD^{4+}T$ cells in neurodegenerative diseases. Front. Neurosci. 12. https://doi.org/10.3389/fncel.2018.00114 (Accessed 15 March 2021).
Varizi, N.D., Rodriguez-Iturbe, B., 2006. Mechanisms of disease: oxidative stress and inflammation in the pathogenesis of hypertension. Nat. Clin. Pract. Nephrol. 2, 582–593.
von Zglinicki, T., 2002. Oxidative stress shortens telomeres. Trends Biochem. Sci. 27 (7), 339–344.
Zeliger, 2016. Predicting disease onset in clinically healthy people. Interdiscip. Toxicol. 9 (2), 101–116.
Zhao, R., Liang, H., Clarke, E., Jackson, C., Xue, M., 2016. Inflammation in chronic wounds. Int. J. Mol. Sci. 17. https://doi.org/10.3390/ijms17122085 (Accessed March 16 2021).

CHAPTER

10

Food

10.1 Introduction

Much of human culture is centered around food preparation and consumption. In the twentieth century, food production and preparation in many parts of the world moved from small farms, home and restaurant kitchens and small food preparation businesses (local bakeries, e.g.) to giant food growing and processing operations. Along with these changes, many chemicals were introduced into food. As a result, numerous exogenous chemicals, all of which raise oxidative stress, are consumed on a daily basis. There are six primary sources for these chemicals (Zeliger, 2011, 2016). These are:

1. Uptake by fruits and vegetables from soil and plant surfaces while growing.
2. Ingestion of contaminated and chemically treated meat and water by cattle, fowl and fish.
3. Food preparation. Deliberate incorporation of chemicals into food.
4. Allowable chemicals in food.
5. Food Packaging.
6. Following diets high in sugar, salt, fats and processed foods.

Though the presence of some toxic chemicals in food is inevitable, given their ubiquitous presence in air, water and soil, others are deliberately added for esthetic, preservation and economic purposes.

10.2 Uptake from soil and plant surfaces

Fertilizers containing nitrates, phosphates and heavy metals are applied to soil to increase yields of crops. Insecticides, herbicides and fungicides are applied to food crops to reduce losses to "pest" species. Plants consumed by humans are further contaminated with toxic chemicals arising from agricultural runoff, industrial runoff, urban runoff and environmental spread of persistent organic pollutants (POPs) into growing fields. As was discussed in Chapter 5, these chemicals may be absorbed into the roots of plants and passed up the food chain to humans by animals that consume such plants and bioaccumulation of these. These chemicals may also be directly consumed by humans who eat the treated plants.

https://doi.org/10.1016/B978-0-323-91890-9.00006-4

10.3 Animal ingestion

Antibiotics are used by meat producers to improve animal production and treat disease. These are administered to beef and dairy cattle, swine, broiler chicks and laying hens (Reyes-Herrera et al., 2005). Prophylactic antibiotics are also widely administered in finfish aquaculture to prevent bacterial infections that result from poor sanitation in fish farming (Cabello, 2006).

The administering of antibiotics to animals and fish leads to the accumulation of residues in the animals' flesh and subsequent ingestion by humans. Such antibiotic residues are due to the development of cytotoxic effects and reduced capacities of detoxification and excretion organs in the treated animals (Reyes-Herrera et al., 2005). In the United States, the federal government has established antibiotic residue tolerances and specifies target tissues for residue monitoring. When muscle tissue is specified as the target tissue for testing, the federal standards do not specify which muscle tissue is to be examined. In poultry, for example, breast muscle accumulates antibiotic residues at higher concentrations than thigh muscle does (Reyes-Herrera et al., 2005).

The effects on humans following consumption of meats with antibiotic residues have not been well studied. It is surmised that such residues help lead to antibiotic-resistant bacteria with obvious human end points.

Anabolic agents are used to increase the rate of growth in livestock (Hoffman and Karg, 1976; Heintzman, 1976). Both steroids that are natural in the body (endogenous steroids) and those that are foreign to the body (exogenous steroids) are used for this purpose. Exogenous steroids include esters of endogenous steroids (e.g., estradiol benzoate and testosterone propionate) or compounds with modified steroidal structures such as trenbolone acetate (Hoffman and Karg, 1976). All these compounds, however, have one characteristic in common, they are transferred through the food chain to humans who consume the treated animals.

Steroids in meat consumed by humans have the propensity to produce endocrine disrupting effects. Estradiol, progesterone and testosterone, which are fed to meat animals, occur naturally in both humans and animals in identical molecular forms. Consuming meat with these compounds in them raises the levels of these hormones in the human body. Very little is known about the effects of such increases in hormone levels in humans. Even less is known of the effects on children (Andersson and Skakkebaek, 1999).

The effects of exogenous steroids can differ from those of endogenous steroids in several ways. Firstly, the biological activity of exogenous sex hormones can be much stronger. Secondly, they may be metabolized differently and thirdly, they may induce effects that are different from those of endogenous steroids (Andersson and Skakkebaek, 1999).

Animals that are hunted and fished for as food often contain bioaccumulated residues of toxic chemicals. These include; POPs (PCBs, dioxins and organochlorine compounds, including pesticides) and mercury.

10.4 Persistent organic pollutants (POPs)

Persistent organic pollutants (POPs) are pervasive in our environment. They are present virtually everywhere in the world including the arctic environs (Braune et al., 2005). Food is the primary source of human exposure to POPs. As was discussed in Chapter 9, they are taken up in plants grown in contaminated soil (Collins et al., 2006). For example, alpha-endosulfan, beta-endosulfan and endosulfan sulfate are absorbed by lettuce growing in soil contaminated with these pesticides (Esteve-Turrillas et al., 2005).

POPs are found in the flesh of animals in all the world's environments. The examples that follow are illustrative.

Fish eaten by people residing in northern Norway are contaminated with PCBs, chlorinated pesticides and their metabolites (Sandanger et al., 2006). Fish living in mangrove habitats in Singapore are contaminated with polybrominated diphenyl ethers (PBDEs) (Bayen et al., 2005). Dioxins and PCBs are found in the eggs of free-range chickens (Schoeters and Hoogenboom, 2006).

POPs are widespread in human food products [7]. An indication of the pervasiveness of POPs is seen from a worldwide study of butter contamination with PCBs, polychlorinated dibenzo-p-dioxins (PCDDs), dibenzofurans (PCDFs), hexachlorobenzene and DDT and its metabolites. These POPs were found to be present in varying degrees in the butter of 37 different nations (Weiss et al., 2005).

PCDDs, PCDFs, PCBs, PBDEs and organochlorine pesticides are contaminants of human breast milk (Nickerson, 2006; Furst, 2006). PCBs and PBDEs are found in human adipose and muscle tissues (Li et al., 2006).

The ubiquitous environmental contaminant 2,3,7,8-tetrachlorodibenzo-p-dioxin (TCDD) accumulates in animal fat and plant tissues. The food chain is the primary source of exposure to humans. TCDD is a multifaceted toxin that regulates the expression of a wide range of drug-metabolizing enzymes and impacts a large number of biological systems. The acute effects of TCDD exposure (including chloracne, porphyria, hepatotoxicity and central and peripheral nervous system toxicity) are well described in the literature. Due to the long-term persistence of TCDD in the human adipose tissue, chronic effects may present as long as several decades after exposure. It is hypothesized that TCDD induced atherosclerosis, hypertension, diabetes and nervous system damage can be present long after initial exposures (Pelclova et al., 2006). As TCDD is slowly released from adipose tissue into the human blood stream over a period of years to decades, mixing with other xenobiotics that are taken up by the body long after the absorption new toxic mixtures are almost constantly being created. These mixtures have unknown effects on humans.

10.5 Mercury in food

Mercury is its elemental form is a liquid that is vaporized upon heating. It is a component of fossil fuel and is found in airborne emissions from fossil fuel burning plants. Mercury travels through the environment via several pathways, including air, water and bioaccumulation routes (Evans et al., 2005).

Inorganic mercury released into the environment is converted by microorganistic activity into methyl mercury (MeHg), a persistent polluter that is soluble in adipose tissue and passed up the food chain. As a result, predator fish species are significantly more contaminated with MeHg than those that feed on plants and benthic organisms. For example, it has been found that in Canada's Arctic and Subarctic regions, benthic feeding fish such whitefish have significantly lower MeHg in their flesh than trout, pike and walleye, which are predatory (Evans et al., 2005).

Methyl mercury is toxic to humans, causing central nervous system and peripheral nervous system injuries (Mahaffey, 2005). Those exposed suffer a degeneration of their nervous systems. Symptoms include numbness in lips and limbs, involuntary movement, constricted vision, slurred speech and hallucinations. The most famous historic example of MeHg poisoning is the Minamata Disease outbreak in Japan (Tsubaki and Irukyama. 1977).

In the 1950s a petrochemical plant dumped as estimated 27 tons of mercury waste into Minamata Bay. A short while after the dumping began, people started noticing that cats in the adjacent town appeared to be going insane and were falling into the sea. Soon thereafter, people in the town became symptomatic as described above. Children of poisoned mothers were born with severe deformities, mental retardation, blindness and deafness. Investigation revealed that the animals and humans of Minamata, who consumed large quantities of locally caught fish, were poisoned by methyl mercury, which had been formed in situ in the bay and passed up the food chain from fish to people. In total, almost 3000 people contracted Minamata Disease (Tsubaki and Irukyama, 1977).

Organic mercury pollution is not limited to salt water environs. This pollution is extensive in lakes and rivers as well. People who consume fish contaminated with MeHg, even at low levels, have reduced neurobehavioral performance (Carta et al., 2002). In the United States, many state agencies recommend limiting the eating of fish taken from these waters, particularly by pregnant women.

10.6 Food preparation. Deliberate incorporation of additive chemicals into food

Food additives are chemical substances, other than basic foods, used in commercial food preparation to achieve preservative, flavor, color, stability and esthetic effects. Though some of the chemicals used are derived naturally, most of the additives used in food production are synthetic and with little or no nutritional value (Goldstein and Goldstein, 2002; Winters, 2004). A glance at almost any cookbook shows that chemical additives are not necessary for food preparation. Their use allows inferior ingredients to be used, extends the shelf lives of many products and exposes people to toxic chemicals.

The basic reasons for using food additives are:

1. Emulsification. Emulsifiers, such as lecithin, are used to keep oil and aqueous phases in salad dressings from separating.
2. Thickening. Thickeners, such as carrageenan and carboxymethyl cellulose, are used to thicken ice cream and jelly and impart texture to bread and cake.
3. Enrichment. Vitamins and minerals, such as vitamin D, thiamin and niacin, are added to fortify milk and flour.
4. Anti-caking. Anti-caking agents, sodium aluminosilicate in salt, and silicon dioxide in powdered milk and non-dairy creamers, e.g., are used to prevent coagulation.
5. Chelation. Chelating agents, such as ethylenediaminetetraacetic acid (EDTA) are added to prevent precipitation of insoluble metal salts. Citric acid and tartaric acid are other chelating agents that are added to prevent discoloration during food processing.
6. Bleaching. Bleaching agents, such as peroxides, are used to whiten flour and cheese.

7. Preservation. Antimicrobial agents, such as methyl paraben, propyl paraben, sodium benzoate and calcium propionate are added to many foods to prevent food spoilage caused by mold, bacteria, fungi or yeast. The use of preservation chemicals extends the shelf lives of many foods and eliminates the need for refrigeration of others.
8. Antioxidant Activity. Antioxidants, such as BHA and BHT, prevent fats and oils from reacting with oxygen and becoming rancid.
9. Coloring. Artificial colors, such as blue # 1, yellow #6 and red #40, are added to impart appealing colors to virtually every type of prepared food.
10. Flavoring. Artificial flavors, such as saccharin for sugar and benzaldehyde for cherry flavor, are used as substitutes for natural flavors. Flavor enhancers, such as monosodium glutamate (MSG), have little or no flavor of their own, but are used to enhance the flavor of other food compounds.

10.7 Chemical impurities in food - allowable xenobiotics

In the United States, The Federal Food, Drug and Cosmetic Act of 1938 (FD&C) gave the Food and Drug Administration (FDA) the authority to oversee and regulate food and food ingredients. The Food Additives Amendment to the FD&C of 1958 requires FDA approval for food additives usage prior to their incorporation into food and also requires the manufacturer to prove an additive's safety for the way it will be used.

Food additives are chemical substances, other than basic foods, used in commercial food preparation to achieve preservative, flavor, color, stability and esthetic effects. Though some of the chemicals used are derived naturally, most of the additives used in food production are synthetic and with little or no nutritional value (Goldstein and Goldstein, 2002; Winter, 2004).

10.8 Allowable chemical impurities in food

In the United States, The Federal Food, Drug and Cosmetic Act of 1938 (FD&C) gave the Food and Drug Administration (FDA) the authority to oversee and regulate food and food ingredients. The Food Additives Amendment to the FD&C of 1958 requires FDA approval for food additives usage prior to their incorporation into food and also requires the manufacturer to prove an additive's safety for the way it will be used. There are, however, exceptions to the regulation. All additives that were determined by the FDA or the U.S. Department of Agriculture (USDA) to be safe for use in specific foods prior to the 1958 amendment were designated as prior-sanctioned substances (an example being nitrites used in luncheon meats). Food additives that had either a history of use in food prior to the 1958 amendment or published scientific evidence of their safe use in food (no matter when published) were also excluded from the requirement that manufacturers prove their safety. Such food ingredients are designated as "generally regarded as safe" (USFDA GRAS, 2020). The GRAS list contains hundreds of chemical additives including monosodium glutamate (MSG), calcium propionate, butylated hydroxy anisole (BHA) and butylated hydroxyl toluene (BHT). It does not contain those chemicals determined by the manufacturers to be covered under the GRAS classification and not requiring notification nor label listing.

The list of chemical additives in foods is not limited to those on the GRAS list. More than 3000 chemicals are permitted to be present in food in the United States. These are listed in an FDA data base referred to as Substances Added to Food Inventory (formally known as Everything Added to Food in the United States – EUFUS) (USFDA SAFI, 2020). This list includes chemicals used in food processing, chemicals arising from machinery used in food processing, extraction solvent residues, pesticide residues, antibiotics, growth hormones and chemicals deliberately added to food. Though the list is long, it does not contain all food additives. Some of these are chemicals added under a GRAS determination made independently of FDA. A partial list of volatile organic SAFI listed additives and the known target organs for each compound is given in Table 10.1. Entries with C after their names are known or suspected carcinogens. Those with T after their names are teratogenic.

The list in Table 10.1 contains numerous lipophiles and hydrophiles. A vast number of mixture combinations are possible. Though most mixtures have not been studied, one possible binary mixture from this list, hexane and methyl ethyl ketone (MEK) has a known synergism. MEK (as well as other ketones on the EAFUS list not shown in Table 10.1) potentiates the neurotoxicity of hexane.

It is not implied here that all the toxic chemicals listed on the EAFUS list are present in all foods, or in any one meal. Many food additives, however, contain large numbers of toxic chemicals.

Food additives number in the thousands. Addressing all of these is beyond the scope of this book. The following sections are illustrative of the toxicants used in commercial food preparation.

Flavor additives are widely used. As an example, Table 10.2 lists the chemicals in artificial strawberry flavoring used in strawberry milk shakes (Feingold Inst, 2020).

The chemicals in Table 10.2 present numerous possible mixtures, the effects of most of which are unknown. An example of a toxic binary mixture of chemicals on this list shows a synergistic effect. When administered together, ethanol inhibits the metabolism of ethyl acetate, resulting in greater toxicity of ethyl acetate (Mortensen et al., 1998).

10.9 Artificial colors

Artificial colors are widely used in commercially prepared food. These are synthetic azo dyes manufactured from petroleum that have no nutritional value. Table 10.3 contains a list of these colors currently approved by the United States Food and Drug Administration (FDA) for use in food (FDA 2021).

Some of the colors listed in Table 10.3 are neurotoxic and others contain carcinogenic components (Feingold, 1979; AAP, 1997; Lancaster and Lawrence, 1999; Kobylewski and Jacobson, 2012). In vitro exposures to a mixture of Blue #1 and yellow #5 was found to induce malignant cell transformation (Price et al., 1978).

An example of the toxicity of artificial colors is provided by a study of children whose diets contained artificial food colors and who were found to exhibit symptoms of attention deficit hyperactivity disorder (ADHD). Removal of the artificial food colors from their diets eliminated their neurotoxic symptoms (Feingold, 1975).

TABLE 10.1 Partial list of SAFI, Substances Added to Food Inventory (formally known as Everything Added to Food in the United States [EUFUS]) data base. All chemicals listed target the skin and eyes.

Chemical	Target organs
Acetaldehyde	RES, CNS
Acetophenone	CNS
Acrolein C, T	Heart, RES,
Amyl alcohol	RES, CNS
Anisole	RES
Benzene C	RES, CNS, blood, bone marrow
Chloroform	CNS, LIV, KID, heart
Cyclohexane	CNS
Diethyl amine	RES, LIV, KID
Ethyl acetate	RES, CNS, RPS
Ethyl acrylate C, T	RES, LIV, KID
Ethylene oxide C	RES, KID, adrenal glands, skeletal muscles, RPS
Glutaraldehyde	RES, CNS, LIV
Hexane	RES, CNS, PNS
Hydrazine C	RES, CNS, KID
Hydrogen sulfide	RES, CNS
Methanol T	RES, CNS, GI
Methyl ethyl ketone	RES, CNS
Methyl methacrylate T	RES, CNS, PNS, KID, LIV
Methylene chloride C	RES, CNS, LIV, blood
Monoethanolamine	RES, CNS, KID, LIV, blood
Phenol	RES, CNS, LIV, KID, heart
Propylene glycol	RES
Pyridine	RES, CNS, LIV, KID
Styrene C	RES, CNS
Trichloroethylene C	RES, CNS, PNS, KID, LIV
Vinyl acetate T	RES, CNS, heart

Abbreviations used are: *CNS*, central nervous system; *GI*, gastrointestinal system; *KID*, kidneys; *LIV*, liver; *PNS*, peripheral nervous system; *RES*, respiratory system; *RPS*, reproductive system; Carcinogens are denoted by *C* after the name, teratogens by *T* after the name.
Toxicology data after Sittig (Pohannish, 2002).

TABLE 10.2 Chemical ingredients in an artificial strawberry flavor used for thick shakes.

Amy acetate
Amyl butyrate
Amyl valerate
Aenthol
Anisyl formate
Benzyl acetate
Benzyl isobutyrate
Butyric acid
Cinnamyl valerate
Cognac essential oil
Diacetyl
Dipropyl ketone
Ethanol
Ethyl acetate
Ethyl amyl ketone
Ethyl butyrate
Ethyl cinnamate
Ethyl heptanoate
Ethyl heptylate
Ethyl lactate
Ethyl methyl phenylglycidate
Ethyl nitrate
Ethyl propionate
Ethyl valerate
Heliotropin
Hydroxyphenyl-2-butanone
a-ionone
Isobutyl butyrate
Lemon essential oil
Maltol
4-methylacetophenone
Methyl anthranilate
Methyl benzoate

TABLE 10.2 Chemical ingredients in an artificial strawberry flavor used for thick shakes.—cont'd

Methyl cinnamate
Methyl heptine carbonate
Methyl naphthyl ketone
Methyl salicylate
Mint essential oil
Nerolin
Neryl isobutyrate
Orris butter
Phenethyl alcohol
Rose
Rum ether
g-undecalactone
Vanillin
Solvent (unspecified)

TABLE 10.3 FDA currently approved artificial colors for use in food.

FD&C blue #1
FD&C blue #2
FD&C green #3
FD&C red #3
FD&C red #40 and its Aluminum lake
FD&C yellow #5
FD&C yellow #6

10.10 Flavor enhancers

Flavor enhancers are also widely added to foods, the most commonly used one being monosodium glutamate (MSG), an excitatory neurotoxin that can freely penetrate certain brain regions and rapidly destroy neurons by hyperactivating N-methyl-D-aspartate (NMDA) receptors. MSG makes the nervous systems of the developing fetus particularly vulnerable to this excitotoxin (Olney, 1994).

Symptoms reportedly developed following ingestion of MSG include; tingling, a burning sensation or a radiating numbness in the back of the neck, forearms and chest; facial pressure tightness, chest pain, headache, nausea, rapid heartbeat, mouth and throat dryness, drowsiness and weakness (Goldstein and Goldstein, 2002; Reit-Lehrer, 1976; Scopp, 1991).

In a blind study, MSG ingestion has been demonstrated to induce asthmatic symptoms in those with this condition. In one study, 41% of those tested developed asthma and other symptoms of MSG ingestion within 1–2 h of eating MSG containing food (Allen et al., 1987).

MSG has also been associated with fibromyalgia syndrome (FM), a painful rheumatologic disorder that is difficult to treat. Patients diagnosed with FM for 2–17 years all had complete or almost complete elimination of their symptoms within months of removing MSG or MSG plus aspartame from their diets. All patients have had recurrence of their symptoms whenever they ingested MSG (Smith et al., 2001).

Aspartame is a non-nutritive sweetener that was first allowed by the FDA for use in dry foods in 1981 and approved for beverages in 1983. It is consumed by tens of millions of people in beverages, instant breakfasts, desserts, breath mints, sugar free chewing gum, vitamins, pharmaceuticals and numerous other products. Though it offers diabetics, dieters and others who must limit sugar intake an alternative sweetening choice, it is toxic. Those consuming it have complained of neurologic, gastrointestinal and allergic reactions (Garriga and Metcalfe, 1998).

Aspartame is completely metabolized in the gut and absorbed as aspartic acid, phenylalanine, methanol and diketopiperazine. Above 86°F, the methanol in aspartame decays, forming formaldehyde and formic acid. When ingested, methanol attacks the eyes, CNS and the GI tract and can damage the liver and kidneys (Pohannish, 2002). Formaldehyde and formic acid are corrosive to mucous membranes and can result in liver and kidney injury and disease when ingested (Pohannish 2002). Formic acid is an established human toxin. Phenylalanine is believed to mediate or exacerbate hepatic encephalopathy (Garriga and Metcalfe, 1998; Hertelendy et al., 1993).

Chronic aspartame ingestion results in an increase of phase I metabolizing enzymes (cytochrome P450) in laboratory animals (Vences-Mejia et al., 2006)] Aspartame is a genotoxin, producing chromosome aberrations (Rencuzogullari et al., 2004). Recent research has shown it to a multipotential carcinogenic agent for laboratory animals, even at a daily dose of 20 mg/kg of body weight, a level that is much less than the current acceptable daily intake (Soffritti et al., 2006).

There is published research on both sides of the toxicity question regarding aspartame, with almost equal numbers of papers claiming toxicity and safety. Most studies address controlled conditions where aspartame exposure could be localized. The foods people eat, however, are never restricted to ingestion of single species and mixture effects may indeed be responsible for the disparities of the test results. The differences in test results are believed to be due the ingestion of aspartame with varying other unidentified chemical species or quantities of other chemicals.

One study considered the toxicity of aspartame when co-administered with the food colorant, Quinoline Yellow (QY). In this study it was shown that synergistic effects are observed when these two additives are given together. Mouse neuroblastoma cells were induced to differentiate and grow neurites when the mixture of aspartame and QY was administered together. Inhibition of neurite outgrowth was found at concentrations of these additives theoretically achievable in plasma by ingestion of a typical snack and drink (Lau et al., 2006).

10.11 Esthetic and storage additives

In addition to their uses as colorants and flavor enhancers, chemical additives are also for esthetic and storage purposes. The Handbook of Food Additives describes more than 8000 trade name and general chemical additives that are used in food products (Ash and Ash, 2002). Those included are listed by category in Table 10.4.

10.12 Volatile organic compounds (VOCs) in food

Volatile organic compounds (VOCs) are present in almost all commercially produced foods consumed in the United States. In a study by FDA scientists, 70 different foods were purchased repeatedly over a 5-year period and analyzed for VOCs. 22 VOCs were found in the foods tested, though no single VOC was found in all foods (Fleming-Jones and Smith, 2003). Table 10.5 lists the foods tested and Table 10.6 lists the 21 VOCs found in foods.

All of the chemicals listed in table in 10.6 are highly toxic and many have been associated with toxic effects of mixtures (Zeliger, 2003, 2011).

10.13 Chemicals in food packaging

Materials used to package food include paper, plastic and rubber food wrappers and containers. These packaging materials contain monomers, oligomers, additives, polymer degradation products, additives and impurities. The chemicals in packaging materials are estimated to contaminate food about 100 times more than pesticides and other POPs (Grob et al., 2006). The contaminants that migrate from packaging into food include allergens, chemicals with estrogenic activity, specific organ toxins, genotoxins and carcinogens. The following examples are illustrative.

1. It is estimated that about one third of all food wrapping contains latex, to which as many as 6% of people are allergic to (Mak et al., 2005).
2. Chemicals with known estrogenic activities contained in plastic and rubber food contact materials include bisphenol A, nonylphenol, benzyl butyl phthalate, styrene oligomers and hydroxylated benzophenones. It has also been found that chemicals in packaging, which either contain a phenol group in their structures or form one easily by hydrolysis

TABLE 10.4 Categories of food additives used in food preparation.

Acidulents
Aerating agents
Alkaline agents
Anticaking agents
Antimicrobials
Antistaling agents
Antioxidants
Antispattering agents
Aromatics
Binders
Bittering agents
Bleaching agents
Bodying agents
Bulking agents
Catalysts
Clouding agents
Coatings
Colorants
Color adjuncts
Color diluents
Color retention aids
Cooling agents
Curing agents
Defoamers/antifoams
Clarifiers
Dietary supplements
Dietary fiber
Dispersants
Dough conditioners
Drying agents
Emulsifiers
Egg replacements
Encapsulants

TABLE 10.4 Categories of food additives used in food preparation.—cont'd

Enzymes
Fat replacements
Fermentation aids
Film-formers
Flavors
Flavor enhancers
Gelling agents
Glazes
Humectants
Instantizing agents
Leavening agents
Masticatory aids
Neutralizers/buffers/pH control agents
Nutrients
Opacifiers
Pickling agents
Preservatives
Propellants
Raising agents
Release agents
Solubilizers
Solvents
Suspending agents
Sweeteners
Synergists
Tenderizers
Texturizers
Thickeners
Vehicles
Vitamins
Viscosity modifiers
Whipping agents [50]

TABLE 10.5 Foods analyzed for VOCs.

Whole milk
Sweet rolls
American cheese
Chocolate chip cookies
Cheddar cheese
Sandwich cookies
Ground beef
Apple pie
Chuck roast
Milk chocolate candy bar
Bacon
Caramels
Hot dogs
Cola
Bologna
Low calorie cola
Salami
Milk-based infant formula
Tuna
Beef, strained/junior
Fish sticks
Carrots, strained/junior
Eggs, scrambled
Apple juice, strained/junior
Peanut butter
Swiss cheese
Corn, cream style
Cream cheese
Popcorn
Chicken nuggets
White bread
Fried chicken, fast food
Blueberry muffins

TABLE 10.5 Foods analyzed for VOCs.—cont'd

Mixed nuts
Corn chips
Graham crackers
Fruit-flavored cereal
Butter crackers
Apples
French fried, fast food
Oranges
Quarter pounder/cheese
Bananas
Taco/tostado
Strawberries
Cheese pizza
Raisins
Cheese/pepperoni pizza
Avocados
Vanilla ice cream
Orange juice
Sherbet
Coleslaw
Popsicles
Tomatoes, raw
Chocolate snack cake
Potato chips
Cake doughnuts
Quarter pounder
Brownies
Meatloaf, homemade
Sugar cookies
Margarine
Sour cream
Butter

(*Continued*)

TABLE 10.5 Foods analyzed for VOCs.—cont'd

Olive/safflower oil
Vanilla ice milk
Fruit-flavored drink
Chocolate cake, commercial
Soy-based infant formula

TABLE 10.6 VOCs found in foods over a 5 year period (Fleming-Jones and Smith, 2003).

Benzene
Bromodichloromethane
n-butylbenzene
Carbon tetrachloride
Chlorobenzene
Chloroform
Cumene
o-dichlorobenzene
p-dichlorobenzene
trans-1,2-dichloroethene
Ethyl benzene
Ethylene dichloride
n-propylbenzene
Styrene
Tetrachloroethylene
Toluene
1,1,1-trichloroethane
Trichloroethylene
1,2,4-trimethylbenzene
m- and/or p-xylene
o-xylene

or metabolism, also display estrogenic activities (Ogawa et al., 2006). Hormone disrupting phthalates also migrate from food packaging into food (Schettler, 2006).

3. Polyethylene terephthalate (PET), which is widely used in water and other beverage containers, contains acetaldehyde and limonene which can migrate into the packaged products (Lissen et al., 1995). PET absorbs organic compounds it comes in contact with and recycling may introduce contaminants upon reuse (Franz et al., 2004). Contamination by recycling is of particular concern when the container has been used to hold cleaners or solvents prior to recycling.
4. Polymerization aids, including initiators and catalysts have been found in plastics intended for food contact. Chemicals detected in these include; methyl benzoate, benzoic acid, biphenyl benzoate, phenyl benzoate and azobisisobutyronitrile (Fordham et al., 2001).
5. Paper and board materials intended for food packaging have been found to contain a large number of toxic organic compounds in them These include; alkyl and aryl aldehydes, BHT, di-tert-butylphenol and substituted benzophenones (Castle et al., 1997).
6. Perfluorochemicals are used in the manufacturing of food packaging materials and cookware. Perfluorooctane sulfonate (PFOS) is a residual impurity in some paper coatings used for food contact. Perfluorooctanoic acid (PFOA) is contained in non-stick cookware coatings. Recent epidemiological studies have shown that both PFOS and PFOA are widely toxic, present in human blood and body organs (Begley et al., 2005; Zeng et al., 2019) and trigger oxidative stress (Chen et al., 2014; Kamendulis et al., 2014).
7. Microwave heat susceptors are packaged with foods intended for microwave oven preparation to generate high temperatures (greater than 300°F) to cook foods such as pizza and French fried potatoes. The metalized polyester film, adhesive and paper packaging materials contained in these products release volatile organic compounds that are absorbed by the foods they package. The VOCs released include; benzene, 1,1,1-trichloroethane and 2-(2-butoxyethoxy)ethanol (McNeal and Hollifield, 1993).
8. Printing inks and adhesives used in packaging frequently release VOCs through packaging layers into foods. Toxic aromatic amines used to catalyze adhesive curing between layers of polyethylene and polypropylene film in food packaging readily permeates through the film and contaminated the food (Pezo et al.,. 2012; Campanella et al., 2015).

10.14 Irradiated food

Food may be preserved by irradiation with beams of ionizing radiation produced by radioactive isotopes. Such treatment kills certain bacteria and molds that induce spoilage. Though irradiated food is not left radioactive, the applied radiation breaks chemical bonds and generates free radicals with, at this time, largely unknown consequences. The safety of irradiated food is being addressed at this time, with some countries allowing it in the marketplace and others banning its sale (Ashley et al., 2004). Unsaturated lipids seem to be particularly vulnerable to irradiation. Irradiated almonds turned rancid upon irradiation with accelerated electrons at a dose of 10 kGy (Sanchez-Bel et al., 2005). Irradiation of PET copolymers intended for food contact significantly increased concentration of acetaldehyde (Komolprasert et al., 2003). To date, the paucity of adequate research suggests caution in accepting irradiated food, particularly since the ingestion of food with increased numbers of free radicals can give rise to higher levels of reactive oxygen species and a higher cancer risk, as was discussed in Chapter 4.

10.15 Chemical preservatives in food

Chemical preservatives are used in almost all packaged foods to maintain stability and promote shelf life. Table 10.7 contains a list of some of these and the food types they are used in (Patnat, 2021).

10.16 Food chemicals and oxidative stress

As discussed in Chapter 2, exposure to all organic and numerous inorganic chemicals results in oxidative stress elevation. All the chemicals listed the sections above fall into this category.

Manufactured foods are contaminated with environmental pollutants, pesticides, growth hormones and antibiotics and chemicals that are incorporated in them for esthetic, manufacturing and storing purposes, rather than for nutritional properties. As a result, people world-wide are constantly ingesting mixtures of OS increasing chemicals of largely unknown composition.

Acute toxic events resulting from ingestion of food that has either been environmentally contaminated or prepared with toxic additives, such as those described in the preceding sections, are less common than chronic exposure effects. Ascribing particular toxic effects to specific agents is often complicated, given the complexities of most diets and most toxic effects of mixtures are difficult to identify. For example, the combined toxic effects of BHA and BHT have been noted where both were known to have been ingested simultaneously. It is quite possible, even probable, that one would unknowingly eat two foods, one containing BHA and the other BHT, develop symptoms and not be able to identify the sources of the toxins. Also, some reactions to foods can easily be misunderstood. For example, some people report an allergic reaction to chocolate following its ingestion. Some so-called chocolate containing products, however, contain no chocolate whatsoever, using extracts and synthetic components to simulate the chocolate taste (Russoff, 1958). In such cases, the symptoms that ensue are responses to unidentified stimuli. Further complicating the equation is the reality that toxic chemical mixtures can also arise from a combination of food ingestion and environmental exposure(s) to different molecular species.

Finally, it should be noted that ingestion of essential foods devoid of chemical additives can also elevate OS. This is the subject of the ensuing sections.

10.17 Dietary choices and oxidative stress

Dietary choices can lead to oxidative stress. Excessive intake of carbohydrates, animal-based proteins, fats and salt result in elevated oxidative stress and chronic inflammation, leading to obesity and other diseases (Tan et al., 2018).

TABLE 10.7 Chemical preservatives and the food types they are used in.

Chemical	Where used
Butylated hydroxy anisole (BHA)	Crackers
	Cereals
	Sausages
	Dried meat
Butylated hydroxy toluene (BHT)	Crackers
	Cereals
	Sausages
	Dried meat
Nitrates and nitrites	Meat
Benzoic acid	Beverages
	Salads
	Bread
	Cheese
	Jams
Sulfites	Fruits
	Vegetables
Calcium propionate	Bread
	Cheese

10.17.1 Carbohydrates

High intake of refined grain and added sugars [Tan refs 11; 32], as well as excessive consumption of added sugars (Ferder et al., 2010; Tappy and Le, 2010; Castro et al., 2015; Rodroguez et al., 2015; Savran et al., 2019) are responsible for sharply increased levels of OS and associated metabolic and cardiovascular diseases.

High fructose intake from sucrose and high fructose corn syrup in sugar-sweetened beverages is a major, if not primary, contributor to elevated oxidative stress, responsible for the onset of the following diseases (Ferder et al., 2010; Chiste et al., 2019; Hernandez-Diazcouder et al., 2019; Masschelin et al., 2020):

Insulin resistance
Metabolic syndrome
Obesity
Hypertension

Atherosclerosis
Hypercholesterolemia
Type 2 Diabetes

10.17.2 Animal protein

Consuming high quantities of red or processed meat is associated with the development of oxidative stress/inflammation-mediated chronic diseases (Van Hecke et al., 2016, 2017; Macho-Gonzales et al., 2020; Ahmad et al., 2020).

These diseases include:

Non-alcoholic fatty liver disease
Colorectal cancer
Insulin resistance
Type 2 diabetes
Cardiovascular disease

Meat and meat products contain salt. The high temperature cooking and processing of meat via curing, salting, smoking, drying and fermentation of meat give rise to the formation of additional OS inducing species, including heterocyclic amines, ammonia and hydrogen sulfide (Tan and Norhaizan, 2019). Iron naturally present in meat gives rise to the formation of ROS during the curing process via the Fenton reaction (Tan and Norhaaizan, 2019).

10.17.3 Nitrates, nitrites and nitrosamines

Nitrates and nitrites are naturally present in plant foods, including lettuce, beets celery and spinach as well as in drinking water. As a result of the increased use of inorganic fertilizer, however, the concentrations of nitrates present in drinking water have increased in many parts of the world. Nitrites, which are endogenously produced in the body, are also added to foods, including processed meats such as bacon, sausage, hot dogs and ham, as well to cheeses, to increase shelf life. Nitrites also are found in humans as a result of nitrate metabolism. Nitrites from both endogenous and exogenous sources react in the gastrointestinal tract to produce N-nitrosamines (Hord et al., 2009; Park et al., 2015; Karwowska and Kononiuk, 2020).

Both nitrites and nitrosamines, which produce reactive nitrogen species (RON), have been associated with numerous acute and chronic diseases, as well as accelerating the normal aging process (Delle-Donne et al., 2006).

10.17.4 Fats

Eating a high fat diet adds significant risk to good health due to increased oxidative stress and mitochondrial dysfunction in several organs (Das and Mandala, 2017; Tan et al., 2018; Lasker et al., 2019). Dietary sources of fat include red meat and dairy products, the overconsumption of which lead to a permanent state of inflammation via the circulation of free fatty acids and generation of white adipose tissue which secretes pro-inflammatory factors (Tan and Norhaizan, 2019).

Animal fats are also a source of cholesterol, a fat which triggers mitochondrial oxidative stress, causes atherosclerosis and has been shown to induce osteoarthritis via increased ROS production (Farnaghi et al., 2017).

In addition to natural fats, trans fatty acids (TFAs), industrially produced fats from partial hydrogenation of vegetable oils with at least one trans configuration double bond are widely used in food preparation. Foods containing TFAs include crackers, bakery products, packaged snacks and margarines. TFAs have been shown to elevate oxidative stress and are associated with accelerating the onset of atherosclerosis (Monguchi et al., 2017; Longhi et al., 2018).

10.17.5 Salt

Large quantities of sodium chloride (salt) are commonly parts of many prepared foods. Excessive salt intake is well known to be associated with cardiovascular disease, including hypertension. The mechanism for this relationship is salt's effect on aortic superoxide (a free radical) production increase (Manning et al., 2003; Banday et al., 2007).

References

AAP, 1997. American Academy of Pediatrics, Committee on Drugs. Interactive ingredients in pharmaceutical products: update. Pediatrics 99 (2), 268–278.

Ahmad, M.I., Ijaz, M.U., Ijas, U.H., 2020. The role of meat protein in generation of oxidative stress and pathophysiology of metabolic syndromes. Food Sci. Anim. Resour. 40 (1), 1–10.

Allen, D.H., Delohery, J., Baker, G., 1987. Monosodium L-glutamate-induced asthma. J. Allergy Clin. Immunol. 80 (4), 530–537.

Andersson, A.M., Skakkebaek, N.E., 1999. Exposure to exogenous estrogens in food: possible impact on human development and health. Eur. J. Endocrinol. 140, 477–485.

Ash, M., Ash, I., 2002. Handbook of Food Additives, second ed. Synapse Information Resources, Inc., Endicott, NY.

Ashley, B.C., Birchfield, P.T., Chamberlain, B.V., et al., 2004. Health concerns regarding consumption of irradiated food. Int. J. Hyg Environ. Health 207 (6), 493–504.

Banday, A.A., Muhammad, A.B., Fazili, F.R., Lokhandwala, M., 2007. Mechanisms of oxidative stress-induced increase in salt sensitivity and development of hypertension in Sprague-Dawley rats. Hypertension 49 (3), 664–671.

Bayen, S., Wurl, O., Karuppiah, S., et al., 2005. Persistent organic pollutants in mangrove food webs in Singapore. Chemosphere 61 (3), 303–313.

Begley, T.H., White, K., Honigfort, P., et al., 2005. Perfluorochemicals: potential sources of and migration from food packaging. Food Addit. Contam. 22 (10), 1023–1031.

Braune, B.M., Outridge, P.M., Fisk, A.T., et al., 2005. Persistent organic pollutants and mercury in marine biota of the Canadian Arctic: an overview of spatial and temporal trends. Sci. Total Environ. 351–352, 4–56.

Cabello, F.C., 2006. Heavy use of prophylactic antibiotics in aquaculture: a growing problem for human and animal health and for the environment. Environ. Microbiol. 8 (7), 1137–1144.

Campanells, G., Ghaani, M., Quetti, G., Farris, S., 2015. On the origin of primary aromatic amines in food packaging materials. Trends Food Sci. Technol. 46 (1), 137–143.

Carta, P., Flore, C., Alinovi, R., et al., 2002. Sub-clinical neurobehavioral abnormalities associated with low level of mercury exposure through fish consumption. Neurotoxicology 24 (4–5), 617–623.

Castle, L., Offen, C.P., Baxter, M.J., Gilbert, J., 1997. Migration studies from paper and board food packaging materials. 1. Compositional analysis. Food Addit. Contam. 14 (1), 35–44.

Castro, M.C., Massa, M.L., Arbelaez, L.G., Schinella, F., Gagliardino, J.J., Francini, F., 2015. Fructose-induced inflammation, insulin resistance and oxidative stress: a liver pathological triad effectively disrupted by lipoic acid. Life Sci. 137, 1–6.

Chen, N., Jia, L., Yang, Y., He, D., 2014. Chronic exposure to perfluorooctane sulfonate induces behavior defects in neurotoxicity through oxidative damages, in vivo and in vitro. PLoS One 9 (11), e 113453.

Chiste, J.A., Pereira, B.P., Porto, M.L., et al., 2019. Worsening of oxidative stress, DNA damage, and atherosclerotic lesions in aged LDLr mice after consumption of guarana soft drinks. Oxid. Med. Cell. Longev. https://doi.org/10.1155/2019/9042526. Article id: 904426526. (Accessed 2 April 2021).

Collins, C., Fryer, M., Grosso, A., 2006. Plant uptake of non ionic organic chemicals. Environ. Sci. Technol. 40 (1), 45–52.

Dalle-Donne, I., Rossi, R., Colombo, R., Giustarini, D., Milzani, A., 2006. Biomarkers of oxidative damage in human disease. Clin. Chem. 52 (4), 601–623.

Des, N., Mandala, A., Bhattacharjee, S., et al., 2017. Dietary fat proportionally enhances oxidative stress and glucose intolerance followed by impaired expression of the genes associates with mitochondrial biogenesis. Food Funct. 8 (4), 1577–1586.

Esteve-Turrillas, F.A., Scott, W.C., Pastor, A., Dean, J.R., 2005. Uptake and bioavailability of persistent organic pollutants by plants grown in contaminated soil. J. Environ. Monit. 7 (11), 1093–1098.

Evans, M.S., Muir, D., Lockhart, W.L., et al., 2005. Persistent organic pollutants and mercury in the freshwater biota of the Canadian Subarctic and Arctic: an overview. Sci. Total Environ. 351–52, 94–147.

Farnaghi, S., Prassadam, I., Cai, G., Friliis, T., Du, Z., Crawford, R., et al., 2017. Protective effects of mitochondria-targeted antioxidants and statins on cholesterol-induced osteoarthritis. FASEB J. 31 (1), 356–367.

FDA, 2021. Color additive status list. https://www.fda.gov/industry/color-additive-inventories/color-additive-status-list. (Accessed 23 March 2021).

Feingold, B.F., 1975. Hyperkinesis and learning disabilities linked to artificial food flavors and colors. Am. J. Nurs. 75 (5), 797–803.

Feingold, B.F., 1979. Dietary management of nystagmus. J. Neural. Transm. 45 (2), 107–116.

Feingold Inst, 2020. Artificial Strawberry Flavoring. www.feingold.org/strawberr.html. (Accessed 23 March 2021).

Ferder, L., Ferder, M.D., Inserra, F., 2010. The role of high-fructose corn syrup in metabolic syndrome and hypertension. Curr. Hypertens. Rep. 12, 105–112.

Fleming-Jones, M.E., Smith, R.E., 2003. Volatile organic compounds in foods: a five year study. J. Agric. Food Chem. 51, 8120–8127.

Fordham, P.J., Gramshaw, J.W., Castle, L., 2001. Analysis for organic residues from aids to polymerization used to make plastics intended for food contact. Food Addit. Contam. 18 (5), 461–471.

Franz, R., Mauer, A., Welle, F., 2004. European survey on post-consumer polyethylene terephthalate (PET) materials to determine contamination levels and maximum consumer exposure from food packages made from recycled PET. Food Addit. Contam. 21 (3), 265–286.

Furst, P., 2006. Dioxins, polychlorinated biphenyls and other organohalogen compounds in human milk. Levels, correlations, trends and exposure during breastfeeding. Mol. Nutr. Food Res. 50 (10), 922–933.

Garriga, M.M., Metcalfe, D.D., 1988. Aspartame intolerance. Ann. Allergy 61 (6 Pt 2), 63–69.

Goldstein, M.C., Goldstein, M.A., 2002. Controversies in Food and Nutrition. Greenwood Press, Westport, CT.

Grob, K., Biedermann, M., Scherbaum, E., 2006. Food contamination with organic materials in perspective: packaging materials as the largest and least controlled source? A view focusing on the European situation. Crit. Rev. Food Sci. Nutr. 46 (7), 529–535.

Heintzman, R.J., 1976. The effectiveness of anabolic agents in increasing rate of growth in farm animals; report on experiments in cattle. Environ. Qual. Saf. Suppl. (5), 89–98.

Hernandez-Diazcouder, A., Romero-Nava, R., Carbo, R., Sanchez-Lozada, L.G., Sanchez-Munoz, F., 2019. High fructose intake and adipogenesis. Int. J. Mol. Sci. 2019. https://doi.org/10.3390/ijms20112787. (Accessed 2 April 2021).

Hertelendy, Z.I., Mendenhall, C.L., Rouster, S.D., et al., 1993. Biochemical and clinical effects of aspartame in patients with chronic, stable alcoholic liver disease. Am J Gastroenterol 88 (5), 737–743.

Hoffman, B., Karg, H., 1976. Metabolic fate of anabolic agents in treated animals and residue levels in their meat. Environ. Qual. Saf. Suppl. (5), 181–191.

Hord, N.G., Tang, Y., Bryan, N.S., 2009. Food sources of nitrates and nitrites: the physiologic context for potential benefits. Am. J. Clin. Nutr. 90, 1–10.

Kamendulis, L.M., Wu, Q., Sandusky, G.E., Hocevar, B.A., 2014. Perfluorooctanoic acid exposure triggers oxidative stress in the mouse pancreas. Toxicol Rep 1, 513–521.

Karawowska, M., Konoiuk, A., 2020. Nitrates/Nitrites in food-risk of nitrosative stress and benefits. Antioxidants 9 (3), 241. https://doi.org/10.3390/antiox9030241. (Accessed 2 April 2021).

Kobylewski, S., Jacobson, M.F., 2012. Toxicology of food dyes. Int. J. Occup. Environ. Health 18 (3), 220–246.

Komolprasert, V., McNeal, T.P., Begley, T.H., 2003. Effects of gamma- and electron beam irradiation on semi-rigid amorphous polyethylene terephthalate copolymers. Food Addit. Contam. 20 (5), 505–517.

Lancaster, F.E., Lawrence, J.F., 1999. Determination of benzidine in the food colours tartrazine and sunset yellow FCF, by reduction and derivatization followed by high-performance liquid chromatography. Food Addit. Contam. 16 (9), 381–390.

Lasker, S., Rahman, M.M., Parvez, F., Zamilla, M., Miah, P., Nahar, K., et al., 2019. High-fat diet-induced metabolic syndrome and oxidative stress in obese rats are ameliorated by yogurt supplementation. Sci. Rep. 9, 20026. https://doi.org/10.1038/s41598-019-56538-0.

Lau, K., McLean, W.G., Williams, D.P., Howard, C.V., 2006. Synergistic interactions between commonly used food additives in a developmental neurotoxicity test. Toxicol. Sci. 90 (1), 178–187.

Li, Q.Q., Loganath, A., Chong, Y.S., et al., 2006. Levels of persistent organic pollutant residues in human adipose and muscle tissues in Singapore. J. Toxicol. Environ. Health 69 (21), 1927–1937.

Lissen, J., Reitsma, H., Cozijnsen, J., 1995. Static headspace gas chromatography of acetaldehyde in aqueous foods and polythene terephthalate. Z. Lebensm. Unters. Forsch. 201 (3), 253–255.

Longhi, R., Almeida, R.F., Pettenuzzo, L.F., Souza, D.G., Machado, L., Quincozes-Santos, A., Souza, D.O., 2018. Effect of trans fatty acid-enriched diet on mitochondrial, inflammatory, and oxidative stress parameters in the cortex and hippocampus of Wistar rats. Eur. J. Nutr. 57 (5), 1913–1924.

Mach-Gonzalez, A., Garcimertin, A., Lopez-Oliva, M.E., Bastida, S., Benedi, J., Ros, G., et al., 2020. Can meat and meat-products induce oxidative stress? Antioxidants 9 (7), 638. https://doi.org/10.3390/antiox9070638. (Accessed 2 April 2021).

Mahaffey, K.R., 2005. Mercury exposure: medical and public health issues. Trans. Am. Clin. Climatol. Assoc. 116, 127–153.

Mak, R.K., O'Gorman-Lalor, O., Croom, A., Wakelin, S.H., 2005. An unusual case of latex allergy: contact urticaria from natural rubber latex in chocolate bar wrappers. Clin. Exp. Dermatol. 30 (2), 190–191.

Manning Jr., R.D., Meng, S., Tian, N., 2003. Renal and vascular oxidative stress and salt-sensitivity of arterial pressure. Acta Physiol. Scand. 179 (3), 243–250.

Masschelin, P.M., Cox, A.R., Chernis, N., Hartig, S.M., 2020. The impact of oxidative stress on adipose tissue energy balance. Front. Physiol. 10, 1638. https://doi.org/10.3389/fphys2019.01638. (Accessed 2 April 2021).

McNeal, T.P., Hollifield, H.C., 1993. Determination of volatile chemicals released from microwave-heat-susceptor food packaging. J. AOAC Int. 76 (6), 1268–1275.

Monguchi, T., Hara, R., Hasokawa, M., Nakajima, H., Mori, K., Toh, R., et al., 2017. Excessive intake of trans fatty acid accelerates atherosclerosis through promoting inflammation and oxidative stress in a mouse model of hyperlipidemia. J. Cardol. 70, 121–127.

Mortensen, B., Osvoll, P.O., Woldbaek, T., 1998. In vitro screening for metabolic interactions among frequently occurring binary mixtures of volatile organic chemicals in Norwegian occupational atmosphere. Pharmacol. Toxicol. 83 (2), 49–56.

Nickerson, K., 2006. Environmental contaminants in breast milk. J. Midwifery Womens Health 51 (1), 26–34.

Ogawa, Y., Kawamra, U., Wakui, C., et al., 2006. Estrogenic activities of chemicals related to food contact plastics and rubbers tested by yeast tow-hybrid assay. Food Addit. Contam. 23 (4), 422–430.

Olney, J.W., 1994. Excitotoxins in foods. Neurotoxicol 15 (3), 535–544.

Park, J.E., Seo, J.E., Lee, J.Y., Kwon, H., 2015. Distribution of seven N-nitrosamines in food. Toxicol. Res. 31 (3), 279–288.

Pohannish, R.P., 2002. Sittig's Handbook of Toxic and Hazardous Chemicals and Carcinogens. Noyes Publications, William Andrew Publishers, Norwich, NY.

Pelclova, D., Urban, P., Preiss, J., et al., 2006. Adverse health effects in humans exposed to 2,3,7,8-tetrachlorodibenzo-p-dioxin (TCDD). Rev. Environ. Health 21 (2), 119–138.

Pez, D., Fedeli, M., Bosetti, O., Nerin, C., 2012. Aromatic amines from polyurethane adhesives in food packaging: the challenge of identification and pattern recognition using Quadrupole-Time of Flight Mass Spectrometry. Anal. Chim. Acta 756 (5), 49–59.

Plantnat, 2021. Food Chemical Preservatives List. https://www.planetnat.com/list-chemical-preservatives-food/. (Accessed 22 March 2021).
Price, P.J., Suk, W.A., Freeman, A.E., et al., 1978. In vitro and in vivo indications of the carcinogenicity and toxicity of food dyes. Int. J. Cancer 21 (3), 361–367.
Reif-Lehrer, L., 1976. Possible significance of adverse reactions to glutamate in humans. Fed. Proc. 35 (11), 2205–2211.
Rencuzogullari, E., Tuylu, B.A., Topaktas, M., et al., 2004. Genotoxicity of aspartame. Drug Chem. Toxicol. 27 (3), 257–268.
Reyes-Herrera, I., Schneider, M.J., Cole, K., et al., 2005. Concentrations of antibiotic residues vary between different edible muscle tissues in poultry. J. Food Prod. 68 (10), 2217–2219.
Rodriguez, L., Otero, P., Panadero, M.I., Rodrigo, S., Alvarez-Millan, J.J., Bocos, C., 2015. Maternal fructose intake induces insulin resistance and oxidative stress in male, but not female, offspring. J. Nutr. Metab. 2015. https://doi.org/10.1155/2015/158091. (Accessed 2 April 2021).
Rusoff, I.J., 1958. Process of Producing an Artificial Chocolate Flavor and the Resulting Product. U.S. Patent No. 2,835,590.
Sanchez-Bel, P., Martinez-Madrid, M.C., Egea, I., Romojaro, F., 2005. Oil quality and sensory evaluation of almond (*Prunus amygdalus*) stored after electron beam processing. J. Agric. Food Chem. 53 (7), 2567–2573.
Sandanger, T.M., Brustad, M., Sandau, C.D., Lund, E., 2006. Levels of persistent organic pollutants (POPs) in a coastal northern Norwegian population with high fish-liver intake. J. Environ. Monit. 8 (5), 552–557.
Savran, M., Asci, H., Ozmen, O., Erzurumlu, Y., Savas, H.B., Sonmez, Y., Sahin, Y., 2019. Melatonin protects the heart and endothelium against high fructose corn syrup consumption-induced cardiovascular toxicity via SIRT-1 signaling. Hum. Exp. Toxicol. 38 (10), 1212–1223.
Schettler, T., 2006. Human exposure to phthalates via consumer products. Int. J. Androl. 29 (1), 134–139.
Schoeters, G., Hoogenboom, R., 2006. Contamination of free-range chicken eggs with dioxins and dioxin-like polychlorinated biphenyls. Mol. Nutr. Food Res. 50 (10), 908–914.
Scopp, A.L., 1991. MSG and hydrolyzed vegetable protein induced headache: review and case studies. Headache 31 (2), 107–110.
Smith, J.D., Terpening, C.M., Schmidt, S.O., Gums, J.G., 2001. Relief of fibromyalgia symptoms following discontinuation of dietary excitotoxins. Ann. Pharmacother. 35 (6), 702–706.
Soffritti, M., Belpoggi, F., Degli-Esposti, D., et al., 2006. First experimental demonstration of the multipotential carcinogenic effects of aspartame administered in the feed to Sprague-Dawley rats. Environ. Health Perspect. 114 (3), A176.
Tan, B.L., Norhaizan, M.E., Liew, W.P.P., 2018. Nutrients and oxidative stress: friend or foe? Oxid. Med. Cell. Longev. 2018. https://doi.org/10.1155/2018/19584. (Accessed 2 April 2021).
Tan, B.L., Norhaizan, M.E., 2019. Effect of high-fat diets on oxidative stress, cellular inflammatory response and cognitive function. Nutrients 11 (11), 2579. https://doi.org/10.3390/nu11112579. (Accessed 2 April 2021).
Tappy, L., Le, K.A., 2010. Metabolic effects of fructose and the worldwide increase in obesity. Physiol. Rev. 90 (1), 23–46.
Tsubaki, T., Irukyama, K. (Eds.), 1977. Minamata Disease. Elsevier, Amsterdam.
USFDA GRAS, 2020. Food and Drug Administration. Food generally regarded as safe list (GRAS). https://www.accessdata.fda.gov/scripts/cdrd/cfdpcs/cfcfr/CFRSearch.cofm?fr=184.1. (Accessed 23 March 2021).
USFDA SAFI, 2020. Food and Drug Administration. Substances added to food inventory (SAFI). Previously known as Everything Added to Food in the United States (EAFUS). https://www.fda.gov/food/food-additives-petitions/substances-added=food-formally-eafus. (Accessed 23 March 2021).
Van Hecke, T., Jakobsen, L.M.A., Vossen, E., Gueraud, F., De Vos, F., Pierre, E., Bertram, H.C.S., De Smet, S., 2016. Short-term beef consumption promotes systematic oxidative stress, TMAO formation and inflammation in rats, and dietary fat content modulates these effects. Food Funct. 7, 3760–3771.
Van Hecke, T., Van Camp, J., DE Smet, S., 2017. Oxidation during digestion of meat: interactions with the diet and *Helicobacter pylori* Gastritis, and implications on human health. Compr. Rev. Food Sci. Food Saf. 16 (2), 214–233.
Vences-Mejia, A., Labra-Ruiz, N., Hernandez-Martinez, N., et al., 2006. The effect of aspartame on rat brain xenobiotic-metabolizing enzymes. Hum. Exp. Toxicol. 25 (8), 453–459.
Weiss, J., Papke, O., Bergman, A., 2005. A worldwide survey of polychlorinated dibenzo-p-dioxins, dibenzofurans, and related contaminants in butter. Ambio 34 (8), 589–597.
Winter, R., 2004. A Consumer's Dictionary of Food Additives. Three Rivers Press, New York, NY.

Zeliger, H.I., 2003. Toxic effects of chemical mixtures. Arch. Environ. Health 58 (1), 23–29.
Zeliger, H.I., 2011. Human Toxicology of Chemical Mixtures, second ed. Elsevier, London.
Zeliger, H.I., 2016. Predicting disease onset in clinically healthy people. Interdiscipl. Toxicol. 9 (2), 101–116.
Zeng, A., Song, B., Xiao, R., Zeng, G., Gong, J., Chen, M., et al., 2019. Assessing the human health risks of perfluorooctane by in vivo and in invitro studies. Environ. Int. 126, 598–610.

CHAPTER

11

Sleep deprivation

11.1 Introduction

Adequate sleep is essential for homeostasis in people of all ages. In the United States, it is estimated that one third of adults and half of older people chronically get less than the recommended amount of sleep. The Centers for Disease Control and Prevention (CDC) recommend that adults get at least 7 h of sleep each night and that younger people require greater amounts of nightly sleep. Table 11.1 shows the CDC recommended hours of sleep by age (CDC, 2017).

Sleep interruption has several causes. These include:

- Temperature extremes
- Breathing difficulties, such as sleep apnea
- Interruptions in the Circadian Cycle
- Prevalent health conditions and lifestyle choices that result in waking frequently during the night.

No matter what the cause, sleep interruption has been associated with increased risk of additional disease (Colten and Altevogt, 2006; Koyanagi et al., 2014). Diseases with increased risk of onset due to sleep deprivation are listed in Table 11.2.

Irrespective of the cause, sleep deprivation is associated with increased oxidative stress and inflammation (Atrooz and Salim, 2020). The following sections examine the most prevalent causes of sleep deprivation and their relationship to increased OS.

11.2 Temperature extremes

Ambient temperature extremes are detrimental to successful sleep. Both excessively high and excessively low ambient temperatures are major contributors to sleep deprivation (Gilbert et al., 2004). Core body temperature naturally decreases slightly during sleep. High ambient temperatures prevent this vital reduction and result in loss of sleep. Excessive body temperature reduction, caused by low ambient temperatures also result in sleep loss (Okamoto-Mizuno and Mizuno, 2012).

Oxidative Stress
https://doi.org/10.1016/B978-0-323-91890-9.00023-4

TABLE 11.1 CDC recommended hours of sleep by age.

Age group		Hours of sleep
Newborn	0–3 months	14–17
Infant	4–12 months	12-16, including naps
Toddler	1–2 years	11-14, including naps
Pre-school	3–5 years	10-13, including naps
School age	6–12 years	9–12
Teen	13–18 years	8–10
Adult	18–60 years	7 or more
	61–64	7–9
	65 and older	7–8

TABLE 11.2 Diseases with increased risk of onset due to sleep deprivation.

Obesity
Systemic hypertension
Angina
Myocardial infarction
Stroke
Type 2 diabetes
Insulin resistance
Depression
Anxiety
Arthritis
Chronic lung disease, including asthma

11.3 Sleep apnea

Sleep apnea, a syndrome with repetitive episodes of partial or total collapse of the upper respiratory airway during sleep, results in sleep deprivation and chronic intermittent hypoxia. It is estimated to affect approximately 20% of the population, with middle aged adults and the elderly primarily impacted by it. Sleep apnea elevates the risk for the onset of several chronic diseases (Zhang and Veasey 2012; Lavie, 2015; Eisele et al., 2015; Zhou et al., 2016; Pinto et al., 2016; Bonsignore et al., 2019). These diseases are listed in Table 11.3.

11.4 Circadian cycle interruption

Circadian rhythms are physical, mental and behavioral changes that follow a 24-h cycle in response to the earth's rotation around its axis and the sun. These rhythms are essential to a myriad of physiological processes that maintain homeostasis. In healthy people, circadian body temperature rhythm maintains a temperature range that fluctuates by approximately 1 °C between day and night, with body temperatures declining at night as metabolism de-ceases (Wilking et al., 2013; Coiffard et al., 2021).

Sleep deprivation, which is triggered by both hot and cold temperature extremes, disrupts the circadian cycle (Okamoto-Mizuno and Mizuno, 2012; Goel et al., 2013; Wu et al., 2019). Chronic sleep deprivation due to temperature extremes has been shown to affect circadian rhythms and result in the onset of Alzheimer's disease, Parkinson's disease and interruption of body metabolism (Wu et al., 2019; Vallee et al., 2020; Moller-Levet et al., 2013).

11.5 Prevalent health conditions and lifestyle choices

Multiple health conditions and diseases cause sleep interruption. These are listed in Table 11.4. A direct linear dose-response relationship exists between the number of chronic diseases and the degree of sleep interruption (Koyanagi et al., 2014).

11.6 Sleep deprivation and oxidative stress

All causes of sleep deprivation listed in Table 11.3 result in increased levels of OS. Extreme heat causes sleep interruption and elevates OS (Slimen et al., 2014; Jacobs et al., 2020).

TABLE 11.3 Diseases at elevated risk for onset in people with sleep apnea.

Hypertension
Coronary disease
Cardiovascular disease
Heart failure
Hyperlipidemia
Obesity
Type 2 diabetes
Neurocognitive effects

TABLE 11.4 Health conditions and lifestyle causes of sleep interruption.

Older age
Increased number of chronic diseases
Urinary tract conditions
Asthma
Chronic lung disease
Diabetes
Chronic pain
Excessive alcohol use
Smoking

Extreme cold induces body heat production. This causes increased respiration that is accompanied by increased oxygen consumption and the production of ROS that raise OS levels (Blagojevic et al., 2011; Alva et al., 2013; Cong et al., 2018). A similar oxidative stress effect is observed from cold weather exercise (Martarelli et al., 2011).

Sleep apnea is a cause of chronic oxidative stress in patients affected by it. Biomarkers in such patients show excessive generation of ROS, increased lipid peroxidation and reduced antioxidant capacity (Zhang and Veasey 2012; Lavie, 2015; Eisele et al., 2015; Zho et al., 2016).

Interruptions in the circadian cycle due to sleep deprivation are also associated with higher levels of ROS, lower levels of antioxidants and elevated levels of oxidative stress (Fanjul-Moles and Lopez-Risquelme, 2016; Teixeira et al., 2019).

All the conditions listed in Table 11.4 (as discussed in previous and ensuing chapters) are associated with elevated OS. It is thus safe to conclude that any source of sleep interruption will elevate OS and that chronic sleep interruption will lead to the onset of new disease.

References

Alva, N., Palomeque, J., Carbonell, T., 2013. Oxidative stress and antioxidant activity in hypothermia and rewarming: can RONS modulate the beneficial effects of therapeutic hypothermia? Oxid. Med. Cell. Longev. 2013. https://doi.org/10.1155/2013/957054. (Accessed 12 April 2021).

Atrooz, F., Salim, S., 2020. Chapter eight – sleep deprivation, oxidative stress and inflammation. Adv. Protein Chem. Struct. Biol. 199, 309–336.

Blagojevic, D.P., Grubor-Lajsic, G.N., Spasic, M.B., 2011. Cold defence responses: the role of oxidative stress. Front Biosci. (Schol. Ed.) 3, 416–427. https://doi.org/10.2741/s161.

Bonsignore, M.R., Baiamonte, P., Mazzuca, E., Castrogiovanni, A., Marone, O., 2019. Obstructive sleep apnea and comorbidities: a dangerous liaison. Respir. Med. 14, 8. https://doi.org/10.1186/s40248-019-0172-9. (Accessed 12 April 2021).

Centers for Disease Control and Prevention (CDC), 2017. Sleep and Sleep Disorders. How much sleep do I need? https://www.cdc.gov/sleep/about_how_much_sleep.html. (Accessed 10 April 2021).

Coiffard, B., Diallo, A.B., Mezouar, S., Leone, M., Mege, J.L., 2021. A tangled threesome: circadian rhythm, body temperature, and the immune system. Biology 10 (1), 65. https://doi.org/10.3390/biology10010065. (Accessed 12 April 2021).

Colten, H.R., Altevogt, B.M. (Eds.), 2006. Sleep Disorders and Sleep Deprivation: An Unmet Public Health Problem. National Academy Press. http://www.nap.edu/catalog/11617.html. (Accessed 10 April 2021).

Cong, P., Liu, Y., Liu, N., Zhang, Y., Tong, C., Lin, S., et al., 2018. Cold exposure induced oxidative stress and apoptosis in the myocardium by inhibiting the Nrf2-Keap1 signaling pathway. Cardiovasc Disord. 18 (1), 36. https://doi.org/10.1186/s12872-018-0748-x. (Accessed 12 April 2021).

Eisele, H.J., Markart, P., Schulz, R., 2015. Obstructive sleep apnea, oxidative stress, and cardiovascular disease: evidence from human studies. Oxid. Med. Cell. Longev. 2015. https://doi.org/10.1155/2015/608438. (Accessed 11 April 2021).

Fanjul-Moles, M.L., Lopez-Riquelme, G.O., 2016. Ox Med Cell Long. https://doi.org/10.1155/2016/7420637. Accessed August 18, 2022.

Gilbert, A.A., Van den Heuval, C.J., Ferguson, S.A., Dawson, D., 2004. Thermoregulation as a sleep signaling system. Sleep Med. Rev. 8 (2), 81–93.

Goel, N., Basner, M., Rao, H., Dinges, D.F., 2013. Circadian rhythms, sleep deprivation and human performance. Prog. Mol. Biol. Transl. Sci. 119, 155–190.

Jacobs, P.J., Oosthuizen, M.K., Mitchell, C., Blount, J.D., Bennett, N.C., 2020. Heat and dehydration induced oxidative damage and antioxidant defenses following incubator heat stress and a simulated heat wave in wild caught four-striped field mice *Rhabdomys dilectus*. PLoS One 15 (11). https://doi.org/10.1371/journal.pone.0242279. (Accessed 12 April 2021).

Koyanagi, A., Garin, N., Olaya, B., Ayuso-Mateos, J.L., Chatterji, S., Leonardi, M., et al., 2014. Chronic conditions and sleep problems among adults aged 50 or over in nine countries: a multi-country study. PLoS One 9 (12), e11742. https://doi.org/10.1371/journal.pone.0114742. (Accessed 11 April 2021).

Lavie, L., 2015. Oxidative stress in obstructive sleep apnea and intermittent hypoxia – revisited – the bad ugly and good: implications to the heart and brain. Sleep Apnea Rev. 20, 27–45.

Martarelli, D., Cocchioni, M., Spataro, A., Pompei, P., 2011. Cold exposure increases exercise-induced oxidative stress. J. Sports Med. Phys. Fitness 51 (2), 299–304.

Moller-Levet, C.S., Archer, S.N., Bucca, G., Laing, E.E., Slak, A., Kabiljo, J.C.Y., et al., 2013. Effects of insufficient sleep on circadian rhythmicity and expression amplitude of the human blood transcriptome. PNAS. www.pnas.org/cgi/doi/10.1073/pnas.1217154110. Accessed August 18, 2022. (Accessed 12 April 2021).

Okamoto-Mizumo, K., Mizumo, K., 2012. Effects of thermal environment on sleep and circadian rhythm. J. Physiol. Anthropol. 31, 14. http://www.jphysiolanthropol.com/content/31/1/14.

Pinto, J.A., Ribeiro, D.K., Cavallini, A.F.S., Duarte, C., Freitas, G.S., 2016. Comorbidities Associated with Obstructive Sleep Apnea: A Retrospective Study. Thieme Publications. https://doi.org/10.1055/s-0036-1579546, 2016. (Accessed 12 April 2021).

Slimen, I.B., Najar, T., Ghram, A., Dabbebi, H., Mrad, M.B., Abdrabbah, M., 2014. Reactive oxygen species, heat stress and oxidative-induced mitochondrial damage. A review. Int. J. Hyperthermia 30 (7), 513–523.

Teixeira, K.R.C., dos Santos, C.P., de Medeiros, L.A., Mendes, J.A., Cunha, T.M., D Angelis, J.M., et al., 2019. Night workers have lower levels of antioxidant defenses and higher levels of oxidative stress damage when compared to day workers. Sci. Rep. 9 (1), 4455. https://doi.org/10.1938/s41598=019-40989-6. (Accessed 12 April 2021).

Vallee, A., Lecarpentier, Y., Guillevin, R., Vallee, J.N., 2020. Circadian rhythms, neuroinflammation and oxidative stress in the story of Parkinson's disease. Cells 9 (2), 314. https://doi.org/10.3390/cells9020314. (Accessed 12 April 2021).

Wilking, M., Ndiaye, M., Mukhtar, H., Ahmad, N., 2013. Circadian rhythm connections to oxidative stress: implications for human health. Antioxid. Redox Signal. 19 (2), 192–208.

Wu, H., Dunnett, S., Ho, Y.S., Chang, R.C.C., 2019. The role of sleep deprivation and circadian rhythm disruption as risk factors for Alzheimer's disease. Front. Neuroendocrinol. 54. https://www.sciencedirect.com/science/article/abs/pii/S0091302219300287. (Accessed 12 April 2021).

Zhang, J., Veasey, S., 2012. Making sense of oxidative stress in obstructive sleep apnea: mediator or distractor? Front. Neurol. 3, 179. https://doi.org/10.3389/fneur.2012.00179. (Accessed 11 April 2021).

Zhou, L., Chen, P., Peng, Y., Ouyang, R., 2016. Role of oxidative stress in the neurocognitive dysfunction of obstructive sleep apnea syndrome. Oxid. Med. Cell. Longev. 2016. https://doi.org/10.1155/2016/9626831.

CHAPTER

12

Pharmaceuticals

12.1 Introduction

Pharmaceuticals, both prescription and over-the-counter types, are widely used to combat disease and alleviate disease symptoms. All pharmaceuticals have adverse drug reactions (ADRs) associated with them and most pharmaceutical formulations contain excipients, which though considered "inactive," may have their own side effects and/or cause additional or heightened ADRs (Zeliger, 2016).

Long term use of medications to treat chronic diseases often result in ADRs which require that additional pharmaceuticals be taken to counter such effects.

All pharmaceuticals and most "inactive" ingredients included in their formulations elevate oxidative stress which leads to the onset of additional disease (Zeliger, 2016).

12.2 Pharmaceutical use

As the prevalence of chronic diseases such as type 2 diabetes, obesity and asthma continue to grow, the number of pharmaceuticals taken grows as well. In the United States, more than half of all people over the age of 20 take prescription drugs with more than 4 billion drug prescriptions dispensed in 2019 (Statista, 2020). The data for the time period of 2015–19 shown in Table 12.1 show the breakdown of prescription drug use by age (Martin et al., 2019).

TABLE 12.1 Prescription drug use by age in the United States, 2015–19.

Age	Prescription drug use percent
0–11	18.0
12–19	27.0
20–59	46.7
60 and over	85.0

Oxidative Stress
https://doi.org/10.1016/B978-0-323-91890-9.00029-5

The number of prescription drugs taken also increases with age. In 2015, among people in their 70s, more than one third used five or more different pharmaceuticals (polypharmacy) (Golchin, et al., 2015). In some diseases, polypharmacy is even higher. In Parkinson's disease patients, for example, more than 70% use five or more prescription drugs (McLean et al., 2017). As the prevalence of chronic diseases such as type 2 diabetes, obesity and asthma and the proportion of the population with multiple comorbidities continue to grow, the number of pharmaceuticals taken also grows.

12.3 Pharmaceutical adverse drug reactions

All pharmaceutical drugs (PDs) have Adverse drug reactions (ADRs) associated with their use and ADRs are a leading cause of morbidity and mortality in those using PDs (Kramer, 1981; Alomar, 2014 and the references contained therein). It is also known that the number and severity of ADRs associated with the use of these medications are elevated by a number of parameters listed in Table 12.2 (Alomar, 2014; Lauschke and Ingelman-Sundberg, 2016; Alhawassi et al., 2014; Gochfeld, 2017; Mendrick, et al., 2018; Kaufmann et al., 2015; Sushko et al., 2012; Grzybowski et al., 2015; Kramer, 1981).

TABLE 12.2 Contributing factors to elevation of ADR prevalence.

Age
Gender
Genetic makeup
Polypharmacy
Exposure to organic solvents
Alcohol use
Tobacco use
Recreational drug use
Diet
Chronic inflammation
Preexisting disease
Chronic emotional stress

12.3.1 ADR effects

All pharmaceuticals cause ADRs and produce symptoms that vary from mild to severe to life threatening. The most common symptoms of ADRs are shown in Table 12.3 (Kramer, 1981).

More severe ADRs include the onset of chronic diseases, representative ones of which are listed in Table 12.4 (Alomar, 2014; Kaufman et al., 2015; Mendrick et al., 2018).

TABLE 12.3 Common ADR symptoms.

Nausea
Vomiting
Diarrhea
Abdominal pain
Rash
Skin itch
Drowsiness
Insomnia
Weakness
Headache
Dizziness
Trembles
Twitching
Fever

TABLE 12.4 Diseases that may result from ADRs.

Type 2 diabetes
Numerous cancers
Cardiovascular diseases
Neurological diseases
Respiratory diseases
Immunological effects

ADRs and corresponding onsets of new diseases increase in number and severity with increased numbers of prevalent diseases, as these further weaken the body and necessitate increases in polypharmacy, the number of pharmaceuticals regularly taken. This cascading effect accounts for the fact that illness is a prominent cause of the onset of further illness (Alomar, 2014; Zeliger, 2016).

12.4 Excipients in pharmaceuticals

Excipients are "inactive" ingredients in pharmaceuticals, vitamin and mineral preparations. These compounds are considered to be inert and not affect the intended functioning of the active ingredients. Excipients have a variety of purposes that include:

- Appearance
- Palatability
- Stability
- Bioavailability enhancement
- Consistency.

Almost 800 chemicals have been approved by the FDA for use as inactive additives in drug products for which labeling regulations do not require that they be listed on product labels (Brown, 1983; FDA, 2021). Table 12.5 contains a representative list of excipients approved by the U.S. Food and Drug Administration (FDA) for incorporation into pharmaceuticals. The list contains alcohols, synthetic dyes, ethers, esters, aldehydes, ketones, quaternary ammonium compounds, pigments, surfactants, carboxylic acids, chlorinated hydrocarbons, synthetic sweeteners and heavy metals.

Excipients are not foods, but many are the identical chemicals used as food additives, leading to multiple sources and increasing doses when foods, prescription drugs and/or over-the-counter medications containing them are ingested at the same time.

The use of excipients in pharmaceuticals is analogous to the use of surfactants, solvents and stabilizing agents used in pesticides to enhance absorption, increase active half-life and produce synergism (Zeliger, 2011).

Many excipients have been associated with adverse reactions in those ingesting drugs and vitamin/mineral formulations containing these compounds (Smith and Dodd, 1982; Weiner and Bernstein, 1989; Kumar, et al., 1991; American Academy of Pediatrics, 1997; Kumar, et al., 1996). Examples of these and their effects follow.

Though considered pharmacologically inert, excipients can react with pharmaceutical compounds, resulting in efficacy reduction. Excipients can also be toxic themselves and/or contain impurities that degrade into compounds that decompose active ingredients (Feingold, 1977; Lockery, 1977; Gordon, 1983; Birbeck, 1989; Gordon, 1972; Gordon, 1975; Stevenson et al., 1986; Schwartz, 1983; Towns and Mellis, 1984; Drake, 1986; Walton, 1986; Supramaniam and Warner, 1986; Koppel et al., 1991; Crowley and Martini, 2001; Wu et al., 2011; Hotha et al., 2016; Pottel et al., 2020; Rouaz et al., 2021). Table 12.6 contains a partial list of pharmaceutical excipients, their intended purposes and adverse effects.

TABLE 12.5 Representative list of excipients FDA approved for incorporation into pharmaceuticals.

Aspartame
Benzaldehyde
Benzalkonium chloride
Benzoic acid
Benzyl alcohol
Butanol
Butylated hydroxy anisole (BHA)
Butylated hydroxy toluene (BHT)
Butylene glycol
Cetyl alcohol
Chlorobutanol
D & C blue, green, red and yellow dyes
Diethylene glycol monoethyl ether
Dipropylene glycol
Ethyl acetate
Ethyl butyrate
Ethyl alcohol
Ethylene glycol monoethyl ether
Ethylene glycol
Ethylene diamine
Ferric oxide
Formaldehyde
Glycerol
Isopropyl alcohol
Limonene
Methyl acrylate
Methyl methacrylate
Methyl chloride
Methyl ethyl ketone
Parabens – methyl, ethyl, propyl, butyl
Phenol

(*Continued*)

TABLE 12.5 Representative list of excipients FDA approved for incorporation into pharmaceuticals.—cont'd

Phthalates
Polysorbates
Polyvinyl acetate
Polyvinyl alcohol
Propyl gallate
Propylene glycol
Rhodamine D
Saccharin
Silicon dioxide
Simethicone
Sodium benzoate
Sodium lauryl sulfate
Sodium sulfite
Sorbitol
Stannous chloride
Stearic acid
Succinic acid
Sucralose
Sulfates
Taratazine
Tert-butyl alcohol
Trichloroethane
Trichloroethylene
Triphenyl methane
Titanium dioxide
Trichloromonofluoromethane
Zinc chloride
Zinc stearate

TABLE 12.6 Pharmaceutical excipients and their adverse effects.

Excipient	Purpose	Adverse effects
Aspartame	Sweetener	Neurological reactions
		Hypersensitivity reactions
Benzalkonium chloride	Preservative	Hypersensitivity reactions
		Bronchoconstriction
Benzoic acid		
Benzyl alcohol	Preservative	Metabolic disorders
		Respiratory depression
		Hypersensitivity reactions
D&C Red No. 28		Hyperactivity
Ethyl alcohol	Solvent	CNS depression
	Preservative	CVS toxicity
Food colors	Colorants	Bronchospasm
FD&C Red No 3		Anaphylaxis
FD&C Yellow No 5		Hyperactivity
D&C Red No. 28		
Glycerol	Solvent	Gastrointestinal disorders
Rheological agent	Electrolyte imbalance	
Parabens	Preservatives	Hypersensitivity to thos allergic to aspirin
Phthalates	Coatings	Teratogenic effects
	Plasticizers	
Polyethylene glycol	Solvent	Nephrotoxic
	Rheological agent	Gastrointestinal disorders
Poly sorbates	Surfactant	Deaths in low-weight neonates
	Dispersant	
Propyl gallate	Antioxidant	Dermatitis and skin allergy in neonates
Propylene glycol		
Saccharin	Sweetener	Gastrointestinal disorders
		Urticaria
		Eczema
		Photosensitization

(Continued)

TABLE 12.6 Pharmaceutical excipients and their adverse effects.—cont'd

Excipient	Purpose	Adverse effects
Sorbitol	Sweetener	Gastrointestinal disorders
	Diluent	
Sucralose	Sweetener	Affects gut microbiome
Sulfites	Antioxidant	Hypersensitivity reactions
		Bronchospasm

12.5 Pharmaceuticals and oxidative stress

All ADRs, be they from active ingredients and/or excipients, elevate oxidative stress in a dose-response relationship and contribute to one's total oxidative stress (Alomar, 2014; Deavall et al., 2012; Zeliger, 2016, Banerjee et al., 2016; Zeliger, 2019). Additional pharmaceuticals are often required to counter symptoms caused by primary prescription drugs. These, accordingly, lead to further increases in oxidative stress promoting cascading OS elevation and the onset of further disease (Zeliger, 2016).

Predicting which ADRs will ensue from a particular pharmaceutical is extremely difficult, given the numbers and concentrations of endogenous and exogenous chemical species present in a person's body due to environmental circumstances and lifestyle choices and the numerous mixtures that can arise from combinations of these. Associations of specific ADRs to particular pharmaceuticals can, at this time, only be made through empirical observations due to the large numbers of OS elevating mixtures potentially present in the body at any given time.

References

Alhawassi, T.M., Krass, I., Bajorek, B.V., Pont, L.G., 2014. A systematic review of the prevalence and risk factors for adverse drug reactions in the elderly in the acute care setting. Clin. Interv. Aging 4 (9), 2079–2086.

Alomar, M.J., 2014. Factors affecting the development of adverse drug effects. Saudi Pharmaceut. J. 22, 83–94.

American Academy of Pediatrics, 1997. Committee on drugs. Am. Acad. Ped. 99 (2), 268–278.

Banaerjee, s, Ghosh, J., Sil, P.C., 2016. Drug metabolism and oxidative stress: cellular mechanism and new therapeutic insights. Biochem. Anal. Biochem. https://doi.org/10.4172/2161-1009.1000255, 2016. (Accessed 16 April 2021).

Birbeck, J., 1989. Saccharin-induced skin rashes. N. Z.Med. J. 102, 24.

Brown, J.L., 1983. Incomplete labeling of pharmaceuticals: a list of "inactive" ingredients. N. Engl. J. Med. 309, 439–441.

Crowley, P., Martini, L.G., 2001. Drug-excipient interactions. Pharmaceut. Technol. 13, 26–34.

Deavall, D.G., Martin, E.A., Horner, J.M., Roberts, R., 2012. Drug-induced oxidative stress and toxicity. J. Toxicol. https://doi.org/10.1155/2012/645460, 2012. (Accessed 16 April 2021).

Drake, M.E., 1986. Panic attacks and excessive aspartame ingestion. Lancet 2, 631.

FDA, 2021. Inactive Ingredients Database Download. U.S. Food and Drug Administration. https://www.fda.gov/drugs. (Accessed 14 April 2021).

Feingold, B.F., 1977. Behavioral disturbances linked to the ingestion of food additives. Del. Med. J. 49 (2), 89–94.

Gochfeld, M., 2017. Sex differences in human and animal toxicology: toxicokinetics. Toxicol. Pathol. 45 (1), 172–189.

Golchin, N., Frank, S.H., Vince, A., Isham, L., Meropol, S.B., 2015. Polypharmacy in the elderly. J. Res. Pharm. Pract. 4 (2), 85–88.

Gordon, H.H., 1972. Untoward reactions to saccharin. Cutis 10, 77–81.

Gordon, H.H., 1975. Episodic urticaria due to saccharin ingestion. J. Allergy Clin. Immunol. 56 (1), 78–79, 975.

Gordon, H.H., 1983. Photosensitivity to saccharin. J. Am. Acad. Dermatol. 8, 565.

Grzybowski, A., Zulsdorff, M., Wilhelm, J., Tonagel, F., 2015. Topic optic neuropathies: an updated review. Acta Ophthamol. 93 (5), 402–410.

Hotha, K.K., Roychowdhury, S., Subramanian, V., 2016. Drug-excipient interactions: case studies and overview of drug degradation pathways. Am. J. Anal. Chem. 7 (1), 107–140.

Kaufmann, C.P., Stampfli, D., Hersberger, K.E., Lampert, M.L., 2015. Determination of risk factors for drug-related problems: a multidisciplinary triangulation process. BMJ Open 5, e006376. https://doi.org/10.1136/bmjopen-2014-006376.

Koppel, B.S., Harden, C.L., Daras, M., 1991. Tergitol excipient-induced allergy. Arch. Neurol. 48, 789.

Kramer, M.S., 1981. Difficulties in assessing the adverse effects of drugs. Br. J. Clin. Pharmacol. 11, 105S–110S.

Kumar, A., Anastassios, T., Atlas, R., Hunter, A.G., Beaman, D.C., 1996. Sweeteners, dyes and other excipients in vitamin and mineral preparations. Clin. Pediatr. https://doi.org/10.1177/000992286003500903. Accessed August 24, 2022.

Kumar, A., Weatherly, M.R., Beaman, D.C., 1991. Sweeteners, flavorings and dyes in antibiotic preparations. Pediatrics 87, 352–360.

Lauschke, V.M., Ingelman-Sundberg, M., 2016. The importance of patient-specific factors for hepatic drug response and toxicity. Int. J. Mol. Sci. 17, 1714. https://doi.org/10.3390/ijms17101714 (Accessed 2 October 2021).

Lockery Sr., S.D., 1977. Hypersensitivity to tartrazine (FD&C Yellow No. 5) and other dyes and additives present in foods and pharmaceutical products. Ann. Allergy 38, 206–210.

Martin, C.B., Hales, C.M., Gu, Q., Ogden, C.L., Centers for Disease Control and Prevention, 2019. In: Prescription Drug Use in the United States, 2015-2016. NHCS Data Brief No. 334, May, 2019.

McLean, G., Hindle, J.V., Guthrie, B., Mercer, S.W., 2017. Co-morbidity and polypharmacy in Parkinson's disease: insights from a large Scottish primary care database. BMC Neurol. https://doi.org/10.1186/s12883-017-0904-4 (Accessed 16 April 2021).

Mendrick, D.L., Diehl, A.M., Topor, L.S., Dietert, R.R., Will, Y., La Merrill, M.A., et al., 2018. Metabolic syndrome and associated diseases: from the bench to the clinic. Toxicol. Sci. 162 (1), 36–42.

Pottel, J., Armstrong, D., Zou, L., Fekete, A., Huang, X.P., Torosyan, H., et al., 2020. The activities of drug inactive ingredients on biological targets. Science 369 (6502), 403–413.

Rouaz, K., Chiclana-Rodriguez, B., Ricard, A.N., Sune-Pou, M., Mercade-Frutos, D., Sune-Negre, J.M., et al., 2021. Excipients in the paediatric population: a review. Pharmaceuticals. https://doi.org/10.3390/phamaceuticals/13030387, 2021. (Accessed 16 April 2021).

Smith, J.M., Dodd, T.R.P., 1982. Adverse reactions to pharmaceutical excipients. Pois. Rev. 1, 93–142.

Statistica, 2020. Total Drug Prescriptions Dispensed in the U.S. 2009-2019. https://www.statistica.com/statistics/238702/us-total-medical-prescriptions-issued/ (Accessed 15 April 2021).

Schwartz, H.J., 1983. Sensitivity to ingested metabisulfite: variations in clinical presentations. J. Allergy Clin. Immunol. 71, 487–489.

Stevenson, D.D., Simon, R.A., Lumry, W.R., Mathison, D.A., 1986. Adverse reactions to tartrazine. J. Allergy Clin. Immunol. 78, 182–191.

Supramaniam, G., Warner, J.O., 1986. Artificial food additive intolerance in patients with angio-oedema and urticaria. Lancet 2, 907–909.

Sushko, I., Salmina, E., Potemkin, V.A., Poda, G., Tetko, I.V., 2012. Tox alerts: a web server of structural alerts for toxic chemicals and compounds with potential adverse reactions. J. Chem. Inf. Model. 52, 2310–2316. (Accessed 2 October 2019).

Towns, S.J., Mellis, C.M., 1984. Role of acetyl salicylic acid and sodium metabisulfite in chronic childhood asthma. Pediatrics 73, 631–637.

Walton, R.G., 1986. Seizure and mania after high intake of aspartame. Psychosomatics 27, 218–220.

Weiner, M., Bernstein, I.L., 1989. Adverse reactions to drug formulation agents: A handbook of excipients. Marcel Dekker, Inc.

Wu, Y., Levons, J., Narang, A.S., Raghavan, K., Rao, V.M., 2011. Reactive impurities in excipients: profiling, identification and mitigation of drug-excipient incompatibility. AAPS Phar. Sci. Tech. 12 (4), 1248–1263.

Zeliger, H.I., 2011. Human Toxicology of Chemical Mixtures, second ed. Elsevier, London.

Zeliger, H.I., 2016. Predicting disease onset in clinically healthy people. Inderdiscip. Toxicol. 9 (2), 39–54.

Zeliger, H.I., 2019. Oxidative stress index: disease onset prediction and prevention. EC Pharmacol. Toxicol. 7 (10), 1022–1036 (Accessed 2 October 2019).

CHAPTER

13

Psychological stress

13.1 Introduction

There is an old, often used admonition, "you'll worry yourself to death." Sadly, it is true. Psychological stress can lead to psychiatric disease, infectious disease, chronic inflammation, non-infectious chronic disease and death (Ramanthan et al., 2002; Aich et al., 2009; Cohen et al., 2015). Chronic psychological stress also induces chronic inflammation and oxidative stress which mediate chronic disease and has been linked to cancer, diabetes, cardiovascular, neurological, respiratory and other diseases (Salzano et al., 2014; Reuter et al., 2010; Khansari et al., 2009; McEwen 2006; Semenkov et al., 2015). Psychological stress and OS are bidirectional with each a cause of the other (Bouayed et al., 2009).

13.1.1 Glossary

Terms that enter a discussion of psychological stress and its effects are often used to impart different meanings. To avoid confusion, a glossary of terms used by clinicians and researchers, and adopted here, follows.

Term	Definition
Anxiety	A state of worry, nervousness or fear.
Depression	A state of sadness, low energy, reduced capacity for pleasure, reduced appetite for food or sex, guilt about the past, pessimistic outlook for the future and a strong selfcritical feeling.
Mental disorder	A condition affecting thinking, feeling, mood and behavior.
Post traumatic stress disorder (PTSD)	A diagnosis given following a traumatic event and a time period of non-recovery.
Psychological stress	A state of mental or emotional strain or tension due to adverse or very demanding circumstances.

Psychological stress, also referred to as anxiety, emotional stress or just stress, is directly responsible for causing physical disease and even death. The causes of such stress, listed in Table 13.1, have been shown to be far ranging and include bereavement, worries about loved ones, concerns about one's health and financial worries (Clay 2011).

Oxidative Stress
https://doi.org/10.1016/B978-0-323-91890-9.00015-5

TABLE 13.1 Causes of psychological stress.

Money concerns
Work
The economy
Family responsibilities
Personal relationships
Housing costs
Job stability
Family health problems
Personal safety
Abuse, neglect and other maltreatment

Stress triggers the release of free radicals that lead to oxidative stress, which as has been discussed, is related to the cause of all disease. Increased oxidative stress leads to emotional stress increase, creating a viscous cycle (Salleh, 2008; Gradus et al., 2015; Zeliger et al., 2016; Leonard, 2018; McGinty et al., 2020). Health effects caused or exacerbated, at least in part, if not entirely, by psychological stress are listed in Table 13.2.

Multiple symptoms are associated with stress (Mayer, 2000; McLeod, 2010; Gradus et al., 2015; Leonard, 2018; Zeliger, 2019). These are listed in Table 13.3.

13.2 Chronic stress and disease

Chronic psychological stress affects the immune, digestive and cardiovascular systems, leading to co-morbid disease. Immune system response to chronic stress results in the constant release of stress hormones, including adrenaline and cortisol, as well as cytokines, all of which increase heart rate, elevate blood pressure, are inflammatory and elevate oxidative stress (Hayashi, 2014). These, in turn, affect the digestive and cardiovascular systems, leading to disease un these systems (Segerstrom and Miller, 2004; Hovatta et al., 2010; Morey et al., 2015).

Chronic psychological stress leads to an increased heart rate and hypertension, major risk factors for coronary heart disease. These lead to increased blood cholesterol level via adrenaline-mediated release of free fatty acids which, in turn, result in the cholesterol particles clumping together and resulting in clots in the blood and artery walls with ultimate artery occlusion. Hypertension leads to the formation of small lesions in artery walls where cholesterol tends to get trapped (Bagheri et al., 2016; Albert et al., 2017).

TABLE 13.2 Health effects caused or exacerbated by psychological stress.

Alzheimer's disease
Anxiety
Arthritis
Asthma
Cancer rates of progression
Chronic fatigue syndrome
Chronic pain
Coronary heart disease
COVID-19
Crohn's disease
Depression
Diabetes
Digestive problems
Early aging
Early death
Gut problems – IBS and gut bacteria influences
Heart disease
HIV infection acceleration
Hypertension
Increased cholesterol levels
Irritable bowel syndrome
Migraines
Myocardial infarction
Obesity
Pain in neck, back and shoulders
Psoriasis
Psychiatric diseases
Skin rashes, eczema and psoriasis
Sleep depravation
Slower healing
The common cold, influenza and other viral diseases
Ulcerative colitis

TABLE 13.3 Symptoms associated with stress.

Symptoms
Feeling nervous, tense or fearful
Feeling restless
Panic attacks
Rapid heart rate
Feeling too hot or too cold
Chest pain
Grinding of teeth
Hair loss or graying
Hyperventilation
Heartburn
Head aches
Muscle and joint aches
Nausea
Digestive problems – aches, diarrhea, butterflies in the gut
Sweating
Dizziness
Shaking
Twitching
Fatigue
Feeling weak
Sleep difficulties
Irregular menstrual periods and severe cramping.

Chronic psychological stress can also indirectly affect illness by leading to bad lifestyle practices such as smoking, excessive alcohol consumption, poor diet, lack of exercise and lack of sleep, all of which affect oxidative stress (Zeliger, 2016), as well as acccelerating telomere shortening, associated with early aging (Peel et al., 2004).

Patients presenting with symptoms that are not explained by physical diseases are often told, "it's all in your head," and dismissed. Indeed, that is often true, but not in the derisive sense often conveyed by diagnosing physicians. The fields of neuropsychiatry and neuroimmunology are rapidly developing disciplines but still in their infancies. These posit that neuroinflammatory response of the immune system include changes in which the brain works, leading to symptoms of depression, reactions to stress, inflammation and hence, oxidative stress (Salim et al., 2012; Burke, 2019; Bernhard, 2019; Mondelli and Pariante, 2021).

13.3 Psychological stress and oxidative stress

Oxidative stress and psychological stress are linked in the manifestation of psychological disease, including anxiety disorders, depression, obsessive-compulsive disorder and PTSD (Salim, 2014; Miller and Sadeh, 2014; das Gracas Fedoce et al., 2018). Oxidative stress is both a cause and a consequence of psychological stress, resulting in numerous co-morbidities that are attributed to psychological stress, as well as psychiatric diseases - including schizophrenia, bipolar disorder, Alzheimer's disease, Parkinson's disease and ALS (Ng et al., 2008; Schiavone et al., 2013; Salim, 2014; Plata et al., 2014; Balmus et al., 2016)

The relationship between oxidative stress and neurological disease is not surprising. The brain is perhaps the most vulnerable body organ to OS damage, due to its high utilization of glucose and oxygen, as well as well as containing high peroxidation-susceptible lipid cells (Floyd andHensley, 2002; Li et al., 2022; Miller and Sadeh, 2014). OS consequences in the brain include blood-brain barrier permeability and neurotransmission interruption and disruption of neurogenesis (Uttara et al., 2009; Miller and Sadeh, 2014).

In conclusion, chronic psychological stress, from whatever cause, elevates oxidative stress and is a major contributor to the onset of numerous co-morbid diseases.

References

Aich, P., Potter, A., Griebel, P.J., 2009. Modern approaches to understanding stress and disease susceptibility: a review with special emphasis on respiratory disease. Int. J. Gen. Med. 2, 19–32.

Albert, M.A., Durazo, E.M., Slopen, N., Zaslavsky, A.M., Buring, J.E., Silva, R., et al., 2017. Cumulative psychological stress and cardiovascular risk in middle aged and older women: rationale, design and baseline characteristics. Am. Heart J. 192, 1–12.

Bagheri, B., Meshkini, F., Dinarvand, K., Alikhani, Z., Haysom, M., Rasouli, M., 2016. Life psychosocial stresses and coronary heart disease. Int. J. Prev. Med. https://doi.org/10.4103/2008-7802.190598. (Accessed 22 April 2021).

Balmus, I.M., Ciobica, A., Antioch, I., Dobrin, R., Timofte, D., 2016. Oxidative stress implications in the affective disorders: Main biomarkers, animal models relevance, genetic perspectives, and antioxidant approaches. Oxid. Med. Cell. Longev. https://doi.org/10.1155/2016/3975101. (Accessed 23 April 2021).

Bernhard, T., 2019. Beyond depression. Psychol. Today 50–88. January/February 2019.

Bouayed, J., Rammal, J., Soulimani, R., 2009. Oxidative stress and anxiety. Oxid. Med. Cell. Longev. 2 (2), 63–67.

Burke, M.J., 2019. "It's all in your head" – medicine's silent epidemic. JAMA Neurol. https://doi.org/10.1001/jamaneurol.209.3043, 2019. (Accessed 22 April 2021).

Clay, R.A., 2011. Stressed in America. Am. Psychol. Assoc. 42 (1). https://www.apa.org/monitor/2011/01/stressed-america. (Accessed 21 April 2021).

Cohen, S., Janicki-Deverts, D., Doyle, W.J., 2015. Self-related health in healthy adults and susceptibility to the common cold. Psychosom. Med. 77 (9), 959–968.

Das Gracias Fedoce, A., Ferreira, F., Bota, R.G., Bonet-Costa, V., Sun, P.Y., Davies, K.J.A., 2018. The role of oxidative stress in anxiety disorder: cause or consequence? Free Rad. Res. 52 (7), 737–750.

Epel, E.S., Blackburn, E.H., Lin, J., Dhabhar, F.S., Adler, N.E., Morrow, J.D., 2004. Proc. Natl. Acad. Sci. USA 101 (49), 17312–17315.

Floyd, R.A., Kensley, K., 2002. Oxidative stress in brain aging. Implications for therapeutics of neurodegenerative diseases. Neurobiol. Aging 23 (5), 795–807.

Gradus, J.L., Antonsen, S., Svensson, E., Lash, T.L., Resick, P.A., Hansen, J.G., 2015. Trauma, comorbidity and mortality following diagnoses of severe stress and adjustment disorders: a nationwide cohort study. Am. J. Epidemiol. 182 (5), 451–458.

Hayashi, T., 2014. Conversion of psychological stress into cellular stress response: roles of the sigma-1 receptor in the process. Psychiatr. Clin. Neurosci. https://doi.org/10.1111/pen.12262, 2014. (Accessed 23 April 2021).

Hovatta, I., Juhila, J., Donner, J., 2010. Oxidative stress in anxiety and comorbid disorders. Neurosci. Res. 68 (4), 261–265.

Khansari, N., Shakiba, Y., Mahmoudi, M., 2009. Chronic inflammation and oxidative stress as a major cause of age-related diseases and cancer. Recent Pat. Inflamm. Allergy Drug Discov. 3 (1), 73–80.

Leonard, J., 2018. The Effects of Anxiety on the Body. Medical News Today, 2018. https://www.medicalnewstoday.com/articles/322510.php. (Accessed 21 April 2021).

Li, Q., Zhao, Y., Deng, D., Yang, J., Chen, Y., Liu, J., Zhang, M., 2022. Aggravating effects of psychological stress on Ligature-induced periodontitis via the onvolvement of local oxidative damage and NF-kB activation. Mediators Inflamm. https://doi.org/10.1155/2022/6447056. (Accessed 18 August 2021).

Mayer, E.A., 2000. The neurobiology of stress and gastrointestinal disease. Gut 47, 861–869.

McEwen, B.S., 2006. Sleep deprivation as a neurobiologic and physiologic stressor: allostasis and allostatic load. Metabolism 10 (Suppl. 2), S20–S23.

McGinty, E.E., et al., 2020. Psychological distress and COVID-19-related stressors reported in longitudinal cohort of US adults in April and July 2020. J. Am. Med. Assn. https://doi.org/10.1001/jama.2020.21231. (Accessed 21 April 2021).

McLeod, S.A., 2010. Stress, illness and the immune system. Simply Psychol. https://www.simplypsychology.org/stress-immune.html. (Accessed 22 April 2021).

Miller, M.W., Sadeh, N., 2014. Traumatic stress, oxidative stress and posttraumatic stress disorder: neurodegeneration and the accelerated-aging hypothesis. Mol. Psychiatr. 19 (11), 1156–1162.

Mondelli, V., Pariante, C.M., 2021. What can neuroimmunology teach us about the symptoms of lon-COVID? Oxford Open Immunol. 2 (1), iqab004. https://doi.org/10.1093/oxfimm/aqab004. (Accessed 22 April 2021).

Morey, J.N., Boggero, I.A., Scott, A.B., Segerstrom, S.C., 2015. Current directions in stress and human immune function. Curr. Opin. Psychol. 1 (5), 13–17.

Ng, F., Berk, M., Dear, O., Bush, A.I., 2008. Oxidative stress in psychiatric disorders: evidence base and therapeutic implications. Int. J. Neuropsychoparmacol. 11, 851–876.

Palta, P., Samuel, L.J., Mill III, E.R., Szanton, S.L., 2014. Depression and oxidative stress: results from a meta-analysis of observational studies. Psychosom. Med. 76 (1), 12–19.

Ramanathan, L., Gulyani, S., Nienhuis, R., Siegel, J.M., 2002. Sleep deprivation decreases superoxide dismutase activity in rat hippocampus and brainstem. Neuroreport 13 (1), 1387–1390.

Reuter, S., Gupta, S.C., Chaturvedi, M.M., Aggarwal, B.B., 2010. Oxidative stress, inflammation, and cancer: how are they linked? Free Rad. Biol. Med. 49 (11), 1603–1616.

Salim, S., 2014. Oxidative stress and psychological disorders. Curr. Neurophamacol. 12, 140–147.

Salim, S., Chugh, G., Asghar, M., 2012. Inflammation in anxiety. Adv. Protein Chem. Struct. Biol. 88, 1–25.

Salleh, M.R., 2008. Life event, stress and illness. Malaysian J. Med. Sci. 15 (4), 9–18.

Salzano, S., Checconi, P., Hanschmann, E.M., Lillig, C.H., Bowler, L.D., Chan, P., et al., 2014. Linkage of inflammation and oxidative stress via release of glutathionylated peroxiredoxin-2, which acts as a danger signal. Proc. Natl. Acad. Sci. USA 111 (33), 12157–12162.

Schiavone, S., Jaquet, V., Trabace, L., Krause, K.H., 2013. Antioxidants Redox Signal. 18 (12), 1475–1490.

Semenkov, V.F., Michalski, A.I., Sapozhnikov, A.M., 2015. Heating and ultraviolet light activate anti-stress gene functions in humans. Front. Genet. https://doi.org/10.3889/fgene2015.00245.

Sergstrom, S.C., Miller, G.E., 2004. Psychological stress and the human immune system: a meta-analytic study of 30 years of inquiry. Psychol. Bull. 130 (4), 601–630.

Uttara, B., Singh, A.V., Zamboni, P., Mahajan, R.T., 2009. Oxidative stress and neurodegenerative diseases: a review of upstream and downstream antioxidant therapeutic options. Curr. Neuropharmacol. 7, 65–74.

Zeliger, H.I., 2016. Predicting disease onset in clinically healthy people. Interdiscip. Toxicol. 9 (2), 39–54.

Zeliger, H.I., 2019. Oxidative stress index: disease onset prediction and prevention. EC Parmacol. Toxicol. 7 (10), 1022–1036.

CHAPTER

14

Genetics and epigenetics

14.1 Introduction

Genetics is well-known to be a factor in most non-communicable diseases (Dato et al., 2013; Jiang et al., 2013; Balmus et al., 2016; Guillaumet-Adkins et al., 2017; Cioffi et al., 2019). Indeed, many diseases, including Alzheimer's disease, Parkinson's disease and cancers, just to name a few, "run in families." Recently, epigenetics, as well, has been shown to lead to heritable diseases (Cencioni et al., 2013; Guillaumet-Adkins et al., 2017). Though all non-communicable diseases are more prevalent in those whose ancestors have suffered from those diseases, parental disease is most closely associated with the likelihood of disease onset in an individual (Awdeh et al., 2006). Both genetic traits and epigenetic effects raise OS, which is associated with all heritable diseases (Cencioni et al., 2013; Dato et al., 2013; Jiang et al., 2013; Guillaumet-Adkins et al., 2017).

14.2 Genetics

Aging itself is the single most responsible factor for the onset of what are known as age-related diseases; metabolic diseases, neurodegenerative diseases, cardiovascular diseases, respiratory diseases, musculoskeletal diseases and cancers some of which are listed in. Table 14.1. All these diseases are triggered by major genetic factors (Sorenssen et al., 1988; Obeidat and Hall, 2011; Johnson et al., 2015; Huggins, 2015), with genetic propensity for a given disease onset caused by oxidative stress in specific genes related to that disease (Balmus et al., 2016; Cioffi et al., 2019).

Genetic susceptibility to specific infectious and environmental diseases need not be related to aging alone. Environmental effects and lifestyle choices are known to trigger disease onset in susceptible individuals and such people can be protected from disease onset by avoiding environmental and lifestyle choice-induced oxidative stress exemplified by the list in Table 14.2, containing references for each.

Disease onset due to genetic factors can also be triggered by adverse drug reactions, both type A (predictable from the known pharmacology) and type B (idiosyncratic and not predictable from the known pharmacology) (Kaufman, 2016; Osaniou et al., 2018).

Oxidative Stress
https://doi.org/10.1016/B978-0-323-91890-9.00022-2

TABLE 14.1 Age-related diseases with major genetic factors.

Metabolic
Type 2 diabetes
Obesity
High fasting glucose levels
High LDL and/or HDL levels
High triglycerides level
Neurodegenerative
Alzheimer's disease
Amyotrophic lateral sclerosis
Multiple sclerosis
Parkinson's disease
Affective disorders
Cardiovascular
Atrial fibrillation
Hypertension
Coronary heart disease
Pulmonary function diseases
Musculoskeletal
Bone mineral density
Osteoarthritis
Rheumatoid arthritis
Respiratory
Asthma
Chronic obstructive pulmonary disease (COPD)
Cancer
Breast
Colorectal
Lung
Melanoma
Pancreatic
Prostate

TABLE 14.2 Examples of genetic susceptibility to specific diseases.

Disease	References
Affective disorders	Balmus et al. (2016)
Allergic rhinitis	Manti et al. (2016)
Asthma	Ober and Yao (2011)
Breast cancer	Goldvaser et al. (2017)
Childhood cancer	Perrone et al. (2016)
Cardiovascular disease	Huggins (2015)
Diabetes (type 2)	Juntarawijit and Juntarawijit (2018)
Obesity	Gunther et al. (2018)
Viral diseases	Clohisey and Baillie (2019); Sorensen et al. (1988)

14.3 Epigenetics

Epigenetics, which means "on top of or in addition to genetics," is the study of heritable changes that do not involve changes in the DNA sequence. Epigenetics can be affected by environmental factors, lifestyle, disease presence and age via four, interconnected pathways. Onset of disease has been linked to each of these factors alone, as well to combinations of these (Romani et al., 2015; Nilsson et al., 2018). These factors are:

1. DNA methylation
2. Histone modification
3. Chromatin remodeling
4. Non-coding RNA (ncRNA)

DNA methylation occurs via the addition of a methyl group to DNA, resulting in gene function modification and expression. Diseases associated with DNA methylation include numerous cancers, autoimmune diseases, metabolic and psychological disorders (Jin and Liu, 2018).

Histone modification is a post-translational modification to histone proteins, including methylation, acetylation, phosphorylation and ubiquitylation. Diseases associated with Histone modification include prostate and breast cancers, lymphoma and intellectual disabilities, including mental retardation (Green and Shi, 2014).

Chromatin remodeling is the rearranging of chromatin from a compact to a relaxed state, allowing for transcription factors or other DNA binding proteins to access DNA and impact gene expression. Chromatin remodeling is associated with several human genetic diseases, including ATR-X syndrome, Juberg-Marsidi syndrome and Coffin-Lowry syndrome (Hendrich and Bickmore, 2001).

A non-coding RNA is a functional one transcribed from DNA but not translated into proteins. Non-coding RNA is associated with breast and prostate cancers, numerous cardiovascular diseases and neurodegenerative diseases that include Alzheimer's disease Parkinson's disease, amyotrophic lateral sclerosis (ALS) and Huntington's disease (Lekka and Hall 2018).

14.4 Chemical environmental and other factors in epigenetic effects

Toxic chemicals and chemical mixtures, environmental factors and lifestyle can induce epigenetic effects (Csoka and Szyf, 2009; Kasevska et al., 2011; Hou et al., 2012; Saavedra-Rodriguez and Feig, 2013; Dato et al., 2013; Caffo et al., 2014; Skinner et al., 2015; Nilsson et al., 2018; Kubsad et al., 2019; Cavalli and Heard, 2019; Xu et al., 2020).Table 14.3 contains a partial list of chemicals and the epigenetically-induced diseases associated with these.

TABLE 14.3 Chemical, as well as environmental and lifestyle factors associated with epigenetically-induced effects.

Factor	Effect
A. Chemicals and Chemical Mixtures	
Air pollution	Asthma, various cancers, diabetic nephropathy
Aluminum	Alzheimer's disease, various cancers, cardiac hypertrophy
Arsenic	Various cancers, schizophrenia
Atrazine	Testicular disease, early puberty
Benzene	various cancers, schizophrenia, psoriasis
Benzo[a]pyrene	Behavioral and psychological deficits
Bisphenol A	Heart disease, reduced fertility, social behavioral changes
Bisphenol A/phthalate mixtures	Various cancers, cardiac hypertrophy, prostate disease, kidney disease, obesity
Cadmium	Various cancers, type 2 diabetes, heart disease, traumatic brain injury
DDT	Obesity
Dioxin (TCDD)	Various cancers, prostate disease kidney disease, reduced fertility
Glyphosate	Prostate disease, obesity, kidney disease, ovarian disease, birth abnormalities
Jet fuel (hydrocarbon mixture)	Prostate disease, kidney disease, obesity, immune negative immune and reproductive impacts
Lead	Various cancers, schizophrenia
Mercury	Various cancers, neurological disorders, behavior change

TABLE 14.3 Chemical, as well as environmental and lifestyle factors associated with epigenetically-induced effects.—cont'd

Factor	Effect
Methoxychlor	Obesity, impaired male fertility, ovary disease, kidney disease
Nickel	Various cancers, heart disease
Permethrin/DEET mixture	Prostate disease, kidney disease
Pharmaceutical drug use (chronic)	Adverse drug reactions
Phthalates	Hormonal and behavioral impacts, disruption in ovarian function
Vinclozolin	Anxiety-like behavior, impaired male fertility, prostate disease, kidney disease, immune system impact.
B. Environmental and Lifestyle	
Caloric restriction	Cardiovascular
Diet high in carbohydrates and red	Accelerated aging meat, poor in antioxidants
Ethanol	Neurological deficits, decreased fertility
High fat diet	Reduced insulin sensitivity, increased breast cancer incidence
Obesity	Accelerated aging
Prediabetes or diabetes	Impaired glucose tolerance, reduced insulin sensitivity
Smoking tobacco	Abnormal pulmonary function
Stress	Reduced social interaction, increased stress resilience, disrupted neural connectivity, increased anxiety

14.5 Role of oxidative stress in genetic and epigenetic effects

Oxidative stress is at the heart of genetic and epigenetic effects. A majority of DNA mutations are related to oxidative stress (Denver et al., 2009) and both genetic and epigenetic alterations are observed with the onset of cancer and other heritable diseases (Nishida and Kudo, 2013; Guillaumet-Adkins et al., 2017; Garcia-Guede et al., 2020). Increases in intracellular OS is detrimental to cells by damaging cell components and activating signaling pathways that damage or alter cellular processes essential for proper cell functioning or by directly leading to cell death (Scandalios, 2002). The relationship between OS and cellular effects is a two-way street as susceptibility to oxidative stress is related to genetic makeup (Gunther et al., 2018). Thus, OS is both a cause and a result of genetics and epigenetics.

The roles of specific oxidative stress genes that either promote OS of deplete antioxidants have been proposed in the onset of numerous diseases (Ruperez et al., 2014; Yoo, 2015; Balmus et al., 2016; Cioffi et al., 2019). A partial list of these is shown in Table 14.4.

Mitochondria are one of the major sites of ROS production, as well as one of the major targets of their action (Perrone et al., 2016) and mitochondrial diseases due to inherited defects

TABLE 14.4 Diseases associated with oxidative stress genes.

ADHD
Alzheimer's disease
ALS
Anxiety
Autism
Bipolar disorder
Depression
Fetal alcohol syndrome
HIV-associated dementia
Obesity
Parkinson's disease
Schizophrenia

in mitochondrially-expressed genes are associated with oxidative stress. OS has been studied in the most common mitochondrial diseases and found to be a factor in all (Hayashi and Cortopassi, 2015). This issue is discussed more thoroughly in part II of this book.

The evidence shown above clearly demonstrates that genetics and epigenetics, not only are predictive of an individual's propensity of diseases, but also of the likelihood of that individual possessing oxidative stress genes that may add to one's total oxidative stress level and thereby help trigger disease onset.

References

Awdeh, Z.L., Yinis, E.J., Audeh, M.J., Fici, D., Pugliese, A., Larsen, C.E., Alper, C.A., 2006. A genetic explanation for the rising incidence of type 1 diabetes, a polygenic disease. J. Immunol. 27, 174–181.

Balmus, I.M., Ciobica, A., Anthioch, I., Dobrin, R., Timofte, I., 2016. Oxidative stress implications in the affective disorders: main biomarkers, animal models relevance. Genetic perspectives, and antioxidant approaches. Oxid. Med. Cell. Longevity. https://doi.org/10.1155/2016/3975101.

Caffo, M., Caruso, G., La Fata, G., Barresi, V., Visalli, M., Venza, M., Venza, I., 2014. Heavy metals and epigenetic alterations in brain tumors. Curr. Genomics 15, 457–463.

Cavall, G., Heard, E., 2019. Advances in epigenetics link genetics to the environment and disease. Nature 571, 489–499.

Cencioni, C., Spalotta, F., Martelli, F., Valente, S., Mai, A., Zeiher, A.M., Gaetano, C., 2013. Oxidative stress and epigenetic regulation in aging and are-related diseases. Int. J. Mol. Sci. 14, 17643–17663.

Cioffi, F., Adam, R.H.I., Broersen, K., 2019. Molecular mechanisms and genetics of oxidative stress in Alzheimer's disease. J. Alzheimer's Dis. 72, 981–1017.

Clohisey, S., Baillie, J.K., 2019. Host susceptibility to severe influenza A virus infection. Crit. Care. https://doi.org/10.1186/s13054-019-2566-7. (Accessed 28 April 2021).

Csoka, A.B., Szyf, M., 2009. Epigenetic side-effects of common pharmaceuticals: a potential new field ion medicine and pharmacology. Med. Hypoth. 73 (5), 770–780.

Dato, S., Crocco, P., D'Aquila, P., de Rango, F., Belllizi, D., Rose, G., Passarino, G., 2013. Exploring the role of genetic variability and lifestyle in oxidative stress response for healthy aging and longevity. Int. J. Mol. Sci. 14, 16443–16472.

Denver, D.R., Dolan, P.C., Wilhelm, L.J., Sung, W., Lucas-Lledo, I., Howe, D.K., et al., 2009. A genome-wide view of Caenorhabditis elegans base-substitution mutation process. Proc. Natl. Acad. Sci. USA 106 (38), 16310–16314.

Garcia-Guede, A., Vera, O., Ibanez-de-Caceres, I., 2020. When oxidative stress meets epigenetics: implications in cancer development. Antioxidants 9 (6), 468. https://doi.org/10.3390/antiox9060468. (Accessed 4 May 2021).

Goldvaser, H., Gal, O., Rizel, S., Hendler, D., Neiman, V., Shochat, T., et al., 2017. The association between smoking and breast cancer characteristic and outcome. BMC Caner. https://doi.org/10.1186/s12885-017-3611-z. (Accessed 28 April 2021).

Greer, E.L., Shi, Y., 2014. Histone methylation: a dynamic mark in health, disease and inheritance. Nat. Rev. Genet. 13 (5), 343–357.

Guillaumet-Adkins, A., Yanez, Y., Preis, M.D., Palancia-Ballester, C., Sandoval, J., 2017. Epigenetics and oxidative stress in aging. Oxid. Med. Cell Longevity. https://doi.org/10.1155/2017/9175806 article No. ID9175806. (Accessed 23 January 2020).

Gunther, M., Nimer, F.A., Piehl, F., Risling, M., Mathiesen, T., 2018. Susceptibility to oxidative stress is determined by genetic background in neuronal cell cultures. eNeuro 5 (2), e0335. https://doi.org/10.1523/ENEURO.0335-17.2018. (Accessed 4 May 2021).

Hayashi, G., Crotopassi, G., 2015. Oxidative stress in inherited mitochondrial diseases. Free Radic. Biol. Med. 88 (0 0). https://doi.org/10.1016/j.freeradbiomed.2015.05.039.

Hendrich, B., Bickmore, W., 2001. Human diseases with underlying defects in chromatin structure and modification. Hum. Mol. Genet. 10 (20), 2233–2242.

Hou, L., Zhang, X., Wang, D., Baccarelli, A., 2012. Environmental chemical exposures and human epigenetics. Int. J. Epidomiol. 41, 79–105.

Huggins, G.S., 2015. Genetic susceptibility to oxidative stress and cardiovascular disease. EBioMedicine 2, 1864–1865.

Jiang, T., Yu, J.T., Tian, Y., Tan, L., 2013. Epidemiology and etiology of Alzheimer's disease. From genetic to non-genetic factors. Curr. Alzheimer Res. 9, 852–867.

Jin, Z., Liu, Y., 2018. DNA methylation in human diseases. Genes Dis. https://doi.org/10.1016/j.jendis.2018.01.002. (Accessed 2 May 2021).

Johnson, S.C., Dong, X., Vijg, J., Suh, Y., 2015. Genetic evidence for common pathways in human age-related diseases. Aging Cell 14, 809–817.

Juntarawijit, C., Juntarawijit, Y., 2018. Association between diabetes and pesticides: a case-control study among Thai farmers. Environ. Health Prev. Med. https://doi.org/10.1186/s12199-018-0692-5. (Accessed 28 April 2021).

Kasevska, M., Ivanov, M., Ingelman-Sundberg, M., 2011. Perspectives on epigenetics and its relevance to adverse drug reactions. Clin. Pharmacol. Ther. https://doi.org/10.1038/clpt.211.21. (Accessed 3 May 2021).

Kaufman, G., 2016. Adverse drug reactions: classification, susceptibility and reporting. Nurs. Stand. 30 (50), 53–63.

Kubsad, D., Nissson, E.E., King, S.E., Sadler-Riggelman, I., Beck, D., Skinner, M.K., 2019. Assessment of glyphosate induced epigenetic transgenerational inheritance of pathologies and sperm epimutations: generational toxicology. Sci. Rep. https://doi.org/10.1038/s41598-019-42860-0. (Accessed 3 May 2021).

Lekka, E., Hall, J., 2018. Noncoding RNAs in disease. FEBS Lett. 592, 2884–2900.

Manti, S., Marseglia, L., D'Angelo, G., Cuppari, C., Cusumoto, E., Arrigo, T., et al., 2016. "Cumulative stress": the effects of maternal and neonatal oxidative stress and oxidative stress-inducible genes on programming of atopy. Oxid. Med. Cell Longevity. https://doi.org/10.1155/2016/8651820. (Accessed 28 April 2021).

Nilsson, E.E., Sadler-Riggleman, I., Skinner, M.K., 2018. Environmentally induced epigenetic transgenerational inheritance of disease. Environ. Epigenet. https://doi.org/10.1093/eep/dvy016. (Accessed 2 May 2021).

Nishida, N., Kudo, M., 2013. Oxidative stress and epigenetic instability in human hepatocarcinogenesis. Dig. Dis. https://doi.org/10.1159/000355243. (Accessed 4 May 2021).

Obeidat, M., Hall, I.P., 2011. Genetics of complex respiratory diseases: implications for pathophysiology and pharmacology studies. Br. J. Pharmacol. 163, 96–105.

Ober, C., Yao, T.C., 2011. The genetics of asthma and allergic disease: a 21st century perspective. Immunol. Rev. 242 (1), 10–30.

Osaniou, O., Pirmohamed, M., Daly, A.K., 2018. Pharmacogenetics of adverse drug reactions. Adv. Pharmacol. https://doi.org/10.1016/bs.apha.2018.03.002.

Perrone, S., Lotti, F., Geronzi, U., Guidoni, E., Longini, M., Buonocore, G., 2016. Oxidative stress in cancer-prone genetic diseases in pediatric age: the role of mitochondrial disfunction. Oxid. Med. Cell Longevity. https://doi.org/10.1155/2016/4782426. (Accessed 4 May 2021).

Romani, M., Pistillo, M.P., Banelli, B., 2015. Environmental epigenetics: crossroad between public health, lifestyle and cancer prevention. BioMed. Res. Intl. https://doi.org/10.1155/587983. (Accessed 18 August 2022).

Ruperez, A.I., Gil, A., Aguilera, C.M., 2014. Genetics of oxidative stress in obesity. Int. J. Mol. Sci. 15, 3118–3144.

Saavedra, L., Feig, L.A., 2013. Chronic social instability induces anxiety and defective social interactions across generations. Biol. Psychiatry. https://doi.org/10.1016/j.biopsych.2012.06035. (Accessed 3 May 2021).

Scandalios, J.G., 2002. Oxidative stress responses – what have genome-scale studies taught us? Gen. Biol. 3 (7), 1019.1–1019.6.

Skinner, M.K., Guerrero-Bosagna, C., Haque, M.M., 2015. Environmentally induced epigenetic transgenerational inheritance of sperm epimutations promote genetic mutations. Epigenetics 10 (8), 762–771.

Sorensen, T.I.A., Nielsen, G.G., Andersen, P.K., Teasdale, T.W., 1988. Genetic and environmental influences on premature death in adult adoptees. N. Eng. J. Med. 318, 727–732.

Xu, Z., Xu, H., Lu, Y., 2020. Genetic link to smoking and breast cancer risk. Clin. Epidomiol. 12, 1145–1148.

Yoo, H.J., 2015. Genetic polymorphism related to oxidative stress in autism. In: Dietrich-Muszalska, A., et al. (Eds.), Studies in Psychiatric Disorders. https://doi.org/10.1007/978-1-4939-0440-1_20. (Accessed 4 May 2021).

CHAPTER

15

Aging

15.1 Introduction

Chronological aging per se is a risk factor for many chronic diseases (Barzilai et al., 2018). Aging has been defined broadly as "a multidimentional irreversible accumulation of physical, environmental and social changes" (Cencioni et al., 2013) and more specifically as "a progressive decline in the physiological functions of an organism after the reproductive phase of its life" (Khansari et al., 2099). By either definition, aging is not a disease, but a natural consequence of living.

Aging is characterized by a progressive decline in physiological processes at the molecular, cellular and tissue levels. In cardiovascular aging, for example, this is manifest by heart vessels and the heart itself becoming more rigid and fibrotic with age, which leads to cardiovascular disease (de Almeida et al., 2020).

Average global-wide life expectancy is increasing at a rapid pace with the world's population of those over the age of 60 anticipated to rise from 12% in 2015 to an estimated 22% in 2050 (Brown, 2014). This projected increase is expected to be paralleled by an increase in age-related disease (de Almeida et al., 2020).

Like disease, however, aging is accelerated by chronic inflammation which leads to excessive oxidative stress. A widely held theory is that OS within mitochondria damages the mitochondria, which in turn leads to the production of increased quantities of ROS that cause further damage. Once it starts, this cycle leads to further damage and corresponding aging (Romano et a. 2010). Aging cannot be prevented, but can be delayed via the adoption of healthy lifestyle choices that limit inflammation and oxidative stress.

15.2 Clinical conditions associated with aging

Aging is associated with numerous clinical conditions (Longo et al., 2019). The most common of these are listed in Table 15.1.

All the conditions listed in Table 15.1 are associated directly or indirectly with chronic oxidative stress.

https://doi.org/10.1016/B978-0-323-91890-9.00025-8

TABLE 15.1 Most common aging-associated clinical conditions.

Bone fractures
Cardiovascular disease
Cognitive decline
Depression
Diabetes (type 2)
Disability
Injurious falls
Polypharmacy
Urinary incontinence

15.3 Lowering age of disease onset

The effects of aging are not limited to the elderly. Increased chronic oxidative stress accelerates aging in the young as well, resulting in the manifestation of age-related diseases at younger ages. Cancers of the kidneys, gallbladder, pancreas, thyroid and leukemias in the young have steadily risen from 1910 through 1990 (Sung et al., 2019). Similar effects have been reported for type 2 diabetes (Wilmot and Idris, 2014), asthma (Veenendaal et al., 2019) and Alzheimer's disease (Zhu et al., 2015).

15.4 Age-related disease and oxidative stress

Age-related diseases are found in all body systems. The elderly are more susceptible to oxidative stress due to declining endogenous antioxidant efficiencies (de Almeida et al., 2017) Examples of the effects of aging include increases in:

- Cardiovascular disease
- Neurodegenerative disease
- Psychological stress
- Immune system impact
- Cancer

15.4.1 Cardiovascular disease

Cardiovascular aging affects the heart and the vascular system. Chronic inflammation and its associated oxidative stress lead to tissue deterioration that ultimately give rise to diseases that include hypertension, heart failure, atherosclerosis, arteriosclerosis, myocardial infarction and stroke (de Almeida et al., 2020).

15.4.2 Neurodegenerative disease

Neurodegenerative diseases are strongly related to aging. This is due to increases in oxidized proteins and oxidized DNA lesions that cannot be accounted for by decreased antioxidant enzymes. It is believed that lipid oxidation product reactions with mitochondrial enzymes result in age-dependent energy generation loss and enhanced neuron susceptibility to apoptosis. It is hypothesized that neuroinflammatory processes which are age-related cause death and/or dysfunction in neurodegenerative diseases such as Alzheimer's disease and Parkinson's disease (Floyd and Hensley, 2002).

15.4.3 Psychological stress

Psychological stress is an important factor in the promotion early onset of age-related diseases. Chronic stress leads to elevation of cortisol and insulin levels, decreased anabolic hormone levels and detremental lifestyle choices such as overeating, poor dietary choices, smoking, alcohol use, to name a few. All of these parameters have been shown to lead to increased inflammation and elevated oxidative stress levels which are associated with poorer cognitive function, loss of autonomy and depressive symptoms and the onset of other age-related diseases (Epel et al., 2004; Palmieri et al., 2006; Maugeri et al., 2004).

15.4.4 Type 2 diabetes

Type 2 diabetes is the most common metabolic disease in older adults. Its prevalence increases with age to the point where almost one third of the elderly in the United States are diabetic and three quarters have diabetes or pre-diabetes (Cowie et al., 2009; Gong and Muzumdar, 2012).

Diabetes is characterized by hyperglycemia, a decline or complete loss of insulin action and insulin resistance. It is associated with accelerated cardiovascular disease, neuropathy and retinopathy, all of which are triggered by mitochondrial overproduction of ROS (Giacco and Brownlee, 2010). Oxidative stress is the driving force behind diabetes pathogenesis via impacts on lipid, protein and DNA damage (Ullah et al., 2016). Thus oxidative stress is the central factor in both the development of and complications associated with diabetes.

15.4.5 Immune system impact

Aging leads to physiological changes that affect all body systems. The impact of aging on the immune system leads to increased susceptibility to infectious, cardiovascular, metabolic, autoimmune and neurodegenerative diseases. The immune system itself is impacted by age-associated changes that affect the endocrine, nervous, digestive, cardiovascular and musculoskeletal systems (Muller et al., 2019), all of which are associated with chronic elevated oxidative stress (Zeliger, 2016).

15.4.6 Cancer

Chronic inflammation and oxidative stress are a major cause of age-related cancer. Inflammation pre-disposes susceptible cells to neoplastic transformations by causing cell injuries and preventing cell repair. A direct relationship between age-related chronic inflammation and cancer is well established (Federico et al., 2007). ROS-induced DNA damage is associated with carcinogenesis via induction of transcriptional arrest or induction/replication errors or by genomic instability (Khansari et al., 2009).

15.5 Hallmarks of aging

Nine hallmarks of aging have been identified (Lopez-Otin et al., 2013; Meiners et al., 2015). These are listed in Table 15.2.

All of these hallmarks have been shown to be negatively impacted by OS (von Zglinicki 2002; Kawanishi and Oikawa 2004; Junqueira et al., 2004; Andriollo-Sanchez et al., 2005; Lopez-Otin et al., 2013; Meiners et al., 2015). Brief discussions of each of these hallmarks follow.

15.5.1 Genomic instability

DNA damage accumulates during aging due to ROS and aging carries with it a lifetime accumulation of genetic damage via exogenous physical, chemical and biological agents, many of which induce ROS production, as well as by endogenous factors including errors in DNA replication and the effects of chronic ROS. DNA is under constant threat by ROS as these highly reactive species modify DNA bases and disrupt genome function (van der Rijt et al., 2020; Poetsch, 2020). ROS lead to oxidative stress that damage both nuclear and mitochondrial DNA leading to numerous premature aging effects, including

TABLE 15.2 Hallmarks of aging.

Hallmarks of aging
Genomic instability
Telomere attrition
Epigenetic alterations
Loss of proteostasis
Deregulated nutrient sensing
Mitochondrial dysfunction
Cellular senescence
Stem cell exhaustion
Altered cellular communication

neurodegenerative diseases, frailty, malnutrition, loss of cognition and cancer (Moskalev et al., 2013; Lopez-Otin et al., 2013; Guerville et al., 2020).

When DNA damage occurs, DNA repair, which triggers apoptosis, repair or senescence is initiated. 10^4 mutations occur in the genome on a daily basis. These mutations are repairable, but as aging occurs the capacity for repair decreases, resulting in mutation accumulation and genomic instability that leads to disease onset with increasing age (Niedenrhoffer et al., 2018).

15.5.2 Telomere attrition

Telomeres are repetitive DNA sequences at the ends of eukaryotic chromosomes that are shortened in each somatic cell division. Reduced telomere length is associated with aging and with the onset of cancer, frailty and early death (Epel et al., 2004; Hou et al., 2015; Barnes et al., 2019, Guerville et al., 2020). Oxidative stress has been shown to shorten telomere length (Kawanishi and Oikawa 2004; von Zglinicki 2002), and there is a strong relationship between short telomeres and the risk of mortality, particularly in younger people (Boonekamp et al., 2013). Lowering of oxidative stress, however, has been shown to delay the shortening of telomere length thus prolonging life and reducing incidences of cancer and other diseases (Crous-Bou et al., 2014).

15.5.3 Epigenetic alterations

Epigenetic alterations affect all cells and tissues throughout life. These involve alterations in DNA methylation, histone modifications and chromatin remodeling (Lopez-Otin et al., 2013; van der Rijt et al., 2020). As discussed in Chapter 14, epigenetic effects are oxidative stress dependent. ROS can influence the methylome by formation of oxidized DNA via hydroxylation of pyrimidines and 5-methylcytosine. ROS also affects DNA methylation by DNA oxidation and ten-eleven transformation-mediated hydroxylation. ROS can also indirectly modulate epigenetic machinery activity by impacting histone-modifying enzymes that are dependent upon intercellular levels of essential metabolites. Thus, oxidative stress can impact epigenetic effects on multiple levels (Guillaumet-Adkins et al., 2017). Epigenetic alterations are associated with loss of cognition, physical function and early death (Lopez-Odin et al., 2013; Guerville et al., 2020).

15.5.4 Loss of proteostasis

Aging is associated with impaired proteostasis, which provides mechanisms for the restoration of mis-folded proteins as well as by their removal and degradation so as to prevent the accumulation of damaged components and assure the continuous intercellular protein renewal (Lopez-Otin et al., 2013; Korovila et al., 2017). Unfolded, misfolded or aggregated proteins, resulting from oxidative stress are contributory to age-related diseases, including coronary artery disease, Alzheimer's disease, Parkinson's disease and cataracts (Powers et al., 2009; Korovila et al., 2017; Guerville et al., 2020).

15.5.5 Deregulated nutrient sensing

The mammalian somatotropic axis is comprised of the growth hormone and insulin-like growth factor (IGF-1). IGF-1 shares the down-stream intracellular pathway with insulin in signaling nutrient abundance and anabolism. The decline of the somatotropic axis, resulting from elevated oxidative stress, is a major hallmark of metabolic aging. Deregulated nutrient sensing is associated with malnutrition, Alzheimer's disease and increased frailty (Guerville et al., 2020).

15.5.6 Mitochondrial dysfunction

Respiratory chain efficacy declines with increasing age, reducing ATP generation and increasing electron leakage. ROS is a stress-generated survival response that initially compensates for age-related deterioration, thus providing a positive impact on mitochondrial homeostasis. As age progresses, however, ROS levels continue to increase until they reach levels where they go beyond their original purpose of providing protection and exacerbate, rather than relieve, age related damage (Hekimi et al., 2011; Giorgi et al., 2018). Mitochondrial dysfunction is associated with diseases in numerous systems, including various cardiovascular diseases, neurodegenerative diseases (Alzheimer's disease, Parkinson's disease and ALS), metabolic disease (diabetes), musculoskeletal disease (muscular dystrophy) and cancer (Giorgi et al., 2018).

15.5.7 Cellular senescence

Cellular senescence is a state of stable arrest of the cell cycle coupled to phenotypic changes. These include the production of several molecules (including matrix metalloproteases and pro-inflammatory cytokines) known as senescence-associated secretory phenotype (SASP), which mediates senescence spreading to adjacent cells, inflammation and tissue dysfunction. Cellular senescence is a compensatory mechanism that aims to avoid proliferation of damaged cells that is induced by telomere attrition, DNA damage and excessive mitogenic signaling. Cellular senescence has a causal effect for osteoporosis, osteoarthritis, frailty, cardiovascular diseases, pulmonary fibrosis, kidney diseases, liver diseases, metabolic dysfunction, neurodegenerative diseases and cancer (Collado et al., 2007 Guerville et al., 2020; Kaur and Farr, 2020), all of which are associated with elevated oxidative stress.

15.5.8 Stem cell exhaustion

After development, adult stem cells are present in every body tissue and organ. During aging the repair and regenerative functions of stem cells decline. Stem cell exhaustion is a process in which stem cells lose their ability to self-renew or differentiate, succumb to senescence or apoptosis and eventually are functionally depleted (Chen et al., 2017). Stem cell exhaustion is a consequence of several of the hallmarks described above. These include DNA damage, epigenetic alterations, telomere shortening, cellular senescence and mitochondrial dysfunction, all of which are associated with increased inflammation and elevated oxidative stress (Chen et al., 2017; Guerville et al., 2020). Effects associated with stem cell exhaustion include cardiovascular diseases, pulmonary diseases, diabetes, neurological disorders, decreased endocrine organ outputs, and cancer (Vijg and Campisi, 2008; Boyette and Tuan, 2014; Chen et al., 2017; Sameri et al., 2020).

15.5.9 Altered cellular communication (ACC)

All complex organisms require intercellular communication to coordinate physiological functions. Aging is associated with altered intercellular communication (ACC) that undermines effective intercellular communication. ACC is mainly driven by chronic low-grade inflammation (termed inflammaging (Ferrucci and Fabbri, 2018)) that results from several aging hallmarks, that include cellular senescence and loss of proteostasis and leads to oxidative stress (Guerville et al., 2020). Several sources of inflammaging have been identified (Lopez-Otin et al., 2013). These include:

1. Accumulation of pro-inflammatory tissue damage
2. Failure of an increasingly dysfunctional immune system to effectively clear pathogens and non-functioning cells.
3. Senescent cell secretion of pro-inflammatory cytokines.
4. Enhanced activation of the NF-kB transcription factor
5. A defective autophagy response.

Consequences associated with altered cellular communication include cardiovascular disease and early death (Guerville et al., 2020).

15.6 Summary

Aging, a natural consequence of living, is a major factor in the onset of numerous diseases. All aspects of aging are associated with inflammation and oxidative stress. Hence, aging can be delayed, but not prevented, by limiting lifestyle choices that lead to increased inflammation and elevated oxidative stress

(Guerville et al., 2020).

References

Andriollo-Sanchez, M., Hininger-Favier, I., Meunier, N., Venneria, E., O'Connor, J.M., Maiani, G., Coudray, C., Roussel, A.M., 2005. Age-related oxidative stress and antioxidant parameters in middle-aged and older European subjects: the ZENITH study. Eur. J. Clin. Nutr. 59 (Suppl. 2), S58–S62.

Barnes, R.P., Fouquerel, E., Opresko, P.L., 2019. The impact of oxidative DNA damage and stress on telomere homeostasis. Mech. Ageing Dev. 177, 37–45.

Barzilai, N., Cuervo, A.M., Austad, S., 2018. Aging as a biological target for prevention and therapy. JAMA 320, 1321–1322.

Boonekamp, J.J., Simons, M.J., Hemerik, L., Verhulst, T., 2013. Telomere length behaves as a biomarker of somatic redundancy rather than biological age. Aging Cell 12, 330–332.

Boyette, L.B., Tuan, R.S., 2014. Adult stem cells and diseases of aging. J. Clin. Med. 3, 88–134.

Brown, G.C., 2014. Living too long: the current focus on medical research on increasing the quantity, rather than the quality of life is damaging our health and harming the economy. EMBO Rep. 16 (2), 137–141.

Cencioni, C., Spallotta, F., Martelli, F., Valente, S., Mai, A., Zeiher, A.M., Gaetano, C., 2013. Oxidative stress and epigenetic regulation in ageing and age-related diseases. Int. J. Mol. Sci. 14, 17643–17663.

Chen, F., Liu, Y., Wong, N.K., Xiao, J., So, K.F., 2017. Oxidative stress in stem cell aging. Cell Transplant. 26 (9), 1483–1495.

Collado, M., Blasco, M.A., Serrano, M., 2007. Cellular senescence in cancer and aging. Cell 130, 223–233.

Cowie, C.C., Rust, K.F., Ford, E.S., Eberhardt, M.S., Byrd-Holt, D.D., Li, C., 2009. Full accounting of diabetes and pe-diabetes in the U.S. population in 1988-1994 and 2005-2006. Diab. Care 32 (2), 287–294.

Crous-Bou, M., Fung, T.T., Prescott, J., Julin, B., Du, M., Sun, Q., et al., 2014. Mediterranean diet and telomere length in Nurses' Health Study: population based cohort study. BMJ 349, g6674. https://doi.org/10.1136/bmj.g6674.

de Almeida, A.J.P.O., Ribeiro, T.P., Medeiros, I.A., 2017. Aging: molecular pathways and implications on the cardiovascular system. Oxid. Med. Cell. Longev. 2017. https://doi.org/10.1155/2017/7941563. (Accessed 19 August 2022).

de Almeida, A.J.P.O., de Almeida, R.M.A., Dantas, S.H., de Lima Silva, S., de Oliveira, J.C.P.L., et al., 2020. Unveiling the role of inflammation and oxidative stress on age-related cardiovascular diseases. Oxid. Med. Cell. Longev. https://doi.org/10.1155/2020/1954398. (Accessed 19 May 2021).

Epel, E.S., Blackburn, E.H., Lin, J., Dhabhar, F.S., Adler, N.E., Morrow, J.D., Cawthon, R.M., 2004. Accelerated telomere shortening in response to life stress. Proc. Nat. Acad. Sci. USA 101 (49), 17312–17315.

Federico, A., Morgillo, P., Tuccillo, C., Ciardiello, F., Loguerico, C., 2007. Ind. J. Cancer 121 (11), 2381–2386.

Ferrucci, L., Fabbri, I., 2018. Inflammaging: chronic inflammation in ageing, cardiovascular disease, and frailty. Nat. Rev. Cardiol. 15 (9), 505–522.

Floyd, R.A., 2002. Oxidative stress in brain aging. Implications for therapeutics of neurodegenerative diseases. Neurobiol. Aging 23 (5), 795–807.

Giacco, F., Browlee, M., 2010. Oxidative stress and diabetic complications. Circ. Res. 107 (9), 1058–1070.

Giorgi, C., Marchi, S., Simous, I.C.M., Ren, Z., Morciano, G., Perrone, M., et al., 2018. Mitochondria and reactive oxygen species in aging and age-related diseases. Int. Rev. Cell Mol. Biol. 340, 209–344.

Gong, Z., Muzundar, R.H., 2012. Pancreatic function, type 2 diabetes and metabolism in aging. Int. J. Endocrinol. 2912. https://doi.org/10.1155/2012/320482. (Accessed 22 May 2021).

Guerville, F., Baretto, P.D.S., Ader, I., Andrieu, S., Casteilla, L., Dray, C., et al., 2020. Revisiting the hallmarks of aging to identify markers of biological age. J. Prev. Alzheimer's Dis. 7 (1), 56–64.

Guillaumet-Adkins, A., Yanez, Y., Peris-Diaz, M.D., Calabria, I., Palanca-Ballester, C., Sandoval, J., 2017. Epigenetics and oxidative stress in aging. Oxid. Med. Cell Longev. https://doi.org/10.1155/2017/9175806. (Accessed 22 May 2021).

Hekimi, S., Lapointe, J., Wen, Y., 2011. Taking a "good" look at free radicals in the aging process. Trends Cell Biol. 21, 569–576.

Hou, L., Joyce, B.T., Gao, T., Liu, L., Zheng, Y., Penedo, F.J., et al., 2015. Blood telomere length attrition and cancer development in the normative aging study cohort. EBioMedicine 13;2 (6), 591–596.

Junqueira, V.B., Barros, S.B., Chan, S.S., Rodrigues, L., Giavarotti, L., Abud, R.L., Deucher, G.P., 2004. Aging and oxidative stress. Mol. Aspect. Med. 25 (1–2), 5–16.

Kaur, J., Farr, J.N., 2020. Cellular senescence in age-related disorders. Transl. Res. 226, 96–104.

Kawashini, S., Oikawa, S., 2004. Mechanism of telomere shortening by oxidative stress. Ann. NY Acad. Sci. 1019, 278–284.

Khansari, N., Shakiba, Y., Mahmoudi, M., 2009. Chronic inflammation and oxidative stresses a major cause of age-related diseases and cancer. Recent Pat. Inflamm. Allergy Drug Discov. 3, 73–80.

Korovila, I., Hugo, M., Castro, J.P., Weber, D., Hohn, A., Grune, T., Jung, T., 2017. Proteostasis, oxidative stress and aging. Redox Biol. 13, 550–567. https://doi.org/10.1016/j.redox.2017.07.008. (Accessed 19 Aug 2022).

Longo, M., Bellastella, G., Maiorino, M.I., Meier, J.J., Esposito, K., Giugliano, D., 2019. Diabetes and aging; from treatment goals to pharmacological therapy. Front. Endocrinol. https://doi.org/10.3389/fendo.2019.00045. (Accessed 22 May 2021).

Lopez-Otin, C., Blasco, M.A., Partridge, L., Serrano, M., Kroemer, G., 2013. The hallmarks of aging. Cell 153 (6), 1194–1217.

Maugeri, D., Santangelo, A., Bonanno, M.R., 2004. Oxidative stress and aging: studies on an East-Sicilian, ultraoctagenarian population living in institutes or at home. Arch. Gerontol. Geriatr. Suppl. 9, 271–277.

Meiners, S., Eickelberg, O., Konigshoff, M., 2015. Hallmarks of ageing lung. Eur. Respir. H. 45 (3), 807–827.

Moskalev, A.A., Shaposhnikov, M.V., Zhavoronkov, A., Budovsky, A., Fraifled, V.E., 2013. The role of DNA damage and repair in aging through the prism of Koch-like criteria. Ageing Res. Rev. 12 (2), 661–684.

Muller, L., Di Benddetto, S., Pawelec, G., 2019. The immune system and its deregulation with aging. Subcell. Biochem. 91, 21–43.

Niedernhofer, L.J., GurkarnAU, Wang, Y., Vijg, J., Hoeijmakers, J.H.J., Robbins, P.D., 2018. Nuclear genomic instability and aging. Annu. Rev. Biochem. 87, 295–322.

Palmieri, V.O., Grattagliano, I., Portincasa, P., Palasciano, G., 2006. Systemic oxidative alterations are associated with visceral adiposity and liver steatosis in patients with metabolic syndrome. J. Nutr. 136, 3022–3026.

Poetsch, A.R., 2020. The genomics of oxidative DNA damage, repair and resulting mutagenesis. Comput. Struct. Biotechnol. J. 18, 207–219. https://doi.org/10.1016/j.csbj.2019.12.013.

Powers, E.T., Morimoto, R.I., Dillin, A., Kelly, J.W., Balch, W.E., 2009. Biological and chemical approaches to diseases of proteostasis deficiency. Annu. Rev. Biochem. 78, 959–991.

Romano, A.D., Serviddio, G., de Matthaeus, A., Bellantil, F., Vendemiale, G., 2010. Oxidative stress and aging. J. Nephrol. 15, S29–S36.

Sameri, S., Samadi, P., Dehghan, R., Salem, E., Fayazi, N., Amini, R., 2020. Stem cell aging in lifespan and disease: a state-of-the-art review. Curr. Stem Cell Res. Ther. 15 (4), 362–378.

Sung, H., Siegel, R.L., Rosenberg, P.S., Jemal, A., 2019. Emerging cancer trends among young adults in the USA: analysis of a population-based cancer registry. Lancer Pub. Health. https://doi.org/10.1016/S2468-2667(18)30267-6. (Accessed 19 May 2021).

Ullah, A., Khan, A., Khan, I., 2016. Diabetes mellitus and oxidative stress – a concise review. Saudi Pharmaceut. J. 24, 547–553.

Van der Rijt, S., Molenaars, M., McIntyre, R.L., Janssens, G.E., Houtkooper, R.H., 2020. Integrating the hallmarks of aging throughout the tree of life: a focus on mitochondrial dysfunction. Front. Cell Dev. Biol. https://doi.org/10.3389/fcell.2020.594416. (Accessed 22 May 2021).

Veenendaal, M., Westerik, J.A.M., Van den Bernt, L., Kocks, J.W.H., Bischoff, E.W., Schermer, T.R., 2019. Age- and sex-specific prevalence of chronic comorbidity in adult patients with asthma: a real-life study. NPJ Prim. Care Respir. Med. https://doi.org/10.1038/s41533-019-127-9. (Accessed 19 May 2021).

Vijg, J., Campisi, J., 2008. Puzzles, promises and a cure for ageing. Nature 454 (7208), 1065–1071.

von Zglinicki, T., 2002. Oxidative stress shortens telomeres. Trends Biochem. Sci. 27 (7), 339–344.

Wilmot, E., Idris, I., 2014. Early onset type 2 diabetes: risk factors, clinical impact and management. Ther. Adv. Chronic Dis. 5 (6), 234–244.

Zeliger, 2016. Predicting disease onset in clinically healthy people. Interdiscipl. Toxicol. 9 (2), 39–54.

Zhe, X.C., Tan, L., Wang, H.F., Jiang, T., Cao, L., Wang, J., et al., 2015. Rate of early onset Alzheimer's disease: a systematic review and meta-analysis. Ann. Transl. Med. 3 (3), 38–44.

CHAPTER

16

Diseases and comorbidities

16.1 Introduction

Previous chapters have discussed factors that elevate oxidative stress and thereby lead to disease. These are listed in Table 16.1. Prevalent disease, which also leads to further disease onset, and in many instances is the premier factor leading to elevated OS and additional disease, is the subject of this chapter.

The incidence of disease world-wide is continually increasing. Though people are living longer, we are also living sicker and with increasing numbers of multi-morbid diseases (Murray et al., 2015; Zeliger, 2014; Wallace & Salive, 2013; Pritchard and Rosenorn-Lanng, 2015). As an example of this phenomenon, in the year 2020, 51.8% of adults in the United States had at least one of the diseases listed in Table 16.2 and of the people diagnosed with these diseases, 27.2% were diagnosed with two or more of these diseases (Boersma et al., 2020).

Numerous diseases have reached epidemic and pandemic proportions in the past two generations. The dramatic increase of environmental disease prevalence with time can be seen from a plot of disease percent versus time, from 1940s to 2020. Such plots produce hyperbolic curves such as that in Fig. 16.1. In addition to those diseases listed in Table 16.2, others that fit this plot include autism and autism spectrum disorders (Zeliger, 2013b), type 2 diabetes (Zeliger, 2013a,c), obesity (Wang and Beyoun, 2007) childhood cancers (Parkin et al., 1988), onset of dementia and other neurological diseases (Zeliger, 2013b), and both male and female infertility (Colborn et al., 1996).

Slopes of these curves for rising disease onset exactly correspond to those of plots for chemical production and use versus time, as exemplified by data for synthetic chemical production (Neel and Sargis, 2011), increased pesticide use (Chen and McCarl, 2001), increased world-wide energy production from combustion of fossil fuel, increases in air and water pollution (U.S. Energy Information Administration, 2014), and increases in pharmaceutical use (Kantor et al., 2015).

Elevated oxidative stress is, directly or indirectly, the cause of virtually all disease. It leads to attack on DNA and proteins, lipid peroxidation, interruptions in cell signaling, induction of cell death via apoptosis or necrosis and structural tissue damage (Pizzino et al., 2017). OS is directly responsible for noninfectious (environmental) disease (Davies, 1995; Zeliger and Lipinski, 2015; Zeliger, 2016), and indirectly responsible for the spread of infectious disease via undermining of the immune system (Zaki et al., 2005; Valyl-Nagy and Dermody, 2005).

Oxidative Stress
https://doi.org/10.1016/B978-0-323-91890-9.00014-3

TABLE 16.1 Disease causing OS factors.

Organic chemicals
Inorganic chemicals
Heavy metals
Chemical mixtures
Particles
Fibers
Air pollutants
Water pollutants
Soil pollutants
Alcohol
Recreational drugs
Tobacco
Ionizing radiation
Nonionizing radiation
Inflammation
Chronic trauma
Saturated fats
Red meat
Fructose and high fructose corn syrup
Processed foods
Artificial food colors
Artificial food flavors
Pharmaceuticals
Hot and cold temperature extremes
Sleep deprivation
Circadian cycle interruption
Psychological stress
Genetics
Epigenetics
Aging
Preexisting disease

TABLE 16.2 Diseases of which 51.8% of adults in the United States had at least one of in the year 2020.

Arthritis
Cancer
Chronic obstructive pulmonary disease
Coronary heart disease
Asthma
Hepatitis
Diabetes
Hypertension
Stroke
Weak or failing kidneys

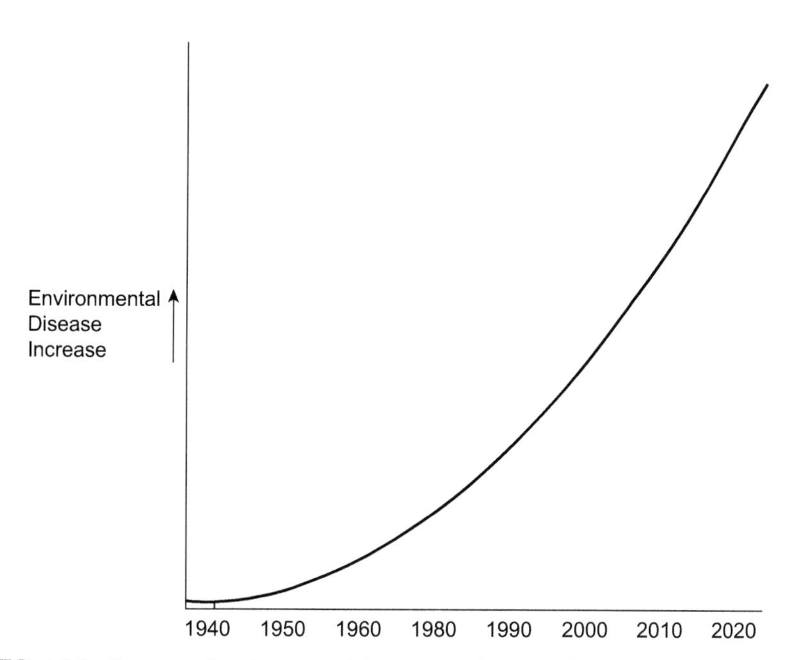

FIG. 16.1 Increase of environmental disease as a function of time, from 1940 to 2020.

The leading causes of death in the United States in 2020 as reported by the Centers for Disease Control and the percentages of each are (Ahmad and Anderson, 2021).

Heart disease	27%
Cancer	23%
COVID-19	13%
Accidents	8%
Stroke	7%
Chronic lower respiratory diseases	6%
Alzheimer's disease	5%
Diabetes	4%
Influenza and pneumonia	3%
Kidney disease	2%
Suicide	2%

TABLE 16.3 Biomarkers of oxidative stress.

Lipid peroxidation
Malondialdehyde (MDA)
4-hydroxy-2-nonenal
Conjugated dienes
Alkane exhalation
Total hydroperoxides
Oxidized LDL
F2-sioprostanes
Advanced lipid peroxidation products
DNA
Nucleotide oxidation
8-hydrdoxy-guanosine
8-hydroxy-desoxyguanosine
Thymidine glycol
7,8-dihydroxy-8-oxo-2′-deoxyguanosine
5-chlorouracil
Deoxyribose oxidation
Proteins
Protein carbonyls
Advanced oxidation protein products

TABLE 16.3 Biomarkers of oxidative stress.—cont'd

Amino acid oxidation
3-nitro-tyrosine
3-chloro-tyrosine
Ischemia-modified albumin
Advanced glycation end products
Oxidants
Hydrogen peroxidetdldtag
Superoxide radical
Nitric oxide
Antioxidant markers
Nonenzymatic
Vitamin A
Vitamin C
Vitamin E
Selenium
Zinc
Glutathione
Cysteine
Uric acid
Enzymatic
Superoxide dismutase
Glutathione peroxidase
Glutathione reductase
Catalase
Nonenzymatic antioxidant capacity
Oxygen radical antioxidant capacity
Total radical-trapping antioxidant parameter
Total antioxidant capacity
Total antioxidant status
Ferric reducing antioxidant potential
Antioxidant gap
Biological antioxidant oxidant potential

TABLE 16.4 Body systems, organs for which disease hallmarks are identified with oxidative stress and references.

Integumentary	
Skin	Wagener et al. (2013)
Hair	Treub (2015)
Subcutaneous tissue	Baek and Lee (2016)
Skeletal	
Bones	Wilson. (2014)
Cartilage	Zahan et al. (2020)
Ligaments	Chen et al. (2020)
Bone marrow	Picou et al. (2019)
Muscular	
Muscles	Sera et al. (2018)
Tendons	Millar et al. (2021)
Nervous	
Brain	Salim (2017)
Spinal cord	Jia et al. (2012)
Nerves	Diaz-Hung and Fragulea (2014)
Eyes	Kruk et al. (2015)
Ears	Celic and Koyuncu (2018)
Endocrine	
Pituitary gland	Sabatino et al. (2018)
Parathyroid gland	Deska et al. (2018), Karbownik-Lewinska and Kokoszo-Bilska (2012)
Thyroid gland	Karbownik-Lewinska and Kokoszo-Bilska (2012)
Thymus	Barbouti et al. (2020)
Pancreas	Robles et al. (2014)
Gonads	Durrani et al. (2012)
Cardivascular	
Heart	Senoner and Dichtl (2019)
Blood	Zidova et al. (2014)
Blood vessels	Sena et al. (2018)
Lymphatic	
Spleen	Khan et al. (1997)

TABLE 16.4 Body systems, organs for which disease hallmarks are identified with oxidative stress and references.—cont'd

Lymph nodes	Gashev et al. (2014)
Thymus	Haroun (2018)
Lymphatic vessels	Thangaswamy et al. (2012)
Respiratory	
Lungs	Park et al. (2009)
Trachea	Chen et al. (2019)
Nasal cavities	Cekin et al. (2009)
Pharynx	Bhattacharyya et al. (2014)
Digestive	
Stomach	Suzuki et al. (2012)
Intestinal tract	Vona et al. (2021)
Liver	Li et al. (2021)
Pancreas	Schulz et al. (1999)
Esophagus	Yoshida (2007)
Salivary glands	Maciejczyk et al. (2019)
Urinary	
Kidneys	Daenen et al. (2019)
Urinary bladder	Miyata et al. (2019)
Urethra	Levy et al. (2019)
Reproductive	
Ovaries	Sulaiman et al. (2018)
Uterus	Lu et al. (2018)
Mammary glands	Calaf et al. (2018)
Testes	Asadi et al. (2017)
Prostate gland	Roumeguere et al. (2017)
Penis	Pailis and Brancato (2012)

Ninety percent of these causes of death, all except accidents and suicide, are attributable to elevated OS. Though no data exist, arguably, a sizable portion of the remaining 10%, accidents and suicide, could be attributed to psychological stress, a known cause of OS.

Oxidative stress levels can be ascertained via biomarkers in human fluids attributable to effects on lipid peroxidation products, DNA, proteins, free radicals, nonenzymatic antioxidant markers, enzymes and nonenzymatic antioxidant capacity (Sanchez-Rodriguez and Mendoza-Nunez, 2019). These are listed in Table 16.3.

Of the biomarkers listed in Table 16.3, those attributable to lipid peroxidation are the most indicative, as cell penetration by toxic agents initially requires breeching lipophilic cell membranes (Zeliger, 2003, 2011; Zeliger and Lipinski, 2015), the effect of which is the production of fatty acid degradation fragments. Of these fragments, malondialdehyde (MDA) is the biomarker most reliably used to indicate the presence of oxidative stress (Nielsen et al., 1997; Lorente et al., 2013; Tangvarasittichai et al., 2009; Sudha et al., 2014). Elevated levels of MDA have been shown to be present in the serum of patients with elevated OS and increased concentration of MDA accordingly is widely used as an indicator of the presence of disease in humans, with the severity of disease being a function of MDA level in a dose response relationship (Nielsen et al., 1997; Zhu et al., 2005; Romeau et al., 2008; Aflanie et al., 2015; Agarwal et al., 1987; Ayala et al., 2014). It has also been shown that serum MDA levels can be used to predict the onset of disease in people who are seemingly clinically healthy (Zeliger, 2016).

16.2 Systems, organs, oxidative stress and disease

Oxidative stress is not only a consequence of disease, but also a cause of disease in all body systems and organs. Table 16.4 lists body systems, a partial list of organs for which disease hallmarks have been identified as OS related and representative references of each.

The immune system, which is the first responder to oxidative stress elevation is composed of six organs which, anatomically, are parts of other systems. These organs, their anatomical locations and references for oxidative stress in their disease are listed in Table 16.5.

16.3 Diseases associated with oxidative stress

Hundreds of diseases in all body systems and organs have been linked to elevated oxidative stress. Table 16.6 contains a representative list of diseases known to be caused by OS (Zeliger, 2016 and the numerous references contained therein).

TABLE 16.5 Immune system organs, their anatomical location and oxidative stress references.

Organ	Location	References
Adenoids	Behind the nose	Yilmaz et al. (2004)
Bone marrow	bone	Picou et al. (2019)
Lymph nodes	lymph nodes	Gashev et al. (2014)
Spleen	Spleen	Khan et al. (1997)
Thymus	thymus	Haroun (2018)
Tonsils	Rear of the pharynx	Cvetkovic et al. (2009)

TABLE 16.6 Diseases associated with oxidative stress.

Metabolic
Type 2 diabetes
Metabolic syndrome
Hyperlipidemia
Obesity
Respiratory
Allergic rhinitis
Asthma
Chemical sensitivity
Chronic obstructive pulmonary disease (COPD)
Neurological
Autism
ADHD
Alzheimer's diesease
Parkinson's disease
Major depression
Motor skill loss
Cognitive loss
Learning disorders
Endocrine
Birth defects
Female infertility
Hypothyroidism
Male infertility
Cardiovascular
Arteriosclerosis
Atherosclerosis
Hypertension
Ischemic heart disease
Myocardial infarction
Stroke

(*Continued*)

TABLE 16.6 Diseases associated with oxidative stress.—cont'd

Gastrointestinal
Cirrhosis of the liver
Crohn's disease
Fatty liver disease
Hepatitis
Irritable bowel disease
Peptic ulcer
Musculoskeletal
Rheumatoid arthritis
Osteoarthritis
Osteoporosis
Rheumatoid arthritis
Urological
Benign prostate hyperplasia
End stage renal disease
Glomeruolsclerosis
Renal vascular disease
Liver
Cirrhosis
Fatty liver disease
Hepatitis
Skin
Acne
Dermatitis
Eczema
Psoriases
SLE (lupus)
Immunological and autoimmune
Acute uticaria
ALS
Chemical sensitivity
Chronic fatigue syndrome
Lupus

TABLE 16.6 Diseases associated with oxidative stress.—cont'd

Sjogram's syndrome
Eye
Cataracts
Glaucoma
Macular degeneration
Corneal and conjunctive diseases
Periodontal
Chronic periodontitis
Cancer
Virtually all cancers and metastasis
Infectious diseases
AIDS
Common cold
Herpes
HIV
Influenza
Tuberculosis

As indicated above, OS does not directly cause infectious diseases, but does so indirectly by undermining the functioning of the immune system via immuno-suppression (Hughes, 1999; Akaike, 2001; Xu et al., 2015; Splettstoesser and Schuff-Werner, 2002). Chemical immuno-suppressants include: polynuclear aromatic hydrocarbons (Jeng et al., 2011), dioxins (Schneider et al., 2008), tobacco smoke (Arcavi and Benowitz, 2004), pesticides (Banerjee et al., 1999) and heavy metals (Liu et al., 2009). Phagocytes generated when the body responds to infectious agents cause further free radical generation via lipid peroxidation, which adds to OS (Bouhafs and Jastrand, 1999; Stossel et al., 1974). Higher levels of OS also further infectious disease in critically ill patients (Andresen et al., 2006).

OS also leads to infectious diseases by impacting the actions of gut microbiota. The following examples illustrate this: diets chronically high in fat (Qiao et al., 2013) increase OS in the gut causing membrane damage that leads to increased permeability and translocation of gut bacteria with ensuing disease (Guarner and Soriano, 2005). Though not a disease, malnutrition is also known to lead to OS and immune system malfunction (Darmon et al., 1993; Katona and Katona-Apte, 2008; Ghone et al., 2013).

16.4 Multimorbidity

As discussed above, disease leads to additional disease, which in turn leads to still more disease. Seemingly unrelated diseases are comorbid with each other. As examples, it has been found that essentially all diseases listed in Table 16.7 are comorbid with all of the others (Zeliger et al., 2012; Zeliger, 2014). Essentially all of the binary combinations of the diseases listed in Table 16.7 have been shown to be comorbid with each other.

As is discussed in Part II of this book, the comorbidities just discussed involve some that can be linked to shared mechanisms of induction and others that cannot. It should also be noted that in pairs of comorbid diseases, either one can precede the other. Examples of such disease pairs are type 2 diabetes and hypertension (Long and Dagogo-Jack, 2011), cardiovascular disease and COPD (Onil and Unwin, 2015) and multiple chemical sensitivity and chronic fatigue syndrome (Zeliger et al., 2015). The only common denominator for these seemingly disparate findings is that oxidative stress is the cause of all disease onset. In the examples just cited as well as in numerous other disease pairs, no other mechanistic explanation can be found. This leads to the conclusion that all disease itself increases oxidative stress in a cumulative manner with multiple morbidities. Thus, oxidative stress caused by the first disease leads to the onset of a second disease, which, in turn leads to yet more total OS that leads to the onset of a third disease, and so on with each additional disease increasing total OS and the likelihood of further disease onset.

The number of multi-morbidities a person can experience has been found to range between 4 and 10 in a study of multi-morbidities in chemically sensitive individuals (Zeliger et al., 2012). Other studies have shown similar results for disease comorbidities (CDC, 2021).

People with noncommunicable diseases not only have high incidences of other noncommunicable diseases, but also with infectious diseases. Examples of comorbidities between noncommunicable and communicable diseases and references for these are shown in Table 16.8.

TABLE 16.7 Diseases, each of which is comorbid with essentially all of the others.

Type 2 diabetes
Atherosclerosis
Myocardial infarction
Hypertension
Stroke
Coronary heart disease
Peripheral heart disease
Ischemic heart disease
Cardiac autonomic function
Cognitive disorders

TABLE 16.7 Diseases, each of which is comorbid with essentially all of the others.—cont'd

Motor disorders
Sensory disorders
Neuropathies
Alzheimer's disease
Parkinson's disease
Amyotrophic lateral sclerosis
Autism
Attention deficit/hyperactivity disorder
Osteoarthritis
Fibromyalgia
Allergic reactions
Multiple chemical sensitivity
Autoimmune diseases
Chronic fatigue syndrome
Asthma
Chronic obstructive pulmonary disease
Obesity
Cancers in multiple organs

TABLE 16.8 Examples of comorbid noncommunicable and communicable diseases and references for these.

Comorbid diseases	References
Influenza and neurological disorders	CDC (2015a), Blanton et al. (2012)
Alzheimer's disease and viral infections	Honjo et al. (2009), Maheshwari and Eslick (2015), Starakis et al. (2011)
HIV and type 2 diabetes	Moni and Lio (2014)
HIV, tuberculosis, malaria and type 2 diabetes	Marais et al. (2013)
Cardiovascular disease and COPD	Onil and Unwin (2015)
chronic infections and heart disease	Madjid et al. (2004)
Type 2 diabetes and hepatitis C	Guo et al. (2011)
COVID-19 and preexisting respiratory and cardio-vascular disease	Liang et al. (2020)

It has also been reported that comorbidities exist between most of the diseases in Table 16.7 and numerous infectious diseases (Zeliger, 2014), as well as between numerous infectious diseases (Andresen et al., 2006).

16.5 Conclusion

All of the numerous of diseases that have been investigated elevate oxidative stress and lead to the onset of additional disease, irrespective of other OS-causing parameters a body may be experiencing. The comorbidities described above add to this hypothesis, as other factors were ruled out in the studies reported. As is discussed in Part III of this book, the presence of prevalent disease is perhaps the greatest indicator of the likelihood of additional disease onset.

References

Aflanie, I., 2015. Effect of heavy metal on malondialdehyde and advanced oxidation protein products concentration: a focus on arsenic, cadmium and mercury. J. Med. Bioengen. 4 (4), 332–337.

Agarwal, S., Ghosh, A., Chatterjee, S.N., 1987. Spontaneous release of malondialdehyde from ultraviolet light exposed liposomal membranes. Z. Naturforsch. 42c, 585–588.

Ahmad, F.B., Anderson, R.N., 2021. JAMA 325 (18), 1829–1830.

Akaike, T., 2001. Role of free radicals in viral pathogenesis and mutation. Rev. Med. Virol. 11 (2), 87–101.

Andresen, H.M., Regueira, H.T., Leighton, F., 2006. Oxidative stress in critically ill patients. Rev. Med. Chile 134 (5), 649–656.

Arcavi, L., Benowitz, N.L., 2004. Cigarette smoking and infection free. Arch. Intern. Med. 164 (20), 2206–2216.

Asadi, N., Bahmani, M., Kheradmand, A., Rafieian-Kopaei, 2017. The impact of oxidative stress on testicular function and the role of antioxidants in improving it. J. Clin. Diagn. Res. 1 (5), IE01–IE05.

Ayala, A., Munoz, M.F., Arguelles, S., 2014. Lipid peroxidation: production, metabolism, and signaling mechanosms of malondialdehyde and 4-hydroxy-2-nonenal. Oxid. Med. Cell. Longev. https://doi.org/10.1155/2014/360438.

Baek, J., Lee, M.G., 2016. Oxidative stress and antioxidant strategies in dermatology. Comm. Free Rad. 21 (4), 164–169.

Banerjee, B.D., Seth, V., Bhattacharya, A., Pasha, S.T., Chakraborty, A.K., 1999. Biochemical effects of some pesticides on lipid peroxidation and free-radical scavengers. Toxicol. Lett. 107 (1–3), 33–47.

Barbouti, A., Vasileiou, P., Evangelou, K., Vlasis, K.G., Papoudou-Bai, A., Gorlgoulia, V.G., et al., 2020. Implications of Oxidative Stress and Cellular Senescense in Age-Related Thymus Involution.

Bhattacharyya, A., Chattopadhyay, R., Mitra, S., Crowe, S.E., 2014. Oxidative stress: an essential factor in the pathogenesis of gastrointestinal mucosal diseases. Physiol. Rev. 94, 329–354.

Blanton, L., Peacock, G., Cox, C., Jhung, M., Finelli, L., Moore, C., 2012. Neurologic disorders among pediatric death associated with the 2009 pandemic influenza. Pediatrics 130 (3), 390–396.

Boersma, P., Black, L.I., Ward, B.W., 2020. Prevalence of multiple chronic conditions among US adults. Prev. Chronic Dis. 2020 (2018), 17–200130. https://doi.org/10.5888/pcd17.200130. (Accessed 12 May 2021).

Bouhafs, R.K., Jastrand, C., 1999. Phagocyte-induced lipid peroxidation of lung surfactant. Pediatr. Pulmonol. 27 (5), 322–327.

Calaf, G.M., Urzua, U., Termini, L., Aguayo, F., 2018. Oxidative stress in female cancers. Oncotarget 9 (1), 23824–23842.

CDC, 2015. Leading Causes of Death. Centers for Disease Control. http://www.cdc.gov/nchs/fastats/leading-causes-of-death.htm.

CDC, Centers for disease control and prevention, 2021. Chronic Diseases in America. www.cdc.gov/chronicdiseases. (Accessed 16 May 2021).

Cekin, E., Ipcioglu, O.M., Erkul, B.E., Ozcan, O., Cincik, H., Gungor, A., 2009. The association of oxidative stress and nasal polyposis. J. Int. Med. Res. https://doi.org/10.1177/147323000903700206. (Accessed 10 May 2021).
Celik, M., Koyuncu, I., 2018. A comprehensive study of oxidative stress in tinnitus patients. Ind J Otolaryngol Head Neck Surg 70 (4), 521–526.
Chen, M., Li, X., Zhang, Z., Xu, S., 2019. Hydrogen sulfide exposure triggers chicken trachea inflammatory injury through oxidative stress-mediated FOS/IL8 signaling. J. Hazard Mater. 368, 243–254.
Chen, C.C., McCarl, B.A., 2001. An investigation of the relationship between pesticide usage and climate change. Clim. Change 50, 475–487.
Chen, Y., Liang, Y., Luo, X., Hu, Q., 2020. Oxidative resistance of leukemic stem cells and oxidative damage to hematopoietic stem cells under pro-oxidative therapy. Cell Death Dis. https://doi.org/10.1038/s41419-020-2488y. (Accessed 10 May 2021).
Colborn, T., Dumanoski, D., Myers, J.P., 1996. Our Stolen Future. Dutton Penguin Books, New York.
Cvetkovic, T., Vlahovic, P., Todorovic, M., Stankovic, M., 2009. Investigation of oxidative stress in patients with chronic tonsillitis. Auris Nasus Larynx 36 (3), 340–344.
Daenen, K., Andries, A., Mekahli, D., Schepdael, A.V., Jouret, F., Bammens, B., 2019. Oxidative stress in chronic kidney disease. Pediatr. Nephrol. 34 (6), 775–791.
Darmon, N., Pelissier, M.A., Heyman, M., Albrecht, R., Desjeux, J.F., 1993. Oxidative stress may contribute to the intestinal dysfunction of weanling rats fed a low protein diet. J. Nutr. 123 (6), 1068–1075.
Davies, K.J.A., 1995. Oxidative stress: the paradox of aerobic life. Biochem. Soc. Symp. 61, 1–31.
Deska, M., Romuk, E., Segiet, O., Polczyk, J., Bula, G., Gawrychowski, J., 2018. Oxidative stress in proliferative lesions of parathyroid gland. Pol. Przegl. Chir. 91 (1), 29–34.
Diaz-Hung, M.L., Fraguela, M.E.G., 2014. Oxidative stress in neurological diseases: cause or effect? Neurologia 29 (8), 451–452.
Durrani, F., Phelps, D.S., Weisz, J., Silveyra, P., Hu, S., Mikerov, A.N., Floros, J., 2012. Gonadal hormones and oxidative stress interaction differentially affects survival of male and female mice after lung *Klebsiella pneumoniae* infection. Exp. Lung Res. 38 (4), 165–172.
Gashev, A.A., Thangaswamy, S., Chatterjee, V., 2014. Lymphatic contractibility and oxidative stress. In: Laher, I. (Ed.), Sysems Biology of Free Radicals and Antioxidants. Springer, Berlin, Heidelberg. https://doi.org/10.1007/978-3-642-30018-9_200. (Accessed 10 May 2021).
Ghone, R.A., Suryakar, A.N., Kulhalli, P.M., Bhagat, S.S., Padalkar, R.K., Karnik, A.C., et al., 2013. A study of oxidative stress biomarkers and effect of oral antioxidant supplementation in severe acute malnutrition. J. Clin. Diagn. Res. 7 (10), 2146–2148.
Guarner, C., Soriano, G., 2005. Bacterial translocation and its consequences in patients with cirrhosis. Eur. J. Gastroenterol. Hepatol. 17, 27–31.
Guo, X., Jin, M., Yang, M., Liu, K., Li, J.W., 2011. Type 2 diabetes mellitus and the risk of hepatitis C virus infection: a systematic review. Sci. Rep. 3, 2981. https://doi.org/10.1038/srep02981.
Haroun, H.S.W., 2018. Aging of thymus gland and immune system. Anat. Physiol. 5 (3), 178–181.
Honjo K van Reekum, R., Verhoeff, N.P., 2009. Alzheimer's disease and infection: do infectious agents contribute to progression of Alzheimer's disease? Alz. Dement. 5 (4), 348–360.
Hughes, D.A., 1999. Effects of dietary antioxidants on the immune function of middle-aged adults. Proc. Nutr. Soc. 58, 79–84.
Jia, Z., Zhu, H., Li, J., Wang, X., Misra, H., Li, Y., 2012. Oxidative stress in spinal cord injury and antioxidant-based intervention. Spinal Cord 50, 264–274.
Jeng, H.A., Pan, C.H., Diawara, N., Chang-Chien, G.P., Lin, W.Y., Huang, C.T., et al., 2011. Polycyclic aromatic hydrocarbon-induced oxidative stress and lipid peroxidation in relation to immunological alteration. Occup. Environ. Med. 68 (9), 653–658.
Kantor, E.D., Rehm, C.D., Haas, J.S., Chan, A.T., Giovannucci, E.L., 2015. Trends in prescription drug use among adults in the United States from 1999-2012. JAMA 314 (17), 1818–1830.
Karbownik-Lewinska, M., Kokoszko-Bilska, A., 2012. Oxidative damage to macromolecules in the thyroid-experimental evidence. Thyroid Res. http://www.thyroidresearchjournal.com/content/5/1/25. (Accessed 11 May 2021).
Katona, P., Katona-Apte, J., 2008. The interaction between nutrition and infection. Clin. Pract. 46, 1582–1588.
Khan, M.F., Boor, P.J., Gu, Y., Alcock, N.W., Ansari, G.A.S., 1997. Oxidative stress in the splenotoxicity of aniline. Fund. Appl. Toxicol. 35, 22–30.

Kruk, J., Kubasik-Kladna, K., Aboul-Enein, H.Y., 2015. The role of oxidative stress in the pathogenesis of eye diseases: current status and a dual role of physical activity. Mini Rev. Med. Chem. 16 (3), 241–257.

Levy, A., Browne, B., Fredrick, A., Stensland, K., Bennett, J., Sullivan, T., et al., 2019. Insights into the pathophysiology of urethral stricture disease to lichen sclerosus: comparison of pathological markers in lichen sclerosus induced strictures vs nonlichen sclerosus induced strictures. J. Urol. 201 (6), 1158–1163.

Li, P., Zhou, H., Tu, T., Lu, H., 2021. Dynamic exacerbation in inflammation and oxidative stress during the formation of peritendinous adhesion resulted from acute tendon injury. J. Orthop. Surg. Res. https://doi.org/10.1186/s13018-021-02445-y. (Accessed 11 May 2021).

Liang, W., Liang, H., Ou, L., Chen, B., Chen, A., Caichen, I., et al., 2020. Development and validation of a clinical risk score to predict the occurrence of clinical illness in hospitalized patients with COVID-19. JAMA Intern. Med. https://doi.org/10.1001/jamainternmed.2020.2033. (Accessed 16 May 2021).

Liu, J., Qu, W., Kadiiska, M.B., 2009. Role of oxidative stress in cadmium toxicity and carcinogenesis. Toxicol. Appl. Pharmacol. 238 (3), 209–214.

Lu, J., Wang, Z., Cao, J., Chen, Y., Dong, Y., 2018. A novel and compact review on the role of oxidative stress in female reproduction. Reproduct. Biol. Endochinol. https://doi.org/10.1186/s12958-018-0391-5. (Accessed 17 May 2021).

Long, A.N., Dagogo-Jack, S., 2011. The comorbidities of diabetes and hypertension: mechanisms and approach to target organ protection. J. Clin. Hypertens. 13 (4), 244–251.

Lorente, L., Martin, M.M., Abreu-Gonzalez, P., Dominguez-Rodriguez, A., Labarta, L., Diaz, C., Sole-Violan, J., et al., 2013. Prognostic value of malondialdehyde serum levels in sever sepsis: a multicenter study. PLoS One 8 (1), e53741.

Maciejczyk, M., Skutnik-Radziszewska, A., Zieniewska, I., Matczuk, J., Domel, E., Danuta, W., et al., 2019. Antioxidant defense, oxidative modification, and salivary gland function in an early phase of cerulein pancreatitis. Ox. Med. Cell Long. https://doi.org/10.1155/2019/8403578. (Accessed 11 May 2021).

Maheshwari, P., Eslicj, G.D., 2015. Bacterial infection and Alzheimer's disease: a metal-analysis. J. Alz. Dis. 43 (3), 957–966.

Madjid, M., Aboshady, I., Awan, I., Litovsky, S., Casscells, S.W., 2004. Influenza and cardiovascular disease. Is there a relationship? Tex. Heart Inst. J. 31 (1), 4–13.

Marais, B.J., Lonroth, K., Lawn, S.D., Migliori, G.B., Mwaba, K., Glaziou, P., et al., 2013. Tuberculosis comorbidity with communicable and nonp-communicable diseases: integrating health services and control efforts. Lancet Infect. Dis. 13 (5), 436–448.

Millar, N.L., Silbernagel, K.G., Thorborg, K., Kirwan, P.D., Galatz, L.M., Abrams, G.D., et al., 2021. Tendinopathy. Nat. Rev. Dis. Primers. https://doi.org/10.1038/s41572-020-00234. Accessed August 23, 2022.

Miyata, Y., Matuso, T., Mitsunari, K., Asai, A., Ohba, K., Sakai, H., 2019. A review of oxidative stress and urinary dysfunction caused by bladder outlet obstruction and treatments using antioxidants. Antioxidants. https://doi.org/10.3390/antiox8050132. (Accessed 11 May 2021).

Moni, M.A., Lio, P., 2014. Network-based analysis of comorbidities risk during an infection: SARS and HIV ase studies. Bioinformatics 15, 333.

Murray, C.J.L., Barber, R.M., Foreman, K.J., Ozgoren, A.A., ABD-Allah, F., Abera, S.F., et al., 2015. Global, regional, and national disability-adjusted life years (DALYs) for 306 diseases and injuries and healthy life expectancy (HALE) for 188 countries, 1990–2013: quantifying the epidemiological transition. Lancet 386 (10009), 2145–2191.

Neel, B.A., Sargis, R.M., 2011. The paradox of progress: environmental disruption of metabolism and the diabetes epidemic. Diabetes 60, 1838–1848.

Nielsen, F., Mikkelsen, B.B., Niesen, J.B., Andersen, H.R., Grandjean, P., 1997. Plasma malondialdehyde as biomarker for oxidative stress: reference interval and efforts of life-style factors. Clin. Chem. 43 (7), 1209–1214.

Onil, T., Unwin, N., 2015. Why the communicable/non-communicable disease dichotom is problematic for public health control strategies: implications of multimorbidity for public health systems in an era of health transition. Int. Health. https://doi.org/10.1093/inthealth/ihv040.

Park, H.S., Kim, S.R., Lee, Y.C., 2009. Impact of oxidative stress on lung diseases. Respirology 14 (1), 27–38.

Parkin, D.M., Stiller, C.A., Draper, G.J., Bieber, C.A., 1988. The international incidence of childhood cancer. Int. J. Cancer 42 (4), 511–520.

Paulis, G., Brancato, T., 2012. Inflammatory mechanisms and oxidative stress in Peyronie's disease: therapeutic "rationale" and related emerging treatment strategies. Inflamm. Allergy - Drug Targets 11 (1), 48–57.

Picou, F., Vignon, C., Debeissat, C., Lachot, S., Kosmider, O., Gallay, N., et al., 2019. Bone marrow oxidative stress and specific antioxidant signatures in myelodysplastic syndromes. Blood Adv. 3 (24), 4271–4279.

Pizzino, G., Irrerea, N., Cucinotta, M., Pallio, G., Mannino, F., Arcoraci, V., et al., 2017. Oxidative Stress: Harms and Benefits for Human Health.

Pritchard, C., Rosenorn-Lanng, E., 2015. Neurological deaths of American adults (55–74) and the over 75's by sex compared with 20 Western countries 1989–2010: cause for concern. Surg. Neurol. Int. 6, 123. https://doi.org/10.4103/2152-7806.161420.

Quio, Y., Sun, J., Ding, Y., Le, G., Shi, Y., 2013. Alterations of the gut microbia i high-fat diet mice is strongly linked to oxidative stress. Appl. Microbil. Biotechnol. 97 (4), 1689–1697.

Robles, L., Vaziri, N.D., Ichii, H., 2014. Role of oxidative stress in pathogenesis of pancreatitis: effect of antioxidant therapy. Pacreat. Disord. Ther. 3 (1), 112–120.

Romieu, I., Barraza-Villarreal, A., Escamilla-Nunez, C., Almstrand, A.C., Diaz-Sanchez, D., Sly, P.D., et al., 2008. Exhaled breath malondialdehyde as a marker of effect of exposure to air pollution in children with asthma. J. Allerge Clin. Immunol. 121 (4), 903–909.

Roumeguere, T., Sfeir, J., Rassy, E., Albisinni, S., Van Antwerpen, P., Boudjeltia, S., et al., 2017. Oxidative stress and prostate disease (Review). Molec. Clin. Oncol. 7, 723–728.

Sabatino, M.E., Grondona, E., Sosa, L.V., Bragato, M., Carreno, L., Juarez, V., et al., 2018. Oxidative stress and mitochondrial adaptive shift during pituitary tumoral growth. Free Radic. Biol. Med. https://doi.org/10.1016/jfreeradbiomed.2018.03.019. (Accessed 11 May 2021).

Salim, S., 2017. Oxidative stress and the central nervous system. J. Pharmacol. Exp. Therapeut. 360, 201–205.

Sanchez-Rodriguez, M.A., Mendoza-Nunez, V.M., 2019. Oxidative stress indexes for diagnosis of health or disease in humans. Oxid. Med. Cell Long. https://doi.org/10.1155/2019/4128152. (Accessed 16 May 2021).

Schneider, D., Manzan, M.A., Crawford, R.B., Chen, W., Kaminski, N.E., 2008. 2,3,7,8-tetrachlorodibenzo-p-dioxin-mediated impairment of B cell differentiation involves dysregulation of paried box 5 (Pax5) isoform, Pax 5a. J. Pharmacol. Exp. Therapeut. 326 (2), 463–474.

Schulz, H.U., Niederau, C., Klonlowski-Stumpe, H., Halangk, W., Luthen, R., Lippert, H., 1999. Oxidative stress in acute pancreatitis. Hepato-Gastroenterology 46 (29), 2736–2750.

Sena, C.M., Leandro, A., Azul, L., Seica, R., Perry, G., 2018. Vascular oxidative stress: impact and therapeutic approaches. Front. Physiol. https://doi.org/10.3389/fphys.2018.01668. (Accessed 11 May 2021).

Senoner, T., Dichtl, W., 2019. Oxidative stress in cardiovascular diseases: still a therapeutic target? Nutrients, 103390/nu11092090. (Accessed 11 May 2021).

Serra, A.J., Prokic, M.D., Vasconsuelo, A., Pinto, J.R., 2018. Oxidative stress in muscle diseases: current and future therapy. Ox. Med. Cell Long. https://doi.org/10.1155/2018/6439138. (Accessed 11 May 2021).

Splettstoesser, W.D., Schuff-Werner, P., 2002. Oxidative stress in phagocytes– "the enemy within". Microsc. Res. Tech. 57 (6), 441–455.

Starakis, J., Panos, G., Koutras, A., Mazokopakis, E.E., 2011. Pathogens and chronic or long-term neurlogic disorders. Cardiovasc. Hematol. Disord. Drug Targets 11 (1), 4052.

Stossel, T.P., Mason, R.J., Smith, A.L., 1974. Lipid peroxidation by human blood phagocytes. J. Clin. Invest. 638–645.

Sudha, S., Sree, J.S., Saranya, R.S., Varun, S., 2014. Evaluation of DNA damage and oxidative stress in road pavement workers occupationally exposed to polycyclic aromatic hydrocarbons. Int. Conf. Adv. Agri. Biol. Environ. Sci. https://doi.org/10.15242/IICBE.C1014023.

Sulaiman, M.A.H., Al-Farsi, Y.M., Al-KHaduri, M.M., Saleh, J., Waly, M.I., 2018. Int. J. Wom. Health 10, 763–771.

Suzuki, H., Nishizawa, T., Tsugawa, H., Mogami, S., Hibe, T., 2012. Roles of oxidative stress inn stomach disorders. J. Clin. Biochem. Nutr. 50 (1), 35–38.

Tangvarasittichai, S., Poonsub, P., Tangvarasittichai, O., Sirigulsatien, V., 2009. Serum levels of malondialdehyde in type 2 diabetes mellitus Thai subjects. Siriraj Med. J. 61, 20–23.

Thangaswamy, S., Bridenbaugh, E.A., Gashev, A.A., 2012. Evidence of increased oxidative stress in aged mesenteric lymphatic vessels. Lymphatic Res. Biol. 10 (2), 53–63.

Trueb, R.M., 2015. The impact of oxidative stress on hair. Int. J. Cosmet. Sci. 37 (Suppl. 2), 25–30.

U.S. Energy and Information Administration. U.S. Annual energy Outlook, 2014. http://www.eia.gov/forcasts/ieo/. (Accessed 6 May 2016).

Valyl-Nagy, T., Dermody, T.S., 2005. Role of oxidative damage in the pathogenesis of viral infections of the nervous system. Histol. Histopathol. 20, 957–967.

Vona, R., Pallotta, L., Cappelletti, M., Severi, C., Matarrese, P., 2021. The impact of oxidative stress in human pathology: focus on gastrointestinal disorders. Antioxidants antiox10020201. (Accessed 11 May 2021).
Wagener, F.A.T.G., Carels, C.E., Lundvig, D.M.S., 2013. Targeting the redox balance in inflammatory skin conditions. Int. J. Mol. Sci. 14 (5), 9226–9167.
Wallace, R.B., Salive, M.E., 2013. The dimentions of multiple chronic conditions: where do we go from here? A commentary on the special collection of preventing chronic disease. Centers for Disease Control and Prevention. Prev. Chronic Dis. 10, 130104. https://doi.org/10.5888/pcd10.130104.
Wang, Y., Beydoun, M.A., 2007. The obesity epidemic in the United States - gender, age, socioeconomic, racial/ethnic, and geographical characteristics: a systemic review and meta-regression analysis. Epidemiol. Rev. 29 (1), 6–28.
Wilson, C., 2014. Oxidative stress and osteoporosis. Nat. Rev. Encocrin. https://doi.org/10.1038/nrendo.2013.225. (Accessed 11 May 2021).
Xu, G.J., Kula, T., Xu, Q., Li, M.Z., Vernon, S.D., Ndung'n, T., et al., 2015. Comprehensive serological profiling of human populations using a synthetic human virome. Science 348, aaa0698. https://doi.org/10.1126/science.aaa0698.
Yilmaz, T., Kocan, E.G., Besler, H.T., 2004. The role of oxidants and antioxidants in chronic tonsillitis and adenoid hypertrophy in children. Int. J. Pediatr. Otorhinolaryngol. 68 (8), 1053–1058.
Yoshida, N., 2007. Inflammation and oxidative stress in gastroesophageal reflux disease. J. Clin. Biochem. Nutr. 40 (1), 13–23.
Zahan, O.M., Serban, O., Gherman, C., Fodor, D., 2020. The evaluation of oxidative stress in osteoarthritis. Med. Pharm. Rep. 93 (1), 12–22.
Zaki, M.H., Akuta, T., Akaike, T., 2005. Nitric oxide-induced nitrative stress involved in microbial pathogenesis. J. Pharmacol. Sci. 98, 117–129.
Zeliger, H.I., 2003. Toxic effects of chemical mixtures. Arch. Environ. Health 58 (1), 23–29.
Zeliger, H.I., 2011. Human Toxicology of Chemical Mixtures, second ed. Elsevier, London.
Zeliger, H.I., Pan, Y., Rea, W.J., 2012. Predicting co-morbidities in chemically sensitive individuals from exhaled breath analysis. Interdiscipl. Toxicol. 5 (3), 123–126.
Zeliger, H.I., 2013a. Lipophilic chemical exposure as a cause of cardiovascular disease. Interdiscipl. Toxicol. 6 (2), 55–62.
Zeliger, H.I., 2013b. Exposure to lipophilic chemicals as a cause of neurological impairments, neurodevelopmental disorders and neurodegenerative diseases. Interdiscipl. Toxicol. 6 (3), 101–108.
Zeliger, H.I., 2013c. Lipophilic chemical exposure as a cause of type 2 diabetes. Rev. Environ. Health 28 (1), 9–20.
Zeliger, H.I., 2014. Co-morbidities of environmental diseases: a common cause. Interdiscipl. Toxicol. 7 (3), 101–106.
Zeliger, H.I., 2016. Predicting disease onset in clinically healthy people. Interdiscipl. Toxicol. 9 (2), 101–116.
Zeliger, H.I., Lipinski, B., 2015. Physiochemical basis of human degenerative disease. Interdiscipl. Toxicol. 8 (1), 15–21.
Zeliger, H.I., POssidente, B., Drake, A.G., 2015. Co-morbidity of multiple chemical sensitivity)MCS) and chronic fatigue syndrome (CGS): which comes first. Pub. Health Prev. Med. 1 (3), 108–111.
Zhu, Q.X., Shen, T., Ding, R., Liang, Z.Z., Zhang, X.J., 2005. Cytotoxicity of trichlorethylene and perchloroethylene on normal human epidermal keratinocytes and protective role of vitamin E. Toxicology 209 (1), 55–67.
Zidova, Z., Garcia-Santos, D., Kapralova, K., Koralkova, P., Mojzikova, R., Dolezal, D., et al., 2014. Oxidative stress and increased destruction of red blood cells contribute to the pathophysiology of anemia caused by DMT1 deficiency. Blood 124 (21), 4027.

CHAPTER

17

Total oxidative stress and disease

17.1 Introduction

Previous chapters to this book have identified causes of oxidative stress and associated disease onset with elevated oxidative stress. These factors are listed in Table 17.1. This chapter addresses the additivity of OS and the use of total oxidative stress as a predictor of the likelihood of disease onset.

17.2 Diseases and oxidative stress

Numerous diseases are caused and/or exacerbated by chronically elevated oxidative stress (Zeliger, 2016). Table 17.2 contains a partial list of these.

17.3 Malondialdehyde as an indicator of oxidative stress

As previously discussed, serum malondialdehyde is the biomarker most often used as an indicator of oxidative stress. This is shown by the data in Table 17.2. Both infectious and noninfectious diseases as well as environmental exposures result in the generation of MDA (Sonnerborg et al., 1988; Zeliger, 2016). The data in this table clearly demonstrate that multiple OS-related diseases, as well as multiple environmental exposures, increase serum MDA levels. The data in Table 17.2 also show that those with multiple MDA increasing sources have higher serum MDA levels and that serum MDA levels are additive, i.e., that there is a dose related relationship, no matter what the OS elevation source(s). The multiple sources can be two or more diseases (for example, type 2 diabetes and cardiovascular disease (Bhutia et al., 2011) or type 2 diabetes and myocardial infarction (Mahreen et al., 2012)); two or more environmental exposures as in (for example, cigarette smoking and road tar fume exposure (Bhutia et al., 2011); or a disease and a toxic exposure (for example, type 2 diabetes and cigarette smoking (Bhutia et al., 2011). The two sepsis data points also show the elevation of MDA with disease severity.

https://doi.org/10.1016/B978-0-323-91890-9.00038-6

TABLE 17.1 Factors that elevate oxidative stress.

Factors
Organic chemicals
Inorganic chemicals
Heavy metals
Chemical mixtures
Particles
Fibers
Air pollutants
Water pollutants
Soil pollutants
Alcohol
Recreational drugs
Smoking
Ionizing radiation
Nonionizing radiation
Inflammation
Chronic trauma
Saturated fats
Red meat
Processed foods
Artificial food colors
Artificial food flavors
Pharmaceuticals
Hot and cold temperature extremes
Sleep deprivation
Circadian cycle interruption
Psychological stress
Genetics
Epigenetics
Aging
Preexisting disease

TABLE 17.2 Diseases caused and/or exacerbated by chronically elevated oxidative stress.

Metabolic
Type 2 diabetes
Metabolic syndrome
Hyperlipidemia
Obesity
Respiratory
Allergic rhinitis
Chronic obstructive pulmonary disease
Asthma
Neurological
Autism
ADHD
Alzheimer's disease
Parkinson's disease
Amyotrophic lateral sclerosis
Major depression
Motor skill decline
Cognitive decline
Learning disorder onset
Endocrine
Male infertility
Female infertility
Hypothyroidism
Birth defects
Cardiovascular
Myocardial infarction
Stroke
Atherosclerosis
Arteriosclerosis
Ischemic heart disease
Hypertension

(*Continued*)

TABLE 17.2 Diseases caused and/or exacerbated by chronically elevated oxidative stress.—cont'd

Gastrointestinal
Irritable bowel disease
Crohn's disease
Peptic ulcer
Musculoskeletal
Rheumatoid arthritis
Osteoarthritis
Osteoporosis
Fibromyalgia
Urinary tract
Benign prostate hypertrophy
Urethritis
Urinary tract infection
Kidney
End stage renal disease
Renal vascular disease
Glomeruolsclerosis
Liver
Cirrhosis
Hepatitis
Fatty liver disease
Skin
Psoriasis
Eczema
SLE (lupus)
Dermatitis
Acne
Immunological and autoimmune
Acute uticaria
Chronic fatigue syndrome
Chemical sensitivity
Lupus
Sjogram's syndrome

TABLE 17.2 Diseases caused and/or exacerbated by chronically elevated oxidative stress.—cont'd

Eye diseases
Cataracts
Glaucoma
Macular degeneration
Corneal and conjunctive diseases
Periodontal
Chronic periodontitis
Cancer
All cancers
Metastasis
Viral and bacterial infectious diseases
Herpes
Influenza
Common cold
TB
Herpes
HIV and AIDS
COVID19

Serum MDA data show that controls for different studies of the same disease may have a range of MDA values for "healthy" individuals. For example, the values for healthy subjects being compared to those with diabetes range from 0.9 to 1.9 mcg/L. This is so because the "healthy" people in the different studies most certainly had different exogenous exposures as well as different lifestyles that would account for the range in MDA levels. It has been reported, for example, that serum MDA levels for a healthy individual was found to vary up and down by as much as 19% over a 6 day period (Nielsen et al., 1997).

*The results shown in Table 17.3, as well as the discussion that follows have, to a great extent, been previously reported by this writer via the following citation: Zeliger, 2016. Predicting disease onset in clinically healthy people. Interdiscip Toxicol; 9 (2):101–116. These are reproduced with permission.

TABLE 17.3 Comparison of serum MDA levels of affected individuals versus healthy controls for those with disease alone, environmental exposure alone, both disease and environmental exposure and multiple environmental exposure. All data are in micromoles per liter (mcmoles/L) reported to two significant figures. Ratios of disease or exposure versus controls are indicative of the relative impact of disease or exposure on elevation of MDA.

	MDA	MDA	Ratio	
Disease/exposure	**Affected**	**Healthy**	**Affected versus healthy[b]**	**References**
Diseases				
Acute COPD	2.4	0.9	2.7	Tug et al. (2004)
Stable COPD	1.2	0.9	0.9	Tug et al. (2004)
COPD	1.3	0.6	2.2	Stupnytska et al. (2014)
Adult ADHD	2.4[a]	0.4[a]	6.0	Bulut et al. (2007)
RI	2.7	1.9	1.4	Kalavacherla et al. (1994)
Sepsis	2.5[a]	1.1[a]	2.3	Lorente et al. (2013)
Sepsis	3.2[a]	1.1[a]	2.9	Lorente et al. (2013a)
CVD	2.8	2.4	1.2	Boaz et al. (1999)
OBS	4.8	2.5	1.9	Olusi (2002)
Psychological stress	4.4	2.5	1.8	Chellappan et al. (2008)
Physical stress (rats)	9.6	19.3	2.0	Nyanatara et al. (2014)
Depression	2.0	0.4	5.0	Baipai et al. (2014)
AR	3.5	2.2	1.6	Alsamarai et al. (2009)
Asthma	4.4	2.2	2.0	Alsamarai et al. (2009)
AR + asthma	7.2	2.2	3.3	Alsamarai et al. (2009)
Met-S	1.0	0.8	1.3	Moreto et al. (2014)
BPH	2.1	1.0	2.1	Merendino et al. (2003)
Stomach CAN	2.6	0.8	3.3	Bitla et al. (2011)
T2D	2.2	1.3	1.7	Tangvarasittichai et al. (2009)
T2D	2.7	0.9	3.0	Bhutia et al. (2011)
T2D + Smoker	3.2	0.9	3.6	Bhutia et al. (2011)
T2D + CVD	3.7	0.9	4.1	Bhutia et al. (2011)
T2D	3.5[a]	1.9[a]	1.8	Mahreen et al. (2012)
T2D + MI	5.5[a]	1.9[a]	2.9	Mahreen et al. (2012)
Obesity	2.0	0.6	3.3	Yesilbersa et al. (2005)
Malnutrition	2.9	1.2	2.4	Ghone et al. (2013)

TABLE 17.3 Comparison of serum MDA levels of affected individuals versus healthy controls for those with disease alone, environmental exposure alone, both disease and environmental exposure and multiple environmental exposure. All data are in micromoles per liter (mcmoles/L) reported to two significant figures. Ratios of disease or exposure versus controls are indicative of the relative impact of disease or exposure on elevation of MDA.—cont'd

	MDA	MDA	Ratio	
Disease/exposure	**Affected**	**Healthy**	**Affected versus healthy[b]**	**References**
E-coli infection	4.2	2.0	2.1	Karaman et al. (2009)
Stomatitis	3.0	2.7	1.1	Khademi et al. (2014)
TB	5.4[a]	2.1[a]	2.6	Kulkarni et al. (2013)
IBS	2.1	1.6	1.3	Mete et al. (2013)
IHD	4.2	2.4	1.8	Metta et al. (2015)
IHD + smoker	6.0	2.4	2.5	Metta et al. (2015)
Environmental exposures				
Smokers	0.9	0.6	1.5	Bloomer (2007)
Smokers	0.7	0.6	1.2	Nielsen et al. (1997)
Smokers	3.8	2.0	1.9	Shah et al. (2015)
Smokers	1.3	0.3	4.3	Bhutia et al. (2011)
Road tar fumes	1.5	0.3	5.0	Sudha et al. (2014)
Smoker + road tar	2.3	0.3	7.7	Sudha et al. (2014)
Artificial food color	2.6[a]	2.1[a]	1.2	Cemek et al. (2014)
Paint thinner	2.0	1.0	2.0	Halifeoglu et al. (2000)
900 MHz radiation	8.5	7.5	1.1	Dasdag et al. (2012)
Healthy only data[b]		0.6[a]		Chakravarty and Rizvi (2011)
		1.3		Hoving et al. (1992)
		0.8		Hu et al. (2006)
		0.9		Bhutia et al. (2011)

[a]*Though MDA concentrations are most commonly reported in micrograms per liter (mcg/L) other units are also reported. Data reported in concentrations other than mcg/L have been converted to mcg/L.*
[b]*For purposes of comparison of affected individuals with healthy or unexposed individuals, all the data in Table 17.2 have been normalized to 1.0 for control, in each instance.*
AR, allergic rhinitis; *BPH*, benign prostate hypertrophy; *CAN*, cancer; *COPD*, chronic obstructive pulmonary disease; *CVD*, cardiovascular disease; *IDH*, ischemic heart disease; *IBS*, irritable bowel syndrome; *Met-S*, metabolic syndrome; *MI*, myocardial infarction; *OBS*, obesity; *RI*, rheumatoid arthritis; *T2D*, type 2 diabetes; *TB*, tuberculosis.

17.4 Oxidative stress additivity - infectious disease

Infectious and noninfectious diseases, result in the generation of MDA (Sonnerborg et al., 1988). Thus, high serum MDA values are predictive of disease prevalence as well as incipient disease onset. Those with infectious diseases have high incidences of other infectious diseases as well as noninfectious diseases, and those with noninfectious diseases have elevated incidences of both other noninfectious as well as infectious diseases.

Elevated MDA concentrations are also associated with all of the OS inducing stimuli presented in Table 17.2: toxic chemicals, particles and fibers, diet and food additives, radiation, pharmaceuticals, psychological stress, physical stress and pre-exising disease

IThese data also show that serum MDA levels can be used to not only predict the likelihood of disease onset, but that there is a dose response relationship between MDA level and disease severity.

17.5 Multimorbidity

Historically, most people who became ill with a single disease perished from it. With the progress made in modern medicine, however, this is no longer the case. Mankind has progressed to where many, if not most, diseases can be treated to prolong life. As a consequence of life prolongation, people are now more likely to have multi-morbidities and more likely to die as a result of a disease other than the first one to ail them (Murray et al., 2016).

The multiple causes of OS predict disease multi-morbidity. Since each disease or toxic exposure raises the level OS, additional morbidity is to be anticipated as the incidence of all the diseases in Table 17.1, as well as all exogenous exposure and the other OS raising parameters shown in this table, raise OS. Accordingly, one ill with virtually any disease or continually exposed to any OS-elevating agent is at increased risk for additional disease. Put it another way, "the sick get sicker."

Comorbidities exist between most of the diseases in Table 17.1 (Zeliger, 2014) as well as between numerous infectious diseases (Andresen et al., 2006). Examples of comorbidities between noninfectious and infectious diseases include:

1. Influenza and neurological disorders (CDC, 2015; Blanton et al., 2012).
2. Alzheimer's disease and viral infections (Honjo et al., 2009; Maheshwari and Eslick, 2015; Starakis et al., 2011).
3. HIV and type 2 diabetes (Moni and Lio, 2014).
4. HIV tuberculosis and malaria with type 2 diabetes (Marais et al., 2013).
5. Cardiovascular disease and COPD in a bidirectional manner (Onil and Unwin, 2015).
6. Chronic infections and heart disease (Madjid et al., 2004).
7. Type 2 diabetes and hepatitis C (Guo et al., 2011).

Multi-morbidities of noninfectious diseases are bidirectional. The following examples are illustrative.

1. In those ill with both with hypertension and type 2 diabetes, the first illness incidence was equally split between the two diseases (Sowers and Epstein, 1995).
2. Diabetes is bidirectional with depression (Mayo, 2015).
3. Depression was bidirectional with myocardial infarction (Chi et al., 2014).
4. A wide range of neurological diseases (including Alzheimer's disease and Parkinson's disease) are di-directional with epilepsy, stroke, cardiac, gastrointestinal diseases and respiratory diseases (Gaitatzis et al., 2012).
5. Asthma and anxiety are bidirectional (Lee et al., 2016).
6. Metabolic syndrome and mental health disorders (including schizophrenia, bipolar disorder, depression, anxiety, attention-deficit/hyperactivity disorder (ADHD) and autism spectrum disorders are bidirectional (Nousen et al., 2013).

As is discussed in Part II of this book, some of the disease-pair combination pairs just mentioned are known to be caused by common pathways, others, however propagate via distinctly different mechanisms. All, however, are due to the effects of elevated oxidative stress.

It follows from the above that a chemical-identified "cause" of a disease may only be part of the overall picture, since it may be only one of several OS contributors. It also follows that the disease that kills need not be the first disease, since an ensuing disease may be a more aggressive killer. For example, chemical sensitivity, which is rarely lethal but is a source of OS, may be followed by type 2 diabetes, which can cause death.

Though high serum MDA is predictive of the onset of additional morbidity, it is not generally possible to predict a specific disease than can ensue based on serum MDA alone. This is so because there are thousands of different lipophilic cell membranes in the human body, all of which produce MDA upon lipid peroxidation. The disease that ensues depends upon which particular membrane type is attacked (Zeliger and Lipinski, 2015).

Noninfectious diseases are accumulation disorders that strike as a result of significant contributory factors. These include: genetic predispositions; sequential absorption of OS-producing agents until toxic levels are reached and/or until all components of toxic mixtures are absorbed in sufficient levels to induce disease; the ability of the body to repair damage is exceeded and its defenses are compromised; or all hallmarks of disease onset are attained (Ayala et al., 2014; Zeliger and Lipinski, 2015). Accordingly, multimorbidity is to be anticipated with increasing age.

17.6 Disease prevention strategies

All disease prevention strategies are aimed at lowering oxidative stress. This can be accomplished via prevention of OS elevation exposures, by administering antioxidants (Kontush et al., 2001; Jain and Chaves, 2011; Coskun et al., 2006) and by treating diseases and their symptoms to cure the disease or reduce severity of symptoms (for example, treating asthma with antiinflammatories (Bartoli et al., 2011), diabetes with insulin (Wang et al., 2014); or depression with therapy to lower oxidative stress (Liu et al., 2015).

Serum MDA data in Table 17.2 show that controls for different studies of the same disease may have a range of MDA values for "healthy" individuals. For example, the values for healthy subjects being compared to those with diabetes range from 0.9 to 1.9 mcg/L. This is so because the "healthy" people in the different studies most certainly had different exogenous exposures as well as different lifestyles that would account for the range in MDA levels. As also discuss, serum MDA levels for a healthy individual can vary up and down by as much as 19% over a 6 day period (Nielsen et al., 1997).

The discussion above strongly suggests that the key to disease prevention is to eliminate as many of the causes of OS as is possible, for it is the total OS from whatever source that causes disease onset. Thus, disease prevention can be accomplished by aggressively treating all diseases, by treating symptoms from conditions with intermittent or occasional manifestations, such as allergic reactions, and by limiting exposures via what are termed here as micro- and macro-preventative measures and the treatment of disease, be it noninfectious or infectious, where possible. It is important that treatment include attention to all morbidities present as well as to all sources of exposures that contribute to OS. The body can generally recover from occasional high-dose acute levels of OS, provided it is not subjected to an instantly lethal dose, but is subject to the onset of disease from chronic elevated levels of OS. Asymptomatic Patients who present with elevated serum MDA levels should be further evaluated to determine the source(s) of their high MDA levels and precautions taken to reduce such sources before the onset of disease, as disease onset will provide additional OS and can lead to further disease.

Disease prevention strategies can be micro- or macro-ones (Zeliger, 2019). Micro-preventative measures include oxidative stress-limiting steps that can be undertaken by the individual.

1. Lifestyle actions such as adherence to a Mediterranean type diet that is high in antioxidant phytochemicals.
2. Carbon-filtering toxic chemical-containing tap water; limiting intake of processed and red meats.
3. Eating foods low in animal fats, sucrose, avoidance of foods and personal care products that contain preservatives such as triclosan, butylated hydroxy anisole (BHA), butylated hydroxy toluene (BHT) and parabens.
4. Limiting exposures to exogenous toxins such as tobacco smoke, pesticides, chemical solvents, and not exercising vigorously in times of high air pollution levels.
5. Avoiding the packaging foods in plastics that exude phthalates and bisphenol-A.
6. Limiting pharmaceutical use to those medically deemed essential.
7. Combating obesity; and seeking prompt medical help when disease strikes.

Macro-preventive measures are societal actions that lead to reducing exposures to oxidative stress elevating parameters. Examples of these are:

1. Educational programs to produce awareness of environmental hazards to good health.
2. Regulatory control of hazardous chemical release, tobacco product sales and pesticide use.

3. Encouragement of organic food production.
4. Mandating strict warning requirements for hazardous chemicals.
5. Stimulation of green energy production to reduce global warming, which increases the release of toxic OS-increasing chemicals, increases the rates of environmental chemical reactions which lead to higher levels of ozone and other air pollutants, increase the risks of wildfires which spew large quantities of pollutants into the air, and greatly facilitates the spread of hazardous toxins throughout the globe, and raises the prevalence of infectious diseases.
6. Promotion of overall healthy living by encouraging physical activity, and making exercise venues available to all.
7. Education of the public to unhealthy lifestyles and the dangers posed by these.
8. Addressing inequalities in all socioeconomic sectors of society, as the poor are generally subjected to living conditions and lifestyle choices that result in elevated oxidative stress.

17.7 Obesity

The danger of obesity is an example of how micro- and macro-preventive measures can act in tandem to lower further oxidative stress-caused disease.

The deleterious effects of obesity linger long after significant weight loss (Stokes and Preston, 2015). White adipose tissue (WAT) serves as a collector and bio-concentrator of exogenous lipophilic chemicals such as PCBs, chlorinated hydrocarbon pesticides, polynuclear aromatic hydrocarbons. other persistent organic pollutants (POPs) and chemicals that are found in air, water, food and everyday products (Arrebola et al., 2013; Brown et al., 2015). These chemicals partition from WAT to blood serum and serve as a constant supplier of OS-inducing toxins to the blood (Yu et al., 2011; Lind et al., 2013). Rapid weight loss, involving drastic reduction in WAT, results in the release of large quantities of toxic lipophiles with resultant significant systemic OS increase, an event that can lead to the onset of diabetes, cardiovascular, renal, liver and other diseases (Lind et al., 2013; Olusi, 2002; Zeliger, 2013; Zeliger, 2013a,b). Accordingly, gradual weight reduction is preferable to rapid weight loss, to allow for the metabolism and elimination of toxic lipophilic chemicals. It is to be noted, however, that many POPs (such as PCBs, dioxins, furans and chlorinated pesticides such as DDT and its metabolite DDE) have very long half lives in the body and may linger up to 30 years or more (Gallo et al., 2011; Yu et al., 2011). This consideration offers an explanation of why the effects of initial obesity linger throughout one's life and the need to prevent obesity throughout life, and particularly during childhood and early adulthood.

17.8 Limits to oxidative stress-mediated disease prevention

There are limitations to disease prevention for several reasons.

1. There are genetic differences which protect some and put others at risk for the onset of disease.
2. Ignorance, socio-economic status, lifestyle, peer pressure and economic interests of chemical manufacturers, mining operations, energy producers and intentional polluters act counteractively to enhance chemical exposures.
3. There are conflicting situations where well meaning people on both sides of an issue can reasonably disagree. Two examples of such situations serve as illustrations of this point and the often-difficult choices that need to be made.
   (a) DDT and its metabolite DDE are persistent organic pollutants that are causative agents of OS and disease (Zeliger, 2011). DDT, however, is still in use in parts of the world to control malaria causing mosquitoes.
   (b) Drinking water is routinely disinfected with chlorine to remove water-borne pathogens. Disinfection byproducts of chlorine treatment, however, are known human toxins that have associated with adverse reproductive effects and diseases including cancer (Windham and Fenster, 2008; Bove et al., 2007).

17.9 Disease onset prediction

Actual serum MDA levels are indicative of the presence of disease or disease promoting oxidative stress. A review of the data in Table 17.2 shows that serum MDA levels are elevated versus controls in those with illness or toxic exposures. Based on the data in this table and numerous other studies, the following scale for serum MDA values in mcg/L as predictors of disease has been proposed (Zeliger, 2016):

These values suggest that asymptomatic individuals with serum MDA levels of 1.20 or greater be evaluated further for disease. The subject of disease onset prediction is the subject of Part III of this book.

17.10 Conclusions

Virtually all human disease is induced by oxidative stress. Total oxidative stress, from whatever source, be it toxic environmental exposure, the presence of disease, lifestyle choices

MDA level	Disease onset likelihood
Less than 1.20	Indicative of a healthy state
1.20–1.40	Disease predicted
1.40–3.00	Disease onset probable
Greater than 3.00	Severe disease likely

or combinations of these, increases the incidence of OS. OS leads to lipid peroxidation of lipophilic cell membranes, which in turn produces MDA. Serum MDA level is an additive parameter resulting from all sources of OS and, therefore, is a reliable indicator of total oxidative stress that can be used to predict the onset of disease in clinically asymptomatic individuals and to suggest the need for further clinical evaluation and treatment that can prevent much human disease. The addition of routine MDA to annual medical checkup blood work could give early warning for a state of elevated oxidative stress and potential for disease onset.

References

Alsamarai, A.M., Alwan, A.M., Ahmad, A.H., Salib, M.A., Salih, J.A., Aldabagh, M.A., et al., 2009. The relationship between asthma and allergic rhinitis in the Iraqi population. Allergol. Int. 58, 649–655.

Andresen, H.M., Regueira, H.T., Leighton, F., 2006. Oxidative stress in critically ill patients. Rev. Med. Chile 134 (5), 649–656.

Arrebola, J.P., Pumarega, J., Gasull, M., Fernandz, M.F., Martin-Olmedo, P., Molina-Molina, J.M., et al., 2013. Adipose tissue concentrations of persistent organic pollutants and prevalence of type 2 diabetes in adults from Southern Spain. Environ. Res. 122, 31–37.

Ayala, A., Munoz, M.F., Arguelles, S., 2014. Lipid peroxidation: production, metabolism, and signaling mechanisms of malondialdehyde and 4-hydroxy-2-nonenal. Oxid. Med. Cell. Longev. https://doi.org/10.1155/2014/360438. (Accessed 27 May 2021).

Baipai, A., Verma, A.K., Srivastava, M., Srivastave, R., 2014. Oxidative stress and major depression. J. Clin. Diagn. Res. 8 (12), CC04–077.

Bartoli, M.L., Novelli, F., Costa, I., Malagrino, L., Melosini, E., Bacci, S., et al., 2011. Malondialdehyde in exhaled breath condensate as a marker of oxidative stress in different pulmonary diseases. Mediat. Inflamm. https://doi.org/10.1155/2011/891752.

Bhutia, Y., Ghosh, A., Sherpa, M.L., Pal, R., Kumar-Mohanta, P., 2011. Aerum malondialdehyde level: surrogate stress marker in the Sikkinmese diabetics. J. Nat. Sci. Biol. Med. 2 (1), 107–112.

Bitla, A.R., Reddy, E.P., Sambasivaih, K., Suchitra, M., Reddy, V.S., Rao, S., 2011. Evaluation of plasma malondialdehyde as a biomarker in patients with carcinoma of the stomach. Biomed. Res. 22 (1), 63–68.

Blanton, L., Peacock, G., Cox, C., Jhung, M., Finelli, L., Moore, C., 2012. Neurologic disorders among pediatric death associated with the 2009 pandemic influenza. Pediatrics 130 (3), 390–396.

Bloomer, R.J., 2007. Decreased blood antioxidant capacity and increased lipid peroxidation in youg cigarette smokers compared to nonsmokers: impact of dietary intake. Nutr. J. 6, 39.

Boaz, M., Matas, Z., Biro, A., Katzir, Z., Green, M., Fainaru, M., Smetana, S., 1999. Serum malondialdehyde and prevalent cardiovascular disease in hemodialysis. Kidney Int. 56, 1078–1083.

Bove Jr., G.E., Rogerson, P.A., Vena, J.E., 2007. Case control study of the geographic variability of exposure to disinfectant byproducts and risk for rectal cancer. Int. J. Health Geogr. 6, 18.

Brown, R.E., Sharma, A.M., Ardern, C.I., Mirdarmadi, P., Mirdarmadi, P., Kuk, J.L., 2015. Secular differences in the association between caloric intake, macronutrient intake, and physical activity with obesity. Obes. Res. Clin. Pract. https://doi.org/10.1016/j.orcp.2015.08.007. (Accessed 27 May 2021).

Bulut, M., Selek, S., Gergerlioglu, H.S., Savas, H.A., Yilmaz, H.R., Yuce, M., Ekici, G., 2007. Malondialdehyde levels in adult attention-deficit hyperactivity disorder. J. Psychiatry Neurosci. 32 (6), 435–438.

CDC, 2015. Leading Causes of Death. Centers for Disease Control 2015. http://www.cdc.gov/nchs/fastats/leading-causes-of-death.htm. (Accessed 27 May 2021).

Cemek, M., Buyukokuroglu, M.E., Hazini, A., Onul, A., Gones, S., 2014. Effects of food color additives on antioxidant functions and bioelement contents of liver, kidney and brain tissues in rats. J. Food Nutr. Res. 2 (10), 686–691.

Chakravarty, S., Rizvi, S.I., 2011. Day and night GSH and MDA levels in healthy adults and effects of different doses of melatonin on these parameters. Int. J. Cell Biol. 5. Article ID 404591.

Chellappan, D., Joseph, J., Shabi, M.M., Krishnamoorthy, G., Ravindran, D., Uthrapathy, S., et al., 2008. Psycho-emotional stress - a cause of coronary artery disease. Acta Sci. Vet. 36 (2), 133–139.

Chi, M.J., Yu, E., Liu, W.W., Lee, M.C., Chung, M.H., 2014. The bidirectional relationship between myocardial infarction and depressive disorders. Int. J. Cardiol. 177 (3), 854–859.

Coscun, O., Yuncu, M., Kanter, M., Buyukbas, S., 2006. Ebelsen protects against oxidative and morphological effects of high concentration chronic toluene exposure on rat sciatic nerves. Eur. J. Gen. Med. 3 (2), 64–72.

Dasdag, S., Akdag, M.Z., Kizil, G., Kizel, M., Cakir, D.U., Yokus, B., 2012. Effect of 900 MHz radiofrequency radiation on beta amyloid protein, protein carbonyl, and malondialdehyde in the brain. Electromag. Biol. Med. 31 (1), 67–74.

Gallo, M.V., Schell, L.M., DeCaprio, A.P., Jacobs, A., 2011. Levels of persistent organic pollutant and their predictors among young adults. Chemosphere 83, 1374–1382.

Gaitatzis, A., Siscodiva, S.M., SAnder, J.W., 2012. The somatic comorbidity of epilepsy: a weighty but often unrecognized burden. Epilepsia 53 (8), 1282–1293.

Ghone, R.A., Suryakar, A.N., Kulhalli, P.M., Bhagat, S.S., Padalkar, R.K., Karnik, A.C., et al., 2013. A study of oxidative stress biomarkers and effect of oral antioxidant supplementation in severe acute malnutrition. J. Clin. Diagn. Res. 7 (10), 2146–2148.

Guo, X., Jin, M., Yang, M., Liu, K., Li, J.W., 2011. Type 2 diabetes mellitus and the risk of hepatitis C virus infection: a systematic review. Sci. Rep. 3, 2981. https://doi.org/10.1038/srep02981. (Accessed 27 May 2021).

Halifeoglu, I., Canatan, H., Ustendag, B., Ilhan, N., Inanc, F., 2000. effect of thinner inhalation on lipid peroxidation of some antioxidant enzymes of people working with paint thinner. Cell Biochem. Funct. 18 (4), 263–267.

Honjo, K., van Reekum, R., Verhoeff, N.P., 2009. Alzheimer's disease and infection: do infectious agents contribute to progression of Alzheimer's disease? Alz. Dement. 5 (4), 348–360.

Hoving, E.B., Laing, C., Rutgers, H.M., Teggeler, M., van Doormaal, J.J., Muskiet, F.A., 1992. Optimized determination of malondialdehyde in plasma lipid extracts using 1,3-diethyl-2-thiobarbituric acid: influence of detection methods and relations with lipids and fatty acids in plasma from healthy adults. Clin. Chem. Acta. 208 (1-2), 63–76l.

Hu, Y., Block, G., Norkus, E.P., Morrow, J.D., Dietrich, M., Hude, M., 2006. Relations of glycemic index and glycemic load with plasma oxidative stress marker. Am. J. Clin. Nutr 84 (1), 70–76.

Jain, S., Chaves, S.S., 2011. Obesity and influenza. CID 53, 422–423.

Kalavacherla, U.S., Ishaq, M., Rao, U.R., Sachindranath, A., Hepsiba, T., 1994. Malondialdehyde as a sensitive marker in patients with rheumatoid arthritis. J. Assoc. Phys. India 42 (10), 775–776.

Karaman, U., Kiran, T.R., Colak, C., Iraz, M., Celik, T., Karabulut, A.B., 2009. Serum malondialdehyde, glutathione and nitric oxide levels in patients infected with entamoeba coli. Int. J. Medicine Med. Sci. 1 (5), 235–237.

Khademi, H., Khozeimeh, F., Tavangar, A., Amini, S., Ghalayani, P., 2014. The serum and salivary level of malondialdehyde, vitamins A, E, and C in patient with recurrent aphthous stomatitis. Adv. Biomed. Res. 3, 246. https://doi.org/10.4103/2277-9175.14636. (Accessed 27 May 2021).

Kontush, A., Mann, U., Arit, S., Ujeyl, A., Luhrs, C., Muller-Thomsen, T., Beisiegel, U., 2001. Influence of vitamin E and C supplementation on lipoprotein oxidation in patients with Alzheimer's disease. Free Radic. Biol. Med. 31 (3), 345–354.

Kulkarni, R., Deshpande, A., Saxena, R., Saxena, K., 2013. A study of serum malondialdehyde and cytokine in tuberculosis patients. J. Clin. Diagn. Res. 7 (10), 2140–2142.

Lee, Y.C., Lee, C.T., Lai, Y.R., Chen, V.C., Stewart, R., 2016. Association of asthma and anxiety: a nationwide population-based study in Taiwan. J. Affect. Discord. 189, 98–105.

Lind, P.M., Lee, D.H., Jacobs, D.R., Salihovic, S., van Bavel, B., Wolff, M.S., Lind, L., 2013. Circulating levels of persistent organic pollutants are related to retrospective assessment of life-time weight change. Chemosphere 90 (3), 998–1004.

Liu, T., Zhong, S., Liao, X., Chen, J., He, T., Lai, S., Jia, Y., 2015. A meta-analysis of oxidative stress markers in depression. PLoS One 10 (10), e0138904.

Lorente, L., Martin, M.M., Abreu-Gonzalez, P., Dominguez-Rodriguez, A., Labarta, L., Diaz, C., Sole-Violan, J., et al., 2013. Prognostic value of malondialdehyde serum levels in sever sepsis: a multicenter study. PLoS One 8 (1), e53741.

Lorente, L., Martin, M.M., Abreu-Gonzalez, P., Dominguez-Rodriguez, A., Labarta, L., Diaz, C., Sole-Violan, J., et al., 2013a. Sustained high serum malondialdehyde levels are associated with severity and mortality in septic patients. Crit. Care 17, R290.

Madjid, M., Aboshady, I., Awan, I., Litovsky, S., Casscells, S.W., 2004. Influenza and cardiovascular disease. Is there a relationship? Tex. Heart Inst. J. 31 (1), 4–13.

Maheshwari, P., Eslick, G.D., 2015. Bacterial infection and Alzheimer's disease: a metal-analysis. J. Alz. Dis. 43 (3), 957–966.

Mahreen, R., Mohsin, M., Ishaq, M., 2012. Significantly increased levels of serum malondialdehyde in type 2 diabetics with myocardial infarction. Int. J. Diabetes Dev. Ctries. 30 (1), 49–51.

Marais, B.J., Lonroth, K., Lawn, S.D., Migliori, G.B., Mwaba, K., Glaziou, P., et al., 2013. Tuberculosis comorbidity with communicable and non-communicable diseases: integrating health services and control efforts. Lancet Infect. Dis. 13 (5), 436–448.

Mayo, P., 2015. Prevention and management of comorbid diabetes and depression. Nurs. Stand. 30 (8), 46–54.

Merendino, R.A., Salvo, F., Saija, A., Di Pasquale, G., Tomaino, A., Minciullo, P.L., et al., 2003. Malondialdehyde in benign prostate hypertrophy: a useful marker? Mediat. Inflamm. 12 (2), 127–128.

Mete, R., Tulubas, F., Oran, M., Yilmaz, A., Avci, B.A., Yildiz, K., et al., 2013. The role of oxidants and reactive nitrogen species in irritable bowel syndrome: a potential etiological explanation. Med. Sci. Mon. Int. Med. J. Exp. Clin. Res. 19, 762–766.

Metta, S., Basalingappa, D.R., Uppala, S., Mitta, G., 2015. Erythrocyte antioxidant defenses against cigarette smoking in ischemic heart disease. J. Clin. Diagn. Res. 9 (6), BC08–BC11.

Moni, M.A., Lio, P., 2014. Network-based analysis of comorbidities risk during an infection: SARS and HIV case studies. Bioinformatics 15, 333.

Moreto, F., de Oliveira, E., Manda, R.M., Burini, R.C., 2014. The higher plasma malondialdehyde concentrations are determined by metabolic syndrome-related glucolipotoxicity. Oxid. Med. Cell. Longev. https://doi.org/10.1155/2014/505368.

Murray, C.J.L., et al., 2016. Global, regional, and national life expectancy, all-cause mortality, and cause-specific mortality for 249 causes of death, 1980-2015: a systematic analysis for the Global Burden of Disease Study 2015. Lancet 388, 1459–1544.

Nielsen, F., Mikkelsen, B.B., Niesen, J.B., Andersen, H.R., Grandjean, P., 1997. Plasma malondialdehyde as biomarker for oxidative stress: reference interval and efforts of life-style factors. Clin. Chem. 43 (7), 1209–1214.

Nousen, E.K., Franco, J.G., Sullivan, E.L., 2013. Unraveling the mechanisms responsible for the comorbidity between metabolic syndrome and mental health disorders. Neuroendocrinology 98 (4), 254–266.

Nyanatara, A.K., Tripathi, Y., Nagaraja, H.S., Jeganathan, P.S., Ganaraja, B.S.R.P., Aron, K.B., 2014. Chronic stress induced tissue malondialdehyde in amygdala nucleus lesioned wister rats. Euro J. Biotech. Biosci. 2 (1), 57–60.

Olusi, S.O., 2002. Obesity is an independent risk factor for plasma lipid peroxidation and depletion of erythrocyte cytoprotic enzymes in humans. Int. J. Obes. 26, 1159–1164.

Onil, T., Unwin, N., 2015. Why the communicable/non-communicable disease dichotom is problematic for public health control strategies: implications of multimorbidity for public health systems in an era of health transition. Int. Health. https://doi.org/10.1093/inthealth/ihv040.

Shah, A.A., Khand, F., Khand, T.U., 2015. Effect of smoking on serum xanthine oxidase, malondialdehyde, ascorbic acid and alphatocopherol levels in healthy male subjects. Pak. J. Med. Sci. 31 (1), 146–149.

Sonnerborg, A., Carlin, G., Akerlund, B., Jarstrand, C., 1988. Increased production of malondialdehyde in patients with HIV infection. Scand. J. Infect. Dis. 20 (3), 287–290.

Sowers, J.R., Epstein, M., 1995. Diabetes mellitus and associated hypertension, vascular disease and nephropathy. Hypertension 26, 869–879.

Starakis, J., Panos, G., Koutras, A., Mazokopakis, E.E., 2011. Pathogens and chronic or long-term neurologic disorders. Cardiovasc. Hematol. Disord.: Drug Targets 11 (1), 40–52.

Stokes, A., Preston, S.H., 2015. Revealing the burden of obesity using weight histories. Proc. Natl. Acad. Sci. USA. https://doi.org/10.1073/pnas.1515472113. (Accessed 27 May 2021).

Stupnyska, G., Fediv, O., 2014. Mechanisms of local and systemic inflammation in chronic obstructive pulmonary disease with comorbid hypertension and obesity. Int. J. Sci. Tech. Res. 3 (5), 218–222.

Sudha, S., Sree, J.S., Saranya, R.S., Varun, S., 2014. Evaluation of DNA damage and oxidative stress in road pavement workers occupationally exposed to polycyclic aromatic hydrocarbons. Int. Conf. Adv. Agri. Biol. Environ. Sci. Oct 15-16m 2014. 1015242/IICBE.C1014023.

Tangvarasittichai, S., Poonsub, P., Tangvarasittichai, O., Sirigulsatien, V., 2009. Serum levels of malondialdehyde in type 2 diabetes mellitus Thai subjects. Siriraj. Med. J. 61, 20–23.

Tug, T., Karatas, F., Terzi, S.M., 2004. Antioxidant vitamins (A,C, and E) and malondialdehyde levels in acute exacerbation and stable periods of patients with chronic obstructive pulmonary disease. Clin. Invertig. Med. 27 (3), 123.

Wang, C.H., Chang, R.W., Ko, Y.H., Tsai, P.R., Wang, S.S., Chen, Y.S., et al., 2014. Prevention of arterial stiffening by using low-dose atorvastatin in diabetes is associated with decreased malondialdehyde. PLoS One 9 (3), e90471.

Windham, G., Fenster, L., 2008. Environmental contaminants and pregnancy outcomes. Fertil. Steril. 89 (Suppl. 1), e111. https://doi.org/10.1016/j.fertnstert.2007.12.041. (Accessed 27 May 2021).

Yesilbursa, D., Serdar, Z., Serdar, A., Sarac, M., Coskun, S., Jale, C., 2005. Lipid peroxides in obese patients and effects of weight loss with orlistat on lipid peroxides levels. Int. J. Obes. 29, 142–145.

Yu, G.W., Laseter, J., Mylander, C., 2011. Persistent organic pollutants in serum and several different fat compartments in humans. J. Environ. Public Health 2011, 417980.

Zeliger, H.I., 2013. Lipophilic chemical exposure as a cause of type 2 diabetes. Rev. Environ. Health 28 (1), 9–20.

Zeliger, H.I., 2013a. Lipophilic chemical exposure as a cause of cardiovascular disease. Interdiscipl. Toxicol. 6 (2), 55–62.

Zeliger, H.I., 2013b. Exposure to lipophilic chemicals as a cause of neurological impairments, neurodevelopmental disorders and neurodegenerative diseases. Interdiscipl. Toxicol. 6 (3), 101–108.

Zeliger, H.I., 2014. Co-morbidities of environmental diseases: a common cause. Interdiscipl. Toxicol. 7 (3), 101–106.

Zeliger, H.I., 2016. Predicting disease onset in clinically healthy people. Interdiscipl. Toxicol. 9 (2), 101–116.

Zeliger, H.I., Lipinski, B., 2015. Physiochemical basis of human degenerative disease. Interdiscp. Toxicol. 8 (1), 15–21.

Zeliger, H.I., 2019. A Pound of Prevention for a Healthier Life. Universal Publishers, Irvine, Boca Raton.

CHAPTER

18

Free radicals

18.1 Introduction

Oxygen is essential for life. Cells use oxygen to generate energy and, for mitochondrial production of adenosine triphosphate (ATP). In the process, free radicals, highly reactive molecules with one or more unpaired electron in their outer orbitals, are generated. Endogenously, free radicals thus created include the reactive oxygen species (ROS) and reactive nitrogen species (RNS) shown in Table 18.1 (Pham-Huy et al., 2015; Lobo et al., 2010; Phaniendra et al., 2015).

Not all ROS and RNS are free radicals. These and their formulas, shown in Table 18.2, readily react to generate free radical reactions in the body.

As noted above, free radicals are produced in the mitochondria as part of the energy generating process. Other sources of free radical production in the human body include both endogenous and exogenous ones (Lobo et al., 2010). These are listed in Table 18.3.

Short term free radical generation via the parameters in Table 18.2 generally do not induce disease onset. Long term, these parameters overwhelm the body's antioxidant defenses and

TABLE 18.1 ROS and RON.

ROS	
Superoxide anion	$O_2{}^{*-}$
Peroxide radical	$O_2{}^{*-2}$
Hydroxyl radical	*OH
Alkoxyl radical	RO*
RON	
Nitric oxide	NO*
Nitrogen dioxide	NO2*

Oxidative Stress
https://doi.org/10.1016/B978-0-323-91890-9.00019-2

TABLE 18.2 Nonfree radicals which easily lead to free radical reactions in the body.

ROS	
Hydrogen peroxide	HOOH
Organic peroxide	ROOH
Ozone	O_3
Singlet oxygen	$^1O_2$
Hypochlorous acid	HOCl
Hypobromous acid	HOBr
RNS	
Nitrous acid	HNO_2
Peroxynitrite	$ONOO^-$
Nitrogen trioxide	N_2O_3

lead to attack on DNA, proteins and lipophilic cell membranes, and also interfere with cell signaling, all of which lead to disease onset.

Although mitochondria are the primary producers of free radicals, ROS are also generated enzymatically. Enzymes which are free radical producers include those listed in Table 18.4 (Cai and Harrison, 2000).

Three of the enzymatic systems, xanthine oxidase, NADA/NAS(P)H oxidase (nicotinamide adenine dinucleotide phosphate), xanthine oxidase, and NO synthase have been widely studied in the cardiovascular system. The free radical producing reactions of the NAD(P)H and xanthine oxydases are shown in Eqs. (18.1–18.4) respectively (Taverne et al., 2013).

NAD(P)H oxidase reaction

$$NAD(P)H + 2O_2 \rightarrow NAD(P)^+ + 2O_2^{*-} + 2\,H^- \quad (18.1)$$

Xanthine oxidase reactions

$$\text{Hypoxanthine} + H_2O + 2O_2 \rightarrow \text{Xanthine} + 2O_2^{*-} + 2\,H^- \quad (18.2)$$

$$\text{Xanthine} + H_2O + 2O_2 \rightarrow \text{Uric acid} + 2O_2^{*-} + 2\,H^- \quad (18.3)$$

$$\text{(Hypo) Xanthine} + H_2O + O_2 \rightarrow \text{Urate} + H_2O_2^{*} \quad (18.4)$$

18.2 Free radical stability

Superoxide, hydroxyl and alkoxyl radicals are very short lived, with half times in the range of 10^{-6} seconds. More complex free radicals, those in species containing double bonds and

TABLE 18.3 Endogenous and exogenous sources of free radical production in the human body.

Endogenous
Mitochondria
Xanthine oxidase
Peroxisomes
Inflammation
Phagocytosis
Arachidonate pathways
Exercise
Trauma
Heat injury
Ischemia/reperfusion injury
Exogenous
Cigarette smoke
Air and water pollutants
Radiation
Ozone
Toxins
Infections
Pesticides
Aromatic hydrocarbons
Chloroaromatics
Complex organic polymers
Polynuclear aromatic hydrocarbons (PAHs)
Persistent organic pollutants (POPs)
Organic solvents
Microplastics
Transition metals
Heavy metals

TABLE 18.4 Enzymatic sources of free radicals.

Enzymatic sources of free radicals
Lipoxygenase
Cyclooxygenase
Cytochrome p450s
NADH/NAD(P)H oxidases
Xanthine oxydase
NO synthase

hetero atoms are more long lived and can persist up to minutes (Kohen and Nyska, 2002; Zipse, 2006; Dellinger et al., 2007; Song et al., 2008; Das and Roychoudhury, 2014; Phanien-dra, et al., 2015; Vejerano et al., 2021). A representative list of species that generate persistent free radicals is listed in Table 18.5.

TABLE 18.5 Representative list of persistent free radical generating species.

Persistent free radical generating species
Benzene
Chlorobenzene
Naphthalene
Phenol
Catechol
Hydroquinone
Benzo[a]pyrene
Pentachloro- and hexachlorobenzene
Pentachlorophenol
Polychlorinated naphthalenes
Polychlorinated biphenyls (PCBs)
Tetra- penta- and hexabromodiphenylether
Polychlorinated dibenzo-p-dioxins (PCDDs)
Polychlorinated dibenzofurans (PCDFs)

Free radicals are stabilized by three factors:

1. Substitution on radical carbon.

The methyl radical is the least stable hydrocarbon radical and stability increases with substitution on the radical carbon atom from primary to secondary to tertiary free radicals. Comparison of the methyl, ethyl, isopropyl and tertiary butyl free radicals is an example of this phenomenon.

2. Resonance stabilization via delocalization.

Free radicals are stabilized where resonance structures obtain, with increasing numbers of resonance structures imparting increased stability. The naphthyl, PCB-derived semiquinone and nitrogen dioxide free radicals are examples of this effect.

3. Free radical geometry.

Geometric configuration increases free radical stability when the radical is alpha to a π bond with which it can overlap. Comparisons of the allyl and n-propyl radicals and the methyl cyclohexyl and benzyl free radicals are examples of this.

These examples are shown in Fig. 18.1

Diesel exhaust particles (DEPs) are ubiquitous in the environment as a result of carbonaceous fuel combustion and tobacco smoking. DEPs serve as absorption sites for PAHs and their corresponding quinones some of which undergo enzymatic and nonenzymatic redox cycling with their derivative hydroquinones via free radical semiquinone intermediates (Chung et al., 2006; Dellinger et al., 2007; Song et al., 2008). This is illustrated by the equation in Fig. 18.2.

Semiquinone radicals, which are quite stable (with half lives of up to several days) readily donate their lone electrons to molecular oxygen to produce the very toxic superoxide radical which accounts for the toxicity of numerous PAHs and other polycyclic chemicals such as PCBs (Asmus et al., 2000; Song et al., 2008).

18.3 The Fenton reaction

ROS are essential for living organisms, including human, to properly function. Iron is an essential element of the body. In excessive quantities, however, iron reacts via the Fenton reaction to overproduce hydroxyl radicals, the most dangerous of free radicals. The Fenton reaction (Eq. 18.5), is the oxidation of divalent iron to its trivalent state by hydrogen peroxide, with the resultant formation of hydroxyl radical and hydroxide ion (Di Meo and Veditti, 2020).

$$Fe^{2+} + H_2O_2 \rightarrow Fe^{3+} + HO^* + OH^- \qquad (18.5)$$

The hydroxyl radical thus formed is associated with oxidative damage to membrane lipids and the onset of numerous diseases, including cancer, atherosclerosis, Alzheimer's disease and Parkinson's disease, diabetes and autoimmune diseases. It is also to be noted that copper also behaves in a Fenton-like manner to trigger free radical formation and disease (Winterbourn, 1995; Brewer, 2007; Lipinski, 2011; Zhao, 2019; Rynkowska et al., 2020).

1. Substitution on carbon radical

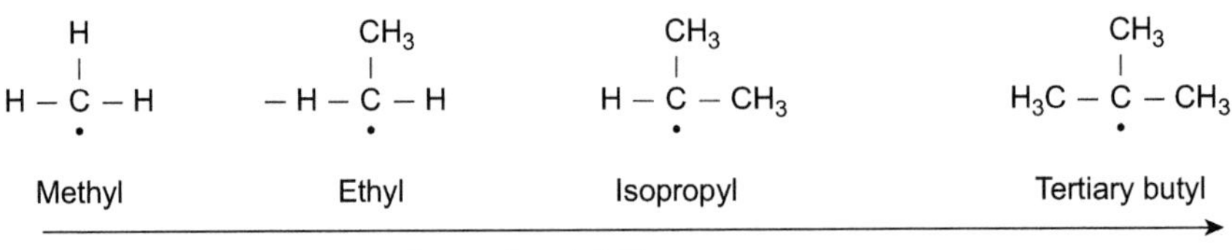

Increasing stability

2. Resonance stabilization via delocalization of the para-semiquinone radical

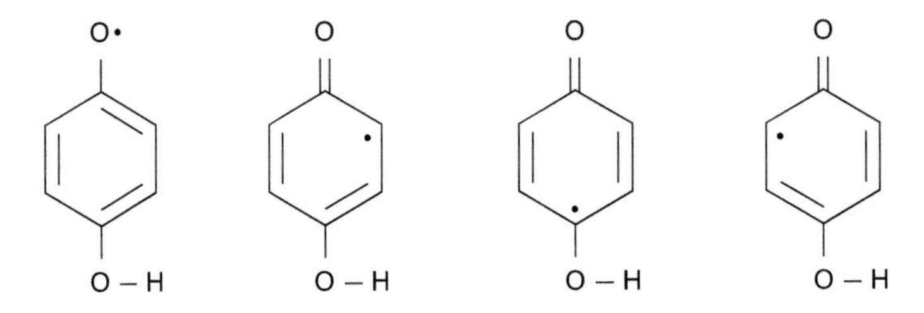

3. Geometric stability of n-propyl versus allyl radicals

n-propyl

allyl

Increasing stability

FIG. 18.1 Free radical stabilization parameters.

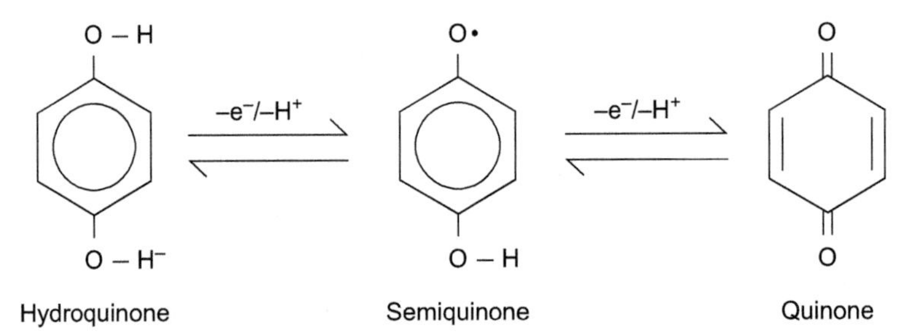

FIG. 18.2 Redox cycling of quinone and hydroquinone via free radical semiquinone intermediate.

Significantly, the trivalent ferric ion produced in the Fenton reaction can be reduced back to the divalent ferrous ion by reaction with hydrogen peroxide to form the hydroxyl and peroxyl free radicals (Eq. 18.6).

$$Fe^{3+} + H_2O_2 \rightarrow Fe^{2+} + HOO^* + H^+ \quad (18.6)$$

18.4 Free radical reactions with DNA

Both endogenous and exogenous sources of free radicals induce DNA damage (Cooke et al., 2003; Rao, 2009; Dizdaroglu and Jaruga, 2012). Hydroxyl radicals react with the DNA base, damaging both the heterocyclic and the sugar components, thus triggering carcinogenesis, mutagenesis and aging. Numerous mechanisms have been identified as causative of these effects examples of which are presented in ensuing chapters. The reaction sequences of hydroxyl radical attack on guanine (Scheme 18.1) and the DNA sugar moiety (Scheme 18.2) are presented as examples in Fig. 18.3 (Nimse and Pal, 2015).

18.5 Free radical reactions with proteins

Super oxide anion, hydroxyl, peroxyl, alkoxyl, hydroperoxyl free radicals and nitrogen dioxide, as well as nonradical species including hydrogen peroxide, ozone, hypochlorous acid, singlet oxygen and peroxynitrite induce free radical protein oxidation. Oxidation of amino acids present in proteins trigger the following effects (Smith et al., 1991; Kikugawa et al., 1994; Dean et al., 1997; Berlett and Stadtman, 1997; Butterfield et al., 1998; Phaniendra et al., 2015).

Protein-protein cross linkages
Loss of enzyme activity
Loss of functioning proteins
Loss of receptor functions
Loss of transport proteins
Conversion of sulfides to disulfides.

Oxidation of different amino acids in proteins produce different oxidation products. Table 18.6 lists protein amino acids and their ROS oxidized products (Phaniendra et al., 2015; Smith et al., 1991; Dean et al., 1997; Berlett and Stadtman, 1997; Butterfield et al., 1988).

The ROS-induced oxidation of lysine, proline, threonine and arginine produce carbonyl derivatives, increased levels of which are found in those with diseases that include Alzheimer's disease, Parkinson's disease, muscular dystrophy, atherosclerosis, diabetes, cataractogenesis and rheumatoid arthritis (Phaniendra et al., 2015).

FIG. 18.3 Reaction sequences of hydroxyl attack on DNA: guanine (Scheme 18.1) and DNA sugar moiety (Scheme 18.2). *Reproduced with permission from S.B. Nimse, D. Pal, Free radicals, natural antioxidants, and their reaction mechanisms, RSC Adv. 5 (27) (2015) 27,986. https://doi.org/10.1039/C4RA13315C.*

Guanine → •OH → C8-hydroxy-adduct radical of guanine

ring opening · reduction · Oxidation

7-Hydro-8-hydroxyguanine

8-hydroxyguanine

reduction · ring opening

2,6-diamino-4-hydroxy-5-formamidopyrimidine

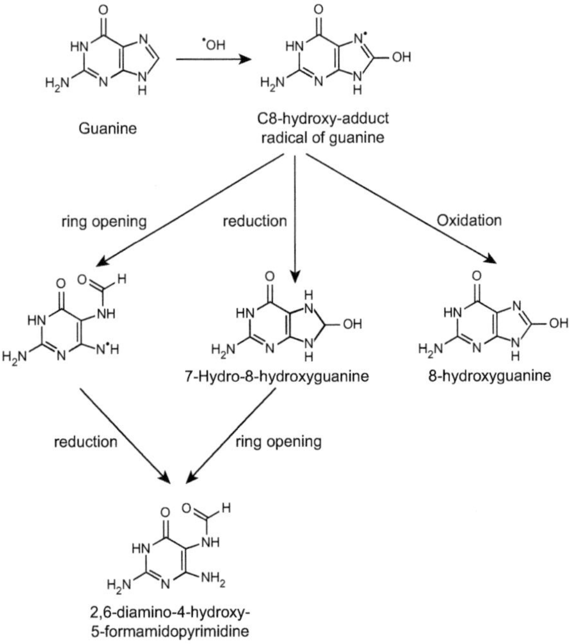

Scheme 1 Reaction of hydroxyl radical with guanine.

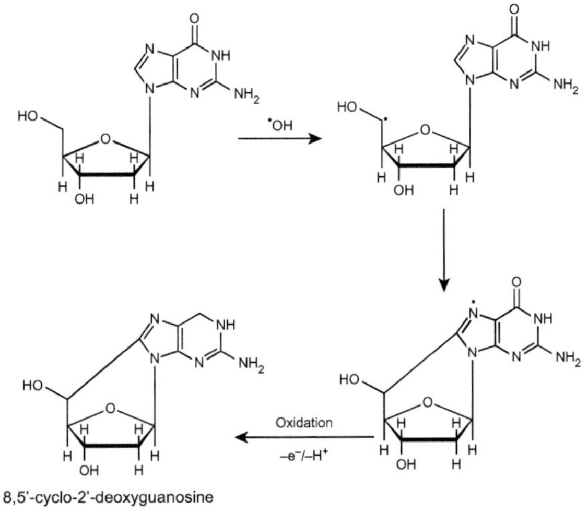

•OH

Oxidation $-e^-/-H^+$

8,5'-cyclo-2'-deoxyguanosine

Scheme 2 Reaction of hydroxyl radical with the sugar moiety of DNA.

TABLE 18.6 Protein amino acids and their ROS oxidized products.

Protein	ROS oxidized products
Tryptophan	Kitrotryptophan
	Kynurenine
	Formylkynurenine
Phenylalanine	2,3-Dihydroxyphenylalanine
	2-,3- and 4-hydroxyphenylalanine
Tyrosene	3,4-Dihydroxyphenylalanine
	Tyrosine-tyrosine cross linkages
	Tyr-O-Tyr
	Cross linked nitro-tyrosine
Histidine	2-Oxohistidine
	Asparagine
	Aspartic acid
Arginine	Glutamic semialdehyde
Lysine	α-aminoadipic semialdehyde
Proline	2-Pyrrolidone
	4-, and 5-hydroxyproline
	Pyroglutamic acid
Threonine	2-Amino-3-ketobutyric acid

18.6 Free radical reactions with lipids

Free radicals (R*) attack body lipids, including the polyunsaturated fatty acids (FAs) contained in cell membranes, breaking these membranes down and ultimately producing aldehydes (RCHO), including malondialdehyde. The sequence starts with the production of an FA radical (Eq. 18.7) and proceeds through the formation of an FA peroxyl radical (Eq. 18.8) to the initiation of a chain reaction that produces FA peroxide and the FA radical (Eq. 18.9) which then generates additional FA radical and FA peroxyl radical that perpetuate the chain reaction that leads to the formation of aldehydes (Eq. 18.10) (Nimse and Pal, 2015). Lipid peroxidation of cell membranes is associated with the onset of numerous diseases, as discussed in the ensuing chapters.

$$FA + R^* \rightarrow FA^* + RH \quad (18.7)$$

$$FA^* + O_2 \rightarrow FAOO^* \quad (18.8)$$

$$FAOO^* + FA \rightarrow FAOOH + FA^* \quad (18.9)$$

$$FOOH \rightarrow FAO^* + FAOO^* + RCHO \quad (18.10)$$

18.7 Free radical signaling

At low concentrations, free radicals have important roles as regulatory monitors in cell signaling processes. At elevated concentrations, however, they are harmful to cell organisms by inactivating cell molecules that are important to signaling. Representative examples of essential signaling functions of free radicals are listed in Table 18.7 (Di Meo and Venditti, 2020).

18.8 Diseases associated with free radicals

Numerous diseases and conditions in which free radicals play a role have been identified (Hallowell, 1994; Southorn, 1988; Sack et al., 2017; Di Meo and Venditti, 2020). Representative examples of these are listed Table 18.8.

Examples of the mechanisms associated with specific free radical-caused disease onset are presented in ensuing chapters.

18.9 Antioxidants

Antioxidants are compounds that modulate free radical reactions by neutralizing excessive free radical concentrations. This subject is explored in detail in part IV of this book.

18.10 Immune system and free radicals

The immune system can be damaged by absorption of exogenous chemicals associated with free radical production in the body, via immunosuppression. Free radicals produced as a result of chronic immune system responses can also induce disease via immunosuppression or by autoimmune attacks on body tissues.

TABLE 18.7 Examples of essential free radical signaling functions.

1. Induce the hydroxyl radical activation of guanylate cyclase, thus acting as a "second messenger"
2. NAD(P)H oxidase stimulated by insulin
3. Serve as mediators in the biosynthesis of prostaglandins
4. Contribute to blood pressure regulation
5. Contribute to cognitive function
6. Induce redox sensitive signal cascades that lead to increased expression of antioxidant enzymes

TABLE 18.8 Representative examples of diseases and conditions in which free radicals play a role.

Adult respiratory distress syndrome
Aging
Allergic encephalomyelitis
Alzheimer's disease
Amyotrophic lateral sclerosis
Asbestosis
Atherosclerosis
Burns
Cancer
Cataracts
Cerebrovascular damage
Contact dermatitis
Depression
Diabetes
Emphysema
Hearing loss
Hypertension
Immune disorders
Insulin resistance
Malaria
Muscular dystrophy
Myasthenia gravis
Pancreatitis
Parkinson's disease
Psoriasis
Psychosis
Renal diseases
Rheumatoid arthritis
Schizophrenia
Sickle cell anemia
Stroke
Systemic lupus
Ulcerative colitis

18.10.1 Exogenous immunocompromising agents

Exposures to exogenous substances which produce free radicals in the body can compromise the immune system by suppressing immune processes. Examples of exogenous immunosuppressant agents are listed in Table 18.9 (National Research Council, 1992; Naidensko et al., 2021).

TABLE 18.9 Exogenous immunosuppressant agents.

Aromatic hydrocarbons
Benzene
Toluene
Polynuclear aromatic hydrocarbons
Naphthalene
Benzo[α]pyrene
Halogenated polynuclear aromatic hydrocarbons
Dioxins
PCBs (polychlorinated biphenyls)
PGAS (polyfluoroalkyl substances
Pesticides
Chlordane
Malathion
Metals
Arsenic
Mercury
Particles
Asbestos
Silica
Food additives
FD&C colorants
Flavorings
Nonnutritive sweeteners
Preservatives

18.10.2 Endogenous immune system free radicals

Free radicals are naturally generated as a result of immune system reactions to physical, chemical, radiation, pathological and psychological challenges. For acute insults, such response is essential to reestablish homeostasis. Chronic attacks from these parameters, however, result in deleterious oxidative stress and increases in frequency and severity of noncommunicable and pathologically induced diseases (Aruoma, 1998; Okopi et al., 2021). Immunotoxins are associated with numerous effects. Table 18.10 lists representative examples of these (Dietert, 2014; Okopi et al., 2021).

Immunotoxins also stimulate free radical involvement in autoimmune diseases.

18.10.3 Autoimmune disease

Autoimmune diseases follow chronic exposures to parameters that are causative of free radical production (Ahsan et al., 2003). Examples are:

Infections
Inflammation
Pharmaceutical use
Environmental chemicals

18.11 Hypothalamus-pituitary-adrenal (HPA) axis

Neuroendocrine response to adverse stress stimuli activates the autonomic sympathetic nervous system, the direct neural innervation of the adrenal cortex and a cascade of hypothalamic hormonal messengers. Together, these lead to a free radical-producing hormone cascade that constitutes the hypothalamic-pituitary-adrenal (HPA) axis. The production and release of Cortisol, well known as the body's stress hormone, is the HPA axis's primary glutcocorticoid hormone. It induces free radical production directly via enhanced

TABLE 18.10 Representative immunotoxic effects.

Bioactivation of cytochrome P450
Lipid peroxidation induction
ATP production inhibition
Apoptosis of hematopoietic stem cells
Apoptosis of hematopoietic immune cells
Immunoglobulin alteration
Cell cycle alteration

mitochondrial activity and oxidative phosphorylation (Colaianna et al., 2013; Spiers et al., 2015; Prevatto et al., 2017).

Cortisol, a steroid hormone, is synthesized by the body from cholesterol. The structures of these are shown in Fig. 18.4. As can be seen from the extensive conjugation in the Cortisol molecule, resonance provides stabilization of a cortisol free radical.

As glucocorticoid receptors are present in almost all body tissues, cortisol affects numerous body systems (Kadmiel and Cidlowski, 2013; Thou et al., 2021). These include:

- Nervous
- Immune
- Cardiovascular
- Reproductive
- Musculoskeletal
- Integumentary

The pervasiveness of Cortisol in a body under chronic psychological stress explains the multitude of diseases attributable to increased oxidative stress due to psychological stress.

FIG. 18.4 Cholesterol and cortisol structures.

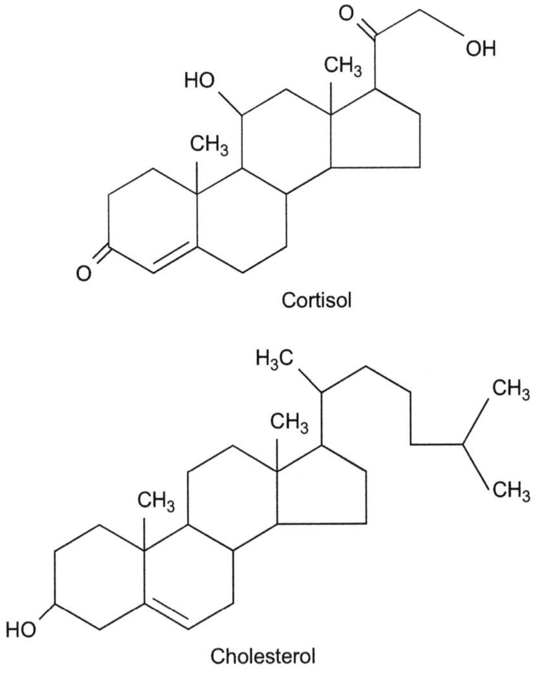

18.12 Summary

Free radicals, formed naturally in the body as part of the energy production process, is essential for life. Excessive quantities of free radicals in the body, no matter how formed, are detrimental to health and causative of numerous illnesses.

References

Ahsani, H., Ali, A., Ali, R., 2003. Oxygen free radicals and systemic autoimmunity. Clin. Exp. Immunol. 131, 398–404.

Arumoma, O.I., 1998. Free radicals, oxidative stress, and antioxidants in human health and disease. JAOCS (J. Am. Oil Chem. Soc.) 75 (2), 199–212.

Asmus, K.D., Bonifacic, M., 2000. Handbook of Oxidants and Antioxidants in Exercise. Elsevier, London.

Berlett, B.S., Stadtman, E.R., 1997. Protein oxidation in aging, disease, and oxidative stress. J. Biol. Chem. 272 (33), 20313–20316.

Brewer, G.J., 2007. Iron and copper toxicity in diseases of aging, particularly atherosclerosis and Alzheimer's disease. Exp. Biol. Med. 232 (2), 323–325.

Butterfield, D.A., Kippal, T., Howard, B., Subramaniam, R., Hall, N., Hensley, K., et al., 1988. Structural and functional changes in proteins induced by free radical-mediated oxidative stress and protective action of the antioxidants N-tert-butyl-alpha- phenylnitrone and vitamin E. Ann. N. Y. Acad. Sci. 854, 448–462.

Cai, H., Harrison, D.G., 2000. Endothelial dysfunction in cardiovascular diseases: the role of oxidant stress. Circ. Res. 87, 840–844.

Chung, S.W., Chung, H.Y., Toriba, A., Kameda, T., Tang, N., Kizu, R., Hayakawa, K., 2007. An environmental quinoid polycyclic aromatic hydrocarbon, ancenaphthenequinone, modulates cyclooxygenase-2 expression through reactive oxygen species generation and nuclear factor kappa B activation in A549 cells. Toxicol. Sci. 95 (2), 348–355.

Colaianna, M., Schiavone, S., Zotti, M., Tucci, P., Morgese, M.G., Backdahl, L., et al., 2013. Neuroendocrine profile in rat model of psychosocial stress: relation to oxidative stress. Antioxidants Redox Signal. 18 (12), 1385–1399.

Cooke, M.S., Evans, M.D., Dizdaroglu, M., Lunec, J., 2003. Oxidative DNA damage: mechanisms, mutation and disease. FASEB J. 17, 1195. https://doi.org/10.1096/fj-0752rev. (Accessed 15 August 2021).

Dazdaroglu, M., Jaruga, P., 2012. Mechanisms of free radical-induced damage to DNA. Free Radic. Res. 46 (4), 382–419.

Das, K., Roychoudhury, A., 2014. Reactive Oxygen Species (ROS) and Response of Antioxidants as ROS-Scavengers during Environmental Stress in Plants.

Dean, R.T., Fu, S., Stocker, R., Davies, M.J., 1997. Biochemistry and pathology or radical-mediated protein oxidation. Biochem. J. 324, 1–18.

Dellinger, B., Lomnicki, S., Khachatrytan, L., Maskos, Z., Hall, R.W., Adounkpe, J., et al., 2007. Proc. Combust. Inst. 31 (1), 521–528.

Di Meo, S., Venditti, P., 2020. Evolution of the Knowledge of Free Radicals and Other Oxidants. Oxidative Med Cell Long. https://doi.org/10.1155/2020/9829176. (Accessed 14 August 2021).

Dietert, R.R., 2014. Developmental immunotoxicity, programming, and noncommunicable diseases: focus on human studies. Adv. Met. Med. https://doi.org/10.1155/2014/867805. (Accessed 23 August 2021).

Hallowell, B., 1994. Free radicals, antioxidants, and human disease: curiosity, cause or consequence. Lancet 344 (8924), 721–724.

Kadmiel, M., Cidlowski, J.A., 2013. Glucocorticoid receptor signaling in health and disease. Pharmacol. Sci. 34 (9), 518–530.

Kikugawa, K., Kato, T., Okamoto, Y., 1994. Damage of amino acids and proteins induced by nitrogen dioxide, a free radical toxin in air. Free Radic. Biol. Med. 16 (3), 373–382.

Kohen, R., Nyska, A., 2002. Toxicol. Pathol. 30 (6), 620–650.

Lipinski, B., 2011. Hydroxyl radical and its scavengers in health and disease. Oxid. Med. Cell. Longev. https://doi.org/10.1155/2011/809696, 2001. (Accessed 12 August 2021).

Lobo, V., Patil, A., Phatak, A., Chandra, N., 2010. Free radicals, antioxidants and functional foods: impact on human health. Pharm. Rev. 4 (8), 118–126.

Naidensko, O.V., Andrews, D.Q., Temkin, A.M., Stoiber, T., Uche, I., Evans, S., Perrone-Gray, S., 2021. Investigating molecular mechanisms of immunotoxicity and the utility of ToxXast for immunotoxicity screening of chemicals added to food. Int. J. Environ. Res. Publ. Health. https://doi.org/10.3390/ijerph18073332, 2021. (Accessed 23 August 2021).

National Research Council, 1992. Subcommittee on Immunotoxicology. Committee on Biologic Markers. Board on Environmental Studies and Toxicology, Commission on Life Sciences. National Research Council. Biologic Markers in Immunotoxicology. National Academmy Press, Washington, D.C. http://nap.edu/catalog/1591.html. (Accessed 23 August 2021).

Nimse, S.B., Pal, D., 2015. Free radicals, natural antioxidants, and their reaction mechanisms. RSC Adv. 5, 27986–28006.

Okapi, A., Isaac, O.T., Agi, A.B., 2021. Oxidative stress in immunotoxity: a biochemical foe or friend? Int. J. Innov. Sci. Res. Technol. 6 (3), 1116–1121.

Pham-Huy, L.I., He, H., Pham-Huy, C., 2015. Free radicals, antioxidants in disease and health. Int. M. Biomed. Sci. 4 (2), 89–96.

Phaniendra, A., Jestadi, D.B., Periyasamy, L., 2015. Free radicals: properties, sources, targets and their implication in various diseases. Ind. J. Clin. Biochem 30 (1), 11–26.

Prevatto, J.P., Torres, R.C., Diaz, B.L., Silva, M.R., Martins, M.A., Carvalho, V.F., 2017. Antioxidant treatment induces hyperactivation of HPA axis by upregulating ACTH receptor in the adrenal and downregulating glucocorticoid receptors in the pituitary. Oxid. Med. Cell. Longev. https://doi.org/10.1155/2017/4156361. (Accessed 23 August 2021).

Rao, K.S., 2009. Free radical induced oxidative stress to DNA: relation to brain aging and neurological disorders. Ind. J Biochem Biophys 46, 9–15.

Rynkowska, A., Stepniak, J., Karbonik-Lewinska, M., 2020. Fenton reaction-induced oxidative damage to membrane lipids and protective effects of 17β-estradiol in porcine ovary and thyroid homogenates. Int. J. Environ. Res. Publ. Health 17, 6841. https://doi.org/10.3390/ijerph17186841. (Accessed 12 August 2021).

Sack, M.N., Fyrquist, F.Y., Saijonmaa, O.J., Fuster, V., Kovacic, J.C., 2017. Basic biology of oxidative stress and the cardiovascular system. J. Am. Coll. Cardiol. 70 (2), 196–211.

Smith, C.D., Carney, J.M., Starke-Reed, P.E., Oliver, C.N., Stadtman, E.R., Floyd, R.A., Markesbery, W.R., 1991. Excess brain protein oxidation and enzyme dysfunction in normal aging and in Alzheimer disease. Proc. Nat. Acad. Sci. USA 88, 10540–10543.

Song, Y., Wagner, B.A., Lehmer, H.J., Buettner, G.R., 2008. Semiquinone radicals from oxygenated polychlorinated biphenyls: electron paramagnetic studies. Chem. Res. Toxicol. 21, 1359–1367.

Southorn, P.A., 1988. Free radicals in medicine. II. Involvement in human disease. Mayo Clin. Proc. 63, 390–408.

Spiers, J.G., Chen, H.S.C., Sernia, C., Lavidis, N.A., 2015. Activation of the hypothalamic-pituitary-adrenal stress axis induces cellular oxidative stress. Front. Neurosci. 8. https://doi.org/10.3389/fnins.2014.00456. (Accessed 23 August 2021).

Taverne, Y.J.H.J., Bogers, A.J.J.C., Duncker, D.J., Merkus, D., 2013. Reactive oxygen species and the cardiovascular system. Oxid. Med. Cell. Longev. https://doi.org/10.1155/2013/862423, 2013. (Accessed 17 August 2021).

Thou, L., Gandhi, J., Sharma, S., 2021. Physiology, Cortisol. National Library of Medicine, National Institutes of Health (Accessed 27 August 2021).

Vejerano, E.P., Ahn, J., Latif, J., Mamun, M., 2021. Environmentally persistent free radicals as sources of POPs. Adv. Environ. Eng. Res. 2 (2), 14. https://doi.org/10.21926/aeer.2102010. (Accessed 12 August 2021).

Winterbourn, C.C., 1995. Toxicity of iron and hydrogen peroxide: the Fenton reaction. Toxicol. Lett. 82–83, 969–974.

Zhao, Z., 2019. Iron and oxidizing species in oxidative stress and Alzheimer's disease. Aging Med. 2, 82–87.

Zipse, H., 2006. Radical stability - a theoretical perspective. Top. Curr. Chem. 263, 163–189.

PART II

Mechanisms of oxidative stress driven disease

CHAPTER

19

Aging mechanism

19.1 Introduction

The world's population is increasing exponentially. Population has increased from one billion in 1800 to more than eight billion currently. The world is also aging and with age comes increased chronic disease. Though a harbinger of chronic disease onset, aging itself is not a disease. Rather, it is a normal function of life that is characterized by a progressive loss of physiological integrity, loss of tissue and organ function with time ultimately leading to the onset of chronic conditions that include cancer, cardiovascular diseases, diabetes, autoimmune diseases, neurodegenerative diseases, chronic obstructive pulmonary disease (COPD), chronic kidney disease, reduced immune function, numerous other diseases and finally death (Flatt, 2012; Montecino-Rodriguez et al., 2013; Aunan et al., 2016; Liguori et al., 2018; Muller et al., 2019).

Oxidative stress accelerates aging, and the oxidative stress theory of aging is based upon the premise that age-associated functional losses ensue following the accumulation of oxidative damage to DNA, proteins and lipids by chronic interactions with reactive oxygen and reactive nitrogen species (ROS and RNS) (Beckman and Ames, 1998; Lopez-Otin et al., 2013; Liguori et al., 2018). As discussed in the earlier chapters, increased oxidative stress from all sources is responsible for increased disease prevalence. This chapter is dedicated to a discussion of the aging process, the mechanisms that drive it and role oxidative stress plays in these.

19.2 Hallmarks of aging

There are nine hallmarks of aging (Lopez-Otin et al., 2013; Guerville et al., 2020). These are listed in Table 19.1.

A discussion of each of the parameters in Table 19.1 follows.

19.2.1 Genomic instability

Genetic damage amassed throughout life is a major contributor to aging. DNA is challenged throughout life by endogenous and exogenous sources, representatives of which are listed in Table 19.2 (Hakem, 2008; Hoeijmakers, 2009; Moskalev et al., 2013; Barnes et al., 2018).

Oxidative Stress
https://doi.org/10.1016/B978-0-323-91890-9.00002-7

TABLE 19.1 Hallmarks of aging.

Hallmarks of aging
Genomic instability
Telomere shortening
Epigenetic alterations
Loss of proteostasis
Deregulated nutrient sensing
Mitochondrial dysfunction
Cellular senescence
Stem cell exhaustion
Altered intercellular communication

TABLE 19.2 Endogenous and exogenous sources of DNA damage.

Endogenous
Hydrolysis
Oxidation
Alkylation
Mismatch of DNA bases
Exogenous
Ionizing radiation
Ultraviolet radiation
Particulate matter
Polynuclear aromatic hydrocarbons
Heterocyclic aromatic amines
Mycotoxins
Nitrosamines
Physical agents
Biological agents

These challenges result in point mutations, translocations, chromosomal gains and losses and other effects (Lopez-Otin et al., 2013). The body repairs these via DNA repair mechanisms, but with time, its ability to do so is overcome, resulting in genomic damage. Compelling evidence exists that excessive genomic damage accelerates aging (Lopez-Otin et al., 2013).

19.2.2 Telomere shortening

With aging, replicative DNA polymerases lose the ability to completely replicate telomeres, the terminal ends of linear DNA molecules. In healthy individuals, this loss is gradual, but in those with exogenously induced DNA damage, the loss is accelerated. Shortened telomeres have been shown to be associated with the onset of age-related disease and shortened lifespan (Armanios et al., 2009; Fumagalli et al., 2012; Hewitt et al., 2012).

Numerous studies have demonstrated that telomere shortening is associated with oxidative stress (Reichert and Stier, 2017; Barnes et al., 2019) and with the onset of numerous diseases, representative examples of which are listed in Table 19.3 (Gavia-Garcia et al., 2021).

19.2.3 Epigenetic alterations

All tissues and systems are continually affected by epigenetic alteration (Talens et al., 2012). These include histone modifications (Sanders et al., 2013; Niu et al., 2015), DNA methylation (Niu et al., 2015; Menezo et al., 2016) and chromatin remodeling, all of these are associated with premature aging and oxidative stress (Lopez-Ortin et al., 2013; Cencioni et al., 2013; Guillaumet-Adkins et al., 2017).

During aging, histones undergo a wide variety of modifications that impact chromatin structure, influence gene expression and affect genome stability and replication Mccauley and Dang (2014).

Aging brings with it a global reduction in DNA methylation and hypermethylation of specific genes (Heyn et al., 2012; Gentilini et al., 2013).

Chromatin remodeling factors are diminished during aging (Pegoraro et al., 2009; Pollina and Brunet, 2011).

TABLE 19.3 Diseases associated with telomere shortening.

Type 2 diabetes
Metabolic syndrome (Met-S)
Cardiovascular diseases
Myocardial infarction
Alzheimer's disease
Osteoporosis
Cataractogenesis

The combination of chromatin remodeling, histone modifications and DNA-methylation alterations are characteristics of aging (Tsurumi and Li, 2012). These oxidative stress effects combine to mediate cell changes which include modulation of gene expression, apoptosis, cell survival and mutagenesis, which are disease driving mechanisms (Kreuz and Fischle, 2016).

19.2.4 Loss of proteostasis

The cellular proteome (all body proteins) is tightly regulated to assure that every protein is correctly synthesized, folded and subcompartmentalized (Balch et al., 2008; Hutt et al., 2009). Several cellular systems assure protein quality control. The failure of these systems negatively impacts cellular homeostasis and functioning via transformation of proteins into cytotoxic products and leading to aging, neurodegenerative diseases, metabolic disorders, myopathies, liver diseases and other systematic disorders (Esser et al., 2004; Morimoto, 2008; Balch et al., 2008).

Proteostasis covers the mechanisms responsible for stabilizing the correct folding of proteins. Numerous studies have identified loss of proteostasis with aging as oxidative stress-driven (Kopito, 2000; Ravikumar et al., 2002; Kourie and Henry, 2001; Munoz, 2003; Olsen et al., 2006 Koga et al., 2011; Korovila et al., 2017).

More than 200 non-enzymatic post-translational protein modifications have been identified, with the degree of oxidative damage both time and dose dependent. High doses and chronic exposures may occur not only via direct ROS action, but also via overwhelming of natural anti-oxidant systems in the body (Korovila et al., 2017).

19.2.5 Deregulated nutrient sensing

Nutrient sensing deregulation is a major contributor in the aging process. First demonstrated in 1935 (McCay et al., 1935), it has been repeatedly demonstrated that reduced caloric intake (without malnutrition) is a major component of nutrient sensing in aging (Fontana et al., 2010; Lopez-Otin et al., 2013; Fontana and Partridge, 2015; Cummings and Lamming, 2017; Falcon et al., 2019). Calorie restriction has also been demonstrated as achievable via diet composition. Hence, nutrient and amino acid sensing mechanisms have been shown to obtain throughout life (Fontana et al., 2010). Interruptions in these mechanisms due to oxidative stress are found in deregulated nutrient sensing, with corresponding impact on longevity (Falcon et al., 2019).

19.2.6 Mitochondrial dysfunction

Mitochondrial function profoundly impacts the aging process. With aging, the efficiency of cellular energy production diminishes, resulting in increased leakage of mitochondrial-generated free radicals which, in turn, lead to ROS production and result in increased oxidative stress (Lopez-Otin et al., 2013; Kudryavtseva et al., 2016).

The mitochondrial free radical theory of aging, first offered in 1965, proposes that the ROS so formed causes further mitochondrial deterioration and global cellular damage. Though numerous studies support the role of ROS in aging, subsequent findings have led to reconsideration of this theory (Hekimi et al., 2016; Ristow and Schmeisser, 2011). Four findings, covering positive, negative and neutral affects of ROS on aging have raised this uncertainty:

1. It has been found that ROS may prolong lifespan in yeast (Mesquita et al., 2010).
2. Genetically manipulated increases of mitochondrial ROS and oxidative damage in mice do not accelerate aging (Van Remmen et al., 2001; Zhang et al., 2009).
3. Increased antioxidant defenses do not increase longevity (Perez et al., 2009).
4. Genetic manipulations that decrease mitochondrial function, but do not increase ROS, do accelerate aging (Vermulst et al., 2008; Hiona et al., 2010).

These seemingly contradictory findings, however, are not mutually exclusive. Other research has demonstrated the role of ROS as a stress-related survival signal aimed at compensating for progressive age-associated mitochondrial decline (Hekimi et al., 2016). Combined, the following two points allow for accommodation of both the mitochondrial free radical theory of aging and the seemingly contradictory results (Lopez-Otin et al., 2013).

1. Aging increases ROS as a survival-maintaining attempt.
2. When ROS levels increase high enough, rather than alleviate, they aggravate Age-associated damage.

19.2.7 Cellular senescence

Cellular senescence (CS) is a state in which cells stop dividing and undergo stereotypical phenotypic alterations that include chromatin and secretome changes as well as tumor suppressor activation (Lopez-Otin et al., 2013; van Deursen, 2014). The term cellular senescence was first introduced in 1961 to describe permanent growth arrest of primary human cells (Hayflick and Moorhead, 1961). Since then, cellular senescence has been shown to be associated with the following parameters:

1. Acting as a safeguard against cancer (Serrano, 1997; Campisi and di Fagagna, 2007).
2. Being positively tied it to embryonic development (Munoz-Espin, 2013; Storer et al., 2013).
3. Participating in wound healing (Jun and Lau, 2010).
4. Participating in tissue repair (Yuh, 2018).
5. Being associated with organismal aging (Baker et al., 2011).

Currently, cellular senescence is being widely investigated as an ameliorating agent for a spectrum of human conditions (Di Micco et al., 2020).

Both endogenous and exogenous sources of oxidative stress have been associated with cellular senescence (Pole et al., 2016; Di Micco et al., 2020).

19.2.8 Stem cell exhaustion

Stem cells, which repair tissue, maintain homeostasis and persist throughout life, are also subject to oxidative stress-driven epigenetic modification (Chen and Kerr, 2019). Adult stem cells are vital in preventing aging or organs and tissues. As aging proceeds, stem cells undergo three detrimental changes. These include:

Micro-environment alterations.
Decline of regenerative capacity.
Loss of function.

Their life-long persistence in the body, leave stem cells vulnerable to accumulation of cellular damage, exhaustion and loss of function (Lopez-Otin, 2013; Oh et al., 2014; Boyette and Tuan, 2014; Sameri et al., 2020).

Studies of human mesenchymal cells have been found to have elevated ROS (Stolzing et al., 2008), and ROS generation has been shown to promote stem cell aging (Pervaiz et al., 2009; Chen et al., 2017).

Stem cell exhaustion is considered one of the ultimate culprits of aging (Lopez-Otin, 2013).

19.2.9 Altered intercellular communication

Intercellular communication is a vital part of homeostasis. With aging, the body's signaling environment becomes more inflammatory in a process known as inflammaging, five causes of which have been identified (Salminen et al., 2012):

1. Accumulation of pro-inflammatory tissue damage.
2. Immune system failure to effectively clear pathogens and dysfunctional host cells.
3. Secretion of pro-inflammatory cytokines.
4. Enhanced activation of the NF-kB transcription factor.
5. Occurrence of a defective autophagy response.

In addition to inflammaging, intercellular communication can be altered by contagious aging, or bystander effects, in which senescent cells induce senescence in neighboring cells via ROS mediation (Nelson et al., 2012). Aging also affects endocrine, neuroendocrine and neuronal intercellular communication (Russell and Kahn, 2007; Lopez-Otin, 2013; Zhang et al., 2013).

19.3 Summary

The nine hallmarks of aging discussed above have been grouped into three categories (Lopez-Otin, 2013):

Primary
Antagonistic
Integrative

The primary hallmarks, all of which are negative, include:

Genomic instability
Telomere shortening
Epigenetic alterations
Loss of proteostasis

The antagonistic hallmarks, which are due to responses to the primary hallmarks, include:

Deregulated nutrient sensing
Mitochondrial dysfunction
Cellular senescence

The integrative hallmarks, which directly affect homeostasis and function, include:

Stem cell exhaustion
Altered intercellular communication

Oxidative stress is detrimental to all nine hallmarks of aging and lowering of OS has been shown to ameliorate their negative effects by delaying aging slowing down the onset of chronic disease. Life can also be prolonged via taking steps to safeguard the body's antioxidants which are vital for homeostasis. These are discussed in Part IV of this book.

References

Armanios, M., Adler, J.K., Parry, E.M., Karim, B., Strong, M.A., Greider, C.W., 2009. Short telomeres are sufficient to cause the degenerative defects associated with aging. Am. J. Hum. Genet. 85 (6), 823–832.

Aunan, J.R., Watson, M.M., Hagland, H.R., Soreide, K., 2016. Molecular and biological hallmarks of ageing. BJS 103, e29–e46. (Accessed 5 September 2021).

Baker, D.J., Wijshake, T., Tchkonia, T., LeBrasseur, N.K., Childs, B.G., van de Sluis, B., et al., 2011. Clearance of p16INK4a – positive senescent cells delays ageing-associated disorders. Nature 479, 232–236.

Balch, W.E., Morimoto, R.I., Dillin, A., Kelly, J.W., 2008. Adapting proteostasis for disease intervention. Science 319 (5865), 916–919.

Barnes, R.P., Fouquerel, E., Opresko, P.L., 2019. Mech. Ageing Dev. 177, 37–45.

Barnes, J.L., Zubair, M., John, K., Poirier, M.C., Martin, F.L., 2018. Carcinogens and DNA damage. Biochem. Soc. Trans. 46 (5), 1213–1224.

Beckman, K.B., Ames, B.N., 1998. The free radical theory of aging matures. Physiol. Rev. 78 (2), 547–581.

Boyette, L.B., Tuan, R.S., 2014. Adult stem cells and diseases of aging. J. Clin. Med. 3, 88–134.

Campisi, J., di Fagnana, F., 2007. Cellular senescence: when bad things happen to good cells. Nat. Rev. Cell. Biol. 8, 729–740.

Cencioni, C., Spallotta, F., Martelli, F., Valente, S., Mai, A., Zeiher, A.M., Gaetano, C., 2013. Oxidative stress and epigenetic regulation in ageing and age-related diseases. Int J Mol Sci. 14, 17643–17663.

Chen, D., Kerr, C., 2019. The epigenetics of stem cell aging comes of age. Trends Cell Biol. 29 (7), 563–568.

Chen, F., Liu, Y., Wong, N.K., Xiao, J., So, K.F., 2017. Oxidative stress in stem cell aging. Cell Transplant. 26 (9), 1483–1495.

Cummings, N.E., Lamming, D.W., 2017. Regulation of metabolic health and aging by nutrient-sensitive signaling pathways. Mol. Cell. Endocrinol. 455, 13–22.

Di Micco, R., Krizanovsky, V., Baker, D., di Fagagna, F., 2020. Cellular senescence in ageing: from mechanisms to therapeutic opportunities. Nat. Rev. Mol. Cell Biol. 22, 75–95.

Esser, C., Alberti, S., Hohfeld, J., 2004. Cooperation of molecular chaperones with the ubiquitin/proteasome system. Biochim. Biosphys. Acta 1695 (1–3), 171–188.

Falcon, P., Escandon, M., Brito, A., Matus, S., 2019. Nutrient sensing and redox balance: GCN2 as a new integrator in aging. Oxid. Med. Cell. Longev. https://doi.org/10.1155/2019/5730532. (Accessed 2 September 2021).

Flatt, T., 2012. A new definition of aging? Front. Genet. 3, 148.

Fontana, L., Partridge, L., 2015. Promoting health and longevity through diet: from model organisms to humans. Cell 161 (1), 106–118.

Fontana, L., Partridge, L., Longo, V.D., 2010. Dietary restriction, growth factors and aging: from yeast to humans. Science 328 (5976), 321–326.

Fumagalli, M., Rossiello, F., Clerici, M., Barozzi, S., Cittaro, D., Kaplunov, J.M., et al., 2012. Telomeric DNA damage is irreparable and causes persistent DNA-damage-response activation. Nat. Cell Biol. 14 (4), 355–365.

Gavia-Garcia, G., Rosado-Perez, J., Arista-Ugalde, L., Aguiniga-Sanchez, I., Santiago-Osorio, E., Mendoza-Nunez, V.M., 2021. Telomere length and oxidative stress and its relation with metabolic syndrome components in the aging. Biology 10, 253. https://doi.org/10.3390/biology/10040253. (Accessed 31 August 2021).

Gentilini, D., Mari, D., Castaldi, D., Remondini, D., Ogliari, G., Ostan, R., et al., 2013. Role of epigenetics in human aging and longevity: genome-wide DNA methylation profile in centenarians and centenarians' offspring. Age 35, 1961–1973.

Guerville, F., Barreto, P.D.S., Ader, I., Andrieu, S., Casteilla, L., Dray, D., et al., 2020. Revisiting the hallmarks of aging to identify markers of biological age. J. Prev. Alzheimer's Dis. 7 (1), 56–64.

Guillaumet-Adkins, A., Yanez, Y., Peris-Diaz, M.D., Calabria, I., Palanca-Ballester, C., Sandoval, J., 2017. Epigenetics and oxidative stress in aging. Oxid. Med. Cell. Longev. https://doi.org/10.1155/2017/9175806. (Accessed 31 August 2021).

Hakem, R., 2008. DNA-damage repair; the good, the bad, and the ugly. Euro. Mol. Biol. Org. J. 27 (4), 589–605.

Hayflick, L., Moorhead, P.S., 1961. The serial cultivation of human diploid cell strains. Exp. Cell Res. 25, 585–621.

Hekimi, S., Wang, Y., Noe, A., 2016. Mitochondrial ROS and the effectors of the intrinsic apoptotic pathway in aging cells: the discerning killers! Front. Genet. 7, 161. https://doi.org/10.3389/fgene.2016.00161.

Hewitt, G., Jurk, D., Marques, F.D.M., Correia-Melo, C., Hardy, T., Gackowska, A., et al., 2012. Telomeres are favoured targets of a persistent DNA damage response in ageing and stress-induced senescence. Nat. Commun. 3, 708. https://doi.org/10.1038/ncomms1708. (Accessed 31 August 2021).

Heyn, H., Li, N., Ferreira, H.J., Moran, S., Pisano, D.G., Gomez, A., et al., 2012. Distinct DNA methylomes of newborns and centenarians. Proc. NY Acad. Sci. 109 (26), 10522–10527.

Hiona, H., Sanz, A., Kujoth, G.C., Pamplona, R., Sao, A.Y., Hofer, T., et al., 2010. Mitochondrial DNA mutations induce mitochondrial dysfunction, apoptosis and sarcopenia in skeletal muscle of mitochondrial DNA mutator mice. PLoS One. https://doi.org/10.1371/journal.pone.0011468. (Accessed 3 September 2021).

Hoeijmakers, J.H.J., 2009. DNA damage, aging and cancer. N. Engl. J. Med. 361, 1475–1485.

Hutt, D.M., Powers, E.T., Balch, W.E., 2009. The proteostasis boundary in misfolding diseases of membrane traffic. FEBS Lett. 583 (16), 2639–2646.

Jun, I.L., Lau, L.F., 2010. Cellular senescence controls fibrosis in would healing. Aging 2 (9), 627–631.

Koga, H., Kaushik, S., Cuervo, A.M., 2011. Protein homeostasis and aging: the importance of exquisite quality control. Ageing Res. Rev. 10 (2), 205–215.

Kopito, R., 2000. Aggresomes, inclusion bodies and protein aggregation. Trends Cell Biol. 10 (12), 523–530.

Korovila, I., Hugo, M., Castro, J.P., Weber, D., Hohn, A., Grune, T., Jung, T., 2017. Proteostasis, oxidative stress and aging. Redox Biol. 13, 550–567.

Kourie, J.I., Henry, C.L., 2001. Protein aggregation and deposition: implications for ion channel formation and membrane damage. Croat. Med. J. 42 (4), 359–374.

Kreuz, S., Fischle, W., 2016. Oxidative stress signaling to chromatin in health and disease. Epigenomics 8 (6), 843–862.

Kudryavtseva, A.V., Krasnov, G.S., Dmitriev, A.A., Alekseev, B.Y., Sadritdinova, A.F., Federova, M.S., et al., 2016. Mitochondrial dysfunction and oxidative stress in aging and cancer. Oncotarget 7 (29), 44879–44905.

Liguori, I., Russo, G., Bulli, G., Aran, L., Della-Morte, D., Gaetano, G., et al., 2018. Oxidative stress, aging and diseases. Clin. Interv. Aging 13, 757–772.

Lopez-Otin, C., Blasco, M.A., Partridge, L., Serrano, M., Kroemer, G., 2013. The hallmarks of aging. Cell 153 (6), 1194–1217.

Mccauley, B.S., Dang, W., 2014. Histone methylation and aging: lessons learned from model systems. Biochim. Biophys. Acta 1839, 1454–1462.

McCay, C.M., Crowell, M.F., Maynard, L.A., 1935. The effect of retarded growth upon the length of life span and upon the ultimate body size. J. Nutr. 10 (1), 63–79.

Menezo, Y.J.R., Silvestris, E., Dale, B., Elder, K., 2016. Oxidative stress and alterations in DNA methylation: two side of the same coin in reproduction. Reprod. Biomed. 33 (6), 668–683.

Mesquita, A., Weinberger, M., Silva, A., Sampaio-Marques, B., Almeida, B., Leao, C., et al., 2010. Caloric restriction or catalase inactivation extends yeast chronological lifespan by inducing H202 and superoxide dismutase activity. Proc. Natl. Acad. Sci. USA 107 (34), 15123–15128.

Montecino-Rodriguez, E., Berent-Maoz, B., Dorshkind, K., 2013. Causes, consequences, and reversal of immune system aging. J. Clin. Invest. 123 (3), 958–965.

Morimoto, R.I., 2008. Proteotoxic stress and inducible chaperone networks in neurodegenerative disease and aging. Genes Dev. 22 (11), 1427–1438.

Moskalev, A.A., Shaposhnikov, M.V., Plyusnina, E.N., Zhavoronkov, A., Budovsky, A., Vanai, H., Fraifeld, V.E., 2013. Ageing Res. Rev. 12 (2), 661–684.
Muller, L., Di Benedetto, S., Pawelec, G., 2019. The immune system and its dysregulation in aging. Sucell. Biochem. 91, 21–43.
Munoz, M.J., 2003. Longevity and heat stress regulation in Caenorhabditis elegans. Mech. Ageing Dev. 124 (1), 43–48.
Munoz-Espin, D., Canamero, M., Maraver, A., Gomez-Lopez, G., Contreras, J., Murillo-Cuesta, S., et al., 2013. Programmed cell senescence during mammalian embryonic development. Cell 155 (5), 1104–1118.
Nelson, G., Wordsworth, J., Wang, C., Jurk, D., Lawless, C., Martin-Ruiz, C., et al., 2012. A senescent cell bystander effect: senescence-induced senescence. Aging Cell 11 (2), 345–349.
Niu, Y., DesMarais, T.L., Tong, Z., Yao, Y., Costa, M., 2015. Oxidative stress alters global histone modification and DNA methylation. Free Rad. Biol. Med. 82, 22–28.
Oh, J., Lee, Y.D., Wagers, A.J., 2014. Stemcell aging: mechanisms, regulators and therapeutic opportunities. Nat Med 20 (8), 870–880.
Olsen, A., Vantipalli, M.C., Lithgow, G.J., 2006. Checkpoint proteins control survival of the postmitotic cells in Caenorhabditis elegans. Science 312 (5778), 1381–1385.
Pegoraro, G., Kubben, N., Wickert, Y., Gohler, H., Hoffmann, K., Misteli, T., 2009. Aging-related chromatin defects via loss of NURD complex. Nat. Cell Biol. 11 (10), 1261–1267.
Perez, V.I., Van Remmen, H., Bokov, A., Epstein, C.J., Vijg, J., Richardson, A., 2009. The overexpression of major antioxidant enzymes does not extend the lifespan of mice. Aging Cell 8 (1), 73–75.
Pervaiz, S., Taneja, R., Ghafarri, S., 2009. Oxidative stress regulation of stem progenitor cells. Antioxid. Redox Signal. 11 (11), 2777–2789.
Pole, A., Dimri, M., Dimri, G.P., 2016. Oxidative stress, cellular senescence and ageing. AIMS Mol. Sci. 3 (3), 300–324.
Pollina, E.A., Brunet, A., 2011. Epigenetic regulation of aging stem cells. Oncogene 30 (28), 3105–3126.
Ravinkumar, B., Duden, R., Rubensztein, D.C., 2002. Aggregate-prone proteins with polyglutamine and polyalanine expansions are degraded by autophagy. Hum. Mol. Genet. 11 (9), 1107–1117.
Reichert, S., Stier, A., 2017. Does oxidative stress shorten telomeres in vivo? A review. Biol. Lett. 13, 201704463. https://doi.org/10.1098/rsbl.2017.0463. (Accessed 31 August 2021).
Ristow, M., Schmeisser, S., 2011. Extending lifespan by increasing oxidative stress. Free Rad. Biol. Med. 51, 327–336.
Russell, S.J., Kahn, C.R., 2007. Endocrine regulation of aging. Nat. Rev. Cell. Biol. 8 (9), 681–691.
Salminen, A., Kaarniranta, K., Kauppinen, A., 2012. Inflammaging: disturbed interplay between autophagy and inflammasomes. Aging 4 (3), 166–175.
Sameri, S., Samadi, P., Dehghan, R., Salem, E., Fayazi, N., Amini, R., 2020. Stem cell aging in lifespan and disease: a state-of-the-art review. Curr. Stem Cell Res. Ther. 15 (4), 362–378.
Storer, M., Mas, A., Robert-Moreno, A., Pecoraro, M., Ortells, M.C., Di Giacomo, V., et al., 2013. Senescence is a developmental mechanism that contributes to embryonic growth and patterning. Cell 155 (5), 1119–1130.
Sanders, Y.Y., Liu, H., Zhang, X., Hecker, L., Bernard, K., Desai, L., et al., 2013. Histone modifications in senescence-associated resistance it apoptosis by oxidative stress. Redox Biol. 1, 8–16.
Serrano, M., Lin, A.W., McCurrah, M.E., Beach, D., Lowe, S.W., 1997. Oncogenic ras provokes premature cell senescence associated with accumulation of p53 and p16INK4a. Cell 88 (5), 593–602.
Stolzing, A., Jones, E., McGonagle, D., Scott, A., 2008. Age related changes in human bone marrow-derived mesenchymal stem cells: consequences for cell therapies. Mech. Ageing Dev. 129 (3), 163–173.
Storer, M., Mas, A., Robert-Moreno, A., Pecoraro, M., Ortells, M.C., Di Giacomo, V., et al., 2013. Senescence is a developmental mechanism that contributes to embryonic growth and patterning. Cell 155 (5), 1119–1130.
Talens, R.P., Christensen, K., Putter, H., Willemsen, G., Christiansen, L., Kremer, D., et al., 2012. Epigenetic variation during the adult lifespan: cross-sectional and longitudinal data on monozygotic twin pairs. Aging Cell 11, 694–703.
Trifunovic, A., Wredenberg, A., Falkenberg, M., Spelbrink, J.N., Bruder, C.E., Bohlooly, Y.M., et al., 2004. Premature aging in mice expressing defective mitochondrial DNA polymerase. Nature 429 (6990), 417–423.
Tsurumi, A., Li, W.X., 2012. Global heterochromatin loss. Epigenetics 7 (7), 680–688.
Van Deursen, J.M., 2014. The role of senescent cells in ageing. Nature 509 (7501), 439–446.
Van Remmen, H., Richardson, A., 2001. Oxidative damage to mitochondria and aging. Exp. Gerentol. 36 (7), 957–968.

Vermulst, M., Wanagat, J., Kujoth, G.C., Bielas, J.H., Rabinovich, P.S., Prolla, T.A., et al., 2008. Nat. Genet. 40 (4), 292–394.

Yuh, M.H., 2018. Cellular senescence in tissue repair: every cloud has a silver lining. Int. J. Dev. Biol. 62 (6–7–8), 591–604.

Zhang, Y., Ikeno, Y., Qi, W., Chaudhari, A., Li, Y., Bokov, A., et al., 2009. Mice deficient in both Mn superoxide dismutase and glutathione peroxidase-1 have increased oxidative damage and greater incidence of pathology but no reduction in longevity. J. Gerentol A Biol. Med. Sci. 64 (12), 1212–1220.

Zhang, G., Li, J., Purkayastha, S., Tang, Y., Zhang, H., Yin, Y., et al., 2013. Hypothalamic programming of systematic ageing involving IKK-beta, NF-kappaB and GnRH. Nature 497 (7448), 211–216.

CHAPTER

20

Obesity

20.1 Introduction

Though long thought to be a condition caused by insufficient willpower, lack of discipline and bad choices, obesity is now known to be a "chronic, relapsing, multifactorial, neurobehavioral disease, wherein an increase in body fat promotes adipose tissue dysfunction and abnormal fat mass physical forces, and resulting in adverse metabolic, biomechanical and psychosocial health consequences" (OMA, 2021). Obesity is an oxidative stress-mediated chronic disease that affects numerous people worldwide and is co-morbid with numerous other diseases.

An understanding of numerous human diseases cannot be complete without first addressing obesity; its mechanisms, comorbidities, impact as a storehouse of disease inducing toxins and association with oxidative stress.

20.2 Statistics

Overweight and obesity are defined by body mass index (BMI), overweight people being those with a BMI of 25–30 and obese individuals those with a BMI greater than 30 (WHO, 2021).

The World Health Organization has prepared the following statistics regarding obesity (WHO, 2021):

- Obesity in the world has nearly tripled since 1975,.
- In 2016, 39% (1.9 billion) of adults aged 18 and up were overweight and 13% of these (over 650 million) were obese.
- In 2020, 39 million children under the age of five were overweight or obese.
- In 2016, over 340 million children and adolescents aged 5–19 were overweight or obese.
- Obesity is largely preventable.

In the United States, the Centers for Disease Control and Prevention have produced the following statistics regarding obesity (CDC, 2021a, 2021b; Fryar et al., 2021).

Oxidative Stress
https://doi.org/10.1016/B978-0-323-91890-9.00004-0

- In the years 2012–18, the obesity prevalence rate was 42.4%.
- In the time span 1999–2000 through 2017–18, obesity prevalence increased from 30.5% to 42.4%. During that same time, severe obesity (BMI greater than 40) increased from 4.7% to 9.2%.
- Obesity affects some groups more than others, with non-Hispanic Black adults having the highest age-adjusted obesity prevalence (49.65), followed by Hispanic adults (44.8%), non-Hispanic White adults (42.2%) and non-Hispanic Asian adults (17.4%).
- Obesity rates in adults are fairly uniform in different age groups. Obesity prevalence in adults is 40.0% in those aged 20–39; 44.8% in those aged 40%–59%; and 42.8% in those aged 60 and older.

20.3 Causes of obesity

The basic cause of being overweight or obese is an imbalance of caloric consumption and expenditure (WHO, 2021). Or simply stated, eating too much, particularly foods high in fats and sugars. A contributory factor is declining rates of physical activity due to increasing inactivity as a consequence of the rising sedentary nature of work, changing transportation modes and urbanization (WHO, 2021). Though the overwhelming cause of obesity is caloric imbalance by choice, this disease is also related to genetic, physical, metabolic, neurological, hormonal and psychological impairments (OMA, 2021).

20.4 Biomarkers of obesity

The biomarkers of obesity include (Vincent and Taylor, 2006)

Hyperglycemia
Hyperleptinemia
Elevated tissue lipid levels
Reduced antioxidant defenses
Increased rates of free radical production
Enzymatic sources in the endothelium
Chronic inflammation

All these biomarkers are associated with elevated levels of oxidative stress.

20.5 Comorbidities

Obese individuals have high risks for developing many co-morbidities, as well as substantial increases in numbers of deaths from all causes. Obesity is associated with at least 60 co-morbidities (Khaodhiar et al., 1999; Pi-Sunyer, 1999; Manna and Jain, 2015; CDC, 2020; CDC, 2021b; Rethink Obesity, 2021). Table 20.1 lists a representative number of these.

TABLE 20.1 Representative list of obesity co-morbidities.

Anxiety
Asthma
Depression
Diabetes (type 2)
Cancer (lung, breast, colorectal, endometrial, esophageal, gallbladder, kidney, ovarian, pancreatic, prostate)
Cardiovascular disease
Dyslipidemia
Female infertility
Gallbladder disease
Gastroesophageal reflux
Gout
Hypertension
Hypouricemia
Kidney disease
Liver disease
Male hypogonadism
Metabolic syndrome
Obstructive sleep apnea
Osteoarthritis
Polycystic ovary syndrome
Stroke
Urinary stress incontinence

It is to be noted that obesity-related co-morbidities follow a similar pattern with differing races and ethnic groups that obesity itself does (Cossrow and Falkner, 2004).

20.6 Obesity, adipose tissue and chemical toxicity

Adipose tissue (AT) is not just a store of fat. It is an integral part of several physiological functions, including metabolic regulation and energy storage. Adipose tissue is also an endocrine organ that releases adipocytokines and adipokines as well as a modulator of the actions of toxic lipophilic chemicals, particularly persistent organic pollutants (POPs) (La Merrill et al., 2013; Marseglia et al., 2015; Lee et al., 2017; Jackson et al., 2018, Castellano et al., 2020; Masschelin et al., 2020).

Adipocytes store glucose as triglycerides in lipid droplets for future energy requirements, such as fasting, via release of fatty acids through lipolytic processes (La Merrill et al., 2013).

Adipocytes secrete several endocrine factors, including leptin and adiponectin, which regulate appetite, metabolic functions and inflammatory functions (Galic et al., 2010). AT also serves as a production and storage site for substances with autocrine, paracrine and neuroendocrine actions that influence behavior, energy regulation, lipid oxidation, immune function, vascular function, hormonal status and its own metabolism (Ouchi et al., 2011).

One of the characteristics of obesity is accumulation of macrophages in AT depots and macrophage secretions that deeply affect adipose cell biology by promoting a proliferative, pro-inflammatory and profibrotic state of preadipocytes and an insulin-resistant state of mature adipocytes (Dalmas et al., 2011). Lymphocytes, natural killer cells and mast cells are found in AT parenchyma in obese individuals, as well as in fibrosis deposits that accumulate in the obese (Divoux et al., 2010). Thus, obesity affects both AT structure and function.

20.6.1 Adipose tissue and persistent organic pollutants

The body acts to eliminate absorbed exogenous lipophilic chemicals by metabolizing these into water soluble species which are easily eliminated. Persistent organic pollutants (POPs),

TABLE 20.2 Stockholm convention original "dirty dozen" and additional POPs.

Original "dirty dozen" POPs
Aldrin
Chlordane
DDT
Dieldrin
Endrin
Mirex
Heptachlor
Hexachlorobenzene
Polychlorinated biphenyls (PCBs)
Toxaphene
Polychlorinated dibenzo-p-dioxins (dioxins)
Polychlorinated dibenzofurans (furans)
NEW POPs
Alpha hexachlorocyclohexane
Beta hexachlorocyclohexane

Chlordecone
Decabromodiphenyl ether
Difocol
Hexabromophenyl
Hexabromocyclododecane
Hexa-, hepta-, and octa-bromodiphenyl ether
Hexachlorobutadiene
Lindane
Pentachlorobenzene
Pentachlorophenol, its salts and esters
Perfluorooctane sulfonic acid (PFOS)
Perfluorooctane sulfonyl fluoride (PFOA), its salts and related
Compounds
Polycholorinated napththalenes
Short-chain chlorinated paraffins
Endosulfan and its related isomers
Tetra- and penta-bromodiphenyl ether

are a group of highly lipophilic chemicals that resist metabolism. A list of 12 of these POPs, dubbed the Stockholm Convention "dirty dozen," were originally identified in 2001 (EPA. 2021). The list has been subsequently expanded to include additional POPs (UN, 2019). Table 20.2 lists all listed to date.

20.6.2 POPs release into the body

The compounds listed in Table 20.2 include pesticides, flame retardants industrial polybrominated, polychlorinated and poyfluoronated species, all of which are environmentally and biologically persistent lipophilic species that bioaccumulate up food chains (Jackson et al., 2018). All resist degradation in the body by xenobiotic metabolizing enzymes and are readily absorbed by, stored in and very slowly released (over years, up to decades) by adipose tissue. Fatty foods (meat, fish and dairy are the human sources of these chemicals (La Merrill et al., 2013). POP storage is believed to be stored primarily in the adipocytes, as adipocyte cytoplasm is almost entirely made up of triglyceride droplets (Bourez et al., 2012).

All POPs are man-made "silent" toxins. Exposures to POPs can occur at concentrations that do not produce symptoms upon exposure, but ultimately can be toxic due to bioaccumulation.

Accordingly, there is no "safe" level for exposures to POPs. Though a zero-exposure level has been set for TCDD, none of the others have yet been so designated (Zeliger, 2011).

Total lipophilic load is a factor in the toxicity of low molecular weight lipophiles. As those are fairly rapidly metabolized and eliminated, their harmful effects are experienced primarily following chronic exposure (Zeliger, 2013; 2013a; 2013b). POPs, however, to not fit into that category as they are not metabolized and eliminated as other lipophiles are. Toluene, for example, a commonly used solvent, is retained in the body by adipose tissue, but has a half-life of one half to 3 days (Filley et al., 2004). POPs, by comparison have half-lives measured in years, not days (Lee et al., 2017).

POPs are also found in the blood, from which they contaminate tissues throughout the body. The rates of POPs release into blood are very slow under normal circumstances, but greatly accelerated by weight loss during which retained POPs are freed from the adipose tissue holding them. This direct relationship between weight loss and the degree of POPs release results in high levels of POPs exposures during periods of drastic weight loss and definitively demonstrates that slow weight loss is a healthier option than crash dieting (Hue et al., 2006; Kim et al., 2012).

20.6.3 Pops as obesogens

POPs have been shown to be obesogens (La Merrill and Birnbaum, 2011). Exposures during prenatal, postnatal and pubertal development can lead to obesity and dyslipidemia in later life via mechanisms involving programming of epigenetic regulation of critical genes (Barouki et al., 2912; Valvi et al., 2012).

20.6.4 Pops as lipotoxins

Following release from AT into the blood, POPs are absorbed in other body areas and lead to accumulation of lipids in non-adipose tissues and induce toxic effects on tissue function, leading to many diseases, including diabetes, hypertension and heart disease (La Merrill et al., 2013; Lee, at al., 2017). Mechanistic pathways attributed to POPs include:

Disruption of AT function and adipocyte differentiation.
Immunotoxicity.
Alteration of the activities of metabolic enzymes via reduction of lipoprotein lipase activity.
Increased expression of inflammatory genes in adipose cells which correlates with insulin resistance in obese individuals.

20.7 Obesity and oxidative stress

Oxidative stress is part of all mechanistic pathways involving the onset and effects of obesity on the human body, as well as the unifying mechanism for co-morbidities including heart disease, type 2 diabetes and numerous cancers (Vincent and Taylor, 2006; Matsuda and Shimomura, 2013; Le Lay et al., 2014; Huang et al., 2015).

Adipose tissue is a significant source of ROS production in the obese from mitochondrial and enzyme sources, as well as from other factors.

20.7.1 Mitochondrial

Increases in uptake of nutrients into adipocytes elevates mitochondrial substrate load, which increases electron transport activity and results in elevated superoxide production by-products (Browlee, 2005).

20.7.2 Enzymes

Several enzymes inside adipocyte cells produce ROS. These include the NAPDH oxidases (Bedard and Krause) and nitric acid synthase (Vasquez-Vivar et al., 1998).

20.7.3 Other factors

Other factors affecting obesity and its associated mechanisms include enhanced ROS production, decreased antioxidant defenses, chronic low-grade inflammation and altered gut microbiota composition, all of which are associated with elevated oxidative stress (Le Lay et al., 2014; Yang et al., 2017; Dao and Clement, 2017; Di Domenico et al., 2019).

20.8 Psychological impact of obesity

In addition to its physiological impacts, obesity carries with it social stigmas that affect psychological health and well being and lead to disorders that include anxiety and depression (Wardle and Cooke, 2005; American Academy of Pediatrics, 2006; Sarwer and Polonsky, 2016; Romain et al., 2017). Psychological impacts of oxidative stress are more fully addressed in chapter 13.

20.9 Summary

Obesity is a very serious disease with numerous health implications. The effects of this disease are compounded by its serving as a reservoir of persistent organic pollutants that are sources of oxidative stress sources which mediate the onset numerous other serious diseases. Obesity as a mechanistic driver for its co-morbidities is addressed in the ensuing chapters.

References

American Academy of Pediatrics, 2006. The emotional toll of obesity. https://www.healthychildren.org/English/health-issues/conditions/obesity/PagesThe-Emotional-Toll-of-Obesity.aspx. (Accessed 6 October 2021).

Barouki, R., Gluckman, P.D., Grandjean, P., Hanson, M., Heindel, J.J., 2012. Development origins of non-communicable disease: implications for research and public health. Environ. Health. https://doi.org/10.1186/1476-069X-11-42. (Accessed 5 October 2021).

Bedard, K., Krause, K.H., 2007. The NOX family of ROS-generating NADPH oxidases: physiology and pathophysiology. Physiol. Rev. 87 (1), 245–313.

Bourez, S., Le Lay, S., Van den Daelin, C., Louis, C., Larondelle, Y., Thome, J.P., et al., 2012. Accumulation of polychlorinated biphenyls in adipocytes: selective targeting to liquid droplets and role of caveolin-1. PLoS One. https://doi.org/10.1371/journal.pone.0031834.

Brownlee, M., 2005. The pathology of diabetic complications: a unifying mechanism. Diabetes 51 (3), 394–397.

Castellano, J.M., Espinosa, J.M., Perona, J.S., 2020. Modulation of lipid transport and adipose tissue deposition by small lipophilic compounds. Front. Cell Dev. Biol. https://doi.org/10.3389/fcell.2020.555359. (Accessed 2 October 2021).

CDC, 2020. The Health Effects of Overweight and Obesity. Centers for Disease Control and Prevention. www.cdc.gov/overweightandobesity. (Accessed 2 October 2021).

CDC, 2021a. Adult Obesity Facts. Centers for Disease Control and Prevention. www.cdc.gov/obesity. (Accessed 2 October 2021).

CDC, 2021b. Adult Obesity Causes & Consequences. Centers for Disease Control and Prevention. www.cdc.gov/adultobesitycausesandconsequences. (Accessed 2 October 2021).

Cossrow, N., Falkner, B., 2004. Race/ethnic issues in obesity-related co-morbidities. J. Clin. Endocrinol. Metab. 89 (6), 2590–2594.

Dalmas, E., Clement, K., Guerro-Millo, M., 2011. Defining macrophage phenotype and function in adipose tissue. Trends Immunol. 32 (7), 307–314.

Dao, M.C., Clement, K., 2017. Gut microbiota and obesity: concepts relevant to clinical care. Eur. J. Intern. Med. 48, 18–24.

Di Domenico, M., Pinto, F., Auagliuto, L., Contaldo, M., Setterbre, G., Romano, A., et al., 2019. The role of oxidative stress and hormones in controlling obesity. Front. Endocrinol. https://doi.org/10.3390/fendo.2019.00540. (Accessed 2 October 2021).

Divoux, A., Lacassa, D., Veyrie, N., Hugol, D., Aissat, A., et al., 2010. Fibrosis in human adipose tissue: composition, distribution and link with lipid metabolism and fat mass loss. Diabetes 59, 2817–2825.

EPA, 2021. Persistent Organic Pollutants: A Global Issue, a Global Response. https://www.epa.gov/international-cooperation/persistent-oragnic-pollutants-global-issue-response#table. (Accessed 5 October 2021).

Filley, C.M., Halliday, W., Kleinschmidt-Demasters, B.K., 2004. The effects of toluene on the central nervous system. J. Neuropathol. Exp. Neurol. 63 (1), 1–12.

Fryar, C.D., Carroll, M.D., Afful, J., 2021. Prevalence of Overweight, Obesity and Sever Obesity Among Adults Aged 20 and Over: United States, 1960–1962 through 2017–2018. CDC. 2020a. Obesity and Overweight. Centers for Disease Control and Prevention. www.cdc.gov/obesityandoverweight. (Accessed 2 October 2021).

Galic, S., Oakhill, J.S., Steinberg, G.R., 2010. Adipose tissue as an endocrine organ. Mol. Cell. Endocrinol. 316, 129–139.

Huang, C.J., McAllister, M.J., Slusher, A.L., Webb, H.E., Mock, J.T., Acevedo, E.O., 2015. Obesity-related oxidative stress: The impact of physical activity and diet manipulation. Sports Med – Open. https://doi.org/10.1186/s40798-015-0031-y. (Accessed 2 October 2021).

Hue, O., Marcotte, J., Berrigan, F., Simoneau, M., Dore, J., Marceau, P., et al., 2006. Increased plasma levels of toxic pollutants accompanying weight loss induced by hypocaloric diet or by bariatric surgery. Obes. Surg. 16, 1145–1154.

Jackson, E., Shoemaker, R., Larian, N., Cassis, L., 2018. Adipose tissue as a site of toxin accumulation. Compr. Physiol. 7 (4), 1085–1135.

Khaodhiar, L., McCowen, K.C., Blackburn, G.L., 1999. Obesity and its comorbid conditions. Clin. Cornerstone 2 (3), 17–31.

Kim, M.J., Pelloux, V., Goyat, E., Tordjman, J., Bui, L.C., Chevallier, A., et al., 2012. Inflammatory pathway genes belong to major targets of persistent organic pollutants in adipose cells. Environ. Health Perspect. 120 (4), 508–514.

La Merrill, M., Birnbaum, L.S., 2011. Childhood obesity and environmental chemicals. Mt. Sinai J. Med. 78 (1), 22–48.

La Merrill, M., Emond, D., Kim, M.J., Antignac, J.P., Le Bizec, B., Clement, K., et al., 2013. Toxicological function of adipose tissue: focus on persistent organic pollutants. Environ. Health Perspect. 121 (2), 162–169.

Lee, Y.M., Kim, K.S., Jacobs Jr., D.R., Lee, D.H., 2017. Persistent organic pollutants in adipose tissue should be considered in obesity research. Obesity 18, 129–139.

Le Lay, S., Simard, G., Martinez, M.C., Andriantsitohaina, R., 2014. Oxidative stress and metabolic pathologies: from an adipocentric point of view. Oxid. Med Cell. Longev. https://doi.org/10.1155/2014/908539. (Accessed 2 October 2021).

Manna, P., Jain, S.K., 2015. Obesity, oxidative stress, adipose tissue dysfunction, and the associated health risks: causes and therapeutic strategies. Metab. Syndr. Relat. Disord. 13 (10), 423–444.

Marseglia, L., Manti, S., D'Angelo, G., Nicotera, A., Parisi, E., Di Rosa, G., et al., 2015. Oxidative stress in obesity: a critical component in human diseases. Int. J. Mol. Sci. 16, 378–400.

Masschelin, P.M., Cox, A.R., Chernis, N., Hartig, S.M., 2020. The impact of oxidative stress on adipose tissue energy balance. Front. Physiol. https://doi.org/10.3389/fphys.2019.01638. (Accessed 2 October 2021).

Matsuda, M., Shimomura, I., 2013. Increased oxidative stress in obesity: implications for metabolic syndrome, diabetes, hypertension dyslipidemia, atherosclerosis, and cancer. Obes. Res. Clin. Pract. 7 (5), e330–e341.

OMA, 2021. Why is obesity a disease? Obesity Association of America. https://obesitymedicine.org/why-is-obesity-a-disease/. (Accessed 2 October 2021).

Ouchi, N., Parker, J.L., Lugus, J.J., Walsh, K., 2011. Adipokines in inflammation and metabolic disease. Nat. Rev. Immunol. 11, 85–97.

Pi-Sunyer, F.X., 1999. Comorbidities of overweight and obesity: current evidence and research issues. Med. Sci. Sports Exerc. 31 (11 Suppl. l), S602–S608.

Rethink Obesity, 2021. Impairments to Health. https://www.rethinkobesity.com/disease-progression/comorbidities-of-obesity.html. (Accessed 2 October 2021).

Romain, A.J., Marleau, J., Baillot, A., 2017. Impact of obesity and mood disorders on physical comorbidities, psychological well-being, health behaviours and use of health services. J. Affect. Disord. https://doi.org/10.1016/j.jad.2017.08.065. (Accessed 6 October 2021).

Sarwer, D.B., Polonsky, H.M., 2016. The psychosocial burden of obesity. Endocrinol Metab. Clin. N. Am. 45 (3), 677–688.

UN, 2019. The New POPs under the Stockholm Convention. www.pops.int/TheConvention/The POPs/TheNewPOPs/tabid/2511/Efault.aspx. (Accessed 5 October 2021).

Valvi, D., Mendez, M.A., Martinez, D., Grimalt, J.O., Torrent, M., Sunyer, J., et al., 2012. Prenatal concentrations of polychlorinated biphenyls, DDE, and DDT and overweight children: a prospective birth cohort study. Environ. Health Perspect. 120 (3), 451–457.

Vasquez-Vivar, J., Kalyanaraman, A.T., Martasek, P., Hogg, N., Masters, B.S.S., Karoui, H., et al., 1998. Superoxide generation by endothelial nitric oxide synthase: the influence of cofactors. Proc. Natl. Acad. Sci. USA 95 (16), 9220–9225.

Vincent, H.K., Taylor, A.G., 2006. Biomarkers and potential mechanisms of obesity-induced oxidative stress in humans. Int. J. Obes. 30 (3), 400–418.

Wardle, J., Cooke, L., 2005. The impact of obesity on psychological well-being. Best Pract. Res. Clin. Endocrinol. Metabol. 19 (3), 421–440.

WHO, 2021. Obesity. World Health Organization. https://who.int/health-topics/obesity#tab=tab_1. (Accessed 2 October 2021).

Yang, B.G., Hur, K.Y., Lee, M.S., 2017. Alterations in gut microbiota and immunity by dietary fat. Yonsei Med. J. 58 (6), 1083–1091.

Zeliger, 2011. Human Toxicology of Chemical Mixtures, second ed. Elsevier, London.

Zeliger, H.I., 2013. Lipophilic exposure as a cause of type 2 diabetes (T2D). Rev. Environ. Health. https://doi.org/10.1515/reveh-2012-0031. (Accessed 15 September 2021).

Zeliger, H.I., 2013a. Lipophilic chemical exposure as a cause of cardiovascular disease. Interdiscipl. Toxicol. 6 (2), 55–62.

Zeliger, H.I., 2013b. Exposure to lipophilic chemicals as a cause of adult neurological impairments, neurodevelopmental disorders and neurodegenerative diseases. Interdiscipl. Toxicol. 6 (3), 101–108.

CHAPTER

21

Cancer

21.1 Introduction

Cancer is not a single disease. It is a group of diseases, all of which cause cells to change and grow uncontrollably anywhere in the body. In the United States, it is estimated that close to two million new cancer cases are diagnosed annually and that 600,000 people die from cancer each year (National Cancer Institute, 2020).

Cancers are named after the body parts in which they originate. The National Cancer Institute has identified 185 different cancer types, the most common of which are listed in descending order in Table 21.1 are (National Cancer Institute, 2020, 2021).

Cancer types are classified by tissue or blood categories (Stanford Healthcare, 2021). These include the following:

1. Sarcoma
2. Carcinoma
3. Lymphoma
4. Leukemia
5. Myeloma
6. Blastoma

1. Sarcoma

Sarcoma is a cancer growing from connective tissue, including cartilage, muscle, bone and fat. Examples of sarcomas include osteosarcoma, soft tissue sarcoma, Ewing's sarcoma and chrondosarcoma.

2. Carcinoma

Carcinomas account for 80%–90% of all cancers. They are malignancies derived from epithelial cells that cover or line the surfaces of organs, glands or body structures. Many target organs associated with secretion, an example being milk-producing breasts. Examples of sarcomas include melanoma, basal cell carcinoma and squamous cell epidermal cancer.

Oxidative Stress
https://doi.org/10.1016/B978-0-323-91890-9.00036-2

TABLE 21.1 Most common cancer types.

Breast cancer
Lung cancer
Prostate cancer
Colorectal cancer
Skin melanoma
Bladder cancer
Non-Hodgkin lymphoma
Kidney cancer
Endometrial cancer
Leukemia
Pancreatic cancer
Thyroid cancer
Liver cancer

3. Lymphoma

Lymphomas originate in nodes or glands of the lymphatic system that are responsible for the production of white cells and the cleansing of body fluids. Lymphomas also originate in the brain and the breast.

4. Leukemia

Leukemia is a cancer of the blood marrow that prevents the production of red blood cells, white blood cells and platelets. Leukemia types include acute lymphocytic leukemia, chronic lymphocytic leukemia, acute myeloid leukemia and chronic myeloid leukemia.

5. Myeloma

Myeloma grows in bone marrow plasma cells. It can occur in one bone forming a single tumor (plasmacytoma), or in multiple bones (multiple myeloma).

6. Blastoma

Blastomas are cancers derived from precursor cells that are most commonly prevalent in children. These include neuroblastoma, nephroblastoma, retinoblastoma, pancreatoblastoma and pleuroblastoma.

A precise understanding of why one person develops cancer and another not still awaits elucidation. There are, however, known risk factors for cancer onset (National Cancer Institute, 2015). These and representative references for each are listed in Table 21.2.

Both intrinsic and non-intrinsic factors present risks for cancer (Wu et al., 2018).

TABLE 21.2 Risk factors for cancer.

Factors	References
Genetics	National Cancer Institute (2017)
Age	White et al. (2014)
Alcohol us	Rumgay et al. (2021)
Environmental exposures to carcinogens	Balali-Mood et al. (2021)
Chronic inflammation	Federico et al. (2007)
Diet	Key et al. (2020)
Hormones	Key (1995)
Immunosuppression	Baniyash (2006)
Mitochondrial dysfunction	Kudryavtseva et al. (2016)
Infectious agents	Masrour-Roudsari and Ebrahimpour (2017)
Obesity	CDC (2021)
Ionizing radiation	National Cancer Institute (2019)
Non-ionizing radiation	Kocaman et al. (2018)
Sunlight exposure	Holick (2014)
Tobacco use	National Cancer Policy Forum, 2016

21.1.1 Intristic risks

Random errors in DNA replication present intrinsic risk for cancer onset. These include single nucleotide errors, deletions and insertions. These cannot be prevented.

21.1.2 Non-instristic risks

Non-intrinsic risk factors include both endogenous and exogenous ones. These include the following, some of which can prevented by altering lifestyle choices:

Endogenous risk factors	Exogenous risk factors
Aging	Chemical exposures
Genetic susceptibility	Radiation
DNA repair failures	Viruses
Hormones	Smoking lack of exercise poor diet
Growth factors	Inflammation

Many exogenous lipophilic chemicals (e.g., polynuclear aromatic hydrocarbons, chlorinated hydrocarbons and persistent organic pollutants) bioaccumulate in adipose tissue, metabolize into DNA-reactive species, form stable DNA adducts, induce free radical formation and/or act via epigenetic mechanisms. These properties, which can make such chemicals highly mutagenic, are believed to be major contributors to human carcinogenesis (Irigaray and Belpomme, 2010).

It is beyond the scope of this bok to thoroughly explore each of these factors. The reader is referred to the literature for follow up. The seemingly different factors listed in Table 21.2, however, have one common trait—all are associated with chronically elevated oxidative stress (Dasari et al., 2020).

As will be shown for each of the individual cancers examined below, redox homeostasis impairment, an essential component of all cancer onset, is a consequence of free radical overproduction and/or insufficient antioxidant defense. Cancer (as well as all oxidative stress-mediated diseases) ensues as a result of free radical attacks on DNA, protein and/or lipid molecules, as well as in the disruptions in signaling that regulate cellular activities.

21.2 Hallmarks of cancer—Hanahan and Weinberg

As discussed above, cancer is not a single disease, but a group of diseases. All cancers, however, share a common characteristic: they involve dynamic genome changes. In 2000, it was proposed that all cancers are manifestations of "six essential alterations in cell physiology that collectively dictate malignant growth" (Hanahan and Weinberg, 2000).

These alterations are:

1. Self-sufficiency in growth signals
2. Insensitivity to growth-inhibitory (antigrowth) signals
3. Evasion of programmed cell death (apoptosis)
4. Limitless replicative potential
5. Sustained angiogenesis
6. Tissue invasion and metastasis

1. Self-sufficiency in growth signals

Normal cells require mitogenic signaling prior to proceeding to a proliferative state, and trans-membrane receptors that bind signaling molecules transmit these signals. Cancerous cells generate their own growth signals, thus bypassing normal growth stimulation signals and allowing cancer cells to accelerate growth.

2. Insensitivity to growth-inhibitory signals

Multiple antiproliferative signals in normal tissues maintain cellular quiescence and tissue homeostasis. Normal growth-inhibitory signals, like proliferative signals, are received via exogenously derived signals. Cancer cells elude proliferative signals.

3. Evasion of programmed cell death

Evasion of programmed cell death is essential to maintain homeostatic cell numbers. Apoptosis is dictated by sensors and effectors. Sensors monitor extra- and intracellular environments for normal and abnormal conditions that dictate whether a cell should live of die. The signals produced by these sensors regulate the effectors which actually affect cellular death. Cancer cells evade programmed death by a number of mechanisms, the most common being the loss of the p53 tumor suppressor gene and the functional inactivation of its product, the p53 protein.

4. Limitless replicative potential

Combined, the three hallmarks just noted; growth signal autonomy, insensitivity to antigrowth signals and resistance to programmed cell death lead to separation of cancer cells' growth programs from normal environmental signals. This results in the fourth cancer hallmark, limitless replicative potential.

5. Sustained angiogenesis

Oxygen and nutrients must be continually supplied to cells for them to function and survive. Once normal tissue is formed, the growth of new blood vessels (angiogenesis) is carefully regulated so as to sustain, but not stimulate excessive growth. Cells contained in aberrant proliferative lesions (cancer cells) initially lack angiogenic ability, but acquire the ability to induce and sustain the growth of new blood vessels without regulation, thus providing the oxygen and nutrients that promote continual tumor growth.

6. Tissue invasion and metastasis

During the development of most types of human cancers, primary tumors invade adjacent tissues and then travel to and invade distant sites (metastasize) where new cancer sites are established. Metastasis, which accounts for 90% of human cancer deaths. is discussed in more detail in Section 21.7.

In their 2000 publication, Hanahan and Weinberg defined the acquired functional capabilities that allow cancer cells to survive, proliferate and disseminate. The functions listed are acquired in different tumor types via distinct mechanisms. In their 2011 paper, Hanahan and Weinberg describe two additional cancer hallmarks, (numbers 7 and 8); reprogramming energy metabolism and evading immune destruction enabling characteristics. Also added to the discussion were two enabling characteristics; genome instability and mutation, and tumor-promoting inflammation reprogramming energy metabolism. These are discussed here.

21.2.1 Emerging hallmarks

7. Reprogramming energy metabolism

Normal cells process glucose aerobically via glycolysis, first to pyruvate in the cytosol and subsequently to carbon dioxide in the mitochondria. Cancer cells, however, produce energy anaerobically via glycolysis, under conditions that do not involve the oxygen-consuming mitochondria. The reprogramming energy metabolism-characteristic of cancer cells has been added as a hallmark of cancer (Hanahan and Weinberg, 2011; Arfin et al., 2021).

8. Evading immune destruction

The immune system is believed to present a significant barrier to tumor formation and progression. Despite this, cancer cells are able to avoid immune system detection, or are able to limit immune system destruction. Either way, these cancer cells evade immune system-eradication. This evasion, though not completely understood at this time, has been labeled as the eighth hallmark of cancer (Hanahan and Weinberg, 2011).

21.2.2 Enabling characteristics

1. Genome instability and mutation

Genome maintenance systems that detect and resolve DNA defects are naturally very efficient. Cancer cells, however, often increase mutation rates via a breakdown in maintenance machinery. Mutation accumulation is also accelerated by cancer cells via the compromising of an array of surveillance systems that normally monitor genomic integrity and force genetically damaged cells into senescence or apoptosis. Genome instability leading to mutation is present in the great majority of human cells and thus labeled as an enabler of hallmarks identified with cancer onset (Hanahan and Weinberg, 2011; Poetsch, 2019).

2. Tumor-promoting inflammation

Immune system responses via inflammatory, oxidative stress elevating cellular activities are the body's attempts to eradicate tumors. Paradoxically, tumor-associated inflammatory responses have the effect of enhancing tumorigenesis and progression. Inflammatory cells release chemicals that include reactive oxygen species which are mutagenic for nearby cancer cells, thereby accelerating their genetic evolution to states of heightened malignancy. Thus, inflammation becomes an enabler of the hallmarks connected to cancer onset (Hanahan and Weinberg, 2011; Greten and Grivennikov, 2019).

21.3 Other hallmarks

Since the initial publication of Hanahan and Weinberg's hallmarks of cancer paper, additional hallmarks of cancer have been proposed. These are addressed here.

21.3.1 Cancer-related inflammation

Inflammatory conditions in some organs, that increase the risk of cancer and inflammation, are also found in the microenvironment of tumors that are not epidemiologically related to inflammation. Rather, an additional mechanism involving cancer-related inflammation that leads to genetic alterations in cancer cells obtains. This has been labeled the seventh hallmark of cancer (Colotta et al., 2009; Greten and Grivennikov, 2019).

21.3.2 Systemic hallmarks

Paul introduced six cancer hallmarks based not only on the cellular and tissue-based hallmarks as Hanahan and Weinberg have, but also on hallmarks that are whole organism-based. This approach takes into account the fact that geographically separated cancer tissues form a system that interacts with the rest of the body via cancer-induced systemic pathways (Paul, 2020). These hallmarks, the result of interaction between and the organism at the macroscopic level are:

1. Primary tumor-metastasis network
2. Global inflammation
3. Immunity inhibition
4. Metabolic changes to cachexia
5. Propensity to thrombosis
6. Distal metastasis

21.3.2.1 Primary tumor-metastasis network

This hallmark has been demonstrated in experimental models where distal tumor resection has shown distal tumor regression in some instances (Flanagan and Yanover, 2001) and accelerated tumor development in others (O'Reilly et al., 1997; Kisker et al., 2001).

21.3.2.2 Global inflammation

Local stromal inflammation is a well-known to be linked to link to cancer progression, with inflammatory cells and molecules involved in almost all aspects of cancer progression and metastasis (Colotta et al., 2009).

21.3.2.3 Immunity inhibition

With cancer, the body's immune system responses, both locally and systemically are inhibited (Chen et al., 2004).

21.3.2.4 Metabolic changes to cachexia

Cancer cells require large amounts of energy to ensure sufficient biomass for their growth. They use oxidative stress to extract nutrients from surrounding stromal cells that are forced to undergo aerobic glycolosis and produce energy-rich nutrients, such as lactate and ketones, to feed cancer cells. resulting in cachexia in multiple organs (Palvides et al., 2010; Palvides et al., 2010a).

21.3.2.5 Propensisty to thrombosis

Cancer is associated with a state of hyper-coaguability. Cancer cells support clot formation and clotting proteins support cancer growth. Cancer cells are also able to adhere to host cells, thereby stimulating additional prothrombotic properties of the thrombosis effector cells (Falanga et al., 2017).

21.3.2.6 Distal metastasis

Both the central nervous system and the lymphatic organs are intimately involved in cancer and associated with metastasis to distal organs (Cao et al., 2010; Burfeind et al., 2016).

21.3.3 Additional cancer hallmarks

As more continues to be learned about cancer, additional hallmarks of it are being identified. Four new ones have recently (2021) been disclosed (Senga and Grose, 2021). These are:

1. Cellular regression from a specialized functional state. Cancer cells can modify normal body mechanisms from specifically prescribed functions and adapt these in order to survive.
2. Epigenetic changes that affect gene expression. Examples include the epigenetic role in the development and progression of several cancers via modification of gene expression such as hypermethylation of tumor suppressor genes and silencing of microRNAs.
3. Role of microorganisms. The composition of microorganisms found in the human gut varies with geographic location, diet and other factors. The microbiome has a multitude of beneficial functions in the human body. It is also linked to 90% of gastric cancers. It is proposed that the microbiome composition, via its immune system influence, can play a pivotal role in cancer.
4. Neuronal signaling. Nerves and neuronal signaling are an indispensable part of tumorigenesis, playing an active role in modulating the tumor microenvironment. These are involved in recruiting blood vessels to tumors, controlling constriction and relaxation of blood vessels, altering the expression of immune checkpoint molecules and providing cues for proliferation to tumor cells.

21.3.4 Cancer hallmark regulation

Recent findings demonstrate that microRNAs and long conceding RNAs regulate most cancer hallmarks via their bonding with DNA, RNA or proteins, or by encoding small peptides via oxidative-stress induced mechanisms (Zhou et al., 2020). These findings further demonstrate the role of oxidative stress in cancer onset.

21.4 Cancer initiation promotion and progression

The hallmarks section above clearly show the involvement of oxidative stress in mutagenesis via DNA damage, protein damage, lipid peroxidation and altered signaling. The ROS that are produced are mutagens that promote tumor initiation. They oxidize guanine in DNA and RNA to yield 8-hydroxyguanine which can pair with adenine during DNA replication, resulting in substitutions that potentially produce mutations leading to cancer initiation (Feig et al., 1994; Upham and Trosco, 2009; Gill et al., 2016).

Oncogene signaling is tied to oxidative stress (Irani et al., 1997; Sattler et al., 2000; Nowicki et al., 2004). The increase in ROS as a result of oncogene signaling is believed to be a contributor to ongoing mutagenesis and genomic instability in cancer cells, thus promoting cancer progression (Gill et al., 2016).

The mutations driving cancer onset may precede diagnosis by years or even decades and once cancer onset has started, it cannot be stopped (Gerstrung et al., 2020). Both endogenous and exogenous agents drive cancer onset.

Cancer is caused by both genetic and environmental sources. Carcinogenesis requires a three-step process:

Initiation
Promotion
Progression

Though all three steps are required, only some sources, complete carcinogens, are causes of all three stages of development. More commonly, different sources are required for one or more of these steps (Kim and Zaret, 2015).

Tumor initiators are carcinogens that induce a first driver mutation in a dividing cell. Tumor promoters are non-genotoxic carcinogens that are able to induce proliferation of mutated cells and prevent these cells from apoptotic loss. Tumor progressors are those carcinogens that allow pre-malignant mutated cells to irreversibly acquire fully malignant cell phenotypes. Contributors to carcinogenesis can come from endogenous or exogenous sources. Endogenous carcinogens are carcinogenic molecules or metabolic intermediates arising as a consequence of respiration and/or from ingestion of food in those living in safe non-polluted environments (Irigaray and Belpomme, 2010).

Examples tumor initiators, promoters and progressors follow (Salnikow and Zhitkovich, 2008; Irigaray and Belpomme, 2010; UC Berkeley, 2021).

Tumor initiators
High molecular weight polycyclic aromatic hydrocarbons (e.g., benzo[α]pyrene)
Nitrosamines
Aromatic amines
Heterocyclic amines
Heterocyclic aromatic amines
Heavy metals and metaloids (e.g. arsenic, chromium, nickel).
Tumor promoters
Phorbol esters (e.g. TPA, O-tetradecanoyl phorbol-13 acetate
Endocrine disruptors (e.g., bisphenol A, estradiol, DES)
Polychlorinated aromatic hydrocarbons (e.g., dioxins and PCBs) are
Redox active xenobiotics
Redox active metals (e.g., iron, copper)
ROS

(*Continued*)

Tumor progressors
Asbestos fibers
Benzene
Peroxides (e.g., benzoyl peroxides)
Inflammation

21.5 Exogenous carcinogens

Numerous exogenous chemical carcinogens (cancer causing agents and co-carcinogens agents that are not by themselves carcinogenic, but may activate carcinogens and/or enhance their carcinogenic effects) are present in the environment. Examples of co-carcinogens are catechol, $C_{10}-C_{14}$ aliphatic hydrocarbons and coal tar derivatives (Van Duuren, 1982; Haverkos et al., 2017). Exposures to carcinogenic/co-carcinogenic mixtures of these occur when both are released together, as is the situation with tobacco smoke, while others of which are formed in the environment from separate emissions.

Multiple exogenous chemicals are present in the environment, numerous ones in low dose concentrations, which alone present little risk of causing cancer. "Cocktail" mixtures of such carcinogens and co-carcinogens, however, have been shown to act together as complete carcinogens (Lutz, 1990).

The list of exogenous carcinogens is a long one. Table 21.3 lists a representative sample (Irigaray and Belpomme, 2010).

Lipophilic carcinogenic chemicals are a particularly worry some class of compounds. Unlike hydrophilic (polar) molecules, lipophilic (hydrophobic) molecules readily penetrate the lipophilic membranes that surround all cells (Zeliger, 2003). Examples of these include benzene, polynuclear aromatic hydrocarbons (PAHs) dioxins, furans, PCBs, organochlorine pesticides. In addition, such lipophiles readily bioaccumulate in adipose tissue which acts as a reservoir from which they are continually released into the blood stream for transport throughout the body. Many carcinogenic lipophiles, including all the persistent organic pollutants (POPs) have long half-lives and can persist in the body for many years or even decades (Zeliger, 2011).

Four factors related to the carcinogenic consequences of chemical absorption of exogeneous toxins are: mixtures, total lipophilic load, sequential absorption of toxic species and resonance stabilization of free radicals produced from carcinogenic chemicals.

21.5.1 Mixtures

Studies on toxicity and carcinogenicity of chemicals are most often carried out in laboratory settings with strict controls over chemical compositions being tested. Exposures in the real world, however, are almost always to mixtures of chemicals, the most studied of which are polluted air and tobacco smoke, both of which have been tied to the onset of cancer.

TABLE 21.3 Representative list of exogenous carcinogens.

Acrolein
Air pollution fine particles
Asbestos
Azo dyes
Benzene and related aromatic compounds
1.3-butadiene
Chromium (hexa-valent)
Dioxins
Formaldehyde
Organochlorine pesticides (several)
Phthalates
Polynuclear aromatic hydrocarbons (PAHs)
Polychlorinated biphenyls (PCBs)
Vinyl chloride

Ninety percent of the world's population breathes air polluted with numerous free radicals and known carcinogens at concentrations that vary from slight to severe. ROS, in addition to those released into the air, are also created in the atmosphere via chemical reactions that are catalyzed by solar radiation. It has been definitively established that those living in areas with high air pollution levels have correspondingly higher incidences of lung and other cancers, including those of the breast, liver and pancreas (AACR, 2021).

Tobacco smoke is composed of thousands of different chemicals, more than 60 of which are established carcinogens. The toxic chemical mix includes, lipophilic and hydrophilic chemicals, PAHs and metals and is associated with at least 13 different cancers (CDC, 2010).

21.5.2 Total lipophilic load

It has been shown that disease onset can be enhanced when mixtures of lipophiles are absorbed, even when components of such mixtures are absorbed at concentrations low enough to have little or no toxicity and that total lipophilic load alone can elevate oxidative stress, serve as promoter of several of the cancer hallmarks and raise disease incidence (Zeliger, 2013; Zeliger, 2013a; Zeliger, 2013b; Raney et al., 2017).

21.5.3 Sequential absorption

As discussed in the previous chapters, it is well established that mixtures of lipophiles and hydrophiles may be more toxic than either of the species alone; are toxic at lower concentration levels than the individual species alone; and exhibit effects not attributable to any of the components of such mixtures (Zeliger, 2011; Raney et al., 2017). It has also been shown that toxic mixture effects can be observed when the lipophilic component(s) of a chemical mixture are absorbed first and demonstrate little or no toxicity until the subsequent absorption of the hydrophilic component(s) (Zeliger and Lipinski, 2015). This is consistent with the propensity for hydrophiles, which are short-lived in the body, to combine with lipophiles, which are stored in adipose tissue and continually released into the body, to combine and produce enhanced toxic effects when mixed together.

21.5.4 Resonance stabilization

As discussed in Chapter 18, free radicals are stabilized by proximity to multiple bonds, the presence of heteroatoms or geometric structure. Carcinogenic PAHs and POPs meet these criteria. Once absorbed through cell membranes and converted to free radicals, these species are stabilized and are orders of magnitude longer-lived than the oxygen free radicals produced via mitochondrial activity. Accordingly, the free radicals so produced are make major contributions to ROS.

21.6 Metals, metalloids and cancer

Several transition metals and metalloids have been identified as established or probable carcinogens (IARC, 2021). These include, but are not limited to arsenic, cadmium, chromium, cobalt and nickel. Metals and metalloids act by an number of mechanisms (Irigaray and Belpomme, 2010) which include:

1. Activating procarcinogens in the liver.
2. Increasing the promoting effect of endogenous steroid hormones (e.g., estrogens).
3. Replacing natural enzyme-associated metals (e.g., zinc) with inactivating substitutes (arsenic, cadmium, nickel).
4. Affecting heterochromatic regions of the genome via an epigenetic mechanism.
5. ROS production leading to mutagenesis (e.g., arsenic, cadmium, chromium, nickel).
6. Metal-mediated free radical formation (asbestos-induced free radical formation due to the iron contained in it via the Fenton reaction).

21.7 Food and cancer

Some foods contain low doses of carcinogenic chemicals, including PAHs, aromatic amines, heterocyclic aromatic amines (HAAs), N-nitroso compounds and aflatoxins. Overcooking meat and other foods at high temperatures results in the productions of HAAs and is also a source of these in the diet (Irigaray and Belpomme, 2010; Saha et al., 2017).

The above are just a few examples of environmental sources of human carcinogens. It is beyond the scope of this book to elaborate further, but the reader is referred to the references cited and the additional references contained in those.

21.8 Metastasis

Metastasis. The ability of cancer to spread to in the body, is a process involving multiple steps (Vanharanta et al., 2013; Gill et al., 2016; National Cancer Institute, 2021). These steps include:

1. Invasion—Growing into, or invading nearby normal tissue.
2. Migration—Penetrating through walls of nearby lymph nodes or blood vessels.
3. Transfer—Travel through the lymphatic system and blood stream to other body parts.
4. Distal invasion—Stopping in small blood vessels at distant locations, invasion of blood vessel walls and moving into surrounding tissue.
5. Seeding—Growing in newly invaded tissue until tiny tumors form.
6. Proliferation—Causing new blood vessels to grow, thus creating a blood supply that allows continual growth of tumors.

Metastasis is vulnerable to interruption at each of the six steps as oxidative stress, which is a promoter of tumor initiation, has also been shown to kill cancer cells at multiple metastatic stages. Indeed, of all tumor cells that enter circulation, only few are able to initiate metastasis at distant organs (Nguyen and Massague, 2007; Piskounova et al., 2015; Peiris-Pages et al., 2015; Gill et al., 2016; Pachmayr et al., 2017). Yet, as discussed above, cancer is a disease that results from accumulations of oxidative stress-driven genomic aberrations that cause its progression and metastasis (Pani et al., 2010; Lee and Kang, 2013; Lee et al., 2017; Liao et al., 2019; Hayes et al., 2020; Fares et al., 2020; Ko et al., 2021). The seemingly conflicting effects of oxidative stress on carcinogenesis and metastasis underline the fact that much yet remains to be learned regarding metastatic mechanisms.

Empirically, however, much is known about metastasis. Cancer is the leading cause of death worldwide and metastasis is responsible for approximately 90% of these deaths (Lee et al., 2017). When primary cancers metastasize, they do so to specific new sites rather than spreading haphazardly, and maintain the characteristics of the primary ones (Welch and Hurst, 2019). Thus, when breast cancer metastasizes to the lung, it is still considered breast cancer not lung cancer. Table 21.4 lists common sites of metastasis (Nguyen and Massague, 2007; Chiang and Massague, 2008; Welch and Hurst, 2019; National Cancer Institute, 2020a).

TABLE 21.4 Common sites of metastasis.

Cancer type	Primary metastasis sites
Bladder	Bone, liver, lung, brain, prostate, colon
Breast	Bone, brain, liver, lung
Colon	Liver, lung, peritoneum
Kidney	Adrenal gland, bone, brain, liver, lung
Lung	Adrenal gland, bone, brain, liver, other lung

(Continued)

TABLE 21.4 Common sites of metastasis.—cont'd

Cancer type	Primary metastasis sites
Melanoma (skin)	Bone, brain, liver, lung, skin, muscle
Melanoma (ocular)	Liver
Ovary	Diaphragm, liver, lung peritoneum
Pancreas	Liver, stomach. lung, peritoneum
Prostate	Adrenal gland, bone, liver, lung
Rectal	Liver, lung, peritoneum
Stomach	Liver, lung, peritoneum
Testes	Lymph nodes, lung, liver
Thyroid	Bone, liver, lung
Uterus	Bone, liver, lung, peritoneum, vagina

The data in Table 21.4 show instances of metastasis of some cancers to new organs due to circulatory patterns. These include colon, ovarian, pancreatic, prostatic, testatic and urinary metastasis. Other cancers; breast, lung, ocular, prostate, stomach and urinary ones, show metastasis to sites that both can and cannot be explained by circulatory patterns (Welch and Hurst, 2019). This further underlines our incomplete knowledge of metastasis.

21.9 Mechanisms associated with specific cancers

The extensive literature published regarding the mechanisms associated with cancer leaves no doubt that oxidative stress is the key factor in cancer initiation, promotion, progression and metastasis. The sections that follow address the specific OS-related mechanistic factors associated with eight of the most common cancer types: lung cancer, breast cancer, prostate cancer, colorectal cancer, melanoma, leukemia, non-Hodgkin lymphoma and kidney cancer.

21.9.1 Lung cancer

Lung cancer is the leading cause of cancer death worldwide. There are several different lung cancer types, the four most common of which are lung nodules, small cell lung cancer, non-small cell lung cancer and mesothelioma (Hopkins Medicine, 2021).

1. Lung nodules. Small tissue masses that can be benign, precancerous or metastatic tumors that have spread from other body parts. As a general rule, the larger the nodule the more likely it is to be cancerous.
2. Small cell lung cancer. The fastest growing lung cancer, with almost all incidences being due to cigarette smoking.

3. Non-small cell lung cancer. The most common type of lung cancer that grows and spreads slower than small cell cancer. The three varieties, adenocarcinoma, large cell carcinoma and squamous cell carcinoma are named for the cell types in these tumors.
4. Mesothelioma. Cancer of the chest lining, most attributable to asbestos exposure. The induction period is long, generally 30–50 years following exposure to asbestos.

Almost any primary cancer from another organ can metastasize to the lungs. The organs from which this occurs most often include the bladder, breast, colon, kidney and prostate.

21.9.1.1 Causes of lung cancer

The causes of lung cancer that have been identified include (Dela Cruz et al., 2011):

1. Tobacco smoking
2. Genetics
3. Diet
4. Obesity
5. Infections
6. Biomass and wood and coal smoke
7. Outdoor Air pollution
8. Occupational chemical exposure
9. Radon

Mechanistic and empirical highlights of these follow.

21.9.1.1.1 Tobacco smoking

Tobacco smoke is a complex mixture of vapors and particulates containing thousands of individual compounds, more than 60 of which are identified carcinogens (IARC, 2021). N-nitrosamines (NNSAs), formed during tobacco processing as well as while smoking, can bind to DNA to create DNA adducts, which, when not repaired, lead to permanent mutations. NNSAs can mediate an array of signaling pathway activation that includes modulation of critical oncogenes and tumor suppressor genes that ultimately result in uncontrolled cancerous cell proliferation and tumorigenesis (Akopyan and Bonavida, 2006).

It is to be noted that those exposed to second hand tobacco smoke, (also called environmental tobacco smoke) also show elevated incidence of lung cancer (Dela Cruz et al., 2001). It has been estimated that 17% of lung cancers in nonsmokers are due to second hand tobacco smoke exposure during childhood and adolescence (Janerich et al., 1990).

21.9.1.1.2 Genetics

The pathogenesis of lung cancer has a genetic factor. Significantly elevated risk associated with having relatives diagnosed with the disease (Matakidou et al., 2005).

21.9.1.1.3 Diet

Dietary factors contribute to the risk for lung cancer. Dietary fat intake is associated with increased risk for lung cancer, while diets rich in the antioxidants beta-carotene and total carotenoids lower lung cancer risk, by protecting against oxidative stress (Nyberg et al., 1998).

21.9.1.1.4 Obesity

There is an obesity paradox in lung cancer. General obesity seemingly protects against lung cancer, while abdominal obesity appears to play a role in its development (Hidayat et al., 2016; Ardesh et al., 2020). Though the precise mechanism for the association between obesity and lung cancer remains unknown, it has been speculated that a mechanism involving complex biological pathways such as hyperinsulinemia, decreased levels of sex hormone binding globulin and increased levels of unbound androgens and estrogens is responsible. All of these pathways are more strongly related to abdominal obesity than to body obesity and several in vitro studies have revealed that small-cell lung cancer cells contain receptors for steroid hormones, including androgens and estrogens (Hidayat et al., 2016).

21.9.1.1.5 Infections

The evidence for connection between infections and lung cancer is, at this point, still not definitive. Human papillomavirus, Epstein–Barr virus and Chlamydia pneumonia are suspected as being causative of lung cancer (Dela Cruz et al., 2001). Lung cancer incidence, however, is significantly increased in individuals with Acquired Immune Deficiency Sydrome (AIDS) (Shebl et al., 2010).

21.9.1.1.6 Biomass, wood and coal smoke

Approximately three billion people worldwide burn wood, biomass (crop residues, twigs and sticks) or coal indoors as their primary sources of energy for cooking and heating. Incomplete combustion of these results in the emission of carcinogenic PAHs and other ROS species (Wilk et al., 2013; Abdulazeez, 2017). Numerous studies have demonstrated the association between the use of these fuels and lung cancer (Kleinerman et al., 2002; Lissowska et al., 2005).

21.9.1.1.7 Outdoor air pollution

As discussed in Chapter 4, air quality has deteriorated to where it is estimated that more than 90% of the world's population regularly breathes contaminated air. The International Agency for Research on Cancer (IARC) has identified 529 agents as either carcinogenic, or potentially carcinogenic to humans, as follows (IARC, 2021):

Group 1	Carcinogenic to humans
Group 2A	Probably carcinogenic to humans
Group 2B	Possibly carcinogenic to humans

Many, if not most of these are components of outdoor air pollution. The Group 1 and Group 2A outdoor air pollutant species specifically identified include metals and fibers, organic chemicals, polynuclear aromatic hydrocarbons and mixtures. Representatives of these, which may be present in air as vapors or particulates, are shown in Table 21.5.

Polluted air varies widely in composition and can be composed of unlimited different mixtures. The pollutants listed in Table 21.5, individually and in mixtures attack all body systems

TABLE 21.5 IARC established or probable outdoor air carcinogens.

Metals and fibers
Arsenic and inorganic arsenic compounds
Asbestos
Beryllium and beryllium compounds
Cadmium and cadmium compounds
Chromium (VI)
Lead compounds
Nickel compounds
Silica dust
Organic chemicals
1,3-butadiene
Benzene
Ethylene oxide
Formaldehyde
Halogenated chemicals
Bis(chloromethyl) ether
Ethylene dibromide
2,3,7,8-tetrachlorodibenzo-p-dioxin
Tetrachloroethylene
Trichloroethylene
1,2,3-trichloropropane
Vinyl bromide
Vinyl chloride
Vinyl fluoride
Polynucear aromatic hydrocarbons
Benzo[a]pyrene
Cyclopenta[cd]pyrene
Dibenz[a,h]anthracene
6-nitrochrysene
Nitropyrene
2-nitrotoluene

(Continued)

TABLE 21.5 IARC established or probable outdoor air carcinogens.—cont'd

Mixtures
Biomass fuel combustion products
Coal combustion products
Coal tar pitch
Coke production products
Creosotes
Diesel engine exhaust
Mineral oils
Polychlorinated biphenyls
Tobacco smoke
Wood dust

and organs via a variety of mechanisms. Combinations of air pollution components with other cancer-causing parameters, further complicate mechanistic considerations.

21.9.1.2 Occupational chemical exposure

Exposures in the workplace are to many of the same species listed in Table 20.5. Indeed, occupational exposures to these are most often to concentrations that are much higher than those contained in polluted air (Zeliger, 2011).

21.9.1.3 Radon

All ionizing radiation is carcinogenic. By definition, ionizing radiation produces free radicals. Alpha-particles, gamma-rays and X-rays are established lung carcinogens (El Ghissassi et al., 2009). Radon-222 is an alpha-particle emitting gas that produces ROS that cause oxidative damage to DNA and has been well documented as both an occupational (in underground structures) and environmental (in buildings such as offices, schools and homes) lung carcinogen (Field and Withers, 2012).

Given the number of different lung cancer types and the numerous causative agents, specific mechanisms still remain to be totally defined. All of the cancer types and all of the causative agents, however, are associated with elevated oxidative stress.

21.9.2 Breast cancer

Breast cancer is the most prevalent of all cancers in women worldwide and it is estimated that about one in eight United States women will develop this disease in her lifetime (Siegel et al., 2020). Table 21.6 lists the known risk factors for cancer (Spivey, 2010; Zeliger, 2011; Feng et al., 2018; White et al., 2018; Majhi et al., 2020; Breast cancer.org, 2021). Several of these factors (the ones starred*) are beyond an individual's ability to control, making the prevalence of breast cancer difficult to address.

TABLE 21.6 Known cancer risk factors.

1. Being a woman*—Breast cancer is about 100 times more prevalent in women than in men.
2. Age*—Aging increases the risk for breast cancer. Most cases are diagnosed in women aged 55 or older.
3. Family history*—The risk of developing breast cancer nearly doubles if a first degree relative (mother, sister or daughter) has been diagnosed with the disease.
4. Genetics*—A woman with a BRCA1 or BRCA2 gene mutation has an 70% chance of developing breast cancer by age 70.
5. Personal breast cancer history*—Those who have been previously diagnosed with breast cancer are 3–4 times more likely to develop cancer in the other breast or a different part of the same breast.
6. Prior chest or face radiation. Radiation treatment prior to age 30 for another cancer (e.g., Hodgkin's disease or non-Hodgkin's lymphoma) elevates the risk of developing breast cancer.
7. Prior breast changes*—Diagnosis with benign breast conditions elevates breast cancer risk.
8. Race/ethnicity*—White woman are more likely to develop breast cancer than Black, Hispanic or Asian Women. Black women, however, are more likely than White women to develop breast cancer at a younger age in a more aggressive form.
9. Being overweight—Being overweight or obese elevates breast cancer risk in women, particularly post menopause. Overweight and obese women also have an increased risk of breast cancer recurrence than those who maintain a healthy weight.
10. Pregnancy history. Having a first full-term pregnancy after the age of 30 raises the risk of breast cancer onset, compared with cohorts who have done so at a younger age.
11. Breastfeeding history. Breastfeeding, particularly if done so for longer than 1 year, lowers breast cancer risk.
12. Menstrual history*—Women who started menstruating prior to the age of 12 or who go through menopause above age 55 have an increased risk of breast cancer.
13. Using hormone replacement therapy. Woman participating in hormone replacement therapy have an elevated breast cancer risk.
14. Drinking alcohol. Chronic alcohol consumption increases a woman's risk of hormone-receptor-positive breast cancer.
15. Smoking. Smoking cigarettes is linked to an elevated breast cancer risk in younger, premenopausal women.
16. Having dense breasts*—Women with dense breasts may be twice as likely to develop breast cancer than those with non-dense breasts.
17. Lack of exercise. Moderate or intense four to 7 h-per-week exercise lowers the risk of developing breast cancer.
18. Low vitamin D levels. There is an inverse relationship between low vitamin D levels and higher breast cancer risk.
19. Light at night. Women who are night workers, or who live in areas with high external light at night (from street and sign lights), have a higher risk of breast cancer than their day worker, or ones not exposed to night time lighting cohorts.
20. Diethylstilbestrol (DES) exposure. Women who took DES in the 1940s–1960s to prevent miscarriage, as well as their daughters who were exposed to DES in utero have elevated breast cancer risks.

(*Continued*)

TABLE 21.6 Known cancer risk factors.—cont'd

21. Poor diet. Eating high fat, high sugar and high quantity grilled, barbecued and smoked meats results to an increased risk of getting breast cancer.
22. Exposure to chemicals in food. Pesticides, antibiotics, hormones and solvents used on crops and livestock feed, as well as non-nutritious chemicals (solvents and processing compounds) raise the risk for breast cancer onset.
23. Exposure to chemicals in cosmetics. Preservatives and other chemicals formulated into cosmetics, sunscreen and personal care products have been shown to raise breast cancer risk.
24. Exposure to air pollution. A number of studies have found an association between chronically breathing polluted air and an elevated breast cancer risk.

There are many types of breast cancers that include a heterogeneous and phenotypically diverse group composed of several biological subtypes that have distinct behaviors and molecular signatures including estrogen receptor expression, human epidermal growth factor 2 expression, proliferation and basal cluster (Feng et al., 2018).

Human mammary gland development is a process that starts in utero, with normal development and cancer progression having molecular level parallels (Macias and Hinck, 2012; Heubner and Ewald, 2014). The drivers of breast carcinogenesis are dysregulation and hijacking of normal signaling pathways that allow cells to communicate with each other and their surrounding environment. Thus cancer is driven by genetic and epigenetic alterations that allow cancer cells to avoid the normal mechanisms that control their proliferation, survival and migration (Sever and Brugge, 2015).

As many as 20% of women with a family history of breast cancer have a mutation in the breast susceptibility genes 1 or 2 (BRCA1 or BRCA2). Mechanistically, BRCA proteins inhibit tumorigenesis by repairing DNA damage. Loss of this function via mutation raises the risk of breast cancer by five to six-fold (Trainer et al., 2010; Narod and Salmena, 2011; Couch et al., 2014).

Epigenetic changes (DNA methylation and histone modification) play important roles in breast cancer causation and progression. Non-coding RNAs (ncRNAs) are key regulators of epigenetic mechanisms and heavily involved in their dysregulation. Most mammalian transcriptions have been shown to encompass ncRNAs that are vital to gene expression and ncRNA dysfunction is associated with breast cancer development (Baylin and Jones, 2011; Suzuki et al., 2016; Lo et al., 2016).

Breast cancers, both intra-tumor in the same individuals and inter-tumor in different individuals are heterogenous, due to the presence of heterogeneous cell populations within individual breast tumors (Ellsworth et al., 2017). Breast cancer is highly metastatic, with 10%–15% of patients developing distant metastases within 3 years of diagnosis with the primary tumor and metastases developing at distant sites 10 years or more after the first diagnosis. The heterogeneity of breast cancer makes it difficult to determine the risk for metastasis (Weigelt et al., 2005; Scully et al., 2012; Yates et al., 2017). It has been found that metastases occurring late after primary tumor diagnosis contain mutations not observed in the primary tumors. This further illustrates that the mechanisms responsible for metastasis are yet to be fully understood.

21.9.3 Prostate cancer

Prostate cancer is the second leading cause of death in men (Testa et al., 2019; ASCO, 2020). Though primarily a cancer in the aged, prostate cancer also strikes younger men. Most prostate cancers cannot be ascribed to a specific cause, but a number of causes have been identified. Twelve of these are listed in Table 21.7 (ASCO, 2020; Facty.com, 2021).

Though no mechanistic pathways have been identified with the items in Table 21.7, all the factors starred (*) are associated with elevated oxidative stress.

Normal prostate cells as well as cancerous prostate cells are androgen-dependent and almost all metastatic prostate cancers require testosterone for growth and proliferation. Accordingly, prostate cancer is treated via surgical or pharmacological androgen deprivation (Pienta and Bradley, 2006; Schrecengost and Knudson, 2013; Weischenfeldt et al., 2013; Testa et al., 2019).

Mechanistically, prostate cancer cells depend upon the androgen receptor as the primary mediator of growth and survival (Debes and Tindall, 2004; Testa et al., 2019). When testosterone enters the cell, it is converted to its active metabolite, dihydrotestosterone, with a five to ten-times greater affinity for the androgen receptor. Dihydrotestosterone binds androgen receptors (ARs) in the cytoplasm, resulting in phosphorylation, dimerization and

TABLE 21.7 Causes of prostate cancer.

1. Aging.* Aging is the primary cause of prostate cancer onset.
2. Weight.* Being overweight or obese is associated with increased odds of getting prostate cancer.
3. Race. Black men of African descent are more like likely than white men to develop prostate cancer.
4. Ethnicity. Prostate cancer is more prevalent in North America or Northern Europe that in other parts of the world, though its prevalence is increasing among Asian men who reside in urban environments.
5. Genetics.* Men with fathers and brothers diagnosed with the disease as well as relatives with breast cancer or BRCA1 and BRCA2 genes are more likely to get prostate cancer than their cohorts without such genetic connections.
6. Lack of exercise.* Prostate cancer is more prevalent in those who lead sedentary lifestyles than in those who exercise regularly.
7. High calcium diet. For unknown reasons, a diet high in calcium is related to prostate cancer onset.
8. Cancer history.* Previous cancers, most notably those of the kidney, bladder or lung, are associated with prostate cancer onset.
9. Hormone factors.* Those with high growth hormone factor IGF-1 are more likely to develop prostate cancer than those with normal levels.
10. Prostate inflammation.* A history of chronic prostate gland inflammation, pre-disposes one to prostate cancer onset.
11. Height. Being taller makes a man more likely to get prostate cancer than being short or of normal height.
12. Agent orange exposure.* Soldiers exposed to agent orange during the Viet Nam war Have been found to be more likely than veterans not so Exposed to develop prostate cancer.

translocation into the nucleus, thus binding to androgen-response elements in the DNA and resulting in activation of genes involved with cell growth and survival (Feldman and Feldman, 2001).

Prostate cancer cells are able to proliferate despite an androgen blockade by developing an ability to use low levels of androgen for growth. One way this is accomplished is via increased expression of the androgen receptor, allowing enhanced ligand binding (Gregory et al., 2001; Chen et al., 2004; Pienta and Bradley, 2006).

A second mechanism for prostate cell proliferation in spite of androgen blockade is via increased sensitivity of the AR to androgens (Gregory et al., 2001). A third mechanism involves increased local production of androgens by prostate cancer cells themselves (Feldman and Feldman, 2001).

21.9.4 Colorectal cancer

Colorectal cancer (CRC) is the second most common cancer in women and the third most common cancer in men, accounting for about 10% of all new cancers worldwide (Tariq and Ghias, 2016; Wong and Yu, 2019). Colorectal cancer is a multifactorial disease caused by a combination of genetic, lifestyle and environmental factors (Aran et al., 2016). Risk factors for this cancer are shown in Table 21.8 (Mayo Clinic, 2018). The factors starred are all associated with elevated oxidative stress.

TABLE 21.8 Risk factors for colorectal cancer.

1. Age.* A majority of colorectal cancers occur in those older than 50, with the risk increasing with age.
2. Race. African-Americans have a greater risk for developing colorectal cancer than those of other races., for unknown reasons.
3. Polyps.* People who have had non-carcinogenic polyps or a previous bout with colorectal cancer have increased risks of future CRC onset.
4. Genetics.* CRC in blood relatives is indicative of a greater likelihood of developing this cancer.
5. Diet.* Eating a low-fiber, high-fat diet characteristic of the Western diet carries a higher risk of CRC onset. Diets rich in red and processed meats have also been found to increase CRC onset risk.
6. Inactivity.* People leading sedentary lifestyles are more likely to develop CRC than those who regularly exercise.
7. Diabetes.* Diabetes or insulin resistance increases the risk for CRC.
8. Weight.* Obese people have a higher risk for CRC than those with normal weight.
9. Smoking.* Smoking increases CRC risk.
10. Alcohol.* Heavy alcoholic use increases the risk for CRC.
11. Radiation.* Radiation therapy for a previous cancer elevates the CRC risk.

Approximately 65% of CRC cases have no family history or apparent genetic predisposition (Burt, 2007). CRC is a heterogenous disease in which one or a combination of three mechanisms are causative (Tariq and Ghias, 2016; Nguyen and Duong, 2018) These are: chromosomal instability, CpG island methylator phenotype and microsatellite instability.

1. Chromasomal instability (CIN)

CIN is the predominant mechanism, accounting for 65%–80% of sporadic CRCs with changes in chromosome number and structure that include gains or losses of chromosomal segments, chromosomal rearrangements and loss of heterozygosity, resulting in gene copy number variations. These changes affect the expression of tumor-associated genes and thereby activation of pathways essential for CR initiation and progression.

2. CpG Island methylator phenotypE (CIMP)

In the CIMP mechanism, hypermethylation in the promoter region leads to transcriptional inactivation of genes that are tumor suppressive or are part of the cell cycle. Mutations in the BRAF gene arise early in the CIMP tumors.

3. Microsatellite instability (MSI)

MSI occurs due to inactivating mutations in the DNA mismatch repair genes that are responsible for correcting DNA replication errors.

CRC is a slow developing cancer in the early stage, progresses moderately in the intermediate stages and rapidly in the final stage. By all mechanisms, the progression of events leading to CRC is shown by the following steps and timeline (Nguyen and Duong, 2018). Non-coding RNAs are implicated in various CRC stages of all three mechanisms (Tariq and Ghias, 2016).

1. Normal Epithelium (over many decades) → Early Adenoma
2. Early adenoma (over 2–5 years) → Intermediate Adenoma
3. Intermediate Adenoma (over 2–5 years) → Late Adenoma
4. Late Adenoma (rapidly) → Cancer

It is thought, but not definitively established, that gut microbiota play a role in colorectal cancer. Gut microbiota interact with host cell to regulate numerous physiological processes, including, but not limited to, energy harvest, metabolism and immune response. Sequencing studies have shown microbial composition and ecological changes in people with CRC and animal studies have demonstrated the role of several bacteria in colorectal carcinogenesis (Wong and Yu, 2019).

Metastasis is the primary cause of death in those with CRC, with the most common sites for CRC metastasis being the liver and peritoneum (Pretzsch et al., 2019). In CRC liver metastasis, ADAM9, which is secreted by hepatic stellate cells, bond to CRC cells and promotes carcinoma invasion through tumor-ECM interaction (Mazzocca et al., 2008; Li et al., 2016).

21.9.5 Melanoma

Skin cancer is the most common type of cancer, with approximately half of the diagnosed cancers in the United States. Though melanomas number only about 4% of all skin cancers, they are the most lethal, accounting for about half of all skin cancer deaths (Skin Cancer Foundation, 2021).

It is estimated that most (60%–70%) of malignant melanomas are caused by ultraviolet (UV) radiation. Sun exposure provides both UVA (315–400 nm) and UVB (280–315 nm) radiation, and indoor tanning beds, which can provide radiation doses of up to 12 times that of the sun, are also a major source of UVA radiation (Koh et al., 1996). Both UVA and UVB have been shown to induce melanoma (Noonan et al., 2012).

Familial genetic predisposition for melanoma is a distant secondary source of this disease, accounting for only an estimated 3%–15% of cases (Debniak, 2004). Familial inherited melanomas most often show changes in tumor suppressor genes (CDKN2A or CDK4) (American Cancer Society, 2020).

A positive association between oxidative stress and the clinical stages of melanoma is well established, as UV contains sufficient energy to break chemical bonds, thus producing free radicals (Bisevac et al., 2018).

Melanocytes produce two types of melanin: eumelanin and pheomelanin. Eumelanin is most common in people with dark skin and dark hair, while pheomelanin predominates in those with light skin, freckles and blond or red hair. Eumelanin lowers the accumulation of UV-induced photoproducts, while pheomelanin is thought to contribute to UV-caused DNA damage by inducing free radical formation following UV exposure (Nasti and Timares, 2015; Raimondi et al., 2008).

Oxidative stress is the primary driver of melanoma, most frequently producing a high variety of somatic mutations, most often involving the BRAF and NRAS genes, but also the c-KIT and PTE genes as well (Venza et al., 2015; Sample and He, 2018; Rossi et al., 2019; Cannavo et al., 2019).

Though primarily a skin disease, melanoma also occurs at other sites, including mucous membranes (hard palate, maxillary gingiva, lip, throat, esophagus, vulva, vagina and perianal region) and the eye (uvea and retina) (Venza et al., 2015). Melanoma may also appear in sun-protected areas of the skin, e.g., palms, soles or subungual regions (acral and mucosal melanomas) (Merkel and Gerami, 2017). The reader is referred to the literature cited for elaboration in these areas.

Vitamins A, C, D, E and K have been found to protect against melanoma onset (Sample and He, 2018).

21.9.5.1 Vitamin A

Vitamin A is tumor suppressive in melanoma, inhibiting growth, invasion and metastasis of melanoma cells.

21.9.5.2 Vitamin C

Vitamin C is thought to show an inverse dose-response relationship in melanoma growth and progression. High vitamin C concentrations inhibit invasion and survival of melanoma cells, and low vitamin C concentrations promote melanoma metastasis.

21.9.5.3 Vitamin D

Vitamin D3 production in the skin is induced in a dose-dependent relationship which depends upon melanin levels and skin type. Dietary intake of vitamin D, however, apparently has no effect upon melanoma risk.

21.9.5.4 *Vitamin E*

Vitamin E has two major forms, tocotrienols and tocopherols. Both inhibit melanoma cell growth and induction of apoptosis.

21.9.5.5 *Vitamin K*

Several forms of vitamin K (including K3 and K5) inhibit proliferation and increase apoptosis of melanoma cells.

UV radiation is a major risk factor for developing melanoma. The risk is elevated by childhood UV exposure as well as by the number of sunburns, or trips to indoor tanning beds throughout life. Oxidative stress is the major driving factor for melanoma onset and plays a major role in the onset, progression and metastasis of the disease. Avoidance of excessive UV exposure coupled with a healthy diet offers protection against melanoma onset.

21.9.6 Leukemia

Leukemia is a hematopoietic malignancy characterized by abnormal proliferation of immature hematopoietic system cells. Different types of leukemia come from different cells as follows (Ntzoachristos et al., 2013):

TABLE 21.9

Leukemia type	Cells of origin
Lymphocytic leukemia	Lymphocytic
Myeloid leukemia	Myeloid
Erythrocytic leukemia	Erythrocytic

Points of origin of other leukemias include the bone marrow, lymph nodes and spleen. Whatever the cell type of origin, leukemia can proceed a via long incubation period (chronic leukemia) or abruptly (acute leukemia). The most common forms of this disease are Chronic myeloid leukemia (CML), acute myeloid leukemia (AML), chronic lymphocytic leukemia (CLL) and acute lymphoblastic leukemia (ALL).

CML: A unique form of leukemia that it is characterized by the presence of the Philadelphia chromosome. AML: The most common form of leukemia overall whose incidence increases with age. CLL: The most common form of adult leukemia that progresses very slowly.

ALL: The most common form of childhood leukemia.

Mechanistically, DNA methylation has been found to fit all three criteria required to demonstrate epigenetics as driver of disease onset and progression (Bonasio et al., 2010; Ntziachristos et al., 2013).

1. A signal must be propagated via DNA replication/cell division.
2. The signal must be transmitted to progeny.
3. The signal must affect gene expression.

Though the exact causes of leukemia remain to be determined, several chemical leukemogens (Adamson and Seiber, 1981) and several other risk factors (Mayo Clinic, 2021; CTCA, 2021) for leukemia have been identified. The chemicals and their uses are listed in Table 21.9 and the other risk factors are shown in Table 21.10.

TABLE 21.10 Chemical leukemogens.

Chemical	Use
Benzene	Industrial solvent Chemical intermediate Gasoline component
Procarbazine	Chemotherapy
Melphalan	Chemotherapy
Thio-TEPA	Chemotherapy
Chlorambucil	Chemotherapy
Cyclophosphamide	Chemotherapy
Chloramphenicol	Antibiotic
Phenylbutazone	Anti-inflammatory

TABLE 21.11 Risk factors for leukemia.

AGE
The risk of most leukemias increase with age. Most cases of ALL occur, however in people under age 20.
Gender
Men have a higher likelihood of developing leukemia than women do.
Blood disorders
Polycythemia vera, idiopathic myelofibrosis, and essential thrombocytopenia elevate AML risk.
Family history
Being a first-degree relative of a CLL patient, or having monozygotic twin with AML or ALL increases the risk for leukemia.
Genetic disorders
Down syndrome, Fanconi anemia, Bloom syndrome, ataxia-telangiectasia and Blackfan-Diamond syndrome elevate the risk for AML.
Previous cancer treatment
Those who have undergone radiation and some types of chemotherapy are at increased risk for leukemia.

TABLE 21.11 Risk factors for leukemia.—cont'd

Smoking
Smoking cigarettes increases the risk for AML.
Radiation
Chronic exposure to electromagnetic radiation from power lines raises the risk for developing ALL.

Of all the known and suspected leukemogens, benzene has been the most extensively studied. The following oxidative stress driven factors have been identified as probable for inducing leukemia (McHale et al., 2011).

1. Targeting critical genes and pathways via induction of genetic, chromosomal or epigenetic abnormalities and genomic instability in hematopoietic stems cells (HSCs).
2. Stromal cell dysregulation.
3. Apoptosis of HSCs and stromal cells.
4. Altered proliferation and differentiation of HSCs.

21.9.7 Non-Hodgkin lymphoma

Non-Hodgkin lymphoma (NHL) is a cancer originating from immature or mature B cells and T cells. There are various types of NHL with differing etiologies, epidemiologies, immunophenotypic, genetic and clinical features. The disease can progress slowly (over many years) or aggressively (Hunt, 2017; Sapkota and Shaikh, 2021).

Several risk factors for NHL onset have been identified. These include infections, pharmaceuticals. environmental exposures, immunodeficiency states, immune disorders, and chronic inflammation (Zhang et al., 2008; Erickson et al., 2008; Anderson et al., 2009; Sapkota and Shaikh, 2021). Table 21.11 lists sources of NHL.

All of the sources in Table 20.9 are associated with oxidative stress (Zeliger, 2016).

TABLE 21.12 Sources of NHL.

Infections
Epstein–Barr virus with Burkitt lymphom
Hepatitis C virus with Splenic marginal zone lymphoma
Human T-cell leukemia virus type 1 with adult T-cell lymphoma
Helicobacter pylori infection with lymphoid tissue lymphomas
Pharmaceuticals
Phenytoin

(*Continued*)

TABLE 21.12 Sources of NHL.—cont'd

Digoxin
Chemotherapy
Immunosuppressants
Environmental exposures
Pesticides
Herbicides
Wood preservatives
Organic chemicals
Solvents
Hair dye
Radiation exposure
Immunodeficiency states
Wiscott-Aldrich syndrome
Severe combined immunodeficiency disease
Induced immunodeficiency states
Immune disorders
Sjogren syndrome
Rheumatoid arthritis
Hashimoto thyroiditis
Celiac disease
Chronic inflammation
Chronic inflammation from any of many multiple sources.

21.9.8 Kidney cancer

There are several types of kidney cancer, the main types being clear cell renal cell carcinoma (ccRCC), transitional cell cancer and Wilms tumor [WIKI # 1]. ccRCC is the most common, accounting for about 80% of all kidney cancer cases Makhov et al. (2017) and the subject of this chapter. The incidence of ccRCC continues to rise, with more than 76,000 new cases predicted in the United States alone for the year 2021 (National Cancer Institute, 2021a).

Risk factors for kidney cancer are shown in Table 21.12 (Hu et al., 2002; ASCO, 2021).

RCC is a heterogenous epithelial malignancy with different subtypes and varied tumor biology (Lim et al., 2007). The origin of ccRCC is thought to arise from proximal tubules, with the VHL gene responsible for it identified on a small locus of the short arm or chromosome 3. This gene acts a loss-of-function tumor-suppressor one that is mutated or a product

TABLE 21.13 Risk factors for kidney cancer onset.

1. Smoking—Tobacco smoking doubles the risk for developing kidney cancer
2. Chemical exposures—Exposures to the following chemicals has been shown to elevate the risk of kidney cancer onset: Benzene Benzidine Cadmium Coal tat Pitch Creosote Asphalt Pitch Soot Herbicides Pesticides Mineral, cutting or lubricating oil Mustard gas Vinyl chloride
3. Gender—Men are two to three times as likely to develop kidney cancer than women are.
4. Race—Black people have higher rates of kidney cancer than white people do.
5. Family history—Those with a strong family history of kidney cancer are more likely to develop the disease than those without such background.
6. Weight—Obesity is an established risk factor for kidney cancer.
7. Hypertension—Men with hypertension are at higher risk for developing kidney cancer
8. Pharmaceuticals—Diuretics, painkillers containing phenacetin, analgesics including aspirin, acetaminophen and ibuprofen have been linked to elevated risk for kidney cancer.
9. Long term dialysis—Those who have been on dialysis for long periods of time are at increased risk of developing cancerous kidney cysts.
10. Previous bladder cancer—Prior bladder cancer raises the risk for kidney cancer onset.
11. Genetic conditions—A number of genetic conditions increase the risk for kidney cancer. These are: Von Hippel-Landau syndrome Hereditary papillary renal cell carcinoma Birt-Hogg-Dube syndrome Hereditary leiomyomatosis Tuberous sclerosis complex syndrome Succinatge dehydrogenase complex syndrome BAP1 tumor predisposition syndrome

of methylation (Herman et al., 1994). Epigenetic regulatory mechanisms (genome-wide methylation and inhibition of protein translation via interaction of microRNA with its target messenger RNA) are central to pathogenesis of ccRCC (Nabi et al., 2018; Braga et al., 2019). Mutation of the genes associated with ccRCC causes dysregulation of the tumor's responses to changes in oxygen, iron, nutrient or energy levels in the tumor's microenvironment (Linehan et al., 2019; Zhang et al., 2021).

21.10 Summary

Oxidative stress is the major factor in the identified or proposed mechanisms of all aspects of cancer initiation, progression and metastasis. Though genetics have been shown to play a role in cancer onset, environmental factors are the primary carcinogens. Cancers in the same organ can have different triggers and proceed by different mechanisms and the same causative effect(s) can induce cancer in multiple body parts. Driver mutations of cancer can often start years or even decades prior to diagnosis. Reducing oxidative stress elevating factors and treatments with antioxidants can prevent or slow some, but not all, cancers. Our understanding of cancer is still in its infancy.

References

AACR, 2021. Air pollution may be associated with many kinds of cancer. Am. Assoc. Cancer Res. https://www.aacr.org/patients-caregivers/progress-against-cancer/air-pollution-associated-cancer/ Accessed September 15, 2021.

Abdulazeez, T.L., 2017. Polycyclic aromatic hydrocarbons. A review. Cogent Environ. Sci. 3, 1. https://doi.org/10.1080/23311843.2017.1339841. Accessed September 18, 2021.

Adamson, R.H., Seiber, S.M., 1981. Chemically induced leukemia in humans. Environ. Health Perspect. 39, 93–103.

Akopyan, G., Bonavida, B., 2006. Understanding tobacco smoke carcinogen NNK and lung tumorigenesis. Int. J. Oncol. 29 (4), 745–752.

American Cancer Society 2020. www.americancancersociety.org/melanoma. Accessed August 23, 2022.

Anderson, L.A., Gadalla, S., Morton, L.M., Landgren, O., Pfieffer, R., Warren, J.L., et al., 2009. Population-based study of autoimmune conditions and the risk of specific lymphoid maliganacies. Ind. J. Cancer 125 (2), 398–405.

Ardesch, F.H., Ruiter, R., Mulder, M., Lahousse, L., Stricker, B.H.C., Kiefte-de Jong, J.C., 2020. The obesity paradox in lung cancer: associations with body size versus body shape. Front. Oncol. https://doi.org/10.3389/fonc.2020.591110. Accessed September 17, 2021.

Arfin, S., Jha, N.K., Jha, S.K., Kesari, K.K., Ruokolainen, J., Roychoudhury, S., et al., 2021. Oxidative stress in cancer cell metabolism. Antioxidants. https://doi.org/10.3390/antiox10050642. Accessed September 17, 2021.

Aron, V., Victorino, A.P., Thuler, L.C., Ferreira, C.G., 2016. Colorectal cancer: epidemiology, disease mechanisms and interventions to reduce onset and mortality. Clin. Colorectal Cancer 15 (3), 195–203.

ASCO, 2020. Prostate cancer: risk factors and prevention. https://www.cancer.net/cancer-types/prostate-cancer/risk-factors-and-prevention. Accessed September 24, 2021.

ASCO, 2021. Kidney cancer: risk factors and prevention. https://www.cancer.net/cancer-types/kidney-cancer/risk-factors-and-prevention. Accessed September 30, 2021.

Balali-Mood, M., Naseri, K., Tahergorabi, Z., Khazdair, M.R., Sadeghi, M., 2021. Toxic mechanisms of five heavy metals: mercury, lead, chromium, cadmium and arsenic. Front. Pharmacol. https://doi.org/10.3389/fphar.2021.643972. Accessed September 8, 2021.

Baniyash, M., 2006. Chronic inflammation, immunosuppression and cancer: new insights and outlook. Semin. Cancer Biol. 16, 80–88.

Baylin, S.B., Jones, P.A., 2011. A decade of exploring the cancer epigenome – biological and translational implications. Nat. Rev. Cancer 11 (10), 726–734.

Bisevac, J.P., Djukic, M., Stanojevic, I., Mijuskovic, Z., Djuric, A., Gobeljic, B., et al., 2018. Association between oxidative stress and melanoma progression. J. Med. Biochem. 37, 12–20.

Hosgood 3rd, H.D., Bofetta, P., Greenland, S., Lee, Y.C.A., McLaughlin, J., Seow, A., et al., 2010. In-home coal and wood use and lung cancer risk: a pooled analysis of the International Lung Cancer Consortium. Environ. Health Perspect. 118 (12), 1743–1747.

Bonasio, R., Tu, S., Reinberg, D., 2010. Molecular signals of epigenetic states. Science 330, 612–616.

Braga, E.A., Fridman, M.V., Loginov, V.I., Dmitriev, A.A., Morozov, S.G., 2019. Molecular mechanisms in clear cell renal cell carcinoma: role of miRNAs and hypermethylated miRNA genes in crucial oncogenic pathways and processes. Front. Genet. https://doi.org/10.3389/fgene.2019,00320. Accessed September 30, 2021.

Breast, 2021. Breast Cancer Risk Factors. cancer.org. https://www.breastcancer.org/risk/factors?gclid=EAIalQobChMltMyxgoaO8wlVTJyzCh0NeAUFEAAYAyAAEgKCmPD_BwE. Accessed September 21, 2021.

Burfeind, K.G., Michaelis, K.A., Marks, D.L., 2016. The central role of hypothalamic inflammation in the acute illness response and cachexia. Semin. Cell Dev. Biol. 54, 42–52.

Burt, R., 2007. Inheritance of colorectal cancer. Drug Discov. Today Dis. Mech. 4, 293–300.

Cannavo, S.P., Tonacci, A., Bertino, L., Casciaro, M., Borgia, F., Gangemi, S., 2019. The role of oxidative stress in the biology of melanoma: a systematic review. Pathol. Res. Pract. 215 (1), 21–28.

Cao, L., Liu, X., Lin, E.J., Wang, C., Choi, R.V., al, at, 2010. Environmental and genetic activation of brain-adipocyte BDNF/leptin axis causes cancer remission and inhibition. Cell 142 (1), 52–64.

CDC, 2010. How Tobacco Smoke Causes Disease. Centers for Disease Control and Prevention. https://www.ncbi.nlm.nih.gov/books/NBK53010/. Accessed September 15, 2021.

CDC, 2021. Centers for Disease Control and Prevention. Obesity and Cancer: 13 Cancers Are Associated with Overweight and Obesity. www.cdc.gov. Accessed September 8, 2021.

Chen, C.D., Welsbie, D.S., Tran, C., Baek, S.H., Chen, R., Vessella, R., et al., 2004. Molecular determinants of resistance to antiandrogen therapy. Nat. Med. 10, 33–39.

Chiang, A.C., Massague, J., 2008. Molecular basis of metastasis. N. Engl. J. Med. 359 (26), 2814–2823.

Colotta F, Allavena P, Sica A, Garlanda C, Mantovani A. Cancer-related inflammation, the seventh hallmark of cancer: links to genetic instability. Carcinogenesis; 30:1073-1081.

Couch, F.J., Nathanson, K.L., Offit, K., 2014. Two decades after BRCA: setting paradigms in personalized cancer care and prevention. Science 343 (6178), 1466–1470.

CTCA, 2021. Risk Factors for Leukemia. https://www.cancercenter.com/leukemia. Accessed October 1, 2021.

Dasari, K., Madu, C.O., Lu, Y., 2020. The role of oxidative stress in cancer. NACS 000584. https://doi.org/10.31031/NACS.2020.04.000585. Accessed September 17, 2021.

Debes, J.D., Tindall, D.J., 2004. Mechanisms of androgen-refractory prostate cancer. N. Engl. J. Med. 351 (15), 1488–1490.

Debniak, T., 2004. Familial malignant melanoma – overview. Hered. Cancer Clin. Pract. 2 (3), 123–129.

Dela Cruz, C.S., TAnoue, L.T., Matthay, R.A., 2011. Lung cancer: epidemiology, etiology, and prevention. Clinics Chest Med. 32, 605–644.

El Gassassi, F., Baan, R., Straif, K., Grosse, Y., Secretan, B., Bouvard, V., et al., 2009. A review of human carcinogens-part D: radiation. Lancet Oncol. 10 (8), 751–752.

Ellsworth, R.E., Blackburn, H.L., Shriver, C.D., Soon-Shiong, P., Ellsworth, D.L., 2017. Molecular heterogeneity in breast cancer: state of the science and implications for patient care. Semin. Cell Dev. Biol. 64, 65–72.

Erickson, M., Hardell, L., Carlberg, M., Akerman, M., 2008. Pesticide exposure as a risk factor for Non-Hodgkin lymphoma including histopathological subgroup analysis. J. Cancer 123 (7), 1657–1663.

Facty.com, 2021. Causes of Prostate Cancer. https://www.facty.om/prostate/cancer. Accessed September 24, 2021.

Falanga, A., Russo, L., Milesi, V., Vignoli, A., 2017. Mechanisms and high risk factors of thrombosis in cancer. Crit. Rev. Oncol.-Hematol. 118, 79–83.

Fares, J., Fares, M.Y., Khachfe, J.J., Salhab, H.A., Fares, Y., 2020. Molecular principles of metastasis: a hallmark of cancer revisited. Signal Transduction Targeted Ther. https://doi.org/10.1038/s41392-020-0134-x. Accessed September 17, 2021.

Federico, A., Morgillo, F., Tuccillo, C., Ciardiello, F., Loguercio, C., 2007. Chronic inflammation and oxidative stress in human carcinogenesis. Int. J. Cancer 121 (11), 2381–2386.

Feig, D.I., Reid, T.M., Loeb, L.A., 1994. Reactive oxygen species in tumorigenesis. Cancer Res. 54 (7 Suppl. l), 1890s–1894s.

Feldman, B.J., Feldman, D., 2001. The development of androgen-independent prostate cancer. Nat. Rev. Cancer 1 (1), 34–45.

Feng, Y., Spezia, M., Huang, S., Yuan, C., Zeng, Z., Zhang, L., et al., 2018. Breast cancer development and progression: risk factors, cancer stem cells, signaling, pathways, genomics, and molecular pathogenesis. Genes Dis. 5, 77–106.

Field, R.W., Withers, B.L., 2012. Occupational and environmental causes of lung cancer. Clin. Chest Med. 33 (4). https://doi.org/10.1016/j.ccm.2012.07.001. Accessed September 20, 2021.

Flanagan, R.C., Yanover, P.M., 2001. The role of radical nephrectomy in metastatic renal cell carcinoma. Semin. Urol. Oncol. 19, 98–102.

Gerstrung, M., Jolly, C., Leshchiner, I., Dentro, S.C., Gonzales, S., Rosebrock, D., et al., 2020. The evolutionary history of 2,658 cancers. Nature 578, 122–128.

Gill, J.G., Piskounova, E., Morrison, S.J., 2016. Cancer, oxidative stress, and metastasis. Cold Spring Harb. Symp. Quant. Biol. https://doi.org/10.1101/sqb.2016.81.030791. Accessed September 13, 2021.

Gregory, C.W., Johnson Jr., R.T., Mohler, J.L., French, F.S., Wilson, E.M., 2001. Androgen receptor stabilization in recurrent prostate cancer is associated with hypersensitivity to low androgen. Cancer Res. 61, 2892–2898.

Greten, F.R., Grivennikov, S.I., 2019. Inflammation and cancer: triggers, mechanisms and consequences. Immunity 51, 27–41.

Hanahan, D., Weinberg, R.A., 2000. The hallmarks of cancer. Cell 100, 57–70.

Hanahan, D., Weinberg, R.A., 2011. Hallmarks of cancer: the next generation. Cell 144, 646–674.

Haverkos, H.W., Haverkos, G.P., O'Mara, M., 2017. Co-carcinogenesis: human Papillomaviruses, coal tar derivatives and squamous cell cervical cancer. Front. Microbiol. https://doi.org/10.3389/fmicb.2017/02253. Accessed September 14, 2021.

Hayes, J.D., Dinkova-Kostova, A.T., Tew, K.D., 2020. Oxidative stress in cancer. Cancer Cell 38, 167–196.

Herman, J.G., Lotif, F., Weng, Y., Lerman, M.I., Zbar, B., Liu, S., et al., 1994. Silencing of the VHL tumor-suppressor gene by DNA methylation in renal sarcoma. Proc. Natl. Acad. Sci. USA 91 (21), 9700–9704.

Heubner, R.J., Ewald, A.J., 2014. Cellular foundations of mammary tubulogenesis. Semin. Cell Dev. Biol. 31, 124–131.

Hidayat, K., Du, X., Chen, G., Shi, M., Shi, B., 2016. Abdominal obesity and lung cancer risk: systematic review and meta-analysis of prospective studies. Nutrients. https://doi.org/10.3390/nu8120810. Accessed September 17, 2021.

Holick, M.F., 2014. Cancer, sunlight and vitamin D. J. Clin. Transl. Endocrinol. https://doi.org/10.1016/j.jcte.2014.10.001. Accessed September 8, 2021.

Hopkins Medicine, 2021. Lung Cancer Types. www.hopkinsmedicine,org.health/conditions-and-diseases/lung-cancer/lung-cancer-types. Accessed September 20, 2021.

Hu, J., Mao, Y., White, K., 2002. Renal cell carcinoma and occupational exposure to chemicals in Canada. Occup. Med. 52 (3), 157–164.

Hunt, J., 2017. Molecular mechanisms in Non-Hodgkin lymphoma. Oncohema Key. https://oncohemakey.com/molecular-mechanisms-in-non-hodgkin-lymphoma/. Accessed September 28, 2021.

IARC, 2021. Agents Classified by AIRC Monographs, Volumes 1-129. https://monographs.iarc.who.int/agents-classified-by-the-iarc/. Accessed September 15, 2021.

Irani, K., Xia, Y., Zweier, J.L., Sollott, S.J., Der, C.J., Fearon, E.R., et al., 1997. Mitogenic signaling mediated by oxidants in Ras-transformed fibroblasts. Science 275 (5306), 1649–1652.

Irigaray, P., Belpomme, 2010. Basic properties and molecular mechanisms of exogenous chemical carcinogens. Carcinogenesis 31 (2), 135–148.

Janerich, D.T., Thompson, W.D., Varela, L.R., Greenwald, P., Chorost, S., Tucci, C., et al., 1990. Lung cancer and exposure to tobacco smoke in the household. N. Engl. J. Med. 323 (10), 632–636.

Key, T.J., 1995. Hormones and cancer in humans. Mutat. Res. 333 (1–2), 59–67.

Key, T.J., Bradbury, K.E., Perez-Cornago, A., Sinha, R., Tsilidis, K.K., Tsugane, S., 2020. Diet, nutrition, and cancer risk: what do we know and what is the way forward? BMJ. https://doi.org/10.1136/bmj.m511. Accessed September 8, 2021.

Kim, J., Zaret, K.S., 2015. Reprogramming of human cancer cells to plripotency for models of cancer progression. EMBO J. 34, 739–747.

Kisker, O., Onizuka, S., Baynard, J., Komiyama, T., Becker, C.M., et al., 2001. Generation of multiple angiogenesis inhibitors by human pancreatic cancer. Cancer Res. 61, 7298–7304.

Kleinerman, R.A., Wang, Z., Wang, L., Metayer, C., Zhang, S., Brenner, A.V., et al., 2002. Lung cancer and indoor exposure to coal and biomass in rural China. J. Occup. Environ. Med. 44 (4), 338–344.

Ko, J., Winslow, M.M., Sage, J., 2021. Mechanisms of small cell lung cancer metastasis. EMBO Mol. Med. 13, e13122.

Kocaman, A., Altun, G., Kaplan, A.A., Deniz, O.G., Yurt, K.K., Kaplan, S., 2018. Genotoxic and carcinogenic effects of non-ionizing electromagnetic fields. Environ. Res. 163, 71–79.

Koh, H.K., Geller, A.C., Miller, D.R., Grossbart, T.A., Lew, R.A., 1996. Prevention and early detection strategies for melanoma and skin cancer – current status. Arch. Dermatol. 132 (4), 436–443.

Kudryavtseva, A.V., Krasnov, G.S., Dmitriev, A.A., Alekseev, B.Y., Kardymon, O.L., Sadritdinova, A.F., et al., 2016. Mitochondrial dysfunction and oxidative stress in aging and cancer. Oncotarget 7 (29), 44879–44905.

Lee, D.J., Kang, S.W., 2013. Reactive oxygen species and tumor metastasis. Mol. Cell. 35 (2), 93–98.

Lee, W.C., Kopetz, S., Wistuba II, Zhang, J., 2017. Metastasis of cancer: when and how? Ann. Oncol. 28 (9), 2045–2046.

Li, J., Ji, Z., Qiao, C., Qi, Y., Shi, W., 2016. Overexpression of ADAM9 promotes color cancer cells invasion. J. Invest. Surg. https://doi.org/10.3109/08941939.2012.728682 Accessed September 26, 2021.

Liao, Z., Chua, D., Tan, N.S., 2019. Reactive oxygen species: a volatile driver of field cancerization and metastasis. Mol. Cancer. https://doi.org/10.1186/s12943-019-0961-y. Accessed September 17, 2021.

Lim, D.L., Ko, R., Pautler, S.E., 2007. Current understanding of the molecular mechanisms of kidney cancer: a primer for urologists. Can. Urol. Assn. J. 1 (2), S13–S20.

Linehan, W.M., Schmidt, L.S., Crooks, D.R., Wei, D., Srinivasan, R., Lang, M., Ricketts, C.J., 2019. The metabolic basis of kidney cancer. Cancer Discov. https://doi.org/10.1158/2159-8290.CD-18-1354. Accessed September 30, 2021.

Lissowska, J., Bardin-Mikolajczak, A., Fletcher, T., Zaride, D., Szeszenia-Dabrowska, N., Rudnai, P., et al., 2005. Lung cancer and indoor pollution from heating and cooking with solid fuels: th IARC international multicentre case-control study in Eastern/Central Europe and the United Kingdom. Am. J. Epidemiol. 162 (4), 326–333.

Lo, P.K., Wolfson, B., Zhou, X., Duru, N., Gernapudi, R., Zhou, Q., 2016. Noncoding RNAs in breast cancer. Briefings Funct. Genomics 15 (3), 200–221.

Lutz, W.K., 1990. Dose-response relationship and low dose extrapolation in chemical carcinogenesis. Carcinogenesis 11 (8), 1243–1247.

Macias, H., Hinck, L., 2012. Mammary gland development. Wiley Interdiscip. Rev. Dev. Biol. 1 (4), 533–557.

Majhi, P., Sharma, A., Roberts, A.L., Daniele, E., Majewski, A.R., Chuong, L.M., et al., 2020. Effects of benzophenone-3 and propylparaben on estrogen receptor-dependent R-loops and DNA damage in breast epithelial cells and mice. Environ. Health Perspect. https://doi.org/10.1289/EHP5221. Accessed September 21, 2021.

Masrour-Roudsari, J., Ebrahimpour, S., 2017. Causal role of infectious agents in cancer: an overview. Caspian J. Intern. Med. 8 (3), 153–158.

Matakidou, A., Eisen, T., Houlston, R.S., 2005. Systematic review of the relationship between family history and lung cancer risk. Br. J. Cancer 93 (7), 825–833.

Mayo Clinic, 2018. Mayo Clinic Family Health Book, fifth ed. Mayo Clinic Press, Rochester, MN.

Mayo Clinic, 2021. Leukemia. https://www.mayoclinic.org/diseases-conditons/leukemia/symptoms-causes.syc-20374373?p=1. Accessed October 1, 2021.

Mazzocca, A., Coppart, R., De Franco, R., Cho, J.Y., Liberman, T.A., Pinzani, M., Toker, A., 2008. A secreted form of ADAM9 promotes carcinoma invasion through tumor-stromal interactions. Cancer Res. 65 (11), 4728–4738.

McHale, C.M., Zhang, L., Smith, M.T., 2011. Current understanding of the mechanism of benzene-induced leukemia in humans: implications for risk assessment. Carcinogenesis 33 (2), 240–252.

Merkel, E.A., Gerami, P., 2017. Malignant melanoma of sun-protected sites: a review of clinical, histological, and molecular features. Lab. Invest. 97, 630–635.

Nabi, S., Kessler, E.R., Bernard, B., Flaig, T.W., Lan, E.T., 2018. Renal cell carcinoma: a review of biology and pathophysiology. F1000Research 7 (F1000 Faculty Rev), 307. https://doi.org/10.12688/f1000research.13179.1. Accessed September 30, 2021.

Narod, S.A., Salmena, L., 2011. BRCA1 and BRCA2 mutations and breast cancer. Discov. Med. 12 (66), 445–453.

Nasti, T.H., Timares, L., 2015. MC1R, eumelanin and pheomelanin: their role in determining the susceptibility to skin cancer. Photochem. Photobiol. 91 (1), 188–200.

National Cancer Institute, 2015. Risk Factors for Cancer. https://www.cancer.gov/about-cancer/causes-prevention/risk. Accessed September 8, 2021.

National Cancer Institute, 2017. The Genetics of Cancer. https://cancer.gov/about-cancer/causes-prevention/genetics. Accessed September 8, 2021.

National Cancer Institute, 2019. Radiation. https://www.cancer.gov/about-cancer/causes-prevention/risk/radiation. Accessed September 8, 2021.

National Cancer Institute, 2020. Cancer Statistics. https://www.cancer.gov/about-cancer/understanding/statistics. Accessed September 8, 2020.

National Cancer Institute, 2020a. Cancer Statistics. https://www.cancer.gov/types/metastatic-cancer. Accessed September 16, 2020.

National Cancer Institute, 2021. Cancer Types. https://cancer.gov/types. Accessed September 8, 2021.

National Cancer Institute, 2021a. Renal Cell Treatment (PDQ) – Health Care Professional. https://www.cancer.gov/types/kidney/hp/kidney-treatment-pdq. Accessed September 30, 2021.

Nguyen, H.T., Duong, H.Q., 2018. The molecular characteristics of colorectal cancer: implications for diagnosis and therapy (review). Oncol. Lett. 16, 9–18.

Nguyen, D.X., Massague, J., 2007. Genetic determinants of cancer metastasis. Nat. Rev. Genet. 8, 341–352.

Noonan, F.P., Raza, Z.M., Wolnicka-Glubisz, A., Anver, M.R., Bahn, J., et al., 2012. Melanoma induction by ultraviolet A but not ultraviolet B radiation requires melanin pigment. Nat. Commun. 3 (884). https://doi.org/10.1038/ncomms1893. Accessed September 27, 2021.

Nowicki, M.O., Falinski, R., Koptyra, M., Slupianek, A., Stoklosa, T., Gloc, E., et al., 2004. BCR/ABL oncogenic kinase promotes unfaithful repair of the reactive oxygen species-dependent DNA double-strand breaks. Blood 104 (12), 3746–3753.

Ntziachristos, P., Mullenders, J., Trimarchi, T., Alfantis, I., 2013. Mechanisms of epigenetic regulation of leukemia onset and progression. Adv. Immunol. https://doi.org/10.1016/B978-0-12-410524-9.00001-3. Accessed September 28, 2021.

Nyberg, F., Agrenius, V., Svartengren, K., Svensson, D., Pershagen, G., 1998. Dietary factors and risk of lung cancer in never-smokers. Ind. J. Can. 78 (4), 430–436.

O'Reilly, M.S., Boehm, T., Shing, Y., Fukai, N., Vasios, G., et al., 1997. Endostatin, an endogenous inhibitor of angiogenesis and tumor growth. Cell 88277–88285.

Pachayr, E., Treese, C., Stein, U., 2017. Underlying mechanisms for distant metastasis – molecular biology. Vis. Med. 33, 11–20.

Pani, G., Galeotti, T., Chiarugi, P., 2010. Metastasis: cancer cell's escape from oxidative stress. Cancer Metastasis Rev. 29 (2), 351–378.

Paul, D., 2020. The systemic hallmarks of cancer. J. Cancer Metastasis 6, 29. https://doi.org/10.20517/2394-4722.2020.63. Accessed September 10, 2021.

Pavlides, S., Tsirigos, A., Migneco, G., Whitaker-Menezes, D., Chiavarina, B., Flomenberg, N., et al., 2010. The autophagic tumor stroma model of cancer: role of oxidative stress and ketone production in fueling tumor cell metabolism. Cell Cycle 9 (17), 3485–3505.

Pavlides, S., Tsirigos, A., Vera, I., Flomenberg, N., Frank, P.G., Casimoro, M.C., et al., 2010a. Loss of stromal caveloin-1 leads to oxidative stress, mimics hypoxia and drives inflammation in tumor microenvironment, conferring the "reverse Warburg effect": a transcriptional informatics analysis with validation. Cell Cycle 9 (1), 2201–2219.

Peiris-Pages, M., Martinez-Outschoom, U.E., Sotgia, F., Lisanti, M.P., 2015. Metastasis and oxidative stress: are antioxidants a metabolic driver of progression. Cell Metabol. 22, 956–958.

Pienta, K.J., Bradley, D., 2006. Mechanisms underlying the development of androgen-independent prostate cancer. Clin. Cancer Res. 12 (6), 1665–1671.

Piskounova, E., Agathocleous, M., Murphy, M.M., Hu, Z., Huddlestun, S.E., Zhao, Z., et al., 2015. Oxidative stress inhibits distant metastasis by human melanoma cells. Nature 527, 186–191.

Poetsch, A.R., 2019. The genomics of oxidative DNA damage, repair and resulting mutagenesis. Comp. Struct. Biotech. https://doi.org/10.1016/jcsbj.2019.12.013. Accessed September 17, 2021.

Pretzsch, E., Bosch, F., Neumann, J., Ganschow, P., Bazhin, A., Guba, M., et al., 2019. Mechanisms of metastasis in colorectal cancer and metastatic organotropism: hematogenous versus peritoneal spread. JAMA Oncol. https://doi.org/10.1155/2019/7407190. Accessed September 26, 2021.

Raimondi, S., Sera, F., Gandini, S., Iodice, S., Caini, S., Maisoneuve, P., Fargnoli, M.C., 2008. MC1R variants, melanoma and red hair color phenotype: a meta-analysis. Int. J. Cancer 122 (12), 2753–2760.

Rainey, N., Saric, A., Leberre, E., Dewailly, D., Slomianny, C., Vial, G., Zeliger, H.I., Petit, P., 2017. Synergistic cellular effects including mitochondrial destabilization, autophagy and apoptosis following low-level exposure to a mixture of lipophilic persistent organic pollutants. Sci. Rep. 7, 4728–4748.

Rossi, A., Roberto, M., Panebianco, M., Botticelli, A., Mazzuca, F., Marchetti, P., 2019. Drug resistance of BRAF-mutant melanoma: review of up-to-date mechanisms of actin and promising targeted agents. Eur. J. Pharmacol. https://doi.org/10.1016/j.ejpharm.2019.172621. Accessed September 26, 2021.

Rumgay, H., Shield, K., Charvat, H., Ferrari, P., Sornpaisarn, B., Obot, I., et al., 2021. Global burden of cancer in 2020 attributable to alcohol consumption: a population-based study. Lancet Oncol. 22 (8), 1071–1080.

Saha, S.K., Lee, S.B., Won, J., Choi, H.Y., Kim, K., Yang, G.M.Y., et al., 2017. Correlation between oxidative stress, nutrition, and cancer initiation. Int. J. Mol. Sci. 18. https://doi.org/10.33390/ijms18071544. Accessed September 15, 2021.

Salnikow, K., Zhitovich, A., 2008. Genetic and epigenetic mechanisms in metal carcinogenesis and cocarcinogenesis; nickel, arsenic, chromium. Chem. Res. Toxicol. 21, 28–44.

Sample, A., He, Y.Y., 2018. Mechanisms and prevention of UV-induced melanoma. Photodermatol. Photoimmunol. Photomed. 34 (1), 13–24.

Sapkota, S., Shaikh, H., 2021. Non-Hodgkin Lymphoma. National Library of Medicine, StatPearls Publishing. https://www.ncbi.nih.gov/books/NBK559328/. Accessed September 28, 2021.

Sattler, M., Verma, S., Shrikhande, F., Byrne, C.H., Pride, Y.B., Winkler, T., et al., 2000. The BCR/ABL tyrosine kinase induces production of reactive oxygen species in hematopoietic cells. J. Biol. Chem. 275 (32), 24273–24278.

Schrecengost, R.S., Knudson, K.E., 2013. Molecular pathogenesis and progression of prostate cancer. Semin. Oncol. 40 (3), 244–258.

Scully, O.J., Bay, B.H., Yip, G., Yu, Y., 2012. Breast cancer metastasis. Cancer Genomics Proteomics 9 (5), 311–320.

Senga, S.S., Grose, R.P., 2021. Hallmarks of cancer – the new testament. Open Biol. 11, 200358. https://doi.org/10.1098/rsob.20.0358.

Sever, F., Brugge, J.S., 2015. Signal transduction in cancer. Cold Spring Harb Perspect. Med. 5 (4). https://doi.org/10.1101/cshpespect.00600098. Accessed September 22, 2021.

Shebl, F.M., Engels, E.A., Goedert, J.J., Chaturvedi, A.K., 2010. Pulmonary infections and risk of lung cancer among persons with AIDS. J. Acquir. Immune Defic. Syndr. 55 (3), 375–379.

Skin Cancer Foundation, 2021. Skin Cancer Facts and Figures. www.skincancer.org. Accessed September 27, 2021.

Spivey, A., 2010. Light at night and breast cancer risk worldwide. Environ. Health Perspect. 118 (12), A525.

Stanford Healthcare, 2021. Cancer Types. https://stanfordhealthcare,org/medical-conditions/cancer/cancer/cancer-types.html. Accessed September 9, 2021.

Suzuki, H., Maruyama, R., Yamamoto, E., Niinuma, T., Kai, M., 2016. Relationship between noncoding RNA dysregulation and epigenetic mechanisms in cancer. Adv. Exp. Med. Biol. 927, 109–135.

Tariq, K., Ghias, K., 2016. Colorectal cancer carcinogenesis: a review of mechanisms. Cancer Biol. Med. https://doi.org/10.28092/j.issn.2095-3941.2015.0103. Accessed September 26, 2021.

Testa, U., Castelli, G., Pelosi, E., 2019. Cellular and molecular mechanisms underlying prostate cancer development: therapeutic implications. Medicine (Basel). https://doi.org/10.3390/medicines6030082. Accessed September 24, 2021.

Trainer, A.H., Lewis, C.R., Tucker, K., Meiser, B., Friedlander, M., Ward, R.L., 2010. The role of BRCA mutation testing in determining breast cancer therapy. Nat. Rev. Clin. Oncol. 7 (12), 707–717.

UC Berkeley, 2021. Chemical Carcinogenesis: Initiation, Promotion and Progression. NST 110 Toxicology, Dept. Nutritional Sciences and Toxicology, University of California, Berkeley. Accessed September 14, 2021.

Upham, B.L., Trosko, J.E., 2009. Oxidative-dependent integration of signal transduction with intercellular gap junctional communication in the control of gene expression. Antioxidants Redox Signal. 11 (2), 297–307.

Van Duuren, B.L., 1982. Cocarcinogens and tumor promoters and their environmental importance. J. Am. Coll. Toxicol. 1 (1), 17–27.

Vanharanta, S., Shu, W., Brenet, F., Hakimi, A.A., Heguy, A., Viale, A., et al., 2013. Epigenetic expansion of VHL-HIF signal output drives multi-organ metastasis in renal cancer.

Venza, M., Visalli, M., Beninati, C., De Gaetano, G., Teti, D., Venza, I., 2015. Cellular mechanisms of oxidative stress and action in melanoma. Oxid. Med. Cell Longev. https://doi.org/10.1155/481782. Accessed September 26, 2021.

Weigelt, B., Peterse, J.L., Van 't Veer, L.J., 2005. Breast cancer metastasis: markers and models. Nat. Rev. Cancer 5 (8), 591–602.

Weischenfeldt, J., Simon, R., Feuerbach, L., Schalangen, K., Weichenhan, D., Minner, S., et al., 2013. Integrative genomic analyses reveal and androgen-driven somatic alteration landscape in early onset prostate cancer. Cancer Cell 23 (2), 159–170.

Welch, D.R., Hurst, D.R., 2019. Defining the hallmarks of metastasis. Cancer Res. 79 (12), 3011–3027.

White, M.C., Holman, D.M., Boehn, J.E., Peipins, L.A., Grossman, S., Henley, S.J., 2014. Age and cancer risk: a potentially modifiable relationship. Am. J. Prev. Med. 46 (3 0 1), S7–S15.

White, A.J., Bradshaw, P.T., Hamra, G.B., 2018. Air pollution and breast cancer: a review. Curr. Epidemiol. Rep. 5 (2), 92–100.

Wilk, A., Waliforski, P., Lassak, A., Vashistha, J., Lirette, D., Tate, D., et al., 2013. Polycyclic aromatic hydrocarbons-induced ROS accumulation enhance mutagenic potential of T-antigen from human polyomavirus JC. J. Cell. Physiol. 228 (11), 2127–2138.

Wong, S.H., Yu, J., 2019. Gut microbiota in colorectal cancer: mechanisms of action and clinical applications. Nat. Rev. Gastroenterol. Hepatol. 16, 690–704.

Wu, S., Zhu, W., Thompson, P., Hannun, Y.A., 2018. Nat. Commun. https://doi.org/10.1038/s41467-018-05467-z. Accessed September 30, 2021.

Yates, L.R., Knappskog, S., Wedge, D., Farmery, J.H.R., Gonzalez, S., Martincorena, I., et al., 2017. Genomic evolution of breast cancer metastasis and relapse. Cancer Cell 32 (2), 169–184.

Zeliger, H.I., 2003. Toxic effects of chemical mixtures. Arch. Environ. Health 58 (1), 23–29.

Zeliger, H.I., 2011. Human Toxicology of Chemical Mixtures, second ed. Elsevier, London.

Zeliger, H.I., 2013. Lipophilic exposure as a cause of type 2 diabetes (T2D). Rev. Environ. Health. https://doi.org/10.1515/reveh-2012-0031. Accessed September 15, 2021.

Zeliger, H.I., 2013a. Lipophilic chemical exposure as a cause of cardiovascular disease. Interdiscipl. Toxicol. 6 (2), 55–62.

Zeliger, H.I., 2013b. Exposure to lipophilic chemicals as a cause of adult neurological impairments, neurodevelopmental disorders and neurodegenerative diseases. Interdiscipl. Toxicol. 6 (3), 101–108.

Zeliger, H.I., 2016. Predicting disease onset in clinically healthy people. Interdiscipl. Toxicol. 9 (2), 39–54.

Zeliger, H.I., Lipinski, B., 2015. Physiochemical basis of human degenerative disease. Interdiscipl. Toxicol. 8 (1), 15–21.

Zhang, Y., Sanjose, S.D., Bracci, P.M., Morton, L.M., Wang, R., Brennan, P., et al., 2008. Personal use of hari dye and the risk of certain subtypes of Non-Hodgkin lymphoma. Am. J. Epidemiol. 167 (11), 1321–1331.

Zhang, Y., Narayanan, S.P., Mannan, R., Raskind, G., Wang, X., Vats, P., et al., 2021. Single-cell analyses or renal cell cancers reveal insights into tumor microenvironment, cell of origin, and therapy response. Proc. Natl. Acad. Sci. USA. https://doi.org/10.1073/pnas.2103240118. Accessed September 30, 2021.

Zhou, L., Zhang, Z., Huang, Z., Nice, E., Zou, B., Huang, C., 2020. Revisiting cancer hallmarks: insights from the interplay between oxidative stress and non-coding RNAs. Molec. Biomed. https://doi.org/10.1186/s43556-020-00004-1. Accessed September 17, 2021.

CHAPTER

22

Atherosclerosis

22.1 Introduction

Cardiovascular disease (CVD) is the leading cause of human death in the world The primary CVD diseases are atherosclerosis, arterial hypertension, heart failure, atrial fibrillation and myocardial infarction. The mechanisms for onset of all are oxidative stress-related (Senoner and Dichtl, 2019).

Of these CVDs, atherosclerosis (ATS) is the major cause of morbidity and mortality worldwide and is the subject of this chapter.

22.2 Hallmark of atherosclerosis

The sole hallmark for ATS is a fibrous plaque on the artery wall. It is preceded by an initial fatty streak that is caused by an accumulation of lipid-laden foam cells in the intimal layer of the artery (Crowther, 2005).

22.3 Atherosclerosis risk factors

Atherosclerosis is a slow, progressive disease whose origin can be traced back to childhood. The known risk factors for ATS are listed in Table 22.1 (Singh et al., 2002; Zeliger, 2013; Rafieian-Kopaei, 2014; Schuett et al., 2015; Stanford Health Care, 2021; Mayo Clinic, 2021).

22.4 Atherosclerosis progression

The pathophysiology of ATS involves four distinct steps.

1. Endothelial cell injury
2. Lipoprotein deposition
3. Inflammatory reaction
4. Smooth cell cap formation

Oxidative Stress
https://doi.org/10.1016/B978-0-323-91890-9.00026-X

TABLE 22.1 Risk factors for atherosclerosis.

Age
Hypertension
Elevated level of cholesterol
Elevated level of triglycerides
Elevated level of lipoprotein
Elevated level of C-reactive protein
Chronic exposure to lipophilic chemicals
Bisphenol A
Phthalates
Low molecular weight hydrocarbons
Chlorinated low molecular weight hydrocarbons
Polynuclear aromatic hydrocarbons
Persistent organic pollutants
Polychlorinated biphenyls (PCBs)
Dioxins
Furans
Organochlorine pesticides
Polybrominated biphenyl ethers (PBDEs)
Esters of perfluorooctanoic acid (PFOEs)
Smoking
Chronic alcohol use
Obesity
Insulin resistance
Type 1 diabetes
Chronic inflammation from other diseases
Arthritis
Lupus
Psoriasis,
Inflammatory bowel disease
Sleep apnea
Sedentary life style
Diet high in saturated fat, trans fat, cholesterol, sugar and sodium
Family history of early heart disease
Chronic psychological stress

22.5 Mechanism of onset

Endothelial dysfunction is the initiating factor for atherosclerotic plaque formation. The endothelium (inner blood vessel lining) is a highly active monolayer that modulates vascular tone cellular adhesion, thromboresistance, smooth muscle cell proliferation and vessel wall inflammation (Rudijanto, 2007; Doran et al., 2008; Basatemur et al., 2019). Central to this role is the production and release of endothelium-derived relaxing factors, including nitrous oxide (NO), and endothelium-derived contracting factors. In endothelial dysfunction, there is reduced production and/or availability of NO, as well as an imbalance between endothelium-derived vasodilators and vasoconstrictors (Forstermann and Munzel, 2006; Deanfield et al., 2007; Favero et al., 2014).

As listed in Table 22.1, risk factors for endothelial dysfunction include aging, hyperglycemia, smoking, sedentary lifestyle, hypercholesterolemia, arterial hypertension and family history of early atherosclerotic disease, all of which are mediated by oxidative stress and chronic inflammation. Indeed, oxidative stress/chronic inflammation are the main drivers of endothelial dysfunction (Hadi et al., 2005).

In the propagation of the atherosclerosis, dyslipidemia, hypertension, and tobacco smoking lead to changes in the endothelium that promote adhesion and increased permeability of macromolecules. This, in turn, promotes the entry and retention of low-density lipoprotein (LDL) which is sequentially oxidatively modified by free radicals (Maiolino et al., 2013). Oxidatively modified LDL particles (OxLDL) are central to the mechanism, leading to the release of bioactive phospholipids which activate endothelial cells that express adhesion of leukocyte adhesion molecules (Hansson et al., 2006; Xu et al., 2013).

After adhesion, LDL is oxidatively modified and attracts inflammatory cells into the arterial wall. Following endothelial injury, inflammatory mediators are released further, increasing leukocyte recruitment. Modified LDL invokes an immune response that results in the incorporation of inflammatory cells into the arterial wall which lead to increased superoxide production, reduced NO bio-availability and plaque formation (Steinbrecher et al., 1984; Ross et al., 1999; Singh et al., 2002; Rafieian-Kopaei et al., 2014).

Macrophage foam cells are critical in the development of ATS as they are associated with the uptake of cholesterol (derived from lipoprotein), that dictates the transformation of macrophages into foam cells, which are the prototypical cells in atherosclerotic cells (Yu et al., 2013).

Following plaque formation, smooth muscle cells migrate to plaque surfaces, creating a fibrous cap. Thick fibrous caps are stable, thinner ones, however, more easily rupture, causing thrombosis (Badimon and Vilahur, 2014; Otsuka et al., 2016) which is the primary cause other cardiovascular diseases, including myocardial infarction and stroke (Grootaert et al., 2015).

22.6 Oxidative stress and Artherosclerosis

Inflammation, mitochondrial dysfunction, autophagy, apoptosis and epigenetics accelerate the onset of atherosclerosis. All induce oxidative stress which is the driving force for all phases of ATS (Pastori et al., 2013; Soeki and Sata, 2016; Yang et al., 2017). The Yang paper identifies the connections between oxidative stress to each of 28 individual parameters associated with onset and propagation of ATS (Yang et al., 2017). These are listed in Table 20.2. The reader is referred to the references cited in the Yang paper for each of these parameters for elaboration.

The factors listed in Table 20.2 highlight the numerous mechanistic factors that are associated with atherosclerosis onset and propagation. All are oxidative stress-mediated. Further evidence for the OS-ATS relationship comes from the findings that antioxidants, including antioxidant vitamins, ACEI and ATLR antagonists, statins and probucol have demonstrated NOx activity suppression and mitigation of oxidative stress in atherosclerosis (Paravicini and Touyz, 2008; Malekmohammad et al., 2019).

TABLE 20.2 Factors associated with oxidative stress in atherosclerosis onset.

1. Primary ROS-producing systems: xanthine oxidase (XO), uncoupled endothelial nitric oxide synthase (eNOS), mitochondrial respiratory chain enzymes and nicotinamide adenine dinucleotide phosphate (NADPH) oxidase (NOXs).
2. Macrophage-induced oxidative stress.
3. Macrophage cellular oxidation.
4. 7-hydroperoxide (7-OOH).
5. Macrophage ATG5 deficiency.
6. Thiol oxidative stress.
7. Inflammation.
8. Oxidized HDL (ox-HDL).
9. NLRP3 inflammasome.
10. Autophagy.
11. Light chain 3 (LC3).
12. Lectin-like ox-LDL Receptor-1 (LOX-1).
13. Autophagy related 7 (ATG7).
14. Apoptosis
15. Granulocyte-macrophage colony stimulating factor (GM-CSF).
16. Protein kinase Cβ (PKCβ).
17. Heptaglobin 2-2 (Hp2-2).
18. B-cell lymphoma-2 (Bcl-2).
19. Superoxide dismutase (SOD).
20. Mitochondria.
21. Retinol-binding protein 4 (RBP4).
22. Macrophage mitochondrial oxidative stress ((mitoOS).
23. Mitochondrial genome (mtDNA).
24. Homocysteine (Hcy).
25. Angiostatin II (ANG II).
26. Epigenetic modifications.
27. Oxidized LDL (Ox-LDL).
28. DNA methylation.

References

Badimon, L., Vilahur, G., 2014. Thrombosis formation on atherosclerotic lesions and plaque rupture. J. Intern. Med. 276 (6), 618–632.

Basatemur, G.L., Jorgensen, H.F., Clarke, M.C.H., Bennett, M.R., Mallat, Z., 2019. Vascular smooth muscle cells in atherosclerosis. Nat. Rev. Cardiol. 16, 727–744.

Crowther, M.A., 2005. Pathogenesis of atherosclerosis. Hematology Am. Soc. Hematol. Educ. Program 1, 436–441.

Deanfield, J.E., Halcox, J.P., Rabelink, T.J., 2007. Endothelial function and dysfunction. Circulation 115 (10), 1285–1295.

Doran, A.C., Meller, N., McNamara, C.A., 2008. Role of smooth muscle cells in the initiation and early progression of atherosclerosis. Arterioscler. Thromb. Vasc. Biol. 28, 8120819.

Favero, G., Paganelli, C., Buffoli, B., Rodella, L.F., Rezzani, R., 2014. Endothelium and its alterations in cardiovascular diseases: lifestyle intervention. BioMed Res. Int. https://doi.org/10.1155/2014/801896, 2014. Accessed October 9, 2021.

Forstermann, U., Munzel, T., 2006. Endothelial nitric oxide in vascular disease. Circulation 113 (13), 1708–1714.

Grootaert, A., da Costa Martins, P.A., Bitsch, N., Pintelon, K., De Meyer, G.R., Martinet, W., et al., 2015. Defective autophagy in vascular smooth muscle cells accelerates senescence and promotes neointima formation and atherogenesis. Autophagy. https://doi.org/10.1080/15548627.2015.1096485, 2004. Accessed October 11, 2021.

Hadi, H.A.R., Carr, C.S., al Suwaidi, J., 2005. Endothelial dysfunction: cardiovascular risk factors, therapy and outcome. Health Risk Manag. 1 (3), 183–198.

Hansson, G.K., Robertson, A.K.L., Soderberg-Naucler, C., 2006. Inflammation and atherosclerosis. Ann. Rev. Pathol. 1, 297–329.

Maiolino, G., Rossitto, G., Caielli, P., Bisogni, V., Rossi, G.P., Calo, L.A., 2013. The role of oxidized low-density lipoproteins in atherosclerosis: the myths and the facts. Mediators Inflamm. https://doi.org/10.1155/2013/714653. Accessed August 24, 2022.

Malekmohammad, K., Sewell, R.D.E., Rafieian-Kopaei, M., 2019. Antioxidants and atherosclerosis: mechanistic aspects. Biomolecules 9 (8), 301. https://doi.org/10.3390/biom9080301. Accessed October 11, 2021.

Mayo Clinic, 2021. Arteriosclerosis/atherosclerosis. www.mayoclinic.org. Accessed October 8, 2021.

Otsuka, F., Yasuda, S., Noguchi, T., Ishibashi-Ueda, H., 2016. Pathology of coronary atherosclerosis and thrombosis. Cardiovasc. Diagn. Ther. 6 (4), 396–408.

Paravicini, T.M., Touyz, R.M., 2008. NADPA oxidases, reactive oxygen species and hypertension. Diabetes Care 31 (Suppl. 2), S170–S180.

Pastori, D., Carnevale, R., Pigntelli, P., 2013. Is there a clinical role for oxidative stress biomarkers in atherosclerotic diseases? Intern. Emerg. Med. 9 (2), 123–131.

Rafieian-Kopaei, M., Setorki, M., Doudi, M., Baradaran, A., Nasri, H., 2014. Atherosclerosis: indicators, risk factors and new hopes. Int. J. Prevent. Med. 5 (8), 927–946.

Ross, R., 1999. Atherosclerosis – an inflammatory disease. N. Eng. J. Med. 340 (2), 115–126.

Rudijanto, A., 2007. The role of vascular smooth muscle cells in the pathogenesis of atherosclerosis. Acta Med. Indones. 39 (2), 86–93.

Scheuett, K.A., Lehrke, M., Marx, N., Burgmaier, M., 2015. High-risk cardiovascular patients: clinical features, cormorbidities, and interconnecting mechanisms. Front. Immunol. 6, 591. https://doi.org/10.3389/fimmu.2015.00591. Accessed October 8, 2021.

Senoner, T., Dichtl, W., 2019. Oxidative stress in cardiovascular diseases: still a therapeutic target. Nutrients 11 (9), 2090. https://doi.org/10.3390/nu11092090. Accessed October 8, 2021.

Singh, R.B., Sushma, S.A., Xu, Y.J., Arneja, A.S., Dhalla, N.S., 2002. Pathogenesis of atherosclerosis: a multifactorial process. Exp. Clin. Cardiol. 7 (1), 40–53.

Soeki, T., Sata, M., 2016. Inflammatory biomarkers and atherosclerosis. Int. Heart J. 57 (2), 134–139.

Stanford Health Care, 2021. Atherosclerosis Causes. https://stanfordhealthcare.org/medical-conditions/blood-heart-circulation/atherosclerosis/causes.html. Accessed October 8, 2021.

Steinbrecher, U.P., Parthasarathy, S., Leake, D.S., Witztum, J.L., Steinberg, D., 1984. Modification of low desity lipoprotein by endothelial cells involves lipid peroxidation and degradation of low density lipoprotein phospholipids. Proc. Natl. Acad. Sci. U.S.A. 81 (12), 3883–3887.

Xu, S., Ogura, S., Chen, J., Little, P.J., Moss, J., Liu, P., 2013. LOX-1 in atherosclerosis: biological functions and pharmacological modifiers. Cell. Mol. Life Sci. 70 (16), 2859–2872.

Yang, X., Yang, L., Ren, X., Zhang, X., Hu, D., Gao, Y., et al., 2017. Oxidative stress-mediated atherosclerosis: mechanisms and therapies. Front. Physiol. 8, 600. https://doi.org/10.3390/fphys.2017.00600. Accessed October 11, 2021.

Yu, X.H., Fu, Y.C., Zhang, D.W., Yin, K., Tang, C.K., 2013. Foam cells in atherosclerosis. Clin. Chim. Acta 424, 245–252.

Zeliger, I., 2013. Lipophilic chemical exposure as a cause of cardiovascular disease. Interdiscip. Toxicol. 6 (2), 55–62.

CHAPTER

23

Alzheimer's disease

23.1 Introduction

Alzheimer's disease (AD) is an age-related neurodegenerative disease that is the most prevalent source of dementia in the elderly. Though primarily afflicting those aged 65 and over, AD also is increasingly being observed in younger people starting from around age 40 (Zhu et al., 2015; NHS 2021). It has been estimated that AD currently strikes one in 20 people over the age of 65 (up to 35 million in 2015) and that the rate doubles about every 5 years beyond age 65 to where about one third of all people aged 85 or older may have AD (NIH 2019; NHS, 2021).

The average duration of AD after diagnosis is 10 years, but as many as 20 years may elapse from initiation of mild cognitive impairment, wherein one does not suffer any significant barriers, to impacted functioning in daily activities or work, to where participating in life's activities becomes problematical (Sutherland et al., 2013). AD is not curable or reversable at this time.

23.2 Hallmarks of Alzheimer's disease

It is well established that misfolding, aggregation and brain accumulation of protein aggregates trigger the pathogenesis of AD (Edwards et al., 2019).

The hallmarks of AD include the following neuropathological markers (Edwards et al., 2019). These are:

Extracellular amyloid-β plaques
Intracellular neurofibrillary tangles
Inflammation
Synaptic impairment
Neuronal loss.

These lead to cognitive and mood changes as well as oxidative stress biomarkers. OS biomarkers cannot be used clinically in AD, due to the need to take serum samples from the source, in this instance, the brain (Collin et al., 2018). Cognitive and mood biomarkers, however, can readily be measured (Alzheimer's Association, 2021). These are listed in Table 23.1.

Oxidative Stress
https://doi.org/10.1016/B978-0-323-91890-9.00020-9

TABLE 23.1 Ten biomarkers (symptoms) of Alzheimer's disease.

1.	Memory loss that disrupts daily life.
2.	Challenges in planning or solving problems.
3.	Difficulty completing familiar tasks.
4.	Confusion with time or place.
5.	Trouble understanding visual images and spatial relationships.
6.	New problems with words in speaking or writing.
7.	Misplacing things and losing the ability to retrace steps.
8.	Decreased or poor judgment.
9.	Withdrawal from work or social activities.
10.	Changes in mood or personality.

23.3 Risk factors for Alzheimer's disease

The risk factors for AD, in addition to age, include genetics, head trauma, lifestyle choices, environmental exposures and pre-existing conditions (Trempe and Lewis, 2018; Duthie et al., 2011; Zeliger, 2019; Edwards et al., 2019; Silva et al., 2019; NIH, 2019a and the numerous references contained in these). A representative sampling of AD risk factors are listed in Table 23.2.

TABLE 23.2 Alzheimer's disease risk factors.

Genetics and epigenetics
AD runs in families. Inheriting one copy of the ApoE4 gene triples the risk of developing AD, while inheriting 2 copies of this gene increases the AD risk by 10–15 fold. AD in blood relatives without the ApoE4 gene (an epigenetic effect) also increases the risk for AD, with the degree of increase being a function of AD prevalence in total numbers of direct relatives (parents, siblings and grandparents).
Head trauma
Traumatic brain injury Chronic traumatic encephalopathy
Lifestyle choices
Not treating type 2 diabetes, if present Smoking Heavy consumption of alcohol Not acting to prevent or treat hypertension Not acting to prevent or reverse overweight and obesity Not maintaining a healthy diet that is low caloric; low in saturated fats, Carbohydrates red and processed meats; and high in fruit, Vegetable, whole grain and antioxidant consumption Playing professional contact sports (football, soccer, rugby, hockey)

Physical inactivity – a sedentary lifestyle
Lack of mental stimulation
Environmental factors
Chronic exposure to polluted air
Chronic exposure to pesticides
Chronic exposure to heavy metals
Chronic exposure to radiation.
Pre-existing conditions
Blood brain barrier breakdown
Traumatic brain injury
Chronic inflammation
Chronic infectious diseases (viral, bacterial, fungal and parasitic)
Chronic psychological stress
Chronic sleep disturbance
Obesity
Chronic migraine headaches
Chronic infectious diseases (viral, bacterial, fungal and parasitic)
Glaucoma
Metabolic syndrome
Lipoprotein disorders
Type 2 diabetes
Hyperthyroidism
Rheumatoid arthritis
Hypertension
Atherosclerosis
Stroke
Parkinson's disease
Asthma
Insomnia
Sleep apnea
Osteoporosis
Psychosis
Neurosis
Depression
Cancer[a]

[a]*The co-morbidity of AD with cancer is, at this point, still open for debate. Some studies have indicated a mechanistic association, while others have reported a protective effect of AD against developing cancer (Duthie et al., 2011; Majd et al., 2019).*

Many of the comorbidities listed in Table 23.2 are to diseases that are far different in nature and mechanisms of onset. Some of these comorbidities are due to common mechanisms, while others may be due to different parameters associated with aging or via pure happenstance. All the risk factors listed above, however, elevate oxidative stress and lead to disease onset (Zeliger, 2016, 2019). The subject of disease co-morbidity is more fully discussed in Chapter 29.

Many studies have demonstrated that the lifestyle choices listed in Table 23.2 are modifiable and addressing these can lower or at least delay the onset of AD (Silva et al., 2019; Edwards et al., 2019).

23.4 Alzheimer's disease co-morbidity data

The average AD patient has between three and four co-morbidities (Poblador-Plou et al., 2014; Vargese et al., 2021; Santiago and Potashkin, 2021). The most prevalent and the approximate percentages of each are shown in Table 23.3 (Doraiswamy et al., 2002).

23.5 Mechanisms of Alzheimer's disease onset

Oxidative stress plays a key role in the development of Alzheimer's disease as well as in other neurodegenerative diseases, evidence coming from the presence of lipid, protein and DNA oxidation products in vivo. Neuronal sensitivity to oxidation is high due to the high polyunsaturated fatty acid content in cell membranes, weak antioxidant defense and high consumption of oxygen by nerve cells (Cheignon et al., 2018; Huang et al., 2016; Collin 2019; Singh et al., 2019; Silva et al., 2019).

The physiological features of Alzheimer's disease are (Collin, 2019):

Deposition of protein aggregates
Extracellular amyloid plaques (Aβ)
Intracellular Tau, or neurofibrillary tangles
Loss of synaptic connections.

TABLE 23.3 Approximate percentage of AD patients with specific disease category comorbidities.

Disease category	Percent
Musculoskeletal	55
Genitourinary	50
Ear, nose and throat	44
Vascular	40
Heart	35
Lower gastrointestinal	30
Endocrine/metabolic/breast	25
Upper gastrointestinal	20
Non-dementia neurological	20
Hemopoietic	15
Respiratory	10
Liver	5
Renal	5

The accumulation of neurotoxic Aβ oligomer peptides and Tau protein sets off a cascade of events, including neurodegeneration, neuroinflammation, synaptic connection impairment, cholinergic denervation, neurotransmitter imbalance, neuronal loss and dendritic alterations (Singh et al., 2019).

In AD, the Aβ peptide is found in aggregated in plaques (composed of Aβ fibrils and metal ions) and Tau neurofibrillary tangles. Copper is present in relatively high levels in the brain and Aβ can chelate metal ions, making Aβ-copper a direct source of ROS in the presence of oxygen via a Fenton-like reaction (Zhao, 2019). Indeed, superoxide has been shown to be generated as an intermediate during H_2O_2 production by Aβ-copper (Rabier et al., 2016; Collin, 2019).

Increased levels of lipid, protein and DNA oxidation products are associated with elevated Aβ levels, with the lipid peroxidation biomarkers malondialdehyde (MDA) and 4-hydroxynonelal (4-HNE) markedly elevated in AD brains (Markesbery and Lovell, 1998). 4-HNE is neurotoxic, causing apoptosis and altering microtubule structure (Neely et al., 1999). 4-HNE also reacts with lipoid acid and forms adducts with proteins found in the human brain by targeting amino acids (cysteine, histidine and lysine) (Sayre et al., 1997; Gegotek and Skrzydkewska, 2019). 4-HNE adducts with Tau have been found to modify Tau conformation and to lead to formation of neurofibrillary tangles (Liu et al., 2005). Protein oxidation in AD is evidenced by the presence of elevated levels of carbonylated proteins in the hippocampus and parietal cortex areas of the brain (those most involved in AD) (Hensley et al., 1995). AD patients with even mild cognitive impairment have been found to have higher oxidative stress levels than their healthy cohorts (Vergallo et al., 2018).

Four basic mechanisms, have been proposed to account for how Aβ causes neurofibrillary tangles. All are mediated by oxidative stress (Blurton-Jones and Laferla, 2006).

1. Aβ promotes the activation of specific kinases (e.g., GSK3β) that catalyze the hyperphosphorylation of tau, leading to its conformational change and formation of neurofibrillary tangles.
2. Neuroinflammation promoted by Aβ leading to the production of pro-inflammatory cytokines that stimulate the phosphorylation of tau.
3. A process induced by Aβ that reduces the capacity of tau degradation by the proteosome.
4. Aβ-promoted defects in axonal transport leading to inadequate localization of and tau and its messenger RNA, which, in turn, leads to hyperphosphorylation and aggregation of neurofibrillary tangles.

The precise mechanism by which Aβ causes neurofibrillary tangles remains to be determined.

References

Alzheimer's Association, 2021. 10 Warning Signs of Alzheimer's. https://www.alz.org.

Blurton-Jones, M., Laferla, F.M., 2006. Pathways by which Abeta facilitates tau pathology. Curr. Alzheimer Res. 3 (5), 437–448.

Cheignon, C., Tomas, M., Bonnefont-Rousselot, D., Fallar, P., Hureau, C., Collin, F., 2018. Oxidative stress and the amyloid beta peptide in Alzheimer's disease. Redox Biol. 14, 450–464.

Collin, F., 2019. Chemical basis of reactive oxygen species reactivity and involvement in neurodegenerative diseases. Int. J. Mol. Sci. 20 (10), 2407. https://doi.org/10.3390/ijms20102407. Accessed October 15, 2021.

Collin, F., Cheignon, C., Hureau, C., 2018. Oxidative stress as a biomarker for Alzheimer's disease. Biomark. Med. 12 (3), 201–203.

Doraiswamy, P.M., Leon, J., Cummings, J.L., Marin, D., Neumann, P.J., 2002. Prevalence and impact of medical comorbidity in Alzheimer's disease. J. Geronol. 57A (3), M173–M177.

Duthie, A., Chew, D., Soiza, R.L., 2011. Non-psychiatric comorbidity associated with Alzheimer's disease. QJ. Med. 104, 913–920.

Edwards III, G.A., Gamez, N., Escobedo Jr., G., Calderon, O., Moreno-Gonzalez, I., 2019. Modifiable risk factors for Alzheimer's disease. Front. Aging Neurosci. 11, 146. https://doi.org/10.3389/fnagi.2019.00146. Accessed October 13, 2021.

Gegotek, A., Skrzydlewska, E., 2019. Biological effect of protein modifications by lipid peroxidation products. Chem. Phys. Lipids 221, 46–52.

Hensley, K., Hall, N., Subramaniam, R., Cole, P., Harris, M., Aksenov, M., et al., 1995. Brain regional correspondence between Alzheimer's disease histopathology and biomarkers of protein oxidation. J. Neurochem. 65 (5), 2146–2156.

Huang, W.J., Zhang, X., Chen, W.W., 2016. Role of oxidative stress in Alzheimer's disease (Review). Biomed. Rep. 4, 519–522.

Liu, Q., Smith, M.A., Avila, J., DeBernardis, J., Kansal, M., Takeda, A., et al., 2005. Alzheimer-specific epitopes of tau represent lipid peroxidation-induced conformations. Free Rad. Biol. Med. 38 (6), 746–754.

Majd, S., Power, J., Majd, Z., 2019. Alzheimer's disease and cancer: when two monsters cannot be together. Front. Neurosci. 13, 155. https://doi.org/10.3389/fnins.2019.00155. Accessed October 15, 2021.

Markesbery, W.R., Lovell, M.A., 1998. Four-hydroxynonenal, a product of lipid peroxidation, is increased in the brain in Alzheimer's disease. Nerobiol. Aging 19 (1), 33–36.

Neely, M.D., Sidell, K.R., Graham, D.C., Montine, T.J., 1999. The lipid peroxidation product 4-hydroxynonenal inhibits neurite outgrowth, disrupts neuronal microtubules, and modifies cellular tubulin. J. Neurochem. 72 (6), 2323–2333.

NIH, 2019. What Causes Alzheimer's Disease? National Institute on Aging. https://www.nia.nih.gov/health/what-causes-alzheimers-disease. Accessed October 13, 2021.

NIH, 2019a. What Do We Know about Diet and Prevention of Alzheimer's Disease? National Institute on Aging. https://www.nia.nih.gov/health/what-do-we-know-about-diet-and-prevention-alzheimers-disease. Accessed October 13, 2021.

NHS, 2021. Causes Alzheimer's Disease. National Health Services, UK. https://www.nha.uk/conditions/alzheimers-disease/causes/. (Accessed 13 October 2021).

Poblador-Plous, B., Calderon-Larranaga, A., Marta-Moreno, J., Hancco-Saavedre, J., Sicras-Mainar, A., Soljak, M., et al., 2014. Comorbidity of dementia: a cross-sectional study of primary care older patients. BMC Psychiatr. 2014. http://www.biomedcentral.com/1471-244X/14/84. Accessed October 15, 2021.

Rabier, K., Ayala, S., Alies, B., Rodrigues, J.V., Rodriguez, S.B., La Penna, G., et al., 2016. Free superoxide is an intermediate in the production of H_2O_2 by copper(I)- Aβ peptide and O_2. Angew Chem. Int. Ed. Engl. 55 (3), 1085–1089.

Santiago, J.A., Potashkin, J.A., 2021. The impact of disease comorbidities in Alzheimer's disease. Front. Aging Neurosci. 13. https://doi.org/10.3389/fnagi.2021.631770.

Sayre, L.M., Zelasko, D.A., Harris, P.L., Perry, G., Salomon, R.G., Smith, M.A., 1997. 4-hydroxynonenal-derived advanced lipid peroxidation end products are increased in Alzheimer's disease. J. Neurochem. 68 (5), 2092–2097.

Silva, M.V.F., Loures, C., Alves, L.C.V., de Souza, L.C., Borges, K.B.G., Carvalho, M., 2019. Alzheimer's disease: risk factors and potentially protective measures. J. Biomed. Sci. 26 (1), 33. https://doi.org/10.1186/s12919-019-0524y. Accessed October 13, 2021.

Singh, A., Kukreti, R., Saso, L., Shrikant, K., 2019. Oxidative stress: a key modulator in neurodegenerative diseases. Molecules 24 (8), 1583. https://doi.org/10.3390/molecules24081583. Accessed October 15, 2021.

Sutherland, G.T., Chami, B., Uoussef, P., Witting, P.K., 2013. Oxidative stress in Alzheimer's disease: primary villain or physiological by-product? Redox Rep. 18 (4), 134–141.

Trempe, C.L., Lewis, T.J., 2018. It's never too early or too late – end the epidemic of Alzheimer's by reversing causation from pre-birth to death. Front. Aging Neurosci. 10, 205. https://doi.org/10.3389/fnagi.2018.00205. Accessed October 13, 2021.

Vargese, S.S., Halonen, P., Raitanen, J., Forma, L., Jylha, M., Aaltonen, M., 2021. Comorbidities in dementia during the last years of life: a register study of patterns and time differences in Finland. Aging Clin. Exper. Res. 33 (12), 3285. https://doi.org/10.1007/s40520-021-01867-2. Accessed October 15, 2021.

Vergallo, A., Giampietri, L., Baldacci, F., Volpi, L., Chico, L., Pagni, C., et al., 2018. Oxidative stress assessment in Alzheimer's disease: a clinic setting study. Am. J. Alzheimers Dis. Other Demen. 33 (1), 35–41.

Zeliger, H.I., 2016. Predicting disease onset in clinically healthy people. Interdiscip. Toxicol. 9 (2), 39–54.

Zeliger, H.I., 2019. Predicting Alzheimer's disease. Eur. J. Med. Health Sci. 1 (1). https://doi.org/10.24018/ejmed.2019.1.1.6. (Accessed 13 October 2021).

Zhao, Z., 2019. Iron and oxidizing species in oxidative stress and Alzheimer's disease. Aging Med. 2 (2), 82–87.

Zhu, X.C., Tan, L., Wang, H.F., Jiang, T., Cao, L., Wang, C., et al., 2015. Rate of early onset Alzheimer's disease: a systematic review and meta-analysis. Ann. Transl. Med. 3 (3), 38. https://doi.org/10.3978/jissn.2305-5839.2015.01.19. Accessed October 13, 2021.

CHAPTER

24

Type 2 diabetes (T2D)

24.1 Introduction

Type 2 diabetes (T2D) is a progressive, chronic metabolic disease characterized by a lowering of β-cell production and decline in insulin resistance. It is a disease in which the risks of myocardial infarction, stoke and microvascular events are associated with hyperglycemia (Fonseca, 2009). T2D is characterized as having a blood fasting glucose level of equal to or greater than 126 mg/dL, or hemoglobin A_{1c} level of 6.5% or greater and generally have blood pressures of 130/80 or greater and low-density lipoprotein cholesterol levels (LDL) that are equal to or greater than 100 mg/dL (Wang et al., 2021). There is a strong relationship between aging and obesity and the risk for developing T2D. Each decade of age increases the likelihood of developing T2D and each additional kilogram of weight elevating the risk of T2D onset by 9% (Oguntibeju, 2019).

T2D generally progresses from pre-diabetes, a condition characterized by a fasting glucose level (FGL) of less than 126 mg/dL the rate of progression from pre-diabetes to T2D has been shown to accelerate with increasing fasting glucose level (FGL), higher body mass index (BMI), higher blood pressure and lower serum high density lipoprotein level (HDL) (Fonseca, 2009). Factors associated with progression of pre-diabetes to diabetes are listed in Table 24.1.

Type 2 diabetes is accompanied by a host of symptoms that serve to alert the need to seek professional care (ADA, 2021). These are listed in Table 24.2.

The primary hallmark of T2D is insulin resistance. Other hallmarks of diabetes are oxidative stress markers of lipid peroxidation (malondialdehyde, 8-iso-prostaglandin F2α, thiobarbituric acid reactive substances and hydroxynonenal (Bigaagli and Lodovici, 2019).

The incidence of diabetes is rapidly rising from a prevalence of 9.8% in 1999–2000 to 14.3% in 2017–18 in the United States. Worldwide, the World Health Organization has reported that the number of people with diabetes has risen from 108 million in 1980 to 422 million in 2014 (Lin et al., 2018; WHO, 2021).

Diabetes prevalence in the world continues to increase dramatically. The rates in the United States have increased from 3.3 cases per 1000 in 1980 to 7.8 cases in 2007 and are projected to increase by 20%–33% by the year 2050. Worldwide, it is estimated that 7.7% of the population will have diabetes with 439 million adults affected by the year 2030 (Wang et al., 2021; WHO, 2021). Many epidemic and pandemic diseases prevalent today, including T2D, have been at least in part attributed to environmental exposures to exogenous toxic chemicals (Zeliger, 2011).

Oxidative Stress
https://doi.org/10.1016/B978-0-323-91890-9.00027-1

TABLE 24.1 Factors associated with progression of pre-diabetes to diabetes.

Younger age
Elevated FGL
Increase in FGL
High plasma insulin
Decreased insulin response to glucose
High BMI
Weight gain
Dyslipidemia
Hypertension
Poor β-cell function
Poor treatment choice

TABLE 24.2 Type 2 diabetes symptoms.

Frequent urination
Severe thirst
Severe hunger, despite eating
Extreme fatigue
Blurry vision
Cuts or bruises that are slow to heal
Tingling, pain or numbness in the extremities

Global diabetes prevalence projections for the years 1999–2045 are shown in Table 24.3 (Saeedi et al., 2019).

T2D is distinguished from type 1 diabetes, an autoimmune disease characterized by β-cell destruction (ADA, 2015). This chapter is devoted to a discussion of T2D.

24.2 Risk factors for type 2 diabetes

Non-modifiable, modifiable and partially modifiable factors put one at risk for developing diabetes (Zeliger, 2013; Wilmot and Idris, 2014, NIH, 2016). Table 24.4 lists these.

TABLE 24.3 Global diabetes prevalence projections 1999–2045.

Year	Percent	Millions of people
1999	9.3	463
2030	10.2	578
2045	10.9	700

Modifiable T2D risk factors are those which may be addressed by lifestyle changes and/or treatment of medical conditions. Chronic chemical exposure is only partially modifiable due to the pervasiveness of chemicals in the environment and the accumulation of persistent organic pollutants in body fat. The following discussion addresses this subject.

24.3 Chemical exposure as a cause of type 2 diabetes

The rapid rise in prevalence of T2D has been linked to exposures to a number of different chemicals (Zeliger, 2013, and the numerous references therein). These are: persistent organic pollutants (POPs) that include organochlorine pesticides (OCs), dioxins, furans, polychlorinated biphenyls (PCBs), polybrominated biphenyls (PBBs) and polybrominated diphenyl ethers used as fire retardants; exudates from plastics, including phthalates and bisphenol-A; polluted air; and tobacco smoke. These chemicals vary widely in structure, chemical properties and composition. Although different mechanisms of action have been suggested for some of the individual species, only one unifying explanation has been shown to explain the effects of all these toxins on T2D - their lipophilicity and their propensity to elevate oxidative stress (Zeliger, 2013; Zeliger 2016).

A review of the medical and toxicological literature reveals that T2D has been shown to present following the accumulation of lipophilic exogenous chemicals in body serum and/or tissues in a dose-response relationship (Lee et al., 2006). It has been shown that it is the lipophilicity of the exogenous chemicals that is responsible for inducing T2D by permeating lipophilic membranes and providing an entry for hydrophilic species (Zeliger, 2013). As discussed previously, the absorption of all lipophilic chemicals results in the elevation of OS.

Exposures to hydrophilic chemicals; arsenic, mercury, nitrates, nitrites and N-nitroso compounds have also been shown to increase the risk of diabetes (Longnecker and Daniels, 2001; Joshi and Shrestha, 2010; Del Razo et al., 2011). The toxicities of these, however, have been shown to increase when co-administered with lipophilic species (Zeliger, 2016).

Mixtures of lipophilic and hydrophilic chemicals are toxic to humans at very low levels of concentration, far below the known toxic levels for each of the components of such mixtures (Zeliger, 2003). It has also been shown that exposures to the lipophilic and hydrophilic species of such mixtures need not occur simultaneously, but can occur sequentially, with the hydrophilic exposure coming sometime after the lipophilic exposure, but provided that the lipophilic species is still present in the body (Zeliger et al., 2012). Such a phenomenon is believed to operate with the induction of T2D. In the case of T2D, the lipophilic species

TABLE 24.4 Risk factors for type 2 diabetes.

Non-modifiable
Age 45 or older
Family history of diabetes
Being African American
Being Alaska Native
Being American Indian
Being Asian American
Being Native Hawaiian
Being Pacific Islander
Being Hispanic/Latino
Have acanthosis nigricans
Have polycyctic ovary syndrome
History of gestational diabetes
History of heart disease
History of stroke
Depression
Modifiable
Obesity
Overweight
Hypertension
High LDL
Low HDL
Physically inactivity
Partially modifiable
Chronic exposure to lipophilic environmental chemicals

can be in the form of long-lived POPs (PCBs, OCs, dioxins, furans, PBBs and PBDEs), which once absorbed are stored in body fat and can remain in the body for up to decades and can continually transfer to serum (Yu et al., 2011; Pocar et al., 2003). The lipophiles can also be intermediate lived ones (phthalates, bisphenol-A and polynuclear aromatic hydrocarbons such as those found in polluted air and tobacco smoke); and short-lived species (C-1 to C-8 alkanes, benzene, ethyl benzene, styrene, toluene and xylenes; whose concentrations in the body remain more or less in a steady state when exposure to these is continual and absorption replaces quantities lost through metabolism and elimination (Kim et al., 2011; Culver et al., 2012; Zeliger, 2012). The lipophilic species can also be a long-use pharmaceutical, which though metabolized or eliminated from the body quickly, is constantly replenished on a regular basis. Statins and antipsychotics are examples of such medications that have been shown to increase T2D prevalence (Kim et al., 2011; Zeliger, 2012).

It is the lipophilicity of the exogenous chemicals that is responsible for inducing T2D either by its own action and/or by facilitating the absorption of hydrophilic species that alone would not penetrate lipophilic membranes. A study of T2D clusters lends credence to this point.

24.3.1 Type 2 diabetes clusters

A disease cluster is an outbreak of a particular disease in a group of individuals in greater than anticipated numbers following a common exposure by that group to a causative agent or agents. Clusters can arise from accidents, occupational exposures or environmental exposures (Zeliger, 2004). Several T2D clusters have been reported in the literature.

1. Sevesco, Italy
   A 1976 accident in Sevesco, an area north of Milan Italy caused more than 278,000 people to be exposed to 2,3,7,8-tetrachlorodibenzo-p-dioxin (TCDD). A follow up study 25 years later showed rates of T2D in those exposed to be as much as 35% higher than the rates in cohorts who were not exposed to TCDD (Consonni et al., 2008).
2. Yucheng, Taiwan
   In the late 1970s, thousands of Taiwanese residents were poisoned by the consumption of rice-bran oil contaminated with PCBs in what has become known as the Yucheng ("oil disease") incident. 1054 of these victims were examined between 1993 and 2003 and compared with unexposed controls. T2D prevalence in the women who exposed to the PCBs was twice as high as that in the control group (Wang et al., 2008).
3. Pesticides, United States
   A study of 1303 Mexican Americans either working on a farm using organochlorine pesticides or in a pesticide processing plant in the Western, United States, found that those with detectable serum levels of organochlorine pesticides had 2.4–3.5 times higher self-reported incidences of diabetes than Hispanics residing in the Metropolitan New York City area or in Dade County, Florida who were not engaged in similar work (Cox et al., 2007).
4. Great Lakes Fishermen
   A cohort study of Great Lakes sport fishermen spanning the early 1990s through 2005, found that those who consumed the fish they caught which contained *p,p* -diphenyldichlroethane (DDE) had higher DDE levels in their serum than those who caught and released the same fish. All subjects were diabetes free at the onset of the study, but 8.4 years later, the fish eaters had a 2.0–2.7 fold increase in T2D prevalence in a dose-related manner compared with controls who did not consume Great Lakes caught fish (Turyk et al., 2009; Upropec et al., 2010).
5. Vietnam Veterans
   United States Air Force veterans of the Vietnam War who participated in Operation Ranch Hand (the spraying of Agent Orange from) 1962–71, were exposed to TCDD. A

study of 989 of these veterans published in 1997 showed that compared with controls (1276 Air Force personnel who served in Southeast Asia at the same time, but not involved in the spraying of herbicides) Operation Ranch Hand veterans had serum TCDD levels more than 3 times higher and T2D incidence rate 1.5 times higher than the incidence of T2D in controls (Henriksen et al., 1997).

6. POPs Mixture

A study of 2016 individuals investigating the relationship between serum POPs concentrations of two dioxins and four organochlorine pesticides with diabetes found strong increased associations between all six POPs and T2D. The study also found strong dose-response relationships between each of the POPs and T2D prevalence (Lee et al., 2006).

24.3.2 Lipophilic chemicals shown to cause T2D

Numerous studies have demonstrated the relationship between lipophilic chemicals in body serum and increased risk of diabetes. These chemicals include POPs, exudates from plastics, polluted air, tobacco smoke and pharmaceuticals (and/or their metabolites).

24.3.2.1 POPs

POPS are compounds that are long lived in the body once absorbed. As discussed in Chapter 20, all are lipophilic compounds that are long-lasting in the body. POPs associated with increased diabetes incidence are listed in Table 24.5 (Zeliger, 2013). In an animal study, it has been definitively established that PCB-126 enhances weight gain, impairs insulin sensitivity, elevates insulin levels, elevates serum triglycerides, raises cholesterol and increases free radical generation. This PCB also modifies the expression of proteins related to oxidative stress on the islets of Langerhans, which is indicative of early β-cell failure (Loiola et al., 2016).

24.3.2.2 Plastic exudates

Semi-volatile organic compounds associated with increased diabetes prevalence include bisphenol-A (BPA) and phthalates.

Bisphenol-A; [2,2-(4,4′-dihydroxydiphenyl)propane]; is very widely used in the production of polycarbonate plastics, epoxy resins, flame retardants, paper products and other specialty chemicals. It is found in food and water containers and in the resin linings of food and beverage cans (Liao and Kannan, 2011). Though most human exposure to BPA is via ingestion, it is also readily absorbed through the skin (Heudorf et al., 2007). BPA is rapidly metabolized in the body and its metabolites are readily analyzable in urine. Elevated BPA metabolites in urine have been associated with T2D independently of other diabetes risk factors (Shankar and Teppala, 2011; Silver et al., 2011).

Phthalates (diesters of phthalic acid) are used as plasticizers in polyvinyl chloride (PVC) plastics and are components of many plastic products including vinyl flooring, house siding, building materials, household furnishings, toys, clothing, food and beverage containers and packaging. Phthalates are also incorporated into cosmetic and personal care products such as nail polish, hair spray, perfumes and deodorants, pharmaceuticals, nutritional supplements medical devices, dentures, oils, lotions, shampoos and body creams (Santodonatto, 1997; Koo and Lee, 2004; Schletter, 2006; Koniecki et al., 2011). Since phthalates are not chemically

TABLE 24.5 Persistent organic pollutants associated with increased diabetes incidence.

Aldrin
Chlordane
DDT and its metabolite DDE
Dieldrin
Endrin
Heptachlor
Hexachlorobenzene
Mirex
Pentachlorobenzene
Toxaphene
Polybrominated biphenyls (PBBs)
Polybrominated biphenyl ethers (PBDEs)
Polychlorinated biphenyls (PCBs)
Polychlorinated dibenzodioxins (PCDDs)
Polychlorinated furans.

bound to PVC resins, they can leach, evaporate, migrate and abrade off plastics into the indoor environment and run off into water sources. Human absorption occurs via ingestion, inhalation and dermal contact (Heudorf et al., 2007).

All of the most widely used phthalates; di(2-ethylhexyl phthalate), dibutyl phthalate, dimethyl phthalate, diethyl phthalate and diisobutyl phthalate; are lipophilic chemicals (Howard et al., 1985). Though these compounds are relative short lived in the body, their ubiquitous nature assures that constant levels are maintained in body serum (van Baal et al., 2011).

Phthalates are endocrine disruptors, with negative impacts on thyroid hormones in children (Boas et al., 2010). They are also directly associated with metabolic disorders in humans, including T2D, obesity, insulin resistance and other chronic diseases in humans (van Baal et al., 2011).

24.3.2.3 *Polluted air*

People breathing polluted air have been shown to have an increased prevalence of T2D (Bhatnagar, 2009; Kramer et al., 2010; Tillett, 2010; Coogan et al., 2012). Polluted air contains numerous volatile and semi-volatile organic compounds, including aliphatic, mononuclear and polynuclear aromatic hydrocarbons, whose primary sources are the incomplete combustion and evaporation of petroleum products. A representative list of these, all of which are lipophilic species, is given in Table 24.6 (Zeliger, 2013).

TABLE 24.6 A representative list of volatile and semi-volatile lipophilic species contained in polluted air.

Volatile
C-1 to C-8 alkanes
Benzene
Ethyl benzene
Styrene
Toluene
Xylenes
Semi-volatile
C9 to C16 alkanes
Naphthalene
Fluorine
Phenanthrene
Anthracene
Fluoanthene
Pyrene
Benz[a]antharacne
Chrysene
Benzofluoanthenes
Dibenz[a,h]anthracene
Benzo[ghi]perylene
Indeno[1,2,3-cd]pyrene

24.3.2.4 Tobacco smoke

Smokers as well as those exposed to second hand (environmental) tobacco smoke have been shown to have an increased prevalence of T2D (Guerin et al., 1992; Willi et al., 2007; Hayashino et al., 2008; Kowall et al., 2010; Ko et al., 2011; Zhang et al., 2012). Many of the same semi-volatile organic compounds present in polluted air that are listed in Table 2, as well as numerous other lipophilic compounds are also present in tobacco smoke (Guerin, 1992).

24.3.2.5 Pharmaceuticals

A study reporting on the T2D prevalence of post-menopausal women has demonstrated the relationship between the use of five different statins and increased prevalence of T2D (de Vries and Cohen, 1993). The five statins; Lovastatin, Simvastatin, Fluvastatin,

Atorvastatin and Pravastatin that are the subject of that study differ significantly from each other in many ways, but all are lipophilic. It has been proposed that these lipophilic statins are behaving in a mechanistic manner similar to other lipophilic exogeneous chemicals in acting to increase the risk of T2D. Though statins are not as long lived in the body as POPs or even PAHs, phthalates and BPA, they are taken on a daily basis and an individual using one of these statins effectively establishes a steady state serum level of lipophile (Culver et al., 2012; Zeliger, 2012). The association between regular use of lipophilic pharmaceuticals and metabolic impacts has also been recently reported for children. Children treated with lipophilic second-generation lipophilic antipsychotics have been shown to have high prevalence of metabolic syndrome, fore runner to T2D (Devlin et al., 2012).

24.4 Mechanisms of type 2 diabetes onset

Type 2 diabetes ensues following pancreatic decline of β-cell function. This decline is associated with insulin resistance and progressively increasing hyperglycemia (Fonseca, 2009). β-cell production of insulin is normally tied to the body's energy needs. Elevated blood glucose level stimulates insulin secretion and decreases glycogen secretion. The reverse occurs when blood glucose levels are depressed. The net effect is the maintenance of normal glycemia, as occurs during consumption of food. In the development of T2D, β-cells fail to produce sufficient insulin to meet such needs and when this failure occurs, individual β-cells and the entire β-cell mass attempt to compensate by producing additional insulin (Weyer et al., 2001).

It is to be noted that though β-cell function is primary, decreasing β-cell mass is also an important factor in T2D progression. β-cell mass is regulated by a number of factors. These are listed in Table 24.6 (Donath et al., 2005). All the items listed in Table 24.7 elevate oxidative stress.

Aging, inflammation and obesity are major contributors to the mechanisms of T2D development.

1. Aging

   Aging raises the incidence of T2D as, it is accompanied by a natural decline in insulin secretion; a reduced response to demands for insulin release; a slowing of glucose

TABLE 24.7 β-cell mass regulating factors.

Hyperglycemia
Dyslipidemia
Leptin
Cytokines
Autoimmunity
Some pharmaceuticals

transport in β-cells; and negative impacts on β-cell proliferation, apoptosis and neogenesis (Gong and Mazumdar, 2012). Aging also affects the accumulation of visceral fat which, in turn, leads to chronic inflammation that promotes T2D development (Chen et al., 1988).

2. Inflammation

   Though not a disease, inflammation exacerbates T2D due to production of ROS and enhances tissue destruction (Donath et al., 2008).

3. Obesity

   Obesity is the third major risk factor for T2D onset. Adipose dysfunction in obesity results in overproduction of pro-inflammatory adipokines, leading to chronic inflammation and oxidative stress (Gregor and Hotamisligil, 2011; Le Lay et al., 2014).

24.5 Effects of ROS

Compared to other tissues, β-cells have fewer antioxidant defense enzymes, making them particularly vulnerable to overproduction of ROS. Excess ROS results in attacks on lipids, proteins and DNA, leading to β-cell dysfunction and death. Excess ROS also leads to insulin resistance and lower insulin secretion (Betteridge, 2000; Bigagli and Lodovici, 2019).

In addition to β-cell dysfunction and insulin resistance, hyperglycemia-induced oxidative stress is also linked to late complications of T2D via the polyol, advanced glycation end product formation, protein kinase C (PKC-) diacylglycerol and hexosamine pathways (Evans et al., 2003; Brownlee, 2001; Giacco and Brownlee, 2010).

The oxidative biochemistry of glucose, which itself generates ROS also results in oxidation and damage to biomolecules (lipids, proteins, DNA and RNA) causing cellular damage in endothelial cells of large and small blood vessels and the myocardium, further accelerate organ dysfunction and ultimately lead to micro- and macrovascular diseases (Schwartz et al., 2017; Kaur et al., 2018; Teodoro et al., 2018; Bigagli and Lodovici, 2019).

24.6 Type 2 diabetes and oxidative stress

As is evident from the discussion above, oxidative stress is a major part of all facets of T2D development, including β-cell dysfunction, insulin signaling and insulin resistance. The individual steps involved in these are shown in Tables 24.8 and 24.9 (Yaribeygi et al., 2020). Free radical production induces insulin resistance through five molecular mechanisms. These are listed in Table 24.10 (Yaribeygi et al., 2020).

24.7 Type 2 diabetes complications

Complications associated with T2D include acute (metabolic) and chronic (systemic) ones. These complications, listed in Table 24.11, are associated with the oxidative stress-driven mechanisms that obtain in T2D (Giacco and Brownlee, 2010; Le Lay et al., 2014; Ullah et al., 2016).

TABLE 24.8 Role of oxidative stress in β-cell dysfunction, leading to insulin.

Resistance
Increases in apoptotic processes
Decreases metabolic pathways in β-cells
Impairs K_{ATP} channels
Inhibits transcription factors as Pdx-1 and MafA
Decreases β-cell neogenesis
Causes mitochondrial dysfunction
Activates toll-like receptors
Induces molecular pathways as nuclear factor kappa b, mitogen-activated protein kinases, and hexosamine pathways

TABLE 24.9 Oxidative stress-impaired insulin signaling pathways.

Increases insulin signal transduction
Modulates, downregulates and induces signal parameters, leading to reduction of insulin signaling pathways which in turn produces insulin resistance.

TABLE 24.10 Free radical-induced mechanistic pathways that induce insulin resistance.

Cause β-cell dysfunction
Depress insulin-regulated glucose transporter (GLUT-4) expression
Negatively impact insulin signaling pathways
Cause inflammation
Causes mitochondrial dysfunction.

24.8 Diabetes outlook

Type 2 diabetes is an incurable, yet treatable disease that, if untreated, can lead to serious morbidity and death. The mechanisms associated with developing T2D are all oxidative stress-driven. Primary risk factors for T2D are age and epigenetic parameters, which are not preventable and obesity, which can be treated. Obesity presents a second danger, as adipose

TABLE 24.11 Acute and chronic complications of T2D.

ACUTE
Infection
Diabetic ketoacidosis
Hyperglycemic, hyperosmolar, non-ketonic coma
Polydipsia
Polyuria
Fatigue
Blurred vision
Stroke
Hypertension
Heart disease
Peripheral vascular disease
Foot sores and other problems
CHRONIC
Infection
Blindness
Retinopathy
Cataracts
Neuropathy
Atherosclerosis
Peripheral vascular disease
Amputation
Cerebrovascular disease
Retinopathy
Cataracts
Renal disease
Foot sores and other problems

tissue, is a reservoir for persistent organic pollutants and other lipophilic chemicals which constantly serves as source of supply of lipophilic chemicals known to increase T2D risk.

Data regarding all T2D causing parameters varies from different nations, races, ethnic groups and geographic areas worldwide. Hence, the following data analyses are meant to be representative and not definitive for all areas.

24.8.1 Aging

Life expectancy in the United States has increased in the past 160 years from 39.4 years in 1860 to 78.9 years in 2020 along a more or less straight-line curve (O'Neill, 2021). It is anticipated that this curve will continue to be followed. With increasingly more available medical care, it is expected that similar results will be noted worldwide.

24.8.2 Obesity rates

In the 1950s, approximately 10% of U.S. adults were classified as obese. In 2011–12, the number obese had risen to 35% and continues to rise (Police, 2020). Similar data has been reported in a U.S. study covering the years 1999 to 2015–16 (CDC, 2017; Wang et al., 2020). Data from Sweden shows that obesity increased from 9.1% in 1995 to 17.0% in in 2017 (Hemmingsson et al., 2021).

24.8.3 Epigenetics

More than 400 genetic loci have been found to be associated with type 2 diabetes, many of which primarily function in β-cells (Mahajan et al., 2018). Circadian rhythm interruption, nutrient and environmental factors have been found to have epigenetic effects in the onset and complications associated with T2D (Rosen et al., 2018). The ongoing increases in quantities and varieties of synthetic chemicals being released into the environment can only continue to raise epigenetic effects.

24.8.4 Diabetes projection

T2D prevalence continues to increase worldwide. The percentage of U.S. adults with T2D is projected to increase from 9.1% in 2014 to 13.9% in 2030 and to 17.9% in 2060 (Lin et al., 2018; WHO, 2021). At this time (2022), it is not possible to attribute the projected rises in T2D numbers to each of the above parameters. The population continues to age and with it so do increases in disease prevalence. Obesity rates also continue to rise with time and this too is contributory. As discussed in Chapter 14, epigenetic effects are major contributors to T2D, and almost entirely due to environmental effects that are continuing to rise as well.

References

ADA, 2015. American Diabetes Association. Classification and diagnosis of diabetes. Diabetes Care 38 (Suppl. 1), S8–S16.

ADA, 2021. Diabetes Symptoms. American Diabetes Association. https://www.diabetes.org/diabetes/type-2/symptoms. Accessed October 18, 2021.

Betteridge, D.J., 2000. What is oxidative stress? Metabolism 49 (2), 3–8.

Bhatnagar, A., 2009. Could dirty air cause diabetes? Circulation 119, 492–494.

Bigagli, E., Lodovici, M., 2019. Circulating oxidative stress biomarkers in clinical studies on type 2 diabetes and its complications. Oxid. Med. Cell. Longev. 2019. https://doi.org/10.1155/2019/59533685. Accessed October 18, 2021.

Boas, M., Frederiksen, H., Feldt-Rasmussen, U., Skakkebaek, N.E., Hegedus, L., Hilsted, L., Juul, A., Main, K.M., 2010. Childhood exposure to phthalates: associations with thyroid function, insulin-like growth factor I and growth. Environ. Health Perspect. 118 (10), 1458–1464.

Brownlee, M., 2001. Biochemistry and molecular cell biology of diabetic complications. Nature 414 (6865), 813–820.

CDC, 2017. Long-term trends in diabetes. Diabetes translation. United States diabetes surveillance system. https://www.cdc.gov/diabetes/data. Accessed October 24, 2021.

Chen, M., Bergman, N., Porte Jr., D., 1988. Insulin resistance and beta-cell dysfunction in aging: the importance of dietary carbohydrate. J. Clin. Endocrin. Metab. 67, 951–957.

Consonni, D., Pesatori, A.C., Zocchetti, C., Sindaco, R., D'Oro, L.C., Rbagotti, M., Bertazzi, P.A., 2008. Mortality in a population exposed to dioxin after the Seveso, Italy accident in 1976: 25 years of follow-up. Am. J. Epidemiol. 167, 847–858.

Coogan, P.F., White, L.F., Jerrett, M., Brook, R.D., Su, J.G., Seto, E., et al., 2012. Air pollution and incidence of hypertension and diabetes mellitus in black women living in Los Angeles. Circulation 125 (6), 767–772.

Cox, S., Niskar, A.S., Narayan, K.M., Marcus, M., 2007. Prevalence of self reported diabetes and exposure to organochlorine pesticides among Mexican Americans: health and nutrition examination survey, 1982-1984. Environ. Health Perspect. 115 (2), 1747–1752.

Culver, Al, Ockene, I.S., Balasubramanian, R., Olenddzki, B., Sepavich, D.M., Wachawski-Wende, J., et al., 2012. Statin use and risk of diabetes mellitus in postmenopausal women in the women's health initiative. Arch. Int. Med. 172 (2), 144–152.

de Vries, A.C.J., Cohen, L.H., 1993. Different effects of hypolipidemic drugs pravastatin and lovastatin on the cholesterol biosynthesis of the human ocular lens in organ culture and on the cholesterol content of the rat lens in vivo. Biochim. Biophys. Acta 1167, 63–69.

Del Razo, L.M., Garcia-Vargas, G.G., Valenzuela, O.L., Castallanos, E.H., Sanchez-Pena, L.C., Currier, J.M., et al., 2011. Exposure to arsenic in drinking water is associated with increased prevalence of diabetes: a cross-sectional study in the Zimapan and Lagunera regions of Mexico. Environ. Health 10, 33.

Devlin, A.M., Ngai, Y.F., Ronsley, R., Panagiotopoulos, C., 2012. Cardiometabolic risk and MTHFR C677T variant in children treated with second-generation antipsychotics. Trans. Psychiatry 2, e71. https://doi.org/10.1638/tp/2011.68. Accessed October 20, 2021.

Donath, M.Y., Ehses, J.A., Maedler, K., Schumann, D.M., Ellingsgaard, H., Eppler, E., et al., 2005. Mechanisms of beta-cell death in type 2 diabetes. Diabetes 54 (Suppl. 2), S108–S113.

Donath, M.Y., Schumann, D.M., Faulengach, M., Ellinsgaard, H., Perren, A., Ehses, J.A., 2008. Islet inflammation in in type 2 diabetes: from metabolic stress to therapy. Diabetes Care 31 (Suppl. 2), S161–S164.

Evans, J.L., Goldfine, I.D., Maddux, B.A., Grodsky, G.M., 2003. Are oxidative stress-activating signaling pathways mediators of insulin resistance and beta-cell dysfunction? Diabetes 52 (1), 1–8.

Fonseca, V.A., 2009. Defining and characterizing the progression of type 2 diabetes. Diabetes Care 32 (Suppl. 2), S151–S156.

Giacco, F., Brownlee, M., 2010. Oxidative stress and diabetic complictions. Circ. Res. 107 (9), 1058–1070.

Gong, Z., Muzumdar, R.H., 2012. Pancreatic function, type 2 diabetes, and metabolism in aging. Int. J. Endocrinol. 2012. https://doi.org/10.1155/2012/320482. Accessed October 21, 2021.

Gregor, M.F., Hotamisligil, G.S., 2011. Inflammatory mechanisms in obesity. Ann. Rev. Immunol. 29, 415–445.

Guerin, M.R., Jenkins, R.A., Tomkins, B.A., 1992. The Chemistry of Environmental Tobacco Smoke: Composition and Measurement. Lewis Publishers, Chelsea, MI.

Hayashino, Y., Fukuhara, S., Okamura, T., Yamato, H., Tanaka, H., Tanaka, K., et al., 2008. A prospective study of passive smoking and risk of diabetes in a cohort of workers: the high-risk and population strategy for occupational health promotion (HIPOP-OHP) study. Diabetes Care 31, 732–734.

Hemminssson, E., Ekblom, O., Kallings, L.V., Andersson, G., Wallin, P., Soderling, J., et al., 2021. Prevalence and time trends of overweight, obesity and severe obesity in 447,925 Swedish adults, 1995–2017. Scand. J. Public Health 49, 377–383.

Henriksen, G.L., Ketchum, N.S., Michalek, J.E., Swaby, J.A., 1997. Serum dioxin and diabetes mellitus in veterans of operation Ranch Hand. Epidemiology 8 (3), 252–258.

Heudorf, U., Mersch-Sundermann, V., Angerer, J., 2007. Phthalates: toxicology and exposure. Int. J. Hyg. Environ. Health 210 (5), 623–634.

Howard, P.H., Banerjee, S., Robillard, K.H., 1985. Measurement of water solubilities, octanol/water partition coefficients and vapor pressures of commercial phthalate esters. Environ. Toxicol. Chem. 4 (5), 653–661.

Joshi, S.K., Shrestha, S., 2010. Diabetes mellitus: a review of its association with different environmental factors. Kathmandu Univ. Med. J. 8 (29), 109–115.

Kaur, R., Kaur, M., Singh, J., 2018. Endothelial dysfunction and platelet hyperactivity in type 2 diabetes mellitus: molecular insights and therapeutic strategies. Cardiovasc. Diabetol. 17 (1), 121.

Kim, M.J., Marchand, P., Henegar, C., Antignac, J.P., Alili, R., Poitou, C., et al., 2011. Fate and complex pathogenic effects of dioxins and polychlorinated biphenyls in obese subjects before and after drastic weight loss. Environ. Health Perspect. 119 (3), 377–383.

Koniecki, D., Wang, R., Moody, R.P., Zhu, J., 2011. Phthalates in cosmetic and personal care products: concentrations and possible dermal exposure. Environ. Res. 111 (3), 329-6.

Ko, K.P., Min, H., Ahn, Y., Park, S.J., Cim, C.S., Park, J.K., Kim, S.S., 2011. A prospective study investigating the association between environmental tobacco smoke exposure and the incidence of type 2 diabetes in never smokers. Ann. Epidemiol. 21 (1), 42–47.

Koo, H.J., Lee, B.M., 2004. Estimated exposure to phthalates in cosmetics and risk assessment. J. Toxicol. Environ. Health 67 (23–24), 1901–1914.

Kowall, B., Rathmann, W., Strassburger, K., Heier, M., Holle, R., Thorand, B., et al., 2010. Association of passive and active smoking with incident type 2 diabetes mellitus in the elderly population: the KORA S4/F4 cohort study. Eur. J. Epidemiol. 25 (6), 393–402.

Kramer, U., Herder, C., Sugiri, D., Strassburger, K., Schikowski, T., Ranft, U., Rathmann, W., 2010. Traffic-related air pollution and incident type 2 diabetes: results from the SALIA cohort study. Environ. Health Perspect. 118 (9), 1273–1279.

Le Lay, S., Simard, G., Martinez, M.C., Ramaroson, A., 2014. Oxidative stress and metabolic pathologies: from an adipocentric point of view. Oxid. Med. Cell. Longev. 2014. https://doi.org/10.1155/2014/908539. Accessed October 22, 2021.

Lee, D.H., Lee, I.K., Song, K., Steffes, M., Toscano, W., Baker, B.A., Jacobs Jr., D.R., 2006. A strong dose-response relation between serum concentrations of persistent organic pollutants and diabetes. Diabetes Care 29 (7), 1638–1644.

Liao, C., Kannan, K., 2011. Widespread occurrence of bisphenol-A in paper and paper products: implications for human exposure. Environ. Sci. Tech. 45 (21), 9372–9379.

Loiola, R.A., dos Anjos, F.M., Shimeda, A.L., Cruz, W.S., Drewes, C.C., Rodriguez, S.F., et al., 2016. Long-term in vivo polychlorinated biphenyl 126 exposure induces oxidative stress and alters proteomic profile on islets of Langerhans. Sci. Rep. 2016. https://doi.org/10.1038/srep27882. Accessed October 22, 2021.

Lin, J., Thompson, T.J., Cheng, Y.J., Zhuo, X., Zhang, P., Gregg, E., et al., 2018. Projection of the future diabetes burden in the United States through 2060. Popul. Health Metrics 2018. https://doi.org/10.1186/s12963-0166-4. Accessed October 24, 2021.

Longnecker, M.P., Daniels, J.L., 2001. Environmental contaminants as etiologic factors for diabetes. Environ. Health Perspect. 109 (Suppl. 6), 871–876.

Mahajan, A., Taliun, D., McCarthy, M.I., 2018. Fine-mapping type 2 diabetes loci to single-varient resolution using high-density imputation and islet-specific epigenome maps. Nat. Genet. 50, 1505–1513.

NIH, 2016. Risk Factors for Type 2 Diabetes. National Institute of Diabetes and Digestive and Kidney Diseases. https://www.niddk.nih.gove/health-information/diabetes/overview/risk-factors-type-2-diabetes. Accessed October 18, 2021.

Oguntibeju, O.O., 2019. Type 2 diabetes mellitus, oxidative stress and inflammation: examing the links. Int. J. Physiol. Pharmacol. 11 (3), 45–63.

O'Neill, A., 2021. Life Expectancy in the United States, 1860–2020. Statistica.com. https://statistica.com/statistics/1040079-expectancy-united-states-all-time. Accessed October 24, 2021.

Pocar, P., Brevini, T.A.L., Fischer, B., Gandolfi, F., 2003. The impact of endocrine disruptors on oocyte competence. Reproduction 125, 313–325.

Police, S., 2020. How much have obesity rates risen since 1950? Livestrong.com article 384722. https://www.livestrong.com/article/384722-how-mucg-hve-obesity-rates-risen-since-1950/. Accessed October 24, 2021.

Rosen, E.D., Kaestner, K.H., Natarajan, R., Patti, M.E., Sallari, R., Sander, M., Sztak, K., 2018. Epigenetics and epigenomics: implications for diabetes and obesity. Diabetes 67, 1923–1931.

Saeedi, P., Petersohn, I., Salpea, P., Malanda, B., Karuranga, N., Colagiuri, S., et al., 2019. Global and Regional Projections for 2030 and 2045: Results from the International Diabees Federation Diabetes Atlas, ninth ed. Diabetes Res. Clin. Pract. https://doi.org/10.1016/j.diabres.2019.107843 Accessed October 18, 2021.

Santodonato, J., 1997. Review of the estrogenic and antiestrogenic activity of polycyclic aromatic hydrocarbons: relationship to carcinogenicity. Chemosphere 34 (4), 835–848.

Schettler, T., 2006. Human exposure to phthalates via consumer products. Int. J. Androl. 29 (1), 181–185.

Schwartz, S.S., Epstein, S., Corkey, B.E., Grant, S.F.A., Gavin, J.R., Aguilar, R.B., Herman, M.E., 2017. A unified pathophysiological construct of diabetes and its complications. Trends Endocrin. Metabol. 28 (9), 645–655.

Shankar, A., Teppala, S., 2011. Relationship between urinary bisphenol A levels and diabetes mellitus. J. Clin. Endocrinol. Metab. 96 (12), 3822–3826.

Silver, M.K., O'Neill, M.S., Sowers, M.R., Park, S.K., 2011. Urinary bisphenol A and type-2 diabetes in U.S. adults: data from MNANES 2-3-2008. PLoS One 6 (10), e26868. https://doi.org/10.1371/journal.pone.0026868. Accessed October 20, 2021.

Tillett, T., 2010. Traffic trouble: study links diabetes to vehicular pollution. Environ. Health Perspect. 118 (9), a399.

Teodoro, J.S., Nunes, S., Rolo, A.P., Reis, F., Palmeira, C.M., 2018. Therapeutic options regarding oxidative stress, mitochondrial dysfunction and inflammation to hinder the progression of vascular complications of diabetes. Front. Physiol. 9, 1–18.

Turyk, M., Anderson, H.A., Knobeloch, L., Imm, P., Persky, V.W., 2009. Organochlorine exposure and incidence of diabetes in a cohort of Great Lakes sport fish consumers. Environ. Health Perspect. 117 (7), 1076–1082.

Ukropec, J., Radikova, Z., Huckova, M., Koska, J., Kocan, A., Sebokova, E., et al., 2010. High prevalence of prediabetes in a population exposed to high levels of an organochlorine cocktail. Diabetologia 53 (5), 899–906.

Ullah, A., Khan, A., Khan, I., 2016. Diabetes mellitus and oxidative stress – a concise review. Saudi Pharm. J. 24, 547–553.

Van Baal, P.H., Engelfriet, P.M., Boshuizen, H.C., van de Kassteel, J., Schellevis, F.G., Hoogenveen, R.T., 2011. Cooccurrence of diabetes, myocardial infarction, stroke and cancer: quantifying age patterns in the Dutch population using health survey data. Popul. Health Metr. 9, 51–60. https://doi.org/10.1186/1478-7954-9-51. Accessed October 20, 2021.

Wang, L., Li, X., Wang, Z., Bancks, M.P., Carnethon, M.R., Greenland, P., et al., 2021. Trends in prevlaene of diabetes and control of risk factors in diabetes among US adults, 1999–2018. JAMA 326 (8), 1–13. https://doi.org/10.1001/jama.2021.9883. Accessed October 18, 2021.

Wang, S.L., Tsai, P.C., Yang, C.Y., Guo, Y., 2008. Increased risk of diabetes and polychlorinated biphenyls and dioxins. Diabetes Care 31 (8), 1574–1579.

Wang, Y., Beydoun, M.A., Min, J., Xue, H., Kaminsky, L.A., Cheskin, L.J., 2020. Has the prevalence of overweight, obesity and central obesity levelled off in the United States? Trend, patterns, disparities, and future projections for the obesity epidemics. Int. J. Epidemiol. 49 (3), 810–823. https://doi.org/10.1093/ije/dyz273. Accessed October 24, 2021.

Weyer, C., Tatarannin, P.A., Bogardus, C., Pratley, R.E., 2001. Insulin resistance and insulin secretory dysfunction are independent predictors of worsening glucose tolerance during each stage of type 2 diabetes development. Diabetes Care 24 (1), 89–94.

WHO, 2021. World Health Organization. Diabetes. https://www.who.who.int/health-topics/diabetes#tab=tab_1. Accessed October 18, 2021.

Willi, C., Bodenmann, P., Ghali, W.A., Faris, P.D., Cornuz, J., 2007. Active smoking and the risk of type 2 diabetes: a systematic review and meta-analysis. JAMA 298, 2654–2664.

Wilmot, E., Idris, I., 2014. Early onset type 2 diabetes: risk factors, clinical impact and management. Ther. Adv. Chronic Dis. 5 (6), 234–244.

Yaribeygi, H., Sathyapalan, T., Atkin, S.L., Sahebkar, A., 2020. Molecular mechanisms linking oxidative stress and diabetes mellitus. Oxid. Med. Cell. Longev. 2020. https://doi.org/10.1155/2020/8609213. Accessed October 22, 2021.

Yu, G.W., Laseter, J., Mylander, C., 2011. Persistent organic pollutants in serum and several different fat compartments in humans. J. Environ. Public Health. https://doi.org/10.1155/2011/417980. Accessed October 19, 2021.

Zeliger, H.I., 2003. Toxic effects of chemical mixtures. Arch. Environ. Health 58 (1), 23–29.

Zeliger, H.I., 2004. Unexplained cancer clusters: common threads. Arch. Environ. Health 59 (4), 172–176.

Zeliger, H.I., 2011. Human Toxicology of Chemical Mixtures. Elsevier, London.

Zeliger, H.I., 2012. Statins use and risk of diabetes. Arch. Intern. Med. 172 (11), 896.

Zeliger, H.I., 2013. Lipophilic chemical exposure as a cause of type 2 diabetes (T2D). Rev. Environ. Health 2013. https://doi.org/10.1515/reveh-2012-0031. Accessed October 18, 2021.
Zeliger, H.I., 2016. Predicting disease onset in clinically healthy people. Interdiscip. Toxicol. 9 (2), 39–54.
Zhang, L., Curham, G.C., Hu, F.B., Rimm EBForman, J.P., 2012. Association between passive and active smoking and incident type 2 diabetes in women. Diabetes Care 34 (4), 892–897.

CHAPTER

25

Rheumatoid arthritis

25.1 Introduction

Rheumatoid arthritis (RA) is an autoimmune inflammatory disease that primarily attacks many joints simultaneously. RA commonly affects hand, wrist and knee joints causing inflammation and damage to joint linings that result in irreversible damage producing chronic pain, unsteadiness and deformity. RA also impacts systems that do not involve joints in about 40% of those with this disease. These are listed in Table 25.1 (CDC, 2020; Mayo Clinic, 2021).

RA should not be confused with osteoarthritis (OA), another arthritic malady which is caused by joint erosion, but unlike RA is not caused by immune system attack. The hallmark of RA is persistent symmetric polyarthritis (synovitis) that affects the hands and feet (Sommer et al., 2005).

TABLE 25.1 Non-joint systems impacted by rheumatoid arthritis.

Lungs
Heart
Blood vessels
Eyes
Kidneys
Nerve tissue
Bone marrow
Salivary glands
Skin

Oxidative Stress
https://doi.org/10.1016/B978-0-323-91890-9.00008-8

25.2 Complications of rheumatoid arthritis

Rheumatoid arthritis increases the risk of developing the complications (Guo et al., 2018; Mayo Clinic, 2021). These are listed in Table 25.2.

25.3 Biomarkers of rheumatoid arthritis

RA activates inflammation, which elevates cytokine, chemokine and inflammatory reactants that include C-reactive protein (CRP). The immune responses triggered by inflammation result in an overproduction of autoantibodies leading to elevations in immunoglobin M and anti-CCP. In RA, serum peptides are citrullinated or subjected to other posttranslational modifications by various environmental stimuli. When this occurs, the altered peptides are presented to immune cells (including T cells and antigens) and antibodies such as anti-cyclic citrallinated peptides (antiCCP) are produced. Thus, CRP, RF and anti-CCP serve as biomarkers for RA (Gavrilla et al., 2016; Atzeni et al., 2017; Mun et al., 2021).

25.4 Risk factors for rheumatoid arthritis

Epidemiological, genetic and environmental factors are associated with the risk for developing RA (Deane et al., 2017; Novella-Navarro et al., 2021). Table 25.3 contains a list of these.

The risk factors listed in Table 25.3 are examined individually.

TABLE 25.2 Complications associated with rheumatoid arthritis.

Osteoarthritis
Rheumatoid nodules
Dry eyes and mouth
Infections
Altered fat to lean mass
Carpal tunnel syndrome
Heart problems
Lung disease
Lymphoma
Reduced cognitive function
Osteoporosis

TABLE 25.3 Risk factors for Rheumatoid Arthritis.

Risk factor
Age
Gender
History of live births
Genetics/familial history
Race/ethnicity
Obesity
Smoking
Dust inhalation
Gut microbiota and mucosal inflammation
Periodontal disease
Viral infections
Diet

25.4.1 Age

Though RA onset can begin at any age, the risk for RA onset increases with increasing age (CDC, 2020; Deane et al., 2017). It is hypothesized that the age factor is associated with increasing inflammation with time (Gavrila et al., 2016).

25.4.2 Gender

Women are about twice as likely as men to develop RA. The role of hormones in RA development is still being debated, but the higher RA frequency in women has been attributed to the stimulatory effects of estrogens on the immune system and the corresponding increase in inflammation (Viatte et al., 2016).

25.4.3 History of live births

It has been empirically observed that rates of first diagnosis of RA in women are seemingly increased 1–2 years post-partum. Reasons for this observation that have been speculated to be due to either changing hormonal levels or via cells and/or DNA transmitted from the fetus to the mother (Wallenius et al., 2010).

25.4.4 Genetics/familial history

Genetics are a major risk factor for RA. There is an increased prevalence of RA, particularly when RA is present in first degree relatives (Frisell et al., 2016). Those born with human leukocyte antigen class II genotypes are likely to experience worse bouts with RA. The risk is highest when the genetically predisposed individuals are also obese or smokers (Padyukov et al., 2004; van Drongelen and Holoshitz, 2017; CDC, 2020).

25.4.5 Race/ethnicity

There is an increased risk for RA in some race and ethnic groups. As examples, elevated prevalence of RA has been reported in American Indian and Alaska Native populations (Tlingit, Yakima, Pima and Chippewa Indians) (Ferucci et al., 2005).

25.4.6 Obesity

Obesity, with body mass index (BMI) as the indicator, is directly related to the risk of developing RA (Novella-Navarro et al., 2021). A likely association between BMI and RA is that white adipose tissue, which is an endocrine organ, can release inflammatory cytokines (including C-reactive protein, leptin, tumor necrosis factor and interleukin 6) which may lead to autoinflammatory diseases, including RA (Stavropoulos-Kalinoglou et al., 2011; Finckh and Turesson, 2014; George and Baker, 2016; Philippou and Nikiphorou, 2018; Dar et al., 2018).

25.4.7 Smoking

Smoking is the primary environmental risk factor for developing RA, accounting for 20%–30% of the environmental risk. Smoking raises inflammatory pathways that are associated with RA. It is yet to be determined whether smoking triggers initial autoimmunity or acts as the driver that propagates autoimmunity (Novella-Navarro et al., 2021). It is clear, however, that the combination of smoking and being overweight further increases the risk of developing RA (de Hair et al., 2013).

25.4.8 Dust inhalation

Studies have reported on the association between exposure to dust from different sources and the risk of developing RA (Novella-Navarro et al., 2021). The following examples demonstrate this.

1. Silica dust (Sluis-Cremer et al., 1986; Turner and Cherry, 2000; Stolt et al., 2010).
2. Dust from the collapse of the World Trade Center in New York (2013) (Webber et al., 2015).
3. Tobacco smoke (Svendsen et al., 2017; Hedstrom et al., 2018).
4. Air pollution (Sun et al., 2016).
5. Textile dust (Too et al., 2016).

25.4.9 Gut microbiotica and mucosal inflammation

Specific gut microbes (Proteus ad Escherichia species) and mucosal inflammation are thought to contribute to the development of RA (Frank and Pace, 2001; Oliver and Silman, 2006; Ebringer and Rashid, 2014). It is hypothesized that interactions between microbes, environmental factors (such as tobacco smoke or dust particles) and host factors produce mucosal inflammation that facilitates local followed by systemic propagation of autoimmunity (Demoruelle et al., 2014).

25.4.10 Peridontal disease

RA development has been empirically related to periodontitis (PD), which has a higher prevalence in those with RA (Smolik et al., 2009; Arkema et al., 2010; de Molon et al., 2019).The bacterium P. gingivalis (the primary cause of PD) expresses the PAD enzyme which drives the process of protein citrullination, and leads to chronic inflammation characterized by production of pro-inflammatory cytokines (Suarez et al., 2020).

25.4.11 Viral infections

Respiratory viral infections have been associated with increased risk of RA and other autoimmune disease development (Li et al., 2014; Nielsen et al., 2017; Arleevskaya et al., 2017; Joo et al., 2019). Viruses have also been implicated as risk factors in the development of RA by virtue of exhibiting tendencies consistent with outbreaks of flu-like disease (Novella-Navarro et al., 2021). As other studies have failed to find such associations (Sandberg et al., 2015) this association is yet to be definitively established.

25.4.12 Diet

Several dietary factors have been reported as risk factors for RA.* These are:

1. High sodium (salt) (Wu et al., 2013; Salgado et al., 2015).
2. High sugar. Particularly in drinks sweetened with sugar and high-fructose corn syrup (Hu et al., 2014; DeChristopher et al., 2017).
3. High coffee (Mikuls et al., 2002; Lee et al., 2014).
4. High red meat (Pattison et al., 2004; Jin et al., 2021).
5. Low vitamin D (Merlino et al., 2004; Song et al., 2012).

*It should be noted that the subject of dietary risk factors remains controversial as studies other than those cited have reached opposing conclusions. This is attributed to the inability to separate the affects of individual dietary components from total diet and from other cofounding effects.

All of the risk parameters in Table 24.3 (with the possible exception of race and ethnicity) are associated with elevated oxidative stress.

25.5 Mechanisms of rheumatoid arthritis onset

There are two subtypes of RA, dependent upon the presence of absence of anti-citrullinated protein antibodies (ACPAs). The two RA subtypes have differing genetic association patterns as well as different responses to citrullinated antigens. The positive subtype is the more aggressive one and the subject of this discussion (Padyukov et al., 2011; Malmstrom et al., 2017).

Rheumatoid arthritis progresses via four distinct stages. These are thoroughly discussed by Guo et al. (2018, and the references therein) and reviewed here.

1. Triggering
2. Maturation
3. Targeting
4. Fulminant

A discussion of the mechanisms associated with each of these stages follows.

25.5.1 Triggering

ACPA, which is related to genetic, environmental, lifestyle and stochastic factors (van der Woude et al., 2010; Svendsen et al., 2013; Hensvold et al., 2015), ensues following an abnormal antibody response to a number of citrullinated proteins that are distributed throughout the body. These include: fibrin, vimentin, fibronectin, Epstein–Barr Nuclear Antigen-1, α-enolase, type II collagen and histones. The presence of ACPA, which can be detected long before the onset of joint symptoms, is widely used in RA diagnosis.

25.5.2 Maturation

The maturation stage is initiated at secondary lymphoid tissue or bone marrow sites and characterized by epitope spreading, the development of an immune response to the part of an antigen molecule to which an antibody attaches itself. Such an immune response can be present for several years before disease onset (van der Woude et al., 2010a).

25.5.3 Targeting

Following immune activation, RA is characterized by synovitis in symmetrical small joints, an indication of synovial membrane inflammation. In this stage, the synovial joint is targeted by leukocytes and the synovial fluid is overwhelmed by an inflammatory cascade which is attributable to both the innate and adaptive immune systems and leads to ACPA-positive RA (Burmester et al., 1983; Lu et al., 2010; Guo et al., 2018, 51,52).

25.5.4 Fulminant

Synovium contains a mixture of bone marrow-derived macrophages and specialized fibroblast-like synoviocytes (FLSs). Under healthy conditions, synovial cells maintain joint steady state via secretion of hyalouronic acid and lubricin. In RA, however, FLS dysfunction leads to hyperplastic synovium that results in production of inflammatory cytokines and proteinases that lead to a micro-environment allowing for survival of T cells and B cells and neutrophil accumulation (Filer et al., 2006). It is hypothesized that hyperplastic synovium is also due to apoptosis resistance (Aupperle et al., 1998). The net effect of these pathways is the resultant cartilage damage and bone erosion.

1. Cartilage damage

   Cartilage is an essential part of healthy synovial joints. It is composed of chondrocytes and a dense, highly organized extracellular matrix (ECM). In RA, the hyperplastic synovium damages cartilage via directed adhesion and invasion Additionally, inflammatory signals, including those released from ECM, further stimulate FLS activity. FLS synthesized enzymes degrade the cartilage matrix, which is unable to regenerate itself sufficiently. The actions of synovial cytokines and reactive nitrogen intermediates promote apoptosis of essential chrondrocytes, resulting in degradation of cartilage (McInnes and Schett, 2011).
2. Bone loss

   Bone loss, a hallmark of RA, results from the induction of osteoclasts and suppression of osteoblasts. Both inflammation and autoimmunity have been proposed as the mechanistic drivers of bone loss (Guo et al., 2018).

The inflammatory theory of bone loss in RA opines that inflammatory cytokines exert pro-osteoclastogenic effects that act as the trigger (Okamoto et al., 2017). The autoimmunity theory opines that bone loss in RA proceeds either via one of two mechanisms. The first is the formation of immune complex and Fc-mediated osteoclast differentiation. The second proceeds via formation of anticitrullinated vimentin antibodies against the most citrullinated protein. This makes osteoclasts the ideal antigenic targets for ACPA and binding ACPA to osteoclast precursors induces osteoclastogenesis, bone resorption and bone loss (Harre et al., 2012).

25.6 Oxidative stress in rheumatoid arthritis

Rheumatoid arthritis is recognized as an inflammatory disease in which free radicals are involved in its pathogenesis (Quinonez-Flores, 2016; Mateen et al., 2016; da Fonseca et al., 2019; Ramani et al., 2020). Those with RA, compared to controls, have been shown to have elevated ROS production and corresponding oxidative stress, as evidenced by increased lipid peroxidation, protein oxidation and DNA damage, as well as significant reduction in antioxidants. Though the precise mechanisms associated with the individual steps in the progression of RA are the subjects of ongoing research it is clear that oxidative stress is central to the initiation and progression of these of these mechanisms (Mateen et al., 2016; Vaselinovic et al., 2014; Nakajima et al., 2014).

References

Arkema, E.V., Karlson, E.W., Costenbader, K.H., 2010. A prospective study of periodontal disease and risk of rheumatoid arthritis. J. Rheumatol. 37 (9), 1800–1804.

Arleevskaya, M.I., Shafigullina, A.Z., Filina, Y.V., Lemerle, J., Renaudineau, Y., 2017. Associations between viral infection history symptoms, granulocyte reactive oxygen species activity, and active rheumatoid arthritis disease in untreated women at onset: results form a longitudinal cohort study of Tatarstan women. Front. Immunol. 2017. https://doi.org/10.3389/fimmu.2017.01725. Accessed October 27, 2021.

Atzeni, F., Talotta, R., Masala, I.F., Bonogiovanni, S., Boccassini, L., Sarzi-Puttini, P., 2017. Biomarkers in rheumatoid arthritis. Isr. Med. Assoc. J. 19 (8), S12–S16.

Aupperle, K.R., Boyle, D.L., Hendrix, M., Seftor, E.A., Zvaifler, N.J., Firestein, G.S., 1998. Regulation of synoviocyte proliferation, apoptosis, and invasion by the p53 tumor suppressor gene. Am. J. Pathol. 152 (4), 1091–1098.

Burmaster, G.R., Locher, P., Koch, B., Winchester, R.J., Dimitri-Bona, A., Kalden, J.R., Mohr, W., 1983. The tissue architecture of synovial membranes in inflammatory and non-inflammatory joint diseases. Rheumatol. Int. 3, 173–181.

CDC, 2020. Center for disease controls. www.cdc.gov/arthritis/basics/rheumatoid-arthritis. Accessed October 26, 2021.

da Fonseca, L.J.S., Nunes-Souza, V., Boulart, M.O.F., Rabelo, L.A., 2019. Oxidative stress in rheumatoid arthritis: what the future night hold regarding novel biomarkers and add-on therapies. Oxid. Med. Cell. Longev. 2019. https://doi.org/10.1155/2019/7536805. (Accessed 31 October 2021). Accessed.

Dar, L., Tiosano, S., Watad, A., Bragazzi, N.L., Zisman, D., Comaneshter, D., Cohen, A., Amital, H., 2018. Are obesity and rheumatoid arthritis interrelated? Int. J. Clin. Pract. 2018. https://doi.org/10.1111/ijcp.13045. Accessed October 28, 2021.

de Hair, M.J., Landewe, R.B., van de Sande, M.G., Schaardenburg, D., Van Baarsen, L.G., Gerlag, D.M., et al., 2013. Smoking and overweight determine the likelihood of developing rheumatoid arthritis. Ann. Rheum. Dis. 72 (10), 1654–1658.

de Molon, R.S., Rossa, C., Thurlings, R.M., Cirelli, J.A., Koenders, M.I., 2019. Linkage of periodontitis and rheumatoid arthritis: current evidence and potential biological interactions. Int. J. Mol. Sci. 20 (18), 4541.

Deane, K.D., Demoruelle, M.K., Kelmenson, L.B., Kuhn, K.A., Norris, J.M., Holers, V.M., 2017. Genetic and environmental risk factors for rheumatoid arthritis. Best Pract. Res. Clin. Rheumatol. 31 (1), 3–18.

DeChristopher, L.B., Uribarri, J., Tucker, K.L., 2017. Intake of high fructose corn syrup sweetened soft drinks, fruit drinks and apple juice is associated with prevalent coronary heart disease. In: U.S. Adults, Ages 45-59 Y. BMC Nutr. https://doi.org/10.1186/s40795-017-0168-9. Accessed October 28, 2021.

Demoruelle, M.K., Deane, K.D., Holers, V.M., 2014. When and where does inflammation begin in rheumatoid arthritis. Curr. Opin. Rheumatol. 26 (1), 64–71.

Ebringer, A., Rashid, T., 2014. Rheumatoid arthritis is caused by a Proteus urinary tract infection. APMIS 122 (5), 363–368.

Ferucci, E.D., Templin, D.W., Lanier, A.P., 2005. Rheumatoid arthritis in American Indians and Alaska Natives: a review of the literature. Semin. Arthritis Rheum. 34 (4), 662–667.

Filer, A., Parsonage, G., Smith, E., Osborne, C., Thomas, A.M.C., Curnow, S.J., et al., 2006. Differential survival of leukocyte subsets mediated by synovial, bone marrow and skin fibroblasts. Arth. Rheum. 54 (7), 2096–2108.

Finckh, A., Turesson, C., 2014. The impact of obesity on the development and progression of rheumatoid arthritis. Ann. Rheu. Dis. 73 (11), 1911–1913.

Frank, D.N., Pace, N.R., 2001. Molecular-phylogenic analysis of human gastrointestinal microbiota. Curr. Opin. Gastroenterol. 17 (1), 52–57.

Frisell, T., Saevarsdottir, S., Askling, J., 2016. Family history of rheumatoid arthritis: an old concept with new developments. Nat. Rev. Rheumatol. 12 (6), 335–343.

Gavrila, B.I., Ciofu, C., Stoica, V., 2016. Biomarkers in rheumatoid arthritis, what is new? J. Med. Life 9 (2), 144–148.

George, M.D., Baker, J.F., 2016. The obesity epidemic and consequences for rheumatoid arthritis care. Curr. Rheumatol. Rep. 2016. https://doi.org/10.1007/s11926-015-0550-z. Accessed October 28, 2021.

Guo, Q., Wang, Y.W., Xu, D., Nossent, J., Pavlos, N.J., Xu, J., 2018. Rheumatoid arthritis: pathological mechanisms and modern pharmacologic therapies. Bone Res. https://doi.org/10.1038/s41413-018-0016-9. Accessed August 24, 2022.

Harre, U., Georgess, D., Bang, H., Bozec, A., Axman, R., Ossipova, E., et al., 2012. Induction of osteoclastogenesis and bone loss by human autoantibodies against citrullinated vimentin. J. Clin. Invest. 122 (5), 1791–1802.

Hedstrom, A.K., Stawiarz, L., Klareskog, L., Alfressson, L., 2018. Smoking and susceptibility to rheumatoid arthritis in a Swedish population-based case-control study. Eur. J. Epidemiol. 33 (4), 415–423.

Hensvold, A.H., Magnusson, P.K.E., Joshua, V., Hansson, M., Israelsson, L., Ferreira, R., et al., 2015. Environmental and genetic factors in the development of anticitrullinated protein antibodies (ACPAs) and ACPA-positive rheumatoid arthritis: an epidemiological investigation in twins. Ann. Rheum. Dis. 74 (2), 375–380.

Hu, Y., Costenbader, K.H., Gao, X., Al-Daabil, M., Sparks, J.A., Solomon, D.H., et al., 2014. Sugar-sweetened soda consumption and risk of developing rheumatoid arthritis in women. Am. J. Clin. Nutr. 100 (3), 959–967.

Jin, J., Li, J., Gan, Y., Liu, J., Zhao, X., Chen, J., et al., 2021. Red meat intake is associated with early onset of rheumatoid arthritis: a cross-sectional study. Sci. Rep. 2021. https://doi.org/10.1038/s41598-021-85035-6. Accessed October 28, 2021.

Joo, Y.B., Lim, Y.H., KimKJ, Park, K.S., Park, Y.J., 2019. Respiratory viral infections and the risk of rheumatoid arthritis. Arthritis Res. Ther. 2019. https://doi.org/10.1186/s13075-019-1977-9. Accessed October 27, 2021.

Lee, Y.H., Bae, S.C., Song, G.G., 2014. Coffee or tea consumption and the risk of rheumatoid arthritis: a meta-analysis. Clin. Rheumatol. 33 (11), 1575–1583.

Li, S., Yu, Y., Yue, Y., Ahang, Z., Su, K., 2014. Microbial infection and rheumatoid arthritis. J. Clin. Cell. Immunol. 2014. https://doi.org/10.4172/2155-9899.1000174. Accessed October 27, 2021.

Lu, M.C., Lai, N.S., Yu, H.C., Huang, H.B., Hsieh, S.C., Yu, C.L., 2010. Anti-citrullinated protein antibodies bind surface-expressed citrullinated Grp78 on monocyte/macrophages and stimulate tumor necrosis factor alpha production. Arthritis Rheum. 62 (5), 1213–1223.

Malmstrom, V., Catrina, A.I., Klareskog, L., 2017. The immunopathogenesis of seropositive rheumatoid arthritis: from triggering to targeting. Nat. Rev. Immunol. 17 (1), 60–75.

Mateen, S., Moin, S., Khan, A.Q., Zatar, A., Fatima, N., 2016. Increased reactive oxygen species formation and oxidative stress in rheumatoid arthritis. PLoS One 11 (4). https://doi.org/10.1371/journal.pone.0152925. Accessed October 31, 2021.

Mayo Clinic, 2021. www.mayoclinic.org/syc-20353648. Accessed October 26, 2021.

McInnis, I.B., Schett, G., 2011. The pathogenesis of rheumatoid arthritis. N. Eng. J. Med. 365 (23), 2205–2219.

Merlino, L.A., Curtis, J., Mikuls, T.R., Cerhan, J.R., Criswell, L.A., Saag, K.G., 2004. Vitamin D intake is inversely associated with rheumatoid arthritis: results from the Iowa women's health study. Arthritis Rheum. 50 (1), 72–77.

Mikuls, T.R., Cerhan, J.R., Criswell, L.A., Merlino, L., Mudano, A.S., Burma, M., et al., 2002. Coffee, tea and caffeine consumption and risk of rheumatoid arthritis: results from the Iowa women's health study. Arthritis Rheum. 46 (1), 83–91.

Mun, S., Lee, J., Park, M., Shin, J., Lim, M.K., Kang, H.G., 2021. Arthritis Res. Therapy 2021. https://doi.org/10.1186/s13075-020-02405-7. Accessed October 26, 2021.

Nakajima, A., Aoki, Y., Shibata, Y., Sonobe, M., Terajima, F., Takahashi, H., 2014. Identification of clinical parameters associated with serum oxidative stress in patients with rheumatoid arthritis. Modern Rheumatol. 24 (6), 926–930.

Nielsen, J., Krause, T.G., Molbak, K., 2017. Influenza-associated mortality from all-cause mortality, Denmark 2010/11-2016/17: the fluMOMO model. Influenza Other Resp. Dis. 12 (5), 591–604.

Novella-Navarro, M., Plasencia-Rodriguez, C., Nuno, L., Balsa, A., 2021. Risk factors for developing rheumatoid arthritis in patients with undifferentiated arthritis and inflammatory arthralgia. Front. Med. 2021. https://doi.org/10.3389/fmed.2021.668898.

Okamoto, K., Nakashima, T., Shinohara, M., Negeshi-Koga, T., Komatsu, N., Terashima, A., et al., 2017. Osteoimmunology: the conceptual framework unifying the immune and skeletal systems. Physiol. Rev. 97 (4), 1295–1349.

Oliver, J.E., Silman, A.J., 2006. Risk factors for the development of rheumatoid arthritis. Scand. J. Rheumatol. 35 (3), 169–174.

Padyukov, L., Sliva, C., Stolt, P., Alfresson, L., Klareskog, L., 2004. A gene-environment interaction between smoking and shared epitope genes in HLA-DR provides a high risk of seropositive rheumatoid arthritis. Arthritis Rheum. 50 (10), 3085–3092.

Padyukov, L., Seielstad, M., Ong, R.T.H., Ronnelid, J., Seddinhzadeh, M., Alfredsson, L., Klareskog, L., 2011. A genome-wide association study suggests contrasting associations in ACPA-positive versus ACPA-negative rheumatoid arthritis. Ann. Rheu. Dis. 70 (2), 259–265.

Pattison, D.J., Symmons, D.P.M., Lunt, M., Welch, A., Luben, R., Bingham, S.A., et al., 2004. Dietary risk factors for the development of inflammatory polyarthritis: evidence for a role of high-level red meat consumption. Arthritis Rheum. 50 (12), 3804–3812.

Philippou, E., Nikiphorou, E., 2018. Are we really what we eat? Nutrition and its role in the onset of rheumatoid arthritis. Autoimmun. Rev. 17 (11), 1074–1077.

Quinonez-Flores, C.M., Gonzalez-Chavez, S.A., Del Rio Najera, D., Pacheco-Tena, C., 2016. Oxidative stress relevance in the pathogenesis of the rheumatoid arthritis: a systematic review. BioMed Res. Int. 2016. https://doi.org/10.1155/2016/6097417. Accessed October 31, 2021.

Ramani, S., Pathak, A., Dalal, V., Paul, A., Biswas, S., 2020. Oxidative stress in autoimmune diseases: an under dealt malice. Curr. Protein Pept. Sci. 21 (6), 611–621.

Salgado, E., Bes-Rastrollo, M., de Irala, J., Carmona, L., Gomez-Reino, J.J., 2015. Obs Study Med (Baltimore) 2015. https://doi.org/10.1097/MD0000000000000924. Accessed October 28, 2021.

Sanberg, M.E., Bengtsson, C., Klareskog, L., Alfredsson, L., Saevarsdottir, S., 2015. Recent infections are associated with decreased risk of rheumatoid arthritis: a population-based case-control study. Ann. Rheum. Dis. 74 (5), 904–907.

Sluis-Cremer, G.K., Hessel, P.A., Hnizdo, E., Chrchill, A.R., 1986. Relationship between silicosis and rheumatoid arthritis. Thorax 41 (8), 596–601.

Smolik, I., Robinson, D., El-Gabalawy, H.S., 2009. Periodontitis and rheumatoid arthritis: epidemiologic, clinical, and immunologic associations. Compend. Cont. Educ. Dent. 30 (4), 188–190.

Sommer, O.J., Kladosek, A., Weiler, V., Czembirek, H., Boeck, M., Stiskal, M., 2005. Rheumatoid arthritis: a practical guide to state-of-the-art imaging, image interpretation, and clinical implications. RadioGraphics 25, 381–398.

Song, G.G., Bae, S.C., Lee, Y.H., 2012. Association between vitamin D intake and the risk of rheumatoid arthritis: a meta-analysis. Clin. Rheumatol. 31 (12), 1733–1739.

Stavropoulos-Kalinoglou, A., Metsios, G.S., Koutedakis, Y., Kitas, G.D., 2011. Obesity in rheumatoid arthritis. Rheumatology 50, 450–462.

Stolt, P., Bengtsson, A.Y.C., Kallberg, H., Ronnelid, J., Lundberg, I., Klarskog, L., Alfredsson, L., 2010. Silica exposure among current smokers is associated with a high risk of developing ACPA-positive rheumatoid arthritis. Ann. Rheum. Dis. 69 (6), 1072–1076.

Suarez, L.J., Garzon, H., Arboleda, S., Rodriguez, A., 2020. Oral dysbiosis and autoimmunity: from local periodontal responses to an imbalanced systemic immunity. A review. Front. Immunol. 2020. https://doi.org/10.3389/fimmu.2020.591255. Accessed October 27, 2021.

Sun, G., Hazlewood, G., Bernatsky, S., Kaplan, G.G., Eksteen, B., Barnabe, C., 2016. Association between air pollution and the development of rheumatic disease: a systematic review. Int. J. Rheumatol. 2016. https://doi.org/10.1155/2016/5356307. Accessed October 27, 2021.

Svendsen, A.J., Kyvik, K.O., Houen, G., Junker, P., Christensen, K., Christiansen, L., et al., 2013. On the origin of rheumatoid arthritis: the impact of environment and genes – a population based twin study. PLoS One 2013. https://doi.org/10.1371/journal.pone.0057304. Accessed October 29, 2021.

Svendsen, A.J., Junker, P., Houen, G., Kyvik, K.O., Nielsen, C., Skytthe, A., Holst, R., 2017. Incidence of persistent rheumatoid arthritis and the impact of smoking: a historical twin cohort study. Arthritis Care Res. 69 (5), 616–624.

Too, C.L., Muhamad, N.S., Ilar, A., Padyukov, L., Alfedsson, L., Klareskog, L., Murad, S., Bengtsson, C., et al., 2016. Occupational exposure to textile dust increases the risk of rheumatoid arthritis: results form a Malaysian population-based case-control study. Ann. Rheum. Dis. 75 (6), 997–1002.

Turner, S., Cherry, N., 2000. Rheumatoid arthritis in workers exposed to silica in the pottery industry. Occup. Environ. Med. 57 (7), 443–447.

Van der Woude, D., Alemayehu, W.G., Verduijn, W., de Vries, R.R.P., Houwing-Duistermaat, J.J., Huizinga, T.W.J., Toes, R.E.M., 2010. Nat. Genet. 42 (10), 814–816.

Van der Woude, D., Rantapaa-Dahlqvist, S., Ioan-Fascinay, A., Onnekink, C., Schwarte, C.M., Verpoort, K.N., et al., 2010a. Epitope spreading of anti-citrullinated protein antibody response occurs before disease onset and is associated with the disease course of early arthritis. Ann. Rheum. Dis. 69 (8), 1554–1561.

van Drongelen, V., Holoshitz, J., 2017. HLA-disease associations in rheumatoid arthritis. Rheum. Dis. Clin. North Am. 43 (3), 363–376.

Veselinovic, M., Barduzic, N., Vuletic, M., Zivkovic, V., Tomic-Lucic, A., Djuric, D., et al., 2014. Oxidative stress in rheumatoid arthritis patients: relationship to diseases activity. Mol. Cell. Biochem. 391 (1–2), 225–232.

Viatte, S., Lee, J.C., Fu, B., Espeli, M., Lunt, M., De Wolf, J.N., 2016. Association between genetic variation in FOXO3 and reductions in inflammation and disease activity in inflammatory polyarthritis. Arthritis Rheumatol. 68 (11), 2629–2636.

Wallenius, M., Skomsvoll, J.F., Irgens, L.M., Salvesen, K.A., Koldingsnes, W., Mikkelsen, K., et al., 2010. Postpartum onset of rheumatoid arthritis and other chronic arthritides: results from a patient register linked to a medical birth registry. Ann. Rheum. Dis. 69 (2), 332–336.

Webber, M.P., Moir, W., Zeig-Owens, R., Glaser, M.S., Jaber, N., Hall, C., et al., 2015. Nested case-control study of selected systemic autoimmune diseases in World Trade Center rescue/recovery workers. Arthritis Rheumatol. 67 (5), 1369–1376.

Wu, C., Yosef, N., Thalhamer, T., Zhu, C., Xiao, S., Kishi, Y., Regev, A., Kuchroo, V.K., 2013. Induction of pathogenic TH17 cells by inducible salt-sensing kinase SGK1. Nature 496 (7446), 513–517.

CHAPTER

26

Asthma

26.1 Introduction

Asthma is a chronic non-communicable condition affecting both adults and children (Pate et al., 2021). It is a heterogenous disease manifest via numerous symptoms brought on by multiple stimuli. Asthma is the most common inflammatory lung disease and is characterized by airway inflammation, variable airflow obstruction, bronchial hyper-responsiveness and narrowing of the lung's small airways whose symptoms include cough, wheezing, dyspnea, chest tightness and increased mucin production, either alone or in any combination of these symptoms (Reddy, 2011; Jia et al., 2013; Holgate et al., 2015; WHO, 2021). It is estimated that, worldwide, asthma affected 262 million people and caused 461,000 deaths in 2019 (WHO, 2021).

Asthma symptom triggers vary from individual to individual. Representative samples of these are listed in Table 26.1 (Holgate et al., 2015; EHHI, 2018; WHO, 2021; EPA, 2021). These are listed in Table 26.1.

As has been discussed in Part I of this book, all the asthma triggers in Table 26.1 elevate oxidative stress.

Though a single direct cause of asthma remains unknown, genetics, environmental factors and combinations of these are believed to be causative. A number of parameters have been linked to its onset (Reddy, 2011; Holgate et al., 2015; WHO, 2021). These are listed in Table 26.2.

26.2 Phenotypes of asthma

A phenotype is defined as "the set of observable characteristics of an individual resulting from the interaction of its genotype with the environment" (Bush, 2019). Many phenotypes of asthma, with varying underlying mechanisms (endotypes) have been described and additional ones continue to be elucidated (Chung, 2014; Holgate et al., 2015; Asthma UK, 2019; Chamberlain, 2020; ACAAI, 2021; Asthma Australia; 2021).

Asthma phenotypes generally fall into two primary categories; eosinophilic inflammation and allergic sensitization. Examples of other asthma phenotypes include exercise induced, obesity caused and smoking related ones Parameters helpful for the differentiation of asthma phenotypes are listed in Table 26.3 (Popovic-Grle et al., 2021).

The parameters listed in Table 26.3 have been applied to asthma phenotyping (Popovic-Grle et al., 2021). These phenotypes are listed in Table 26.4.

https://doi.org/10.1016/B978-0-323-91890-9.00010-6

TABLE 26.1 Asthma symptom triggers.

Viral respiratory infections
Extreme cold and hot ambient temperatures
Smoke
Chemical fumes
Grass
Grain
Tree pollen
Mold
Dust
Dust mites
Animal Fur
Feathers
Food additives
Medications
Gastric reflux
Strong odors
Stress and anxiety

TABLE 26.2 Parameters linked to asthma onset.

Asthma in primary relatives (parents and siblings)
Allergic conditions
Urban living
Tobacco smoke
Air pollution
Burning wood smoke
Low birth weight
Frequent viral respiratory infections
Chronic exposure to dust and dust mites
Occupational exposure to chemicals and fumes
Obesity

TABLE 26.3 Parameters helpful for the differentiation of asthma phenotypes.

Disease symptoms
Triggers
Body shape and weight
Age of onset
Etiology
Atopic status
Smoking
Irritant exposures
Clinical laboratory results
Biomarkers
Lung function
Bronchodilator reversibility effectiveness
Response to therapy and level of asthma control
Fractional exhaled nitric oxide
Drug reactions
Need for hospitalization and/or intensive care
Remission duration
Skin involvement (urticaria, eczema, atopic dermatitis)
Digestive system involvement (e.g., esophagitis)
Upper respiratory airway involvement (nasal polyps)

Adapted and reproduced with permission from Popovic-Grle, Stajduhar, A., Lampalo M., Rnjak., 2021. Biomarkers in different asthma phenotypes. Genes 12, 801. https://dx.doi.org/10.3390/genes12060801. Under Creative Commons Attribution (CC BY) licence (https://creativecommons.org/licences/by/4.0/).

Asthma phenotypes according to etiology
Allergic asthma
Non-allergic asthma
Aspirin exacerbated respiratory disease
Exercise-induced asthma

(*Continued*)

Occupational asthma
Asthma phenotypes according to clinical characteristics
Obesity-related asthma
Smoking-associated asthma
Cough variant asthma
Persistent asthma
Intermittent asthma
Premenstrual asthma
Preschool asthma
Post-puberty asthma
Early-onset asthma
Infantile asthma
Late-onset asthma
Very late onset asthma
Exacerbations-prone asthma
Atypical asthma
Classic asthma
Asthma phenotypes with underlying diseases
Eosinophilic granulomatosis with poly angiitis (Churg-Strauss syndrome)
Allergic bronchopulmonary mycosis
Asthma with bronchiectasis
Asthma with immunodeficiency
Asthma with α-1 antitrypsin deficiency
Asthma phenotypes according to pulmonary function tests
Reversible asthma
Asthma with fixed airway obstruction
Asthma with non-reversible airway obstruction
Restrictive ventilatory disorders such as asthma
Airway hyperresponsiveness
Asthma with high inflammatory component (via fractional exhaled nitric oxide)
Asthma with low inflammatory component (via fractional exhaled nitric oxide)
Brittle asthma (wide variation of peak expiratory flow)
Asthma phenotype according to cellular composition of airway inflammation

Eosinophilic asthma
Neutrophilic asthma
Mixed asthma
Paucigranulocytic asthma
Asthma phenotypes based on treatment response and level of asthma control
Severe asthma
Difficult -to-treat asthma
Refractory asthma
Treatment-resistant asthma
Problematic asthma
Uncontrolled asthma
Steroid-resistant asthma
Steroid-dependent asthma
Asthma with a history of respiratory failure and/or intubation and mechanical ventilation
Mild asthma
Benign asthma
Asthma phenotypes based on the level of type 2 cytokine profile (modern approach)
T2 high asthma
T2-low (or non T2 high)
Asthma phenotypes according to etiology
Allergic asthma—previously extrinsic
Non-allergic asthma—previously intrinsic
Aspirin exacerbated respiratory disease (usually connected to nasal polyposis, Samter's triad of syndrome de Widal)
Exercise-induced asthma
Occupational asthma

Reproduced with permission from Popovic-Grle, Stajduhar, A., Lampalo M., Rnjak., 2021. Biomarkers in different asthma phenotypes. Genes 12, 801. https://dx.doi.org/10.3390/genes12060801. Under Creative Commons Attribution (CC BY) licence (https://creativecommons.org/licences/by/4.0/).

Alternate ways to describe asthma include those identified by frequency and severity have been proposed. An example follows (Asthma Australia, 2021).

1. Mild Intermittent. Symptoms are mild and don't interfere with daily activities. Duration of symptoms is less than 2 days per week or two nights per month.
2. Mild Persistent. Symptoms occur more than twice per week, though not daily, or as many as four nights per month.

TABLE 26.4 Primary risk factors for asthma.

Primary risk factors for asthma
Gender
Ethnicity
Family history
Genetics
Epigenetics
Exposure to animals
Atopy
Allergic sensitization
Smoking
Air pollution
Diet
Microbial respiratory infections
Antibiotics
Occupational exposure

3. Moderate Persistent. Symptoms are experienced daily and at least one night per week, but not every night. Ability to carry on with daily activities may be limited.
4. Severe Persistent. Symptoms occur several times per day and most nights. Daily activities are extremely impacted.

26.3 Biomarkers of asthma

A perfect biomarker for asthma is yet to be determined. Though a number of different biomarker types have been proposed, it is generally agreed that biomarkers for asthma generally fall into two types, T2-high inflammatory and T2-low inflammatory (Wan and Prescott, 2016; Tiotiu, 2018; Popovic-Grle et al., 2021). Type 2 inflammation (T2I) is characterized by immune dysregulation and epithelial barrier dysfunction (ACAAI, 2021). T2I biomarkers of asthma are those defined as being driven by the Th2-cytokines Interleukin (IL)-4, IL-5 and IL-13 (Wan and Prescott, 2016; Tiotiu, 2018).

The major biomarkers of asthma are sputum eosinophil count, fractional exhaled nitric oxide (FeNO) concentration, blood eosinophil count and serum periostin.

1. Sputum eosinophils arise from sputum and are expressed as a percentage of inflammatory cells. Induction of sputum eosinophil count is increased in people exhibiting symptoms of asthma and increases in it may be indicative of asthma exacerbation (Jatakanon et al., 2000).

2. Fractional exhaled nitric oxide (FeNO) concentration.
   Nitric oxide (NO) is synthesized by NO synthases. The exhaled breath of asthmatics contains high levels of NO, produced by inducible NO synthases in airway epithelial cells resulting from airway inflammation (Dweik et al., 2011).
3. Blood eosinophil count.
   Blood eosinophil count is a potential surrogate biomarker for eosinophilic inflammation, as it is less invasive than sputum eosinophil collection. Though not as accurate as sputum eosinophil count, sputum eosinophil count has been shown to be applicable in determining treatment options for asthma (Hastie et al., 2013; Wagener et al., 2015).
4. Periostin.
   Periostin is a protein secreted by bronchial epithelial cells and lung fibroblasts in response to the Th2 cytokines IL4 and IL 13. Though an elevated periostin level is associated with asthma, its contribution to asthma pathogenesis, if any, remains to be determined (Takayama et al., 2006).

Asthma biomarkers unrelated to type 2 inflammation include elevated immunoglobin E (IgE), sputum neutrophil, interleukin-17 (IL-17), IL-6 and C-reactive protein. These and other T2-high inflammatory and T2-low inflammatory biomarkers for asthma are described in the literature (Subbarao et al., 2009; Wan and Prescott, 2016; Popovic-Grle et al., 2021).

26.4 Risk factors for asthma

Asthma is a heterogenous disease manifest via numerous symptoms that result from multiple stimuli. As a result, there are multiple risk factors for asthma, the primary ones of which are listed in Table 26.4, discussed and referenced below.

26.4.1 Gender

The risk for asthma onset is higher for boys than for girls up to age 13–14 but higher for female adolescents and young adults (Bjornson and Mitchell, 2000). At around age 20, the incidence of asthma is the same for men as for women, but the ratio increases for women above age 40. This phenomenon is believed to be attributable to the use of hormone replacement therapy by women (Troisi et al., 1995).

26.4.2 Ethnicity

Asthma risk is more prevalent among some ethnic groups, with the risk greater for African Americans than for Caucasian Americans serving as an example (CDC, 2011).

26.4.3 Family history

Having a parent with asthma, one is at 3–6 times higher risk for developing this disease than if both parents are asthma free (American Lung Association, 2020)

26.4.4 Genetics

Genetic factors have been shown to be associated with risk for asthma or related phenotypes including bronchial hyperresponsiveness, eosinophilia or elevated IgE levels. Studies in multiple ethnic populations have identified over 25 genetic loci, including immune system genes in people with asthma. More than 600 candidate genes for asthma have been identified, but the complexity of the disease has prevented precise connections (Ober and Hoffjan, 2006; Toskala and Kennedy, 2015).

26.4.5 Epigenetics

Epigenetic factors regulate many processes. Allergic phenotype asthma is regulated by DNA methylation and histone modifications and asthma development is associated with the interplay between inherited and environmental factors. Chemically modified DNA and histone proteins are thought to play a crucial role in the translation of environmental interactions in the expression of asthma. Whether epigentic changes are the cause or consequence of the disease, however, remains an open question (Relton and Smith, 2012; Begin and Nadeau, 2014; Toskala and Kennedy, 2015).

26.4.6 Atopy and allergic sensitizations

Atopy (the tendency to produce specific IgE antibodies as a reaction to exposure to an allergen) in childhood is a key factor in for the risk of developing asthma (Illi et al., 2006; Sly et al., 2008). The presence of these specific IgE antibodies in blood or via skin prick tests has been epidemiologically associated with asthma (Oksel and Custovic, 2018; Sonntag et al., 2019).

Sensitization to airborne allergens, (e.g., house dust mites, cat and cockroach allergens) further elevates IgE antibody production and greatly elevates the risk for developing asthma. The development of allergic sensitization in very early childhood is a major risk factor for the onset of asthma at school age (Kusel et al., 2007). Indeed, the combination of atopy and allergic sensitization greatly increases the risk for developing asthma (Comberiati et al., 2017; Di Cicco et al., 2020). The subject of allergic sensitization is further discussed in Section 26.6 below.

26.4.7 Tobacco smoke exposure

Smoking tobacco presents a high risk for asthma onset in both sexes, with women at greater risk than men. Children of mothers who smoked during pregnancy or who have been exposed to second hand smoke are also at risk for developing asthma (Burke et al., 2012; Toskala and Kennedy, 2015).

26.4.8 Air pollution

Both outdoor and indoor air pollution exposures produce a major risk factor for asthma development when air is polluted with fine particulates (PM 2.5–2.10) and gaseous compounds resulting from fuel combustion (most prominently, nitrogen dioxide) (Toskala and Kennedy, 2015)

26.4.9 Obesity

Obesity is a significant risk factor for asthma. Epidemiological studies have shown that asthma onset is more likely in obese children and adults and that symptoms of the disease are greater in the obese than in those of normal weight (Sutherland, 2014). It is thought that immunological mechanisms and elevated airway inflammation common to both diseases connect the two illnesses (Sutherland et al., 2012).

26.4.10 Diet

Diet has been indirectly identified as a risk factor for asthma. Asthma is more prevalent in Western societies in which the major dietary components consist of refined grains, high sugar content, high fat content, processed and red meats (foods that have pro-inflammatory effects), and in which the minor components are fruits and vegetables (foods high in antioxidant-reducing that reduce inflammation-reducing antioxidants). Chronic airway inflammation is a key component of asthma. Diets high in fats, processed foods and sugary sodas and fruit drinks have been shown to be associated with the risk for developing asthma (Tromp et al., 2012; Patel et al., 2014; Berentzen et al., 2015; DeChristopher et al., 2016; Guilleminault et al., 2017) Whether foods are a direct or indirect cause of asthma is still an open question as Western diet consumers tend to have a higher obesity prevalence than those who eat healthier foods, such as those in the Mediterranean diet, and that it is obesity that is the asthma driver (Garcia-Larsen et al., 2016; Guilleminault et al., 2017).

26.4.11 Microbial respiratory infections

Viral and bacterial infections during infancy and early childhood cause wheezing and some children who experience viral respiratory infections develop chronic asthma (American Lung Association, 2020). However, whether lower respiratory tract infections promote sensitization to airborne allergens that leads to persistent asthma is an open question as childhood viral infections have been found to be pathogenic to some but protective for others (Martinez, 1994; Friedlander et al., 2005; Subbarao et al., 2009).

26.4.12 Antibiotics

Several studies have addressed the question of whether early-life antibiotic use is a risk factor for asthma. The general consensus is that administering antibiotics is associated with a slight risk of developing asthma, with risk rising as courses of antibiotic prescription increase (Marra et al., 2009; Ahmadizar et al., 2017; Slob et al., 2020).

26.4.13 Occupational exposure

Occupational asthma (OA) is defined as follows: "A disease characterized by variable airflow limitation and/or nonspecific bronchial hyperresponsiveness due to causes and conditions attributable to a particular occupational environment and not to stimuli encountered outside the workplace" (Chan-Yeung and Malo, 1994).

OA includes both immunological (occurring following latent exposure to an immune sensitizing agent), and non-immunological, in which onset is not associated with sensitization and with or without a latent exposure period. Most OA causing agents are immunological (Chan-Yeung and Malo, 1994; Bakerly et al., 2008; Tan and Bernstein, 2014; Toskala and Kennedy, 2015). Hundreds of individual immunological and non-immunological sensitizing agents of occupational asthma have been documented. Examples of both types are listed in Table 26.5 (World Allergy Organization, 2021).

TABLE 26.5 Common causes of occupational asthma.

Immunological
Flour dust
Latex
Animal products (cow dander, eggs, pork)
Shellfish (clams, lobster, shrimp)
Arthropods (beetles, fruit flies, gipsy moth)
Microorganisms (amoebas, molds, mushrooms)
Plants (baby's breath, cauliflower and broccoli pollen, grass)
Enzymes (lactase, proteases, trypsin)
Gums
Rubber-derived proteins
Woods (pine, beach, western red cedar)
Tobacco
Pharmaceuticals
Diisocyanates (toluene diisocyanate, methylene diphenyl diisocyanate, hexamethylene diisocyanate)
Acid anhydrides
Amines
Epoxy resins
Metals (chromium, cobalt, nickel, platinum, zinc)
Plastics monomers (cyanoacrylates, styrene, vinyl chloride)
Dyes (basic blue 99, henna red, textile dyes)
Ethylene diamine tetraacetic acid (EDTA)
Freon
Gluteraldehyde
Triclosan

TABLE 26.5 Common causes of occupational asthma.—cont'd

Non-immunological
Machining fluids
Acetic acid
Chlorine
Formaldehyde
Ozone
Perchloroethylene
Pesticides
Sulfur dioxide

26.5 Cross sensitization

Exposures to respiratory sensitizing agents can occur via inhalation and also via dermal absorption. Hundreds of respiratory sensitizers and thousands of dermal sensitizers have been identified (Kimber et al., 2018; World Allergy Organization, 2021). Some sensitizers affect both the respiratory system and the skin. Exposure of such chemicals can lead to cross-sensitization, whereby inhalation can lead to dermal sensitization and allergic contact dermatitis and dermal absorption can lead to respiratory sensitization and asthma (Bello, 2007; Redlich and Herrick, 2008; Redlich, 2010; Zeliger, 2011; Kimber, 2010). Primary examples of cross-sensitizing chemicals are isocyanates (used in wood floor finishes, adhesives and plastics) and amines (used in epoxy resins) (Redlich and Herrick, 2008; Redlich, 2010).

26.6 Mechanisms of asthma and oxidative stress

Asthma is a chronic disease causing intermittent bronchial hyperresponsiveness. Airway inflammation is one of the two primary factors in asthma onset. It is caused by a combination of genetic and environmental factors including the involvement of multiple genetic loci, increased Th2 cytokines leading to recruitment of inflammatory cells to the airway, an increase in reactive oxygen species and mitochondrial dysfunction in activated inflammatory cells leading to tissue injury in the bronchial epithelium, which is central to asthma onset (Levine and Reinhaldt, 1983; Busse and Rosenwasser, 2003; Reddy, 2011; Holgate et al., 2015; Bush, 2019).

Oxidative stress (OS), the second major factor in asthma, is elevated in asthmatics compared to their healthy cohorts (Nadeem et al., 2003). It is well established that endogenous and exogenous ROS and RNS species affect macrophage function and promote airway inflammation and asthma severity. Oxidative stress-promoted airway inflammation occurs via induction of diverse pro-inflammatory mediators, enhancement of bronchial hyperresponsiveness, stimulation of bronchospasm and increases in mucin production (Misso and

Thompson, 2005; Cho and Moon, 2010; Holguin and Fitzpatrick, 2010; Sahiner et al., 2011 Reddy, 2011; Manti et al., 2016; de Groot et al., 2019; van der Vliet et al., 2018).

26.6.1 Endogenous oxidative stress

Inflammatory cells, including activated eosinophils, neturophils, monocytes and macrophages, as well as resident epithelial and smooth muscle cells generate ROS in asthmatics (Fahn et al., 1998; Evans et al., 1996; Dworsky, 2000; Lavinskiene et al., 2015). In asthma, ROS attack proteins and lipids, cause DNA mutations, alter signaling and significantly lower antioxidant levels in the body (Misso and Thompson, 2005; van der Vliet et al., 2018; Sahiner, 2011, de Groot et al., 2019).

26.6.2 Exogenous oxidative stress

Inhalation is the major pathway for absorption of exogenous OS raising species. The large surface area of the lungs, provides a huge quantity of absorption sites for volatile compounds and particulates. Ultrafine particles, such as those present in tobacco smoke, readily induce mitochondrial damage in the lungs (Li et al., 2003; de Groot et al., 2019).

26.7 Summary

Asthma is a multi-phenotype, mechanistically diverse disease that strikes in childhood and/or adulthood. It can wax and wane with age or time and can be caused by genetic and/or environmental factors. No matter which phenotype of asthma is being considered, be it endogenously or exogenously caused, the one mechanistic constant is the influence of elevated oxidative stress in its onset and progression.

References

ACAAI, 2021. American College of Allergy, Asthma and Immunology. Types of asthma. https://acaai.org/asthma/types-of-asthma. Accessed November 5, 2021.

Ahmadizar, F., Vijerberg, S.J.H., Arets, H.G.M., de Boer, A., Turner, S., Devereux, G., et al., 2017. Early life antibiotic use and the risk of asthma an asthma exacerbations in children. Pediatr. Allergy Immunol. 28 (5), 430–437.

American Lung Association, 2020. Asthma risk factors. https://www.lung.org/lung-health-diseases/lung-disease-lookup/asthma/asthma-symptoms-causes-risk-factors/asthma-risk-factors. Accessed November 8, 2021.

Asthma Australia, 2021. Types of Asthma. https://asthma.org.au. Accessed November 5, 2021.

Asthma, U.K., 2019. https://www.asthma.org.uk/advice/understanding-asthma/types/. Accessed November 5, 2021.

Bakerly, N.D., Moore, V.C., Vellore, A.D., Jaakkola, M.S., Robertson, A.S., Burge, P.S., 2008. Fifteen-year trends in occupational asthma: data from the Shield surveillance scheme. Occup. Med. 58, 169–174.

Begin, P., Nadeau, K.C., 2014. Epigenetic regulation of asthma and allergic disease. Allergy Asthma Clin. Immunol. 2014. https://doi.org/10.1186/1710-1492-10-227. Accessed November 8, 2021.

Bello, D., Herrick, C.A., Smith, T.J., Woskie, S.R., Streicher, R.P., Cullen, M.R., et al., 2007. Skin exposure to isocyanates: reasons for concern. Environ. Health Perspect. 115 (3), 328–335.

Berentzen, N.E., van Stokkom, V.L., Koppelman, G.H., Schaap, L.A., Smit, H.A., Wijga, A.H., 2015. Associations of sugar-containing beverages with asthma prevalence in 11-year-old children: the PIAMA birth cohort. Eur. J. Clin. Nutr. 69 (3), 303–308.

Bjornson, C.L., Mitchell, I., 2000. Gender differences in asthma in childhood and adolescence. J. Gend. Specif. Med. 3 (8), 57–61.

Burke, H., Leonardi-Bee, J., Hashim, Pine-Abata, H., Chen, Y., Cook, D.G., et al., 2012. Prenatal and passive smoke exposure and incidence of asthma and wheeze: systematic review and meta-analysis. Pediatrics 129 (4), 735–744.

Bush, A., 2019. Pathophysiological mechanisms of asthma. Front. Pediatr. 2019. https://doi.org/10.3389/fped.2019.00068. Accessed November 11, 2021.

Busse, W.W., Rosenwasser, L.J., 2003. Mechanisms of asthma. J. Allergy Clin. Immunol. 111 (3), S799–S804.

CDC, 2011. Centers for Disease Control and Prevention. CDC vital signs. Asthma in the US. Growing Every Year. http://www.cdc.gov/Vital/Signs/Asthma. Accessed November 8, 2021.

Chamberlain, M., 2020. Types of asthma, complete list – with symptoms and severity. https://prescriptionhope.com/types-of-asthma. Accessed November 5, 2021.

Chan-Yeung, M., Malo, J.L., 1994. Aetiological agents in occupational asthma. Eur. Respir. J. 7, 346–371.

Cho, Y.S., Moon, H.B., 2010. The role of oxidative stress in the pathogenesis of asthma. Allergy Asthma Immunol. Res. 2 (3), 183–187.

Chung, K.F., 2014. Defining phenotypes in asthma: a step towards personalized medicine. Drugs 74 (7), 719–728.

Comberiati, P., Di Cicco, M.E., D'Elios, S., Peroni, D.G., 2017. How much asthma is atopic in children? Front. Pediatr. 2017. https://doi.org/10.3389/fped.2017.00122. Accessed November 9, 2021.

DeChristopher, L.R., Uribarri, J., Tucker, K.L., 2016. Intakes of apple juice, fruit drinks and soda are associated with prevalent asthma in US children aged 2–9 years. Public Health Nutr. 19 (1), 123–130.

de Groot, L.E.S., van der VeenTA, Martinez, F.O., Hamann, J., Lutter, R., Melgert, B.N., 2019. Oxidative stress and macrophages: driving forces behind exacerbations of asthma and chronic obstructive pulmonary disease. Am. J. Physiol. Lung Cell. Mol. Physiol. 316, L369–L384.

Di Cicco, M., D'Elios, S., Diego, P., Comberiati, P., 2020. Current Opin. Allergy Clin. Immunol. 20 (2), 131–137.

Dweik, R.A., Boggs, P.B., Erzurum, S.C., Irvin, C.G., Leigh, M.W., Lundberg, J.O., et al., 2011. An official ATS clinical guideline: interpretation of exhaled nitric oxide levels (FENO) for clinical applications. Am. J. Respir. Crit. Care Med. 184 (5), 602–615.

Dworsky, R., 2000. Oxidant stress in asthma. Thorax 5 (Suppl. 2), S51–S53.

EHHI, 2018. The Harmful Effects of Wood Smoke and Growth of Recreational Wood Burning. Environment and Hyman Health, North Haven, CT. www.ehhi.org. Accessed November 8, 2021.

EPA, 2021. Wood Smoke and Your Health. U.S. Environmental Protective Agency. https://www.epa.gov/burnwise/wood-smoke-and-your-health. Accessed November 8, 2021.

Evans, D.J., Lindsay, M.A., O'Connor, B.J., Barnes, P.J., 1996. Priming of circulating human eosinophils following late response to allergen challenge. Eur. Resp. J. 9 (4), 703–708.

Fahn, H.J., Wang, L.S., Kao, S.H., Chang, S.C., Huang, M.H., Wei, Y.H., 1998. Smoking-associated mitochondrial DNA mutations and lipid peroxidation in human lung tissues. Am. J. Respir. Cell Mol. Biol. 19 (6), 901–909.

Friedlander, S.L., Jackson, D.J., Gangnon, R.E., Evans, M.D., Li, Z., Roberg, K.A., et al., 2005. Viral infections, cytokine dysregulation and the origins of childhood asthma and allergic diseases. Pediatr. Infect. Dis. J. 24 (11 Suppl. l), S170–S176.

Garcia-Larsen, V., Del Giacco, S.R., Moreira, A., Bonini, M., Charles, D., Carlsen, K.H., et al., 2016. Asthma and dietary intake: an overview of systemic reviews. Allergy 71 (4), 433–442.

Guilleminault, L., Williams, E.J., Scott, H.A., Berthon, B.S., Jensen, M., Wood, L.G., 2017. Diet and asthma: is it time adapt our message. Nutrients 2017. https://doi.org/10.3390/nu9111227. Accessed November 11, 2021.

Hastie, A.T., Moore, W.C., Li, H., Rector, B.M., Ortega, V.E., Pascual, R.M., et al., 2013. Biomarker surrogates do not accurately predict sputum eosinophil and neutrophil percentages in asthmatic subjects. J. Allergy Clin. Immunol. 132 (1), 72–80.

Holgate, S.T., Wenzel, S., Postma, D.S., Weiss, S.T., Renz, H., Syl, P.D., 2015. Asthma. Nat. Rev. Dis. Primers 2015. https://doi.org/10.1038/nrdp.2015.25. Accessed November 5, 2021.

Holgun, F., Fitzpatrick, A., 2010. Obesity, asthma, and oxidative stress. J. App. Physiol. 108, 754–759.

Illi, S., von Mutius, E., Lau, S., Niggerman, B., Gruber, C., Wahn, U., et al., 2006. Perennial allergen sensitization early in life and chronic asthma in children: a birth cohort study. Lancet 368 (9537), 763–770.

Jatakanon, A., Lim, S., Barnes, P.J., 2000. Changes in sputum eosinophils predict loss of asthma control. J. Respir. Crit. Care Med. 161 (1), 64–72.

Jia, C.E., Zhang, H.P., Lv, Y., Liang, R., Jiang, Y.Q., Powell, H., et al., 2013. The asthma control test and asthma control questionnaire for assessing asthma control: systematic review and meta-analysis. J. Allergy Clin. Immunol. 131 (3), 695–703.

Kimber, I., Basketter, D.A., Dearman, R.J., 2010. Chemical allergens—what are the issues? Toxicology 268, 139–142.

Kimber, I., Poole, A., Basketter, D.A., 2018. Skin and respiratory chemical allergy: confluence and divergence in a hybrid adverse outcome pathway. Toxicol. Res. 7, 586–605.

Kusel, M.M.H., de Klerk, N.H., Kebadze, T., Vohma, V., Holt, P.G., Johnston, S., et al., 2007. Early-life respiratory viral infections, atopic sensitization and risk of subsequent development of persistent asthma. J. Allergy Clin. Immunol. 119 (5), 1105–1110.

Lavinskiene, S., Malakauskas, K., Jeruch, J., Hoppenot, D., Sakalauskas, R., 2015. Functional activity or peripheral blood eosinophils in allergen-induced late-phase airway inflammation in asthma patients. J. Inflamm. 12. https://doi.org/10.1186/s12950-015-0065-4. Accessed November 12, 2021.

Levine, S.A., Reinhardt, J.H., 1983. Biochemical-pathology initiated by free radicals, oxidant chemicals, and therapeutic drugs in the etiology of chemical hypersensitivity disease. Orthomol. Psychiatry 12 (3), 166–183.

Li, N., Sioutas, C., Cho, A., Schmitz, D., Misra, C., Sempf, J., Wang, M., et al., 2003. Ultrafine particulate pollutants induce oxidative stress and mitochondrial damage. Environ. Health Perspect. 111 (4), 455–460.

Manti, S., Marseglia, L., D'Angelo, G., Cuppari, C., Cusumano, E., Arrigo, T., et al., 2016. "Cumulative stress": the effectds of maternal and neonatal oxidative stress and oxidative stress-inducible genes on programming of atopy. Oxid. Med. Cell. Longev. 2016. https://doi.org/10.1155/2016/8651820. Accessed November 11, 2021.

Marra, F., Marra, C.A., Richardson, K., Lynd, L.D., Kozyrskyj, A., Patrick, D.M., et al., 2009. Antibiotic use in children is associated with increased risk of asthma. Pediatrics 123 (3), 1003–1010.

Martinez, F.D., 1994. Role of viral infections in the inception of asthma and allergies during childhood: Could they be protective? Thorax 49 (12), 1189–1191.

Misso, N.L.A., Thompson, P.J., 2005. Oxidative stress and antioxidant deficiencies in asthma: potential modification by diet. Redox Rep. 10 (5). https://doi.org/10.1179/135100005X70233. Accessed November 11, 2021.

Nadeem, A., Chhabra, S.K., Masood, A., Raj, H.G., 2003. Increased oxidative stress and altered levels of antioxidants in asthma. J. Allergy Clin. Immunol. 111 (1), 72–78.

Ober, C., Hoffjan, 2006. Asthma genetics 2006: the long and winding road to gene discovery. Genes Immun. 7 (2), 95–100.

Oksel, C., Custovic, A., 2018. Development of allergic sensitization and its relevance to paediatric asthma. Curr. Opin. Allergy Clin. Immunol. 18 (2), 109–116.

Pate, C.A., Zahran, H.S., Qin, X., Johnson, C., Hummelman, E., Malilay, J., 2021. Asthma surveillance – United States, 2006-2018. MMWR 70 (5). Accessed September 17, 2021.

Patel, S., Custovic, A., Smith, J.A., Simpson, A., Kerry, G., Murray, C.S., 2014. Cross-sectional association of dietary patterns with asthma and atopic sensitization in childhood—in a cohort study. Pediatr. Allergy Immunol. 25 (6), 565–571.

Popovic-Grle, Stajduhar, A., Lampalo, M., Rnjak, 2021. Biomarkers in different asthma phenotypes. Genes 12, 801. https://doi.org/10.3390/genes12060801. Accessed November 8, 2021.

Reddy, P.H., 2011. Mitochondrial dysfunction and oxidative stress in asthma: implications for mitochondria-targeted antioxidant therapeutics. Pharmaceuticals 4, 429–456.

Redlich, C.A., Herrick, C.A., 2008. Lung/skin connections in occupational lung disease. Curr. Opin. Allergy Clin. Immunol. 8, 115–119.

Redlich, C.A., 2010. Skin exposure and asthma. Proc. Am. Thorax Soc. 7, 134–137.

Relton, C.L., Smith, G.D., 2012. Two-step epigenetic Mendelian randomization: a strategy for establishing the causal role of epigenetic processes in pathways to disease. Int. J. Epidemiol. 41 (1), 161–176.

Sahiner, U.M., Birben, E., Erzurum, S., Sacksen, C., Kalayci, O., October 2011. Oxidative stress in asthma. WAO J. 151–158.

Slob, E.M.A., Brew, B.K., Vijverberg, J.H., Kats, C.J.A.R., Longo, C., Pijnenburg, M.W., et al., 2020. Early-life antibiotic use and risk of asthma and eczema: results of a discordant twin study. Eur. Resp. J. https://doi.org/10.1183/13993003.02021-2019. Accessed November 9, 2021.

Sly, P.D., Boner, A.L., Bjorksten, B., Bush, A., Custovic, A., Eigenmann, P.A., et al., 2008. Early identification of atopy in the prediction of persistent asthma in children. Lancet 372 (9643), 1100–1106.

Sonntag, H.J., Filippi, S., Pipis, S., Custovic, A., 2019. Blood biomarkers of sensitization and asthma. Front Pediatrics 2019. https://doi.org/10.3389/fped.2019.00251. Accessed November 8, 2021.

Subbarao, P., Mandhane, P.J., Sears, M.R., 2009. Asthma: epidemiology, etiology and risk factors. CMAJ 181 (9), E181–E190.

Sutherland, E.R., Goleva, E., King, T.S., Lehman, E., Stevens, A.D., Jackson, L.P., et al., 2012. Cluster analysis of obesity and asthma phenotypes. PLoS One 2012. https://doi.org/10.1371/journal.pone.0036631. Accessed November 9, 2021.

Sutherland, E.R., 2014. Linking obesity and asthma. Ann. N. Y. Acad. Sci. 1311, 31–41.

Takayama, G., Arima, K., Kanaji, T., Toda, S., Tanaka, H., Shoji, S., et al., 2006. Periostin: a novel component of subepithelial fibrosis of bronchial asthma downstream of IL-4 and IL-13 signals. J. Allergy Clin. Immunol. 118 (1), 98–104l.

Tan, J., Bernstein, J.A., 2014. Occupational asthma: an overview. Curr. Allergy Asthma Rep. 14 (5), 431. https://doi.org/10.1007/s11882-014-0431-y. Accessed November 10, 2021.

Tiotiu, A., 2018. Biomarkers in asthma: state of the art. Asthma Res. Pract. 2018. https://10.1186/s40733-018-0047-4.

Toskala, E., Kennedy, D.W., 2015. Asthma risk factors. Int. Forum Allergy Rhinol. 5 (S1), S11–S16.

Troisi, R.J., Speizer, F.E., Willett, W.C., Trichopoulus, D., Rosner, 1995. Menopause, postmenopausal estrogen preparations and the risk for adult-onset asthma. Am. J. Crit. Care Med. 152 (4 pt. 1), 1183–1188.

Tromp, I.I., Kiefte-de Jong, J.C., de Vries, J.H., Jaddoe, V.W., Raat, H., Hofman, A., et al., 2012. Dietary patterns and respiratory symptoms in pre-school children: the generation R study. Eur. Resp. J. 40 (3), 681–689.

Van der Vliet, A., Janssen-Heininger, M.W., Anathy, V., 2018. Oxidative stress in chronic lung disease: from mitochondrial dysfunction to dysregulated redox. Mol. Aspects Med. 63, 59–69.

Wagener, A.H., de Nijs, S.B., Lutter, R., Sousa, A.R., Weersink, E.J.M., Bel, E.H., Sterk, P.J., 2015. External validation of blood eosinophils FE(NO) and serum periostin as surrogates for sputum eosinophils in asthma. Thorax 70 (2), 115–120.

Wan, X.C., Prescott, P.G., 2016. Biomarkers in severe asthma. Imminol. Allergy Clin. North Am. 36 (3), 547–557.

World Allergy Organization, 2021. Sensitizing Agents Inducers of Occupational Asthma, Hypersensitivity Pneumonitis and Eosinophilic Bronchitis. www.worldallergy.org. Accessed November 10, 2021.

World Health Organization, 2021. Asthma. https://www.who.int/news-room/fact-sheets/detail/asthma. Accessed November 5, 2021.

Zeliger, H.I., 2011. Human Toxicology of Chemical Mixtures. Elsevier, London.

CHAPTER

27

Liver cirrhosis

27.1 Introduction

Cirrhosis is "an advanced stage of liver fibrosis that is accompanied by distortion of hepatic vasculature," and as "the histological development of regenerative nodules surrounded by fibrous bands in response to chronic liver injury, that leads to portal hypertension and end stage liver disease" (Schuppan and Afdahl, 2008). Cirrhosis can be caused by many factors, including chronic alcoholism, chronic exposure to environmental chemicals and viral disease. No matter the cause, liver cirrhosis is oxidative stress-mediated.

The liver has more than 500 functions in the human body that include the following.

Bile production
Detoxification
Metabolism
Immune system activities
Cholesterol production
Protein synthesis
Blood sugar balance
Micronutrient storage
Blood volume regulation
Endocrine control of blood signaling pathways

These and examples of them are listed in Table 27.1 (Corless and Middleton, 1983; Dabrowska et al., 2009; Eipel et al., 2010; Robinson et al., 2016; Trefts et al., 2017; Hopkins Medicine, 2021; Bowen, 2021; Columbia Surgery, 2021).

Oxidative stress is a critical factor in each of the functions listed in Table 27.1.

27.2 Liver fibroproliferative diseases

The liver is subject to onset numerous chronic fibroproliferative diseases that ultimately lead to cirrhosis, the final stage of fibrosis that ultimately leads to hepatic failure. It is characterized by scar accumulation and nodule formation, as the wound-healing response

Oxidative Stress
https://doi.org/10.1016/B978-0-323-91890-9.00003-9

TABLE 27.1 Functions of the liver.

Bile Production
Essential for digestion
Gastrointestinal system anti-microbial
Detoxification
Alcohol
Drugs
Steroid hormones
Conversion of ammonia to urea for removal
Bacterial removal from the blood stream
Removal of environmental allergens
Removal of toxic environmental chemicals
Metabolism
Carbohydrate metabolism
Fat metabolism
Protein metabolism
Immune system activities
Pathogen control
Hepatic and portal vein immune system maintenance
Immune system transport
Immune system homeostasis
Cholesterol production
Sex hormones' precursor
Vitamin D precursor
Protein synthesis
Blood level amino acid regulation
Prothrombin synthesis
Lipoprotein synthesis
Globulin synthesis
Albumin synthesis
Ceruloplasmin synthesis
Blood sugar balance
Glycogen storage

Micronutrient storage
Vitamins (A, D, E, K, B12)
Minerals (copper, iron, magnesium, zinc)
Blood volume regulation
Endocrine control of blood signaling pathways
Gene expression regulation
Anti-viral response
Inflammation regulation

to hepatocyte injury in which the overproduction of collagen I plays a causative role (Cichoz-Lach and Michalak, 2014; Zhou et al., 2014; Micu et al., 2019).

Liver proliferative diseases include alcoholic liver cirrhosis, non-alcoholic fatty liver disease, hepatitis B, hepatitis C, obstructive cholestasis and hepatic encephalopathy (Cichoz-Lach and Michalak, 2014; Ramachandran and Jaeschke, 2018; Micu et al., 2019; Jarvis et al., 2020).

Oxidative stress plays a critical role in the development and progression of all fibroproliferative liver diseases, either via oxidative reactions or by depletion of antioxidants (Zhu et al., 2012; Cichoz-Lach and Michalak, 2014; Zhou et al., 2014; Li et al., 2015; Ramachandran and Jaeschke, 2018).

There are numerous individual causes of liver fibroproliferative diseases. Five primary ones are addressed here: alcoholic liver disease, drugs, chemical solvents and pollutants, non-alcoholic fatty liver disease and hepatitis C.

27.2.1 Alcoholic liver disease (ALD)

Alcohol metabolism in the liver is related to ROS production, mitochondrial injury and steatosis. Drinking of ethanol initially leads to steatosis, the first step in alcoholic liver disease (ALD). Heavy chronic ethanol consumption leads to progression of ALD from steatosis to hepatitis, fibrosis and cirrhosis. Metabolism of alcohol occurs via initial oxidation to acetaldehyde and ultimately to acetate via free radical mediated mechanisms (Schuppam and Afdhal, 2008; Li et al., 2015).

27.2.2 Drugs

Numerous drugs are known to be hepatotoxic, with many of these having known clinical signatures (phenotypes). These drugs include those of many different classifications (Bjornsson, 2016; Livertox, 2021), with examples listed in Table 27.2. A current complete list of these is available on-line in Livertox.

TABLE 27.2 Examples of hepatotoxic drugs and their classifications.

Drug	Classification
Amoxicillin	Antibiotic
Anabolic steroids	Body building
Halothane	Anesthetic
Ibuprofen	NSAID
Infliximab	Immunosuppressive
Interferon beta	Multiple sclerosis
Isoniazid	Antituberculosis
Phenytoin	Antiepileptic
Simvastatin	Lipid lowering

Drug induced liver injury generally results via a three-step cascade of events that lead to elevation of oxidative stress, activation of stress signaling pathways, impairment of mitochondrial functioning and ultimate cell death (Jaeschke et al., 2002; Yuan and Kaplowitz, 2013; Teschke and Dunan, 2020). The three steps are:

1. Generation of ROS from drugs or their metabolites from oxidation by cytochrome P450.
2. Triggering of immune reactions.
3. Mitochondrial function impairment that initiates apoptosis or necrosis, leading to cell death.

27.2.3 Chemical solvents and pollutants

The liver is the primary organ responsible for metabolism of environmental and occupational toxic chemicals. These and their metabolites are the main cause of liver damage (Franco et al., 1986; Franco, 1991). A broad spectrum of chemicals adversely impacts the liver (Malaguarnera et al., 2012; Haz-Map, 2019). These types and their exemplars are listed in Table 27.3.

Exogenous chemicals cause organic and functional damage to the liver via eight different pathways, each of which is associated with elevated oxidative stress (Malaguanera et al., 2012; Luster et al., 1999). These are:

1. Hepatocyte disruption causing a decrease in ATP levels.
2. Protein transport disruption causing bile flow interruption.
3. Cytolytic T-cell activation.
4. Hepatocyte apoptosis.
5. Bile duct injury.

TABLE 27.3 Categories and examples of chemicals that adversely affect the liver.

Alcohols (ethanol, n-propanol)
Aliphatic hydrocarbons (cyclohexane, cyclopentadiene)
Aliphatic nitro compounds (nitroethane, 2-nitropropane)
Amides (dimethylformamide)
Aromatic amines (nitrobenzene, 2,4,6-trinitrotoluene)
Aromatic hydrocarbons (benzene, toluene, xylenes, ethyl benzene, styrene)
Aromatic amines (benzidine, 4,4′-methylenedianaline)
Brominated solvents (carbon tetrabromide, ethylene dibromide)
Chlorinated solvents (chloroform, trichloroethylene, carbon tetrachloride, 1,2-dichloropropane, methylene chloride.)
Chlorofluorocarbons (dichlorotetrafluoroethane)
Esters (ethyl acetate, vinyl acetate)
Ethers (ethyl ether, dioxane)
Fungicides (pentachlorophenol)
Glycol ethers (2-butoxyethanol)
Halogenated aromatics (chlorobenzene, o-dichlorobenzene)
Halogenated polycyclic aromatic hydrocarbons (PCBs, dioxins, furans, hydrazines (hydrazine, phenylhydrazine))
Metals (arsenic, copper, chromates, thalium)
Pesticides (chlordane, DDT, endrin, mirex, toxaphene)
Refined petroleum (kerosine, stoddard solvent, VM&P naphtha)
Vinyl halides (fluoro-, chloro- and bromo-)

6. Activation of Kupffer cells causing production of pro-inflammatory cytokines—IL-1, IL-6, tumor necrosis factor (TNF) α—and chemokine receptor chemokines and resulting in increased inflammation.
7. Reduction of Cytochrome P450, the most common liver enzyme for detoxification of organic chemicals.
8. Mitochondrial dysfunction via inhibition of fatty acid β-oxidation (causing steatosis) and electron transfer in the respiratory chain and leading to superoxide formation.

27.2.4 Nonalcoholic fatty liver disease (NAFLD)

Nonalcoholic fatty liver disease (NAFLD) is the most common liver disease. It is caused by disrupted uptake, synthesis. oxidation and export of fatty acids and most often due to pre-existing conditions that include obesity, diabetes and dyslipidemia (Jarvis et al., 2020). Oxidative stress is central to NAFLD and particularly harmful to mitochondria, resulting in altered protein synthesis, reduced mitochondrial content and impaired mitochondrial β-oxidation, as well as in impaired gene expression (Gawrieh et al., 2004;Musso et al., 2010; Koek et al., 2011; Podrini et al., 2013; Santos et al., 2013; Cichoz-Lach and Michalak, 2014; Bovi et al., 2021).

27.2.5 Hepatitis C

The World Health Organization estimates that 58 million people worldwide have chronic hepatitis C virus (HCV) infection, with about 1.5 million new infections occurring annually (WHO, 2021). The immune system can clear the virus from some of those infected, but fails to do so in most of those infected (70%–85%).

Liver damage from HCV starts with fibrosis. As scar tissue builds up, disease can progress to cirrhosis and ultimately to end stage liver disease. HCV can also lead to hepatocellular carcinoma and non-Hodgkin lymphoma (Jin, 2007; Hartridge-Lambert et al., 2012).

The connection between HCV and oxidative stress has been well established, with reduced anti-oxidant function mechanistically central (Li et al., 2015). In HCV, glutathione, a primary antioxidant, is decreased; the activity of antioxidant enzymes (superoxide dismutases (SOD),

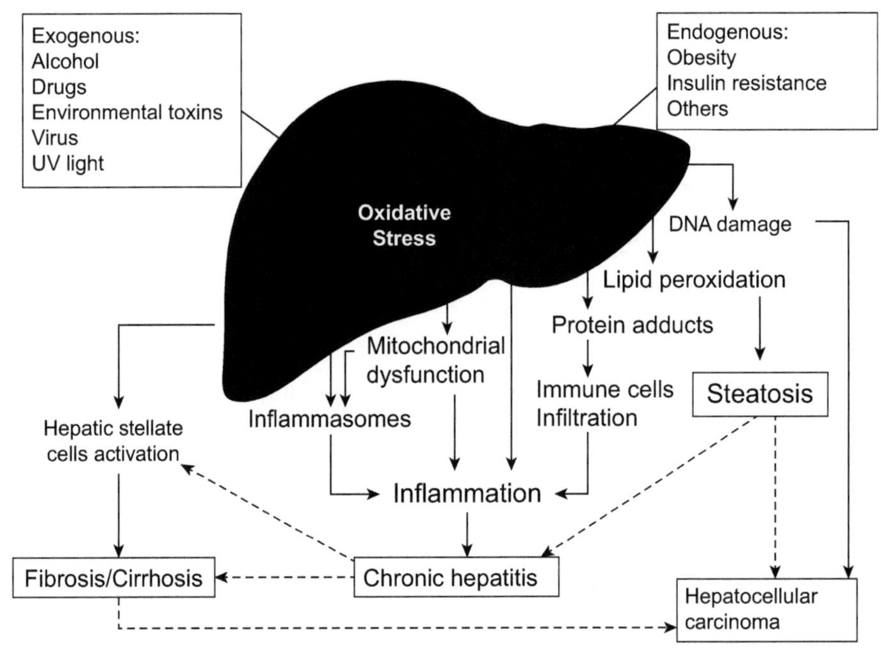

FIG. 27.1 General mechanism scheme of oxidative stress induced by various factors on liver disease. *Reproduced with permission from: Li, S., Tan, H.Y., Wang, N., Zhang, Z.J., Lao, L., Wong, C.W., Feng, Y. 2015. The role of oxidative stress and antioxidants in liver diseases. Int. J. Mol. Sci. https://doi.org/10.3390/ijms161125942. Under Creative Commons Attribution (CC BY) license (https://creativecommons.org/licences/by/4.0/).*

glutathione reductase and glutathione peroxidase) is reduced; lipids and proteins are oxidized; and mitochondrial alterations followed by apoptosis occur—all of which are oxidative stress-mediated (Dabrowska et al., 2009; Ivanov et al., 2013; Cichoz-Lach and Michalak, 2014).

27.3 Oxidative stress and liver disease

Liver disease can be caused by both endogenous and exogenous factors, with oxidative stress central to the development and progression of all (Zhu et al., 2012; Cichoz-Lach and Michalak, 2014; Zhou et al., 2014; Li et al., 2015; Ramachandran and Jaeschke, 2018). These are schematically shown in Fig. 27.1 (reproduced with permission from Li et al., 2015).

As is shown in Fig. 27.1, both endogenous and exogenous sources raise oxidative stress in the liver, the resultant effects of which being chronic hepatitis, steatosis, fibrosis, cirrhosis and hepatocellular carcinoma.

References

Bjornsson, E.S., 2016. Hepatotoxicity by drugs: the most common implicated agents. Int. J. Mol. Sci. https://doi.org/10.3390/ijms17020224. Accessed November 16, 2021.

Bovi, A.P.D., Marciano, F., Mandato, Siano, M.A., Savoia, M., Vajro, P., 2021. Oxidative stress in non-alcoholic fatty liver disease. An updated mini review. Front. Med. https://doi.org/10.3389/fmed.2-21/595371. Accessed November 19, 2021.

Bowen, R., 2021. Metabolic Functions of the Liver. www.vivo.colostate.edu/hbooks/pathways/digestion/liver/metabolic.html/. Accessed November 15, 2021.

Cichoz-Lach, H., Michalak, A., 2014. Oxidative stress as a crucial factor in liver disease. World J. Gastroenterol. 20 (25), 8082–8091.

Columbia Surgery, 2021. The Liver and Its Functions. https://columbiasurgery.org/liver/liver-and-its-functions. Accessed November 15, 2021.

Corless, J.K., Middleton 3rd, H.M., 1983. Norman liver function. A basis for understanding hepatic disease. Arch. Intern. Med. 143 (12), 2291–2294.

Dabrowska, M.M., Panasiuk, A., Flisiak, R., 2009. Signal transduction pathways in liver and influence of hepatitis C virus infection on their activities. World J. Gastroenterol. 15 (18), 2184–2189.

Eipel, C., Abshagen, K., Vollmar, B., 2010. Regulation of hepatic blood flow: the hepatic arterial buffer response revisited. World J. Gastroenterol. 16 (48), 6046–6057.

Franco, G., 1991. New perspectives in biomonitoring liver function by means of serum bile acids: experimental and hypothetical biochemical basis. Br. J. Ind. Med. 43 (2), 139.

Franco, G., Fonte, R., Candura, F., 1986. Hepatotoxicity of organic solvents. Br. J. Ind. Med. 48, 557–561.

Gawrieh, S., Opara, E.C., Koch, T.R., 2004. Oxidative stress in nonalcoholic fatty liver disease: pathogenesis and antioxidant therapies. J. Invest. Med. 52 (8), 506–514.

Hartridge-Lambert, S.K., Stein, E.M., Markowitz, A.J., Portlock, C.S., 2012. Hepatitis C and non-Hodgkin lymphoma: the clinical perspective. Hepatology 55 (2), 634–641.

Haz-Map, 2019. Toxic Hepatitis. https://hax-map.com/heptox1.htm. Accessed November 16, 2021.

Hopkins Medicine, 2021. Liver: Anatomy and Functions. https://hopkinsmedicine.org/health/conditions-and-diseases/liver-anatomy-and-functions. Accessed November 15, 2021.

Ivanov, A.V., Bartosch, B., Smirnova, O.A., Isagullants, M.G., Kochetkov, S.N., 2013. HCV and oxidative stress in the liver. Viruses 5, 439–469.

Jaeschke, H., Gores, G.J., Cederbaum, A.I., Hinson, J.A., Pessayre, D., Lemasters, J.J., 2002. Mechanisms of hepatotoxicity. Toxicol. Sci. 65, 166–176.

Jarvis, H., Craig, D., Barker, R., Spiers, G., Stow, D., Anstee, Q.M., Hanratty, B., 2020. Metabolic risk factors and incident advanced liver disease in non-alcoholic fatty liver disease (NAFLD): a systematic review and meta-analysis of population-based observational studies. PLoS Med. https://doi.org/10.1371/journal.pmed.1003100. Accessed November 16, 2021.

Jin, D.Y., 2007. Molecular pathogenesis of hepatitis C virus-associated hepatocellular carcinoma. Front. Biosci. 12, 222–233.

Koek, G.H., Liedorp, P.R., Bast, A., 2011. The role oxidative stress in non-alcoholic steatohepatitis. Clin. Chim. Acta 412 (15–16), 1297–1305.

Li, S., Tan, H.Y., Wang, N., Zhang, Z.J., Lao, L., Wong, C.W., Feng, Y., 2015. The role of oxidative stress and antioxidants in liver diseases. Int. J. Mol. Sci. https://doi.org/10.3390/ijms161125942. Accessed November 16, 2021.

Livertox, 2021. https://www.ncbi.nlm.nih.gov. Accessed November 19, 2021.

Luster, M.I., Simeonova, P.P., Gallucci, R., Matheson, J., 1999. Tumor necrosis alpha factor and toxicology. Crit. Rev. Toxicol. 29 (5), 491–511.

Malaguarnera, G., Cataudella, E., Giorano, M., Nunnari, G., Chisari, G., Malaguarnera, M., 2012. Toxic hepatitis in occupational exposure to solvents. World J. Gastroenterol. 18 (22), 2756–2766.

Micu, S.I., Manea, M.E., Popoiag, R., Nikolic, D., Andrada, D., Patti, A.M., et al., 2019. Alcoholic liver cirrhosis, more than an simple hepatic disease – a brief review of the risk factors associated with alcohol abuse. J. Mind Med. Sci. 6 (2), 232–236.

Musso, G., Gambino, R., Cassader, M., 2010. Non-alcoholic fatty liver disease from pathogenesis to management: an update. Obes. Rev. 11 (6), 430–445.

Podrini, C., Borghesan, M., Greco, A., Pazienza, V., Mazzoccoli, G., Vinciguerra, M., 2013. Redox homeostasis and epigenetics in non-alcoholic fatty liver disease (NAFLD). Curr. Pharmaceut. Des. 19 (15), 2737–2746.

Ramachandran, A., Jaeschke, H., 2018. Oxidative stress and acute hepatic injury. Curr. Opin. Toxicol. 7, 17–21.

Robinson, M.W., Harmon, C., O'Farrelly, C., 2016. Liver immunology and its role in inflammation and homeostasis. Cell. Mol. Immunol. 13, 267–276.

Santos, J.C.F., Valentim, I.B., Araujo, O.R.P., Ataide, T.R., Goulart, M.O.F., 2013. Development of nonalcoholic hepatopathy: contributions of oxidative stress and advanced glycation end products. Int. J. Mol. Sci. 14, 19846–19866.

Schuppam, D., Afdhal, N.H., 2008. Liver cirrhosis. Lancet 371 (9615), 838–851.

Teschke, R., Danan, G., 2020. Drug induced liver injury: mechanisms, diagnosis, and clinical management. In: Radu-Ionits, F., Pyrsopoulos, N., Jinfa, M., Tintoiu, I., Sun, Z., Bontas, E. (Eds.), Liver Diseases. Springer, Cham. https://doi.org/10.1007/978-3-030-24432-3_9. Accessed November 19, 2021.

Trefts, E., Gannon, M., Wasserman, D.H., 2017. The liver. Curr. Biol. 27 (21), R1147–R1151.

WHO, 2021. Hepatitis C. https://www.who.int/news-room/fact-sheets/hepatitis-c/. Accessed November 20, 2021.

Yuan, L., Kaplowitz, N., 2013. Mechanisms of drug induced liver injury. Clin. Liver Dis. 17 (4), 507–518.

Zho, W.C., Zhang, Q.B., Qiao, L., 2014. Pathogenesis of liver cirrhosis. World J. Gastroenterol. 20 (23), 7312–7324.

Zhu, R., Wang, Y., Zhang, L., Guo, Q., 2012. Oxidative stress and liver disease. Hepatol. Res. 42 (8), 741–749.

CHAPTER

28

Chronic kidney disease (CKD)

28.1 Introduction

The kidneys are filtering organs that remove toxic wastes from the blood stream. Kidney disease, which prevents essential cleansing, can be acutely or chronically caused. Acute kidney disease, in which the kidneys suddenly stop functioning, may be caused by reduced blood flow to the kidneys, direct kidney damage, or sudden urine flow back up. The subject of this chapter is chronic kidney disease (CKD), a condition in which the kidneys are gradually damaged over time to where these organs can no longer provide their blood filtering function.

CKD may be caused by diabetes and other disorders and complications from it can include anemia, weak bones, poor nutritional health, nerve damage, heart disease and blood vessel disease. Hypertension is both a cause and effect of CKD.

28.2 Causes of chronic kidney disease

The primary causes of CKD are diabetes and hypertension. These and other causes are listed in Table 28.1 (Gorenjak, 2009; Shankar et al., 2021; NHS, 2021; Kidney.org, 2021; Mayo Clinic, 2021).

28.3 Hallmarks of chronic kidney disease

The hallmarks of CKD are:

1. Fibrosis.
2. Tubular atrophy.
3. Interstitial inflammation.
4. Biomarkers of oxidative stress.
5. Onset of symptoms over time.

Oxidative Stress
https://doi.org/10.1016/B978-0-323-91890-9.00024-6

TABLE 28.1 Causes of chronic kidney disease.

1.	Diabetes Elevated blood sugar in diabetes damages kidney filters.
2.	Hypertension High blood pressure increases the pressure of the blood against small kidney blood vessels, impeding their normal functioning.
3.	Glomerulonephritis A group of diseases that inflame and damage the kidneys' filtering capability.
4.	Polycyctic kidney disease An inherited disease that causes the formation of large cysts in the kidneys and damages surrounding tissue.
5.	Malformation A congenital narrowing that prevents normal urine outflow, resulting in urine backup and infection.
6.	Autoimmune disease Lupus and other autoimmune diseases.
7.	Obstructions Obstructions to urine flow caused by kidney stones or an enlarged prostate gland.
8.	Infections Chronic microbial urinary tract infections can worsen renal function.
9.	High cholesterol Causes a build-up of fatty deposits in blood vessels supplying the kidneys.
10.	Chronic plharmaceutical use Long-term use of medications, e.g., non-steroidal anti-inflammatories.

The first three of these are pathological ones which are unique to CKD (Yamaguchi, et al., 2015). The fourth, though a biomarker of CKD, is also a biomarker of other oxidative stress related disease and, therefore, not a definitive hallmark of CKD alone (Ling and Kuo, 2018). The individual symptoms in the fifth hallmark, listed in Table 28.2 (Kidney.org, 2021; Mayo Clinic, 2021; NHS, 2021) also cannot serve as definitive biomarkers, as each can be associated with numerous diseases and conditions.

28.4 Risk factors for chronic kidney disease

Genetic and phenotype factors are associated with CKD development risk (Ramos et al., 2008; Jacobson et al., 2020; Zhao et al., 2020; Tsia et al., 2021; Kidney.org, 2021; Mayo clinic, 2021). These factors, listed in Table 28.3, are all associated with elevated oxidative stress.

28.5 Mechanisms of chronic kidney disease and oxidative stress

The parameters that dictate the onset of CKD are: (1) oxidative stress-mediated and characterized by increased ROS production and its associate inflammation; and (2) impaired antioxidant defense (Cachofeiro et al., 2008; Kao et al., 2010; Daenen et al., 2018; Yonova et al., 2018; Ramos et al., 2008; Gyuraszova et al., 2020; Podkowinska and Formanowicz, 2020).

TABLE 28.2 Symptoms of chronic kidney disease.

Hypertension
Dyspnea (if liquid builds up in the lungs)
Chest pain (if liquid builds up around the lining of the heart)
Muscle cramps
Muscle twitches
Swelling of feet and ankles
Puffiness around the eyes (particularly in the morning
Persistent dry, itchy skin
Nausea
Vomiting
Loss of appetite
Metallic taste in the mouth
Fatigue
Weakness
Increase in urination frequency (particularly at night)
Sleep problems
Decreased mental acuity

The mechanisms of ROS overproduction and impaired antioxidant defense system in CKD include the following[1]

28.5.1 Excess ROS production

1. Up-regulation/activation of ROS-producing enzymes, including nicotine amide dinucleotide phosphate – NAD(P)H oxidase, cyclooxygenase, lipoxygenase and others.
2. Uncoupling of NO synthase (via monomerization, depletion of tetrahydro-biopterin, and accumulation of asymmetric dimethylamine.
3. Impairment of mitochondrial electron transport chain.
4. Activation of leukocytes and resident cells.
5. Dissemination of oxidative stress by circulating low density lipoproteins (LDL) and phospholipids via oxidation.

[1] Reproduced with permission from Yonova, D., Trendafilow, I., Georgieva, P., Dimitrova, V., Arabadjieva, D., Verkove, N., 2018. Oxidative stress (OS) in chronic kidney disease (CKD): a mini review. Nephrol Renal Dis 3 (3), 1–3.

TABLE 28.3 Risk factors for developing chronic kidney disease.

1. Genetics
2. Family history
3. Gender
4. Ethnicity (African American, Native American, Asian American, Hispanics)
5. Age
6. Obesity
7. Diabetes
8. Hypertension
9. Cardiovascular disease
10. Kidney injury
11. Chronic use of kidney-injuring pharmaceuticals
12. Abnormal kidney structure
13. Smoking
14. Exposure to nephrotoxic environmental chemicals: Melamine, phthalates, bisphenol A, heavy metals (arsenic, cadmium, lead, mercury, uranium) air pollution

28.5.2 Impaired antioxidant defense system

1. Reduced production of endogenous antioxidants (antioxidant enzymes, glutathione sulfate, ApoA1, albumin, lecithin-cholesterol acyltransferase, melatonin and others.
2. Impaired activation of the regulator genes encoding antioxidant/detoxification molecules.
3. Depletion of antioxidant molecules by ROS.
4. Diminished antioxidant activity of high density lipoproteins (HDL).
5. Reduced intake of fresh fruits and vegetables (due to potassium restriction, causing vitamins deficit.
6. Removal of water-soluble antioxidants by dialysis.
7. Anemia – decreased red blood cells antioxidants: glutathione sulfate, glutathione peroxidase, platelet activating factor – acetyl-hydrolase, phospholipids.

It is clear from the discussion above, that oxidative stress, which is associated with all the CKD risk factors and mechanisms, is the driver of CKD. The kidneys are vital filtering organs without which the human body cannot cleanse itself of toxins. Not all CKD is preventable, given the genetic predisposition of some to it. Avoiding environmental risk factors, however, can prevent CKD for many.

References

Cachofeiro, V., Goicochea, M., Vinuesa, S.G., Oubina, P., Lahera, V., 2008. Oxidative stress and inflammation, a link between chronic kidney disease and cardiovascular disease. Kidney Int. 74 (Suppl. 111), S4–S9.

Daenen, K., Andries, A., Mekahli, D., Schepdael, A.V., Joiret, F., Bammens, B., 2018. Oxidative stress in chronic kidney disease. Pediatr. Nephrol. https://doi.org/10.1007/s00467-018-4005-4 (Accessed 23 November 2021).

Gorenjak, M., 2009. Kidneys and Autoimmune Disease. eJIFCC. http://ifcc.org (Accessed 22 November 2021).

Gyuraszova, M., Gurecka, R., Babichova, J., Tothova, L., 2020. Oxidative stress in the pathophysiology of kidney disease: implications for noninvasive monitoring and identification biomarkers. Oxidative Med. Cell Long. https://doi.org/10.1155/2020/5478708 (Accessed 22 November 2021).

Jacobson, M.H., Wu, Y., Liu, M., Attina, T.M., Naidu, M., Karthikraj, R., et al., 2020. Serially assisted bisphenol A and phthalate exposure and association with kidney function in children with chronic kidney disease in the US and Canada: a longitudinal cohort study. PLoS Med. 1003384. https://doi.org/10.1371/journal.pmed (Accessed 23 November 2021).

Kao, M.P.C., Ang, D.S.C., Pall, A., Struthers, A.D., 2010. Oxidative stress in renal dysfunction: mechanisms, clinical sequelae and therapeutic options. J. Hum. Hypertens. 24, 1–8.

Kidney.org, 2021. Chronic Kidney Disease (CKD) Symptoms and Causes. https://www.kidney.org/atoz/content/about-chronic-kidney-disease (Accessed 22 November 2021).

Ling, X.C., Kuo, K.L., 2018. Oxidative stress in chronic kidney disease. Renal Repl. Ther. https://doi.org/10.1186/s41100-018-0195-2 (Accessed 22 November 2021).

Mayo Clinic, 2021. Chronic Kidney Disease. https://www.mayoclinic.org/diseases-conditons/chronic-kidney-disease/symptoms-causes/syc-20354521 (Accessed 22 November 2021).

NHS, 2021. Overview Chronic Kidney Disease. https://www.nhs.uk/conditons/kidney-disease/ (Accessed 22 November 2021).

Podkowinska, A., Formanowicz, D., 2020. Chronic kidney disease as oxidative stress- and inflammatory-mediated cardiovascular disease. Antioxidants 2020. https://doi.org/10.3390/antiox9080752 (Acccessed November).

Ramos, L.F., Shintani, A., Ikizler, A., Himmerfarb, J., 2008. Oxidative stress and inflammation are associated with adiposity in moderate to severe CKD. J. Am. Nephrol. 19, 593–599.

Shankar, M., Narasimhappa, S., Madhura, N.S., 2021. Urinary tract infection in chronic kidney disease population: a clinical observational study. Cureus 13 (1). https://doi.org/10.7759/cureus.12486 (Accessed November 22, 2021).

Tsai, H.J., Wu, P.Y., Huang, J.C., Chen, S.C., 2021. Environmental pollution and chronic kidney disease. Int. J. Med. Sci. 18 (5), 1121–1129.

Yamaguchi, J., Tanaka, T., Nangaku, M., 2015. Recent advances in understanding kidney disease. F1000 Research. https://doi.org/10.12688/f1000research.6970.1 (Accessed 22 November 2021).

Yonova, D., Trendafilow, I., Georgieva, P., Dimitrova, V., Arabadjieva, D., Verkove, N., 2018. Oxidative stress (OS) in chronic kidney disease (CKD): a mini review. Nephrol. Renal Dis. 3 (3), 1–3.

Zhao, J., Hinton, P., Chen, J., Jiang, J., 2020. Causal inference for the effect of environmental chemicals on chronic kidney disease. Comput. Struct. Biotechnol. J. https://doi.org/10.1016/j.csbj.2019.12.001 (Accessed 22 November 2021).

CHAPTER

29

Disease comorbidities

29.1 Introduction

The onset of a chronic noncommunicative disease (NCD) generally leads to comorbidity. In the United States, for example, more than a quarter of the adult population suffers from two or more chronic diseases (van Oostrom et al., 2012; Jakovljevic et al., 2013; Bauer et al., 2014). As discussed below, comorbidity is observed in diseases with common mechanistic pathways as well in diseases with seemingly different mechanisms. A common thread in many NCDs is the presence of exogenous lipophilic toxic chemicals in the bodies of those affected (Zeliger, 2013a, 2015b,c).

The onset of disease comorbidity is oxidative stress-mediated and primarily caused by age, genetic factors, preexisting diseases, absorption of exogenous toxic chemicals or combinations of these, all of which elevate oxidative stress and thereby induce further disease (Nguyen et al., 2019). Though little can be done to minimize age, genetic and preexisting disease effects, the effects of toxic chemicals can be lowered. This chapter is devoted to comorbidities induced by exogenous toxic chemicals.

29.2 Comorbidity

A study of 400 chemically sensitive individuals exposed to low molecular weight hydrocarbons (LMWHCs) has demonstrated the presence of disease comorbidities in multiple body systems (Zeliger et al., 2012). These are shown in Table 29.1.

Comorbidity of NCD is not limited to just the chemically sensitive people. Exposure to and retention of lipophilic persistent organic pollutants (POPs), semivolatile and volatile exogenous lipophilic chemicals has been associated with increased prevalence of type 2 diabetes (T2D) (Carpenter, 2008; Lee et al., 2010; Zeliger, 2013c), cardiovascular disease (Zeliger, 2013a), and neurological disease (Zeliger, 2013b). Many other NCDs, that affect virtually all body systems, are also associated with exposure to and retention of exogenous lipophilic chemical species. Affected systems include: immunological (Marie et al., 2013), musculoskeletal (Al-Bashri et al., 2013; Strujis et al., 2006); and respiratory diseases (Cazzola et al., 2013; Varela et al., 2013); as well as numerous cancers (Habib et al., 2013; Sorensen, 2013; van Baal et al., 2011).

Oxidative Stress
https://doi.org/10.1016/B978-0-323-91890-9.00032-5

TABLE 29.1 Percent of chemically sensitive people with comorbid diseases.

System	Percent of comorbidity
Nervous	100
Respiratory	83
Cardiovascular	77
Gastrointestinal	77
Musculoskeletal	33

A unifying explanation for induction of NCD by absorbed exogenous lipophilic chemicals has been presented (Zeliger, 2013a, b, c). Review of the medical and toxicological literature shows that the onset of these diseases is associated with the accumulation of exogenous lipophilic chemicals in body serum (Gallo et al., 2011; Lee et al., 2007, 2011; Philibert et al., 2009; Cortu et al., 2007). A dose dependent relationship between persistent organic pollutants (POPs) serum levels and type 2 diabetes (T2D), for example, has been shown to exist (Cortu et al., 2007; Lee et al., 2006). Lipophilic cell membranes are not permeable to most hydrophilic chemicals. Lipophilic chemicals act as solvents and carriers for impermeable organic, inorganic and metallic hydrophiles to facilitate absorption of species which would not otherwise permeate through the cells' lipophilic barriers (Zeliger, 2003; Zeliger and Lipinski, 2015).

It has also been previously shown that mixtures of toxic lipophilic and hydrophilic species produce enhanced toxicities, low-level effects and attacks on organs and/or systems not known to be impacted by either species alone (Zeliger, 2003, 2011). Such effects have been observed following simultaneous exposures to mixtures of lipophilic and hydrophilic chemicals, but can also be triggered by the initial absorption and retention of either long-persisting lipophilic species (POPs) or chronic absorption of more readily cleared lipophiles (lower molecular weight hydrocarbons) followed by sequential uptake of hydrophilic species. Such combinations can then act together as a toxic mixture, with the absorption of different hydrophiles accounting for the onset of different diseases (Zeliger et al., 2012; Zeliger, 2013a,; Zeliger and Lupinski, 2015).

Though different diseases involve attacks on widely disparate organs and systems, comorbidity rates are high when individuals are exposed to environmental lipophilic toxins (Zeliger et al., 2012). The onsets of comorbid diseases do not follow set patterns, with published studies showing that individuals with two comorbid diseases, e.g., T2D and hypertension, are just as likely to become ill with one first as the other first (Sowers et al., 2001; Zeliger et al., 2015). The wide prevalence of NCDs and the lack of a pattern of onset strongly suggests the common cause for these diseases that has been previously reported on (Zeliger, 2013a, 2016).

Previous chapters here discussed individual NCDs. All the diseases addressed in those chapters have numerous comorbidities. References for these are listed in Table 29.2 and Fig. 29.1 schematically shows these comorbid relationships.

TABLE 29.2 References for NCD comorbidities.

Disease pairs	References
ATS-CAN	Suzuki et al. (2017)
OBS-RA	Crowson et al. (2013)
LIV-CAN	Mu et al. (2020)
CAN-RA	Chen et al. (2011)
AD-CAN	Ospina-Romero et al. (2020)
CKD-LIV	Wong et al. (2019)
AST-CAN	Salameh et al. (2021)
CKD-RA	Chiu et al. (2015)
LIV-RA	Selmi et al. (2011)
RA-AST	Kim et al. (2019)
CKD-AD	Xu et al. (2021)
LIV-T2D	Jepsen (2014)
LIV-CAN	Jepsen (2014)
LIV-CKS	Jepsen (2014)
T2D-RA	Albrecht et al. (2018)
AST-T2D	Cazzola et al. (2011)
ATS-LIV	Mukthinuthalapati et al. (2018)
ATS-AST	Gurgone et al. (2020)
T2D-ATS	Iglay et al. (2021)
T2D-AD	Santiago and Potashkin (2021)
ATS-CKD	Valdivielso et al. (2019)
OBS-CKD	Kovesdy et al. (2017)
OBS-LIV	Glass et al. (2019)
AST-OBS	Fedele et al. (2014)
OBS-AD	Alford et al. (2018)
OBS-ATS	Apovian and Gokce (2012)

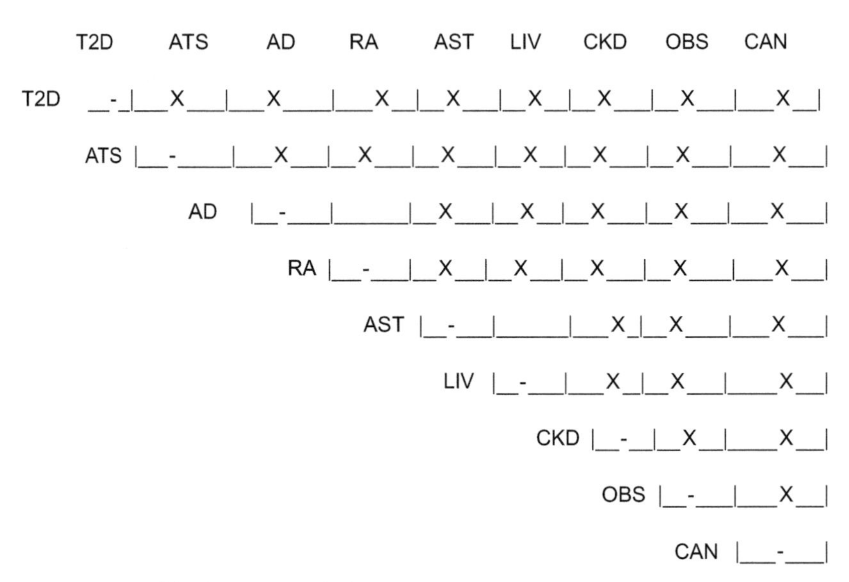

FIG. 29.1 Noncommunicable disease comorbidities.

As can be seen from Fig. 29.1, 32 of the 35 possible binary combinations (94%) do indeed occur.

29.3 Oxidative stress, the common thread

NCDs are late onset diseases that generally follow decades of living during which physiological breakdowns of many systems occur (Wright et al., 2003) and all are triggered by a combination of aging, genetic susceptibility and environmental exposure (Zhang et al., 2010). The mechanisms have been proposed the account for these breakdowns include:

1. Oxidative stress (Uttara et al., 2009; Bolanos et al., 2009; Zeliger, 2016), epigenetic effects (Jakovcevski et al., 2012; Urdinguio et al., 2009; Baccarelli et al., 2009).
2. Low intensity inflammation (Miller et al., 2008; Leonhard et al., 2006).
3. Endocrine disruption (Mostafalou et al., 2013; Colborn et al., 1997).
4. Phenotype connections between diseases, i.e., that comorbidities exist between diseases that are connected to other diseases by a Phenotype Disease Network (Hidalgo et al., 2009; Zhang et al., 2010; Lee et al., 2008).

Low density inflammation, the second side of the oxidative stress coin (see Chapter 9), and endocrine disruption (Meli et al., 2020; Gao et al., 2021) are both associated with chronic oxidative stress, and phenotype connections between disease types do not account for all comorbidities (Hidalgo et al., 2009). Accordingly, as discussed in previous chapters, only chronically elevated oxidative stress can account for all disease comorbidities.

The connection between disease onset and exposures to all the lipophilic chemical types (POPs, semivolatile organic compounds, and volatile organic compounds), which have been independently been associated with each of the diseases listed in Table 29.2 (Zeliger, 2013a) is also consistent with oxidative stress as the driver of disease, as the absorption of all lipophilic species trigger systemic responses that raise oxidative stress and chronic exposure to these induces a steady state of oxidative stress (Zeliger, 2016).

Since all the diseases listed in 29.2 may be related to exogenous lipophilic adsorption and dose-response related (Zeliger, 2013a), it is to a great extent predictable that individuals ill with one of these diseases will eventually ail with at least one other of these diseases (Zeliger et al., 2012). It is also predictable that ailing with a disease in a given system predisposes one to a second disease in that system, examples being the comorbidity of neurodegenerative diseases (Alzheimer's and Parkinson's diseases (Rajput et al., 1993)) and cardiovascular diseases (hypertension with stroke (Cipolla et al., 2018)). There are numerous studies in the literature showing, that where individuals are comorbid with two of these diseases, the comorbidities are independent of the order of onset of the two diseases, i.e., that either of the diseases can precede the other. The following serve as examples of these studies.

1. Individuals with autoimmune disease show higher than expected comorbidities with musculoskeletal disease and type 2 diabetes and that in both instances either of the diseases could precede the other (Somers et al., 2009).
2. In people comorbid with metabolic syndrome and mental health disorders, either condition can precede the other (Nousen et al., 2013).
3. Obesity and depression are common comorbid conditions and either one can come first (Luppino et al., 2010).
4. Hypertension is about twice as common in diabetics as in those without diabetes and either disease can precede the other (Sowers et al., 1995, 2001).
5. There is a high degree of comorbidity between chronic fatigue syndrome, a condition that affects multiple body systems and multiple chemical sensitivity, in which sensitized individuals react to a host of chemical species at concentration levels which do not affect those not sensitized, with an equal probability of either one occurring first (Zeliger et al., 2015).
6. Comorbidity is found between rosacea, an inflammatory skin disease, and Parkinson's disease, a neurodegenerative disorder (Egeberg et al., 2016).
7. Systemic lupus erythematosis is co-morbid with dementia (Park et al., 2021).
8. Those with type 2 diabetes have higher incidences of psychiatric disorders than those with normal metabolic health (Medved et al., 2009).
9. An increased risk of stroke and myocardial infarction has been found following Herpes Zoster virus (Minassian et al., 2015).

Credence for the chronic oxidative stress connection to disease comorbidities is provided by the following considerations:

1. All the diseases in Table 29.1 are late-onset ones, coming mostly after decades of exposures (Wright, et al., 2003).
2. One in four individuals with one of the diseases in Table 29.2 is likely to be stricken with at least one more of these diseases (Bauer et al., 2014; Jakovljevic et al., 2013; van Oostrom et al., 2012).

3. Eight types of environmental diseases are listed in Table 29.2. Of the 35 possible binary disease combinations, 32 (94%), have been shown to be comorbid (see Fig. 29.1).
4. Not only does absorption of exogenous lipophiles cause these diseases by inducing oxidative stress, but as discussed in Chapter 20, one of these diseases, obesity, has also been shown to cause retention of lipophilic POPs in white adipose tissue.

A consideration of obesity is in order here. Obesity is a key to co-morbidity. Body Mass Index (BMI) of 30 or greater is considered obese (Luppino et al., 2010). BMI is also a predictor of human adipose tissue concentration of POPs (Vaclavik et al., 2006). This is consistent with the fact that obesity is usually associated with CVS, T2D and other diseases, as adipose tissue continually releases the lipophiles it holds to the blood stream, making for steady-state chronic impact. Obesity is itself caused by POPs, phthalates, bisphenol A, polynuclear aromatic hydrocarbons, low molecular weight hydrocarbons and other exogenous lipophiles (Dirinick et al., 2014; Choi et al., 2014; Langer et al., 2014; Lee et al., 2014; Simmons et al., 2014). Being obese and having high serum endogenous lipophiles (cholesterol and tryglycerides) also contributes to the absorption of these exogenous lipophiles (Wang et al., 2007; Vaclavik et al., 2006).

Finally, it is not implied that lipophilic exposure is the only chemical cause of oxidative stress-mediated NCD. Chronic exposure to heavy metals and continually breathing polluted air are also contributing factors to NCD and multi-morbidity. Heavy metals (including arsenic, cadmium, chromium, cobalt, copper, mercury and nickel) are known to elevate OS and trigger NCDs, including type 2 diabetes, cardiovascular and neurological diseases (Carocci et al., 2014; Caciari et al., 2013; Kuo et al., 2013; Baccarelli et al., 2009; Khan et al., 2014; Agarwal et al., 2011; Mates et al., 2010). Oxides of nitrogen (NO_2, NO_X) and fine particulates ($PM_{2.5}$) have been shown to be associated with increased incidence of multi-morbidity in the aged (Autenrieth et al., 2018; Lee et al., 2020).

Though the focus of this chapter is the effects of chronic exposure to toxic chemicals, two other factors must be kept in mind. Firstly, as discussed in previous chapters, all oxidative stress-raising factors cause disease onset. Secondly, all disease and associated symptoms increase oxidative stress. Combined, these account for the onset of unexplained and unpredictable disease comorbidities.

References

Agarwal, S., Zaman, T., Tuzcu, E.M., Kapadia, S.R., 2011. Heavy metals and cardiovascular disease: results from the National Health and Nutrition examination survey (NHANES) 1999–2006. Angiology 62 (5), 422–429.

Albrecht, K., Ramos, A.L., Hoffmann, F., Redeker, I., Zink, A., 2018. High prevalence of diabetes in patients with rheumatoid arthritis: results from a questionnaire survey linked to claims data. Rheumatology (Oxford) 57 (2), 329–336.

Al-Bishri, J., Attar, S.M., BAssuni, N., Al-Yofaiey, Qutbuddeen, H., Al-Harthi, S., et al., 2013. Comorbidity profile among patients with rheumatoid arthritis and the impact on prescription trend. Clin. Med. Insights Arthritis Musculoskelet. Disord. 6, 11–18.

Alford, S., Patel, D., Perakakis, N., Mantzoros, C.S., 2018. Obesity as a risk factor for Alzheimer's disease: weighing the evidence. Obes. Rev. 19 (2), 269–280.

Apovian, C.M., Gokce, N., 2012. Obesity and cardiovascular disease. Circulation 125 (9), 1178–1182.

Autenrieth, C., Hampel, R., Wolf, K., Cyrys, J., Doring, A., Pershagen, G., Peters, A., 2018. Long term exposure to air pollution and risk of systemic inflammation-induced multimorbidity among the elderly: results from the

population-based kora-age study. Circulation. https://www.ahajournals.org/doi/1161/circ.suppl_AP181. (Accessed 29 November 2021).
Baccarelli, A., Bollati, V., 2009. Epigenetics and environmental chemicals. Curr. Opin. Pediatr. 21 (2), 243–251.
Bauer, U.E., Briss, P.A., Goodman, R.A., Bowman, B.A., 2014. Prevention of chronic disease in the 21st century: elimination of the leading preventable causes of premature death and disability in the USA. Lancet 384 (9937), 45–52.
Bolanos, J.P., Moro, M.A., Lizasoain, I., Almeida, A., 2009. Adv. Drug Deliv. Rev. 61 (14), 1299–1315.
Caciari, T., Sancini, A., Fioravanti, M., Capozzella, A., Casale, T., Montuori, L., et al., 2013. Cadmium and hypertension in exposed workers: a meta-analysis. Int. J. Occup. Med. Environ. Health 26 (3), 440–456.
Carocci, A., Rovito, N., Sinicropi, M.S., Genchi, G., 2014. Mercury toxicity and neurodegenerative effects. Rev. Environ. Contam. Toxicol. 229, 1–18.
Carpenter, D.O., 2008. Environmental contaminants as risk factors for developing diabetes. Rev. Environ. Health 23 (1), 59–74.
Cazzola, M., Calzetta, A., Betoncelli, G., Novelli, L., Cricelli, C., Rogliani, P., 2011. Asthma and comorbid medical illness. Eur. Respir. J. 38, 42–49.
Cazzola, M., Segreti, A., Calzetta, L., Rogliani, P., 2013. Comorbidities of asthma: current knowledge and future research needs. Curr. Opin. Pulm. Med. 19 (1), 36–41.
Chen, Y.J., Chang, Y.T., Wang, C.B., Wu, C.Y., 2011. The risk of cancer in patients with rheumatoid arthritis: a nationwide cohort study in Taiwan. Arthritis Rheum. 63 (2), 352–358.
Chiu, H.Y., Huang, H.L., Li, C.H., Chen, H.A., Yeh, C.L., Chiu, S.H., et al., 2015. Increased risk of chronic kidney disease in rheumatoid arthritis associated with cardiovascular complications – a national population-based cohort study. PLoS One. https://doi.org/10.1371/journal.pone.0136508. (Accessed 29 November 2021).
Choi, J., Eom, J., Kim, J., Lee, S., Kim, Y., 2014. Association between some endocrine-disrupting chemicals and childhood obesity in biological samples of young girls: a cross-sectional study. Environ. Toxicol. Pharmacol. 38 (1), 51–57.
Cipolla, M.J., Liebeskind, D.S., Chan, S.L., 2018. The importance of comorbidities in ischemic stroke: impact of hypertension on the cerebral circulation. J. Cerebr. Blood Flow Metabol. 38 (12), 2129–2149.
Colborn, T., Dumanoski, D., Myers, J.P., 1997. Our stolen future. Penguin Books, New York.
Cordu, N., Schymura, M.J., Nogotia, S., Rej, R., Carpenter, D.O., 2007. Diabetes in relation to serum levels of polychlorinated biphenyls and chlorinated pesticides in adult native Americans. Environ. Health Perspect. 115 (10), 1442–1447.
Crowson, C.S., Matteson, E.L., Davis III, J.M., Gabriel, S.E., 2013. Contribution of obesity to the rise in incidence of rheumatoid arthritis. Arthritis Care Res. 65 (1), 71–77.
Dirinck, E.L., Dirtu, A.C., Govidian, M., Covaci, A., Van Gaal, L.F., Jorens, P.G., 2014. Exposure to persistent organic pollutants: relationship with abnormal glucose metabolism and visceral adiposity. Diabetes Care 37 (7), 1951–1958.
Egeberg, A., Hansen, P.R., Gislason, G.H., Thyssen, J.P., 2016. Exploring the association between rosacea and Parkinson disease. A Danish nationwide cohort study. JAMA Neurol. 73 (5), 529–534.
Fedele, D.A., Janicke, D.M., Lim, C.S., Abu-Hasan, M., 2014. An examination of comorbid asthma and obesity: assessing differences in physical activity, sleep duration, health-related quality of life and parental distress. J. Asthma 51 (3), 275–281.
Gallo, M.V., Schell, L.M., DeCaprio, A.P., Jacobs, A., 2011. Levels of persistent organic pollutant and their predictors among young adults. Chemosphere 83, 1374–1382.
Gao, C., He, H., Qiu, W., Zheng, U., Chen, Y., Hu, S., Zhao, X., 2021. Oxidative stress, endocrine disturbance, and immune interference in humans showed relationships to serum bisphenol concentrations in a dense industrial area. Environ. Sci. Technol. 55 (3), 1953–1963.
Glass, L.M., Hunt, C.M., Fuchs, M., Su, G.L., 2019. Comorbidities and nonalcoholic fatty liver disease: the chicken, the egg, or both? Fed. Pract. 64–71, 2019.
Gurgone, D., McShane, L., McSherry, C., Guzik, T.J., Maffia, P., 2020. Cytokines and the interplay between asthma and atherosclerosis. Front. Pharmacol. https://doi.org/10.3389/fphar.2020.00166, 2020. (Accessed 29 November 2021).
Habib, S.L., Rojna, M., 2013. Diabetes and risk of cancer. ISRN Oncology. https://doi.org/10.1155/2013/583786 (Accessed 29 November 2021).

Hidalgo, C.A., Blumm, N., Barabasi, A.L., Christakis, N.A., 2009. A dynamic network approach for the study of human phenotypes. PLoS Comput. Biol. 5 (4), e1000353.

Iglay, K., Hannachi, H., Engel, S.S., Li, X., O'Connell, D., Moore, L.M., Rajpathak, S., 2021. Comorbidities in type 2 diabetes patients with and without atherosclerotic cardiovascular diseases: a retrospective database analysis. Curr. Med. Res. Opin. 37 (5), 743–751.

Jakovcevski, M., Akbarian, S., 2012. Epigenetic mechanisms in neurological disease. Nat. Med. 18 (8), 1194–1204.

Jakovljevic, M., Ostojic, L., 2013. Comorbidity and multimorbidity in medicine today: challenges and opportunities for bringing separated branches of medicine closer to each other. Psychiatr. Danub. 25 (Suppl. 1), 18–28.

Jepsen, P., 2014. Comorbidity in cirrhosis. World J. Gastroenterol. 20 (23), 7223–7230.

Khan, A.R., Awan, F.R., 2014. Metals in the pathogenesis of type 2 diabetes. J. Diabetes Metab. Disord. 13, 16. https://doi.org/10.1186/2251-6581-13-16. (Accessed 29 November 2021).

Kim, S.Y., Min, C., Oh, D.J., Choi, H.G., 2019. Increased risk of asthma in patients with rheumatoid arthritis: a longitudinal follow-up study using a national sample cohort. Sci. Rep. https://doi.org/10.1038/s41598-019-43481-3, 2019. (Accessed 29 November 2021).

Kovesdy, C.P., Furth, S.L., Zocali, C., 2017. Obesity and kidney disease: hidden consequences of the epidemic. Can. J. Kidney Health 4, 1–10.

Kuo, C.C., Moon, K., Thayer, K.A., Navas-Acien, A., 2013. Environmental chemicals and type 2 diabetes: an updated systematic review of the epidemiologic evidence. Curr. Diabetes Rep. 13 (6), 831–849.

Langer, P., Ukropec, J., Kocan, A., Drobna, B., Radikova, Z., Huckova, M., et al., 2014. Obesogenic and diabetogenic impact of high organochlorine levels (HCB, p,p-DDE, PCBs) on inhabitants in the highly polluted Eastern Slovakia. Endocr. Regul. 48 (1), 17–24.

Lee, D.H., Porta, M., Jacobs Jr., D.R., Vandenberg, L.N., 2014. Chlorinated persistent organic pollutants, obesity, and type 2 diabetes. Endocr. Rev. 35 (4), 557–601.

Lee, D.H., Lind, P.M., Jacobs Jr., D.R., Salihovic, S., van Bavel, B., Lind, L., 2011. Polychlorinated biphenyls and organochlorine pesticides in plasma predict development of type 2 diabetes in the elderly: the prospective investigation of the vasculature in Uppsala seniors (PIVUS) study. Diabetes Care 34 (8), 1778–1784.

Lee, D.S., Park, J., Kay, K.A., Christakis, N.A., Oltavi, Z.N., Barabasi, A.L., 2008. The implications of human metabolic network topology for disease comorbidity. Proc. Natl. Acad. Sci. USA 105 (29), 9880–9885.

Lee, D.H., Steffes, M.W., Sjoden, A., Jones, R.S., Needham, L.L., Jacobs Jr., D.R., 2010. Low dose of some persistent organic pollutants predicts type 2 diabetes: a nested case-control study. Environ. Health Perspect. 118 (9), 1235–1242.

Lee, D.H., Lee, I.K., Porta, M., Steffes, M., Jacobs Jr., D.R., 2007. Relationship between serum concentrations of persistent organic pollutants and the prevalence of metabolic syndrome among non-diabetic adults: results from the National Health and Nutrition Examination Survey 1999–2002. Diabetologia 50, 1841–1851.

Lee, D.H., Lee, I.K., Song, K., Steffes, M., Toscano, W., Baker, B.A., Jacobs Jr., D.R., 2006. A strong dose-response relation between serum concentrations of persistent organic pollutants and diabetes. Diabetes Care 29 (7), 1638–1644.

Lee, W.J., Liu, L.N., Lin, H.P., Chen, L.K., 2020. $PM_{2.5}$ air pollution contributes to the burden of frailty. Sci. Rep. https://doi.org/10.1038/s41598-020-71408-w, 2020. (Accessed 29 November 2021).

Leonhard, B.E., Myint, A., 2006. Inflammation and depression: is there a causal connection with dementia? Neurotox. Res. 10, 149–160.

Luppino, F.S., de Wit, L.M., Bouvy, P.F., Stijnen, T., Cuijpers Penninx, B.W.J.H., et al., 2010. Overweight, obesity, and depression. Arch. Gen. Psychiatr. 67 (3), 220–229.

Marie, R.A., Hanwell, H., 2013. General health issues in multiple sclerosis: comorbidities, secondary conditions and health behaviors. Continuum 19 (4), 1046–1057.

Mates, J.M., Segura, J.A., Alonso, F.J., Marquez, J., 2010. Roles of dioxins and heavy metals in cancer and neurological diseases using ROS-mediated mechanisms. Free Radic. Biol. Med. 49 (9), 1328–1341.

Medved, V., Joanovic, N., Knapic, V.P., 2009. The comorbidity of diabetes mellitus and psychiatric disorders. Psychiatr. Danub. 21 (4), 585–588.

Meli, R., Monnolo, A., Annunziata, C., Pirozzi, C., Ferrante, M.C., 2020. Oxidative stress and BPA toxicity: an antioxidant approach for male and female dysfunction. Antioxidants. https://www.mdpi.com/2076-3921/9/5/405/pdf. (Accessed 29 November 2021).

Miller, A.H., Ancoli-Israel, S., Bower, J.E., Capuron, L., Irwin, M.R., 2008. Neuroendocrine-immune mechanisms of behavioral comorbidities in patients with cancer. J. Clin. Oncol. 26, 971–982.

Minassian, C., Thomas, S.L., Smeeth, L., Douglas, I., Brauer, R., Langan, S.M., 2015. Acute cardiovascular events after Herpes Zoster: a self-controlled case series analysis in vaccinated and unvaccinated older residents of the United States. PLoS One. https://journals.plos.org/plosmedicine/article?id=10.1371/journal.pmed.10011919. (Accessed 29 November 2021).

Mostafalou, S., Abdollahi, M., 2013. Pesticides and human chronic diseases: evidences, mechanisms and perspectives. Toxicol. Appl. Pharmacol. 268 (2), 157–177.

Mu, X.M., Wang, W., Jiang, Y.Y., Feng, J., 2020. Patterns of comorbidity in hepatocellular carcinoma: a network perspective. Int. J. Environ. Res. Publ. Health. https://doi.org/10.3390/ijerph17093108. (Accessed 29 November 2021).

Mukthinutalapati, P.K., Syed, M., Salazar, M., Akinyeye, S., Fricker, Z., Ghabril, M., et al., 2018. Prevalence of comorbidities in patients with cirrhosis admitted to U.S. safety net hospitals and their impact on clinical phenotype of admission. Am. J. Gastroentrol. 113, S566–S567.

Nguyen, H., Manolova, G., Daskalopoulou, C., Vitoratou, S., Prince, M., Prina, A.M., 2019. Prevalence of multimorbidity in community settings: a systematic review and meta-analysis of observational studies. J. Comorbidity 9, 1–15.

Nousen, E.K., Franco, J.G., Sullivan, E.L., 2013. Unraveling the mechanisms responsible for the comorbidity between metabolic syndrome and mental health disorders. Neuroendocrinology 98 (4), 254–266.

Ospina-Romero, M., Glymour, M.M., Hayes-Larson, E., Mayeda, E.R., Graff, R.E., Brenowitz, W.D., et al., 2020. Association between Alzheimer disease and cancer with evaluation of study biases. A systematic review and meta-analysis. JAMA Netw. Open. https://doi.org/10.1001/jamanetworkopen.2020.25515. (Accessed 30 November 2021).

Park, H., Yim, D.H., Ochirpurev, B., Eom, S.Y., Choi, I.A., Kim, J.H., 2021. Association between dementia and systemic rheumatic disease: a nationwide population-based study. PLoS One. https://doi.org/10.1371/journal.-pone.0248395, 2021. (Accessed 29 November 2021).

Philibert, A., Schwartz, H., Mergler, D., 2009. An exploratory study of diabetes in First Nation community with respect to the serum concentrations of p,p'-DDE and PCBs and fish consumption. J. Environ. Res. Public Health 6 (12), 3179–3189.

Rajput, A.H., Rozdilsky, B., Rajput, A., 1993. Alzheimer's disease and idiopathic Parkinson's disease coexistence. J. Geriatr. Psychiatr. Neurol. 6 (3), 170–176.

Salameh, L., Mahboub, B., Khamis, A., Alsharhan, M., Tirmazy, S.H., Dairi, Y., et al., 2021. Asthma severity as contributing factor to cancer incidence: a cohort study. PLoS One. https://doi.org/10.1371/journal.pone.0250430, 2021. (Accessed 30 November 2021).

Santiago, J.A., Potashkin, J.A., 2021. The impact of disease comorbidities in Alzheimer's disease. Front. Aging Neurosci. https://doi.org/10.3389/fnagi.2021.631770, 2021. (Accessed 30 November 2021).

Selmi, C., De Santis, M., Gershwin, M.E., 2011. Liver involvement in subjects with rheumatic disease. Arthritis Res. Ther. 2011 http://arthritis-research.com/content/13/3/226. (Accessed 30 November 2021).

Simmons, A.L., Schlezinger, J.J., Corkey, B.E., 2014. What are we putting in our food that is making us fat? Food additives, contaminants, and other putative contributors to obesity. Curr. Obes. Res. 3 (2), 273–285.

Somers, E.C., Thomas, S.L., Smeeth, L., Hall, A.J., 2009. Are individuals with an autoimmune disease at higher risk of a second autoimmune disorder? Am. J. Epidemiol. 169, 749–755.

Sorensen, H.T., 2013. Multimorbidity and cancer outcomes: a need for more research. Clin. Epidemiol. 5 (Suppl. 1), 1–2.

Sowers, J.R., Epstein, M., 1995. Diabetes mellitus and associated hypertension, vascular disease, and nephropathy. Hypertension 26, 869–879.

Sowers, J.R., Epstein, M., Frolich, E.D., 2001. Diabetes, hypertension, and cardiovascular disease an update. Hypertension 37, 1053–1059.

Strujis, J.N., Baan, C.A., Schellevis, F.G., Westert, G.P., van den Bos, G.A.M., 2006. Comorbidity in patients with diabetes mellitus: impact on medical health care utilization. BMC Health Serv. Res. 6, 84–93.

Suziki, M., Tomoike, H., Sumiyoshi, T., Nagatomo, Y., Hosoda, T., Nagayama, M., et al., 2017. Incidence of cancer in patients with atherosclerotic cardiovascular disease. IJC Heart Vasc. https://doi.org/10.1016/j.ijcha.2017.08.004, 2017. (Accessed 30 November 2021).

Urdinguio, R.G., Sanchez-Mut, J.V., Esteller, M., 2009. Epigenetic mechanisms in neurological diseases: genes, syndromes, and therapies. Lancet Neurol. 8 (11), 1056–1072.

Uttara, B., Singh, A.V., Zamboni, P., Mahajan, R.T., 2009. Oxidative stress and neurodegenerative diseases: a review of upstream and downstream antioxidant therapeutic options. Curr. Neuropharmacol. 7 (1), 65–74.

PART III

Disease onset prediction

CHAPTER

30

Oxidative stress index

30.1 Introduction

Disease onset prediction is the subject of Part III of this book. As discussed in Chapter 17, total oxidative stress from all sources has been shown to be a predictor of disease onset in clinically asymptomatic individuals. The OS biomarker serum malondialdehyde (MDA) has been shown to be a reliable indicator of total oxidative stress and is indicative of the likelihood of disease onset. The following serum MDA scale is predictive of disease onset likelihood (Zeliger, 2016).

MDA level	Disease onset likelihood
Less than 1.20	Indicative of a healthy state
1.20–1.40	Disease predicted
1.40–3.00	Disease onset probable
Greater than 3.00	Severe disease likely

These values suggest that asymptomatic individuals with serum MDA levels of 1.20 or greater be evaluated further for disease.

30.2 Oxidative Stress Index (OSI)

MDA measurement, as well as measurement for other biomarkers for oxidative stress are invasive and not universally available. It has been shown, however, that oxidative stress measurement can be accomplished by the Oxidative Stress Index (OSI), an alternative, noninvasive method that considers all endogenous and exogenous OS raising parameters (Zeliger, 2017). As demonstrated in Parts I and II of this book, these parameters are listed in Table 30.1.

Oxidative Stress
https://doi.org/10.1016/B978-0-323-91890-9.00018-0

TABLE 30.1 Parameters that raise oxidative stress and lead to increased likelihood of disease onset.

Environmental exposure
Chemicals
Fibers and particles
Air pollution
Water and soil pollution
Radiation
Extreme temperature exposures
Lifestyle choices
Tobacco and vaping smoke
Alcohol and other recreational drugs
Diet high in animal fat, sugar, salt and processed foods
Overeating
Genetics
Genetic and epigenetic background
Birth defects and abnormalities
Pre-existing conditions
Inflammation
Chronic noninfectious disease
Chronic infectious chronic disease
Pharmaceuticals regularly taken
Symptoms chronically present
Psychological stress
Chronic trauma
Sleep deprivation
Obesity
Age
Natural increase with age

30.3 OSI basis

The OSI is based upon the following:

1. As each of the parameters listed in Table 30.1 elevate oxidative stress, it is valid to conclude that the presence of each contributes to total oxidative stress.
2. The presence of each of the parameters in Table 30.1 can be determined via a yes/no patient questionnaire.
3. It is valid to use a patient questionnaire to determine one's medical status, as questionnaires, are routinely used to solicit patient background information to predict the likelihood of onset and severity of impending illness. Examples of such applications are the Charlson Comorbidity Index (Charlson et al., 1987); the Alzheimer's Questionnaire (Sabbagh et al., 2010); the asthma control questionnaire (Jia et al., 2013), The Danish Comorbidity Index (Albertsen et al., 2020); and the COVID-19 Clinical Risk Score (Liang et al., 2020).
4. The contributions of each of the parameters in Table 30.1 are additive, with their sum equal to one's oxidative stress Index (Zeliger, 2017).
5. One point is scored for each check on the questionnaire.
6. A direct relationship between the OSI point score and the likelihood of disease onset obtains.

30.4 Oxidative Stress Index questionnaire

The OSI questionnaire is made up of nine sections. These are:

Personal physical data
Genetics
Prevalent illnesses and conditions
Symptoms
Diagnostic data
Pharmaceuticals regularly taken
Lifestyle
Tobacco, alcohol and recreational drug use
Home, work and school environmental exposures

30.4.1 Physical data

Age, gender, height, weight and residence location.

(a) Oxidative stress increases with age. This is reflected in the OSI by scoring one point for each decade or decade part starting with age 40. Accordingly, for example, one would score 2 OSI points for an individual aged 54.

(b) Height and weight information can be used to determine a score for weight status by consulting the NIH body mass index (BMI) table as shown in Fig. 30.1 (NIH, 2021).

BMI	19	20	21	22	23	24	25	26	27	28	29	30	31	32	33	34	35
Height (inches)	Body Weight (pounds)																
58	91	96	100	105	110	115	119	124	129	134	138	143	148	153	158	162	167
59	94	99	104	109	114	119	124	128	133	138	143	148	153	158	163	168	173
60	97	102	107	112	118	123	128	133	138	143	148	153	158	163	168	174	179
61	100	106	111	116	122	127	132	137	143	148	153	158	164	169	174	180	185
62	104	109	115	120	126	131	136	142	147	153	158	164	169	175	180	186	191
63	107	113	118	124	130	135	141	146	152	158	163	169	175	180	186	191	197
64	110	116	122	128	134	140	145	151	157	163	169	174	180	186	192	197	204
65	114	120	126	132	138	144	150	156	162	168	174	180	186	192	198	204	210
66	118	124	130	136	142	148	155	161	167	173	179	186	192	198	204	210	216
67	121	127	134	140	146	153	159	166	172	178	185	191	198	204	211	217	223
68	125	131	138	144	151	158	164	171	177	184	190	197	203	210	216	223	230
69	128	135	142	149	155	162	169	176	182	189	196	203	209	216	223	230	236
70	132	139	146	153	160	167	174	181	188	195	202	209	216	222	229	236	243
71	136	143	150	157	165	172	179	186	193	200	208	215	222	229	236	243	250
72	140	147	154	162	169	177	184	191	199	206	213	221	228	235	242	250	258
73	144	151	159	166	174	182	189	197	204	212	219	227	235	242	250	257	265
74	148	155	163	171	179	186	194	202	210	218	225	233	241	249	256	264	272
75	152	160	168	176	184	192	200	208	216	224	232	240	248	256	264	272	279
76	156	164	172	180	189	197	205	213	221	230	238	246	254	263	271	279	287

BMI	36	37	38	39	40	41	42	43	44	45	46	47	48	49	50	51	52	53
Height (inches)	Body Weight (pounds)																	
58	172	177	181	186	191	196	201	205	210	215	220	224	229	234	239	244	248	253
59	178	183	188	193	198	203	208	212	217	222	227	232	237	242	247	252	257	262
60	184	189	194	199	204	209	215	220	225	230	235	240	245	250	255	261	266	271
61	190	195	201	206	211	217	222	227	232	238	243	248	254	259	264	269	275	280
62	196	202	207	213	218	224	229	235	240	246	251	256	262	267	273	278	284	289
63	203	208	214	220	225	231	237	242	248	254	259	265	270	278	282	287	293	299
64	209	215	221	227	232	238	244	250	256	262	267	273	279	285	291	296	302	308
65	216	222	228	234	240	246	252	258	264	270	276	282	288	294	300	306	312	318
66	223	229	235	241	247	253	260	266	272	278	284	291	297	303	309	315	322	328
67	230	236	242	249	255	261	268	274	280	287	293	299	306	312	319	325	331	338
68	236	243	249	256	262	269	276	282	289	295	302	308	315	322	328	335	341	348
69	243	250	257	263	270	277	284	291	297	304	311	318	324	331	338	345	351	358
70	250	257	264	271	278	285	292	299	306	313	320	327	334	341	348	355	362	369
71	257	265	272	279	286	293	301	308	315	322	329	338	343	351	358	365	372	379
72	265	272	279	287	294	302	309	316	324	331	338	346	353	361	368	375	383	390
73	272	280	288	295	302	310	318	325	333	340	348	355	363	371	378	386	393	401
74	280	287	295	303	311	319	326	334	342	350	358	365	373	381	389	396	404	412
75	287	295	303	311	319	327	335	343	351	359	367	375	383	391	399	407	415	423
76	295	304	312	320	328	336	344	353	361	369	377	385	394	402	410	418	426	435

FIG. 30.1 Body mass index table. *Reprinted with permission, U.S. Department of Health & Human Services. National Heart, Lung, and Blood Institute.*

The four BMI weight categories are scored as follows.

BMI range	Classification	OSI score
19–24	Normal	0
25–29	Overweight	1
30–39	Obese	2
40 or more	Extremely obese	3

(c) Location of residence either via city and state or zip code can to be used to determine air pollution levels using the EPA air pollution data (EPA, 2020), with the OSI point scores for air pollution as shown in Table 30.2.

30.4.2 Genetics

Diseases are more likely to strike in those with genetic and epigenetic predispositions. Those whose parents, siblings and grandparents are or have been afflicted with a particular disease are presumed to carry the genetic markers that make onset of the diseases carried by these relatives more likely. One point is scored for each disease prevalent in a parent, sibling or grandparent.

30.4.3 Prevalent diseases and conditions

All diseases and conditions that interfere with homeostasis elevate oxidate stress. One point is scored for each prevalent disease or condition. Clearly, some diseases and conditions are more serious than others and result in higher OS levels. So why score all identically? The unit score is justified because more serious illness generally results greater numbers of symptoms, diagnostic test results and the use of larger numbers of pharmaceuticals, all of which are also scored on the OSI.

30.4.4 Symptoms

Symptoms are associated with all diseases and all symptoms illicit an immune system response that elevates OS. Though some symptoms are higher OS-elevating than others, all chronic diseases and conditions have multiple symptoms associated with them. As shown in Chapter 9, oxidative stress and inflammation are tied together. Hence, included in the symptoms section are all those symptoms associated with inflammation. These are

TABLE 30.2 EPA air pollution value.

EPA air quality classification	OSI points
Good	1
Moderate	2
Unhealthy for sensitive groups	3
Unhealthy	4
Very unhealthy	5
Hazardous	6

listed in Table 30.3 (Pahwa et al., 2020). One OSI point is associated with each checked symptom.

30.4.5 Diagnostic data

Diagnostic data are routinely collected in annual health examinations or when patients report disease symptoms. In either event, data points higher or lower what are considered "normal" are indicators of prevailing OS. One OSI point is scored for each "abnormal" diagnostic data point.

30.4.6 Pharmaceuticals regularly taken

All pharmaceuticals regularly taken result in chronic elevation of OS (Zeliger, 2016). In addition, all pharmaceuticals produce adverse drug reactions (ADRs) which are not only OS raising, but often lead to added symptoms and the need for additional pharmaceuticals to counter ADRs. Each pharmaceutical regularly taken is scored one OSI point.

TABLE 30.3 Symptoms associated with inflammation.

Anxiety
Body pain, arthralgia, myalgia
Brain fog
Depression
Elevated sedimentation rate
Excessive mucous production
Frequent infections
Gastrointestinal
Bloating
Abdominal pain
Constipation
Diarrhea
Weight loss
High C-Reactive protein
High white cell count
Low energy

30.4.7 Diet

Diet, be it unhealthy components or just over eating is a major contributor to elevated OS. This is particularly so in Western societies. One OSI point is scored for each poor dietary component.

30.4.8 Lifestyle

Lifestyle is a major contributor to OS in many. Lifestyle parameters include locations at which major quantities of time are spent. These include places of residence, employment and/or school. One OSI point is scored for each OS-raising lifestyle parameter.

30.4.9 Tobacco and alcohol use

As discussed in Chapter 17, tobacco and alcohol use elevate OS in a dose response relationship. One OSI point is scored for each 10 cigarettes smoked per day and 1 OSI point is scored for consumption of each alcoholic drink, greater than one, regularly consumed on a daily basis.

30.5 OSI score and disease

Empirical data analysis has found the following relationship between the OSI score and the likelihood of disease onset (Zeliger, 2017).

OSI	Disease onset likelihood
0–15	Indicative of good health
16–30	Disease predicted
31–45	Disease onset probable
46 or greater	Disease onset likely

The items in the OSI are listed alphabetically in the questionnaire to assure that the person completing it maintains his/her focus. The OSI is long and listing similar items sequentially could cause an individual to lose interest and check or ignore items. By alphabetizing the items, the person is made to constantly consider a fresh item prior to responding.

The OSI can be administered as a paper document and also lends itself to an electronic format. The electronic format offers the opportunity to easily collect and analyze data from multiple individuals. As is shown in subsequent chapters, data collection and analysis offers opportunities to use the OSI for many other applications.

The Oxidative Stress Index questionnaire is shown in Table 30.4.

TABLE 30.4 The Oxidative Stress Index questionnaire.

OXIDATIVE STRESS INDEX

DATE ________________ NUMBER ___________ (leave blank)

1. PHYSICAL DATA

HEIGHT _____ WEIGHT______

BMI (Body Mass Index)

Find BMI from chart or from height and weight chart (table 37.1)

BMI	OSI POINTS
19–24	0
25–29	1
30–39	2
40 or more	3

BMI SCORE _______

RESIDENCE City _____________ State____________ Zip code ______

RESIDENCE EPA AIR POLLUTION

Find EPA air pollution score from table 37.2

EPA AIR QUALITY CLASSIFICATION	OSI POINTS
Good	0
Moderate	1
Unhealthy for sensitive groups	2
Unhealthy	3
Very unhealthy	4
Hazardous	5

RESIDENCE AIR POLLUTION SCORE ____

TABLE 30.4 The Oxidative Stress Index questionnaire.—cont'd

AGE

Check all age boxes that apply. If, for example, 55 years old, check the first two boxes. If you're 82, check all 5 of these boxes

____ 40 or older
____ 50 or older
____ 60 or older
____ 70 or older
____ 80 or older

TOTAL PHYSICAL DATA ____ (leave blank)

2. GENETICS

List all chronic diseases each parent, sibling or grandparent has/had.

DISEASE	CHECKS	DISEASE	CHECKS
____________	____	____________	____
____________	____	____________	____
____________	____	____________	____
____________	____	____________	____
____________	____	____________	____
____________	____	____________	____

For each disease any one of these people have/had, score one check. If, for example, one grandparent, one parent and one sibling has/had diabetes, enter the number 3 next to diabetes.

TOTAL GENETIC CHECKS ____

ILLNESSES and CONDITIONS

Check all illnesses or conditions that you have been diagnosed with and currently have.

____ Acne

(Continued)

TABLE 30.4 The Oxidative Stress Index questionnaire.—cont'd

____ ADHD (Attention Deficit Hyperactivity Disorder)
____ AIDS or HIV
____ Alcohol addiction
____ Allergic rhinitis (sinus inflammation)
____ ALS (Lou Gehrig's disease)
____ Anemia
____ Anorexia
____ Anxiety disorder
____ Appetite loss
____ Arthritis
____ Asthma
____ Autism and ASD
____ Autism or autism spectrum disorder (ASD)
____ Benign prostate hyperplasia (enlargement - BPH)
____ Bipolar disorder
____ Bronchitis (chronic)
____ Bulimia
____ Bulging or herniated disc
____ Carpal tunnel syndrome

Cancer - Check all that apply. If stage three, for example, check first three

____ Cancer - stage 1
____ Cancer - stage 1 or 2
____ Cancer - stage 1, 2 or 3
____ Cancer - stage 1, 2, 3 or 4
____ Cardiovascular disease
____ Chronic Fatigue Syndrome (CFS)
____ Crohn's disease
____ Common cold (frequent)
____ COVID-19
____ COPD (chronic obstructive pulmonary disease)
____ Crohn's disease
____ Dementia
____ Dengue fever
____ Dental abscess (frequent)
____ Depression
____ Diabetes (type 1 or type 2)
____ Diarrhea (frequent)
____ Diverticulitis
____ Drug addiction
____ Eczema
____ Emphysema
____ Endometriosis

TABLE 30.4 The Oxidative Stress Index questionnaire.—cont'd

____ Epilepsy
____ Fibromyalgia (FM)
____ Frequent common colds
____ Frequent headaches
____ Frequent indigestion
____ Frequent infection
____ Frequent itching
____ Frequent rashes
____ Frequent sinus infections (sinusitis)
____ Glaucoma
____ Gout
____ Gum disease
____ Heart attack
____ Heart Disease or heart problems
____ Hemorrhoids
____ Hepatitis (chronic)
____ High blood pressure (hypertension)
____ High cholesterol
____ Herpes
____ Inflammatory bowl disease
____ Irritable bowel syndrome (IBS)
____ Kidney disease
____ Leukemia
____ Liver cirrhosis
____ Liver disease
____ Lupus
____ Lyme disease
____ Macular degeneration
____ Malaria
____ Metabolic syndrome (pre-diabetes)
____ Middle ear infection (frequent)
____ Migraine headaches
____ Multiple chemical sensitivity (MCS)
____ Multiple sclerosis
____ Obesity
____ Osteoarthritis
____ Osteoporosis
____ Parkinson's disease
____ Periodontal disease (swollen or bleeding gums)
____ Post-traumatic stress disorder (PTSD)
____ Rocky Mountain spotted fever
____ Psoriasis
____ Rosacea
____ Schizophrenia

(Continued)

TABLE 30.4 The Oxidative Stress Index questionnaire.—cont'd

____ Seizures
____ Sexually transmitted disease (STD)
____ Shingles
____ Sjogren's syndrome
____ Sleep apnea
____ Stroke
____ TB (Tuberculosis)
____ Thyroid disease
____ Tourette syndrome
____ Tremors
____ Ulcers
____ Varicose veins
____ West Nile Fever
____ Yellow fever
____ Zika

Write in the names of any other chronic illnesses not listed above and check those.

____ ____________________
____ ____________________
____ ____________________

TOTAL DISEASE CHECKS ____

SYMPTOMS

Check all chronic symptoms currently experienced.

____ Abdominal pain (frequent)
____ Allergic reactions to chemicals
____ Allergic reactions to any foods
____ Allergic reactions to insects
____ Allergic reactions to medications
____ Allergic reactions to plants (Hay fever)
____ Ankle pain
____ Attention span decline
____ Anxiety
____ Bleeding gums
____ Bloating
____ Blood in stool
____ Blood in urine
____ Blurred or cloudy vision
____ Body pain, arthralgia, myalgia
____ Brain fog

TABLE 30.4 The Oxidative Stress Index questionnaire.—cont'd

____ Bruise easily
____ Burning when urinating
____ Butterflies in your stomach often
____ Change in skin color
____ Chest pain
____ Constant chills
____ Constipation
____ Cough that is persistent
____ Coughing or spitting up blood
____ Decision making difficulties
____ Decline in learning ability
____ Decreased eye sight
____ Decreased sex drive
____ Depression
____ Diarrhea
____ Difficulty completing familiar tasks
____ Difficulty concentrating
____ Difficulty getting warm
____ Difficulty maintaining balance
____ Difficulty solving problems
____ Difficulty swallowing
____ Difficulty walking
____ Difficulty concentrating or finding words
____ Dizziness
____ Drained of energy
____ Dreams that are bizarre and recurring
____ Excessive mucous production
____ Excessive thirst
____ Eye discomfort or pain
____ Eye redness
____ Fatigue
____ Feel depressed a lot
____ Feel less alert or fuzzy headed
____ Fever
____ Food allergies
____ Foot pain
____ Foot swelling
____ Frequent or persistent infection
____ Frequent urination
____ Fungal infection such as athlete's food that persists
____ Graying of hair
____ Hair loss (not due to chemotherapy)
____ Have itchy scaly skin rashes
____ Headaches frequently

(*Continued*)

TABLE 30.4 The Oxidative Stress Index questionnaire.—cont'd

____	Hear voices inside you
____	Hearing loss that comes on suddenly
____	Heart palpitations (throbbing)
____	Heartburn
____	Hip pain
____	Hoarseness
____	Increased susceptibility to infections
____	Indigestion (frequent)
____	Insomnia
____	Irregular periods
____	Itchy hands
____	Itchy skin other than hands
____	Jaw pain
____	Leg swelling
____	Learning new things more difficult
____	Light headedness
____	Long recovery time from infections
____	Losing track of time
____	Loss of coordination
____	Loss of muscle tone
____	Loss of taste
____	Low energy
____	Lower back pain
____	Memory loss
____	Mood swings from very high to very low and vice versa
____	Mouth sores that don't go away quickly
____	Muscle aches that last a long time
____	Muscle cramps
____	Muscle spasms
____	Nasal congestion
____	Nausea
____	Neck pain
____	Nervousness
____	Nightmares regularly
____	Nose bleeds
____	Knee pain
____	Numbness or tingling in hands or feet
____	Pain in joints
____	Heart palpitations
____	Pelvic pain
____	Perspire (sweat) profusely
____	Post nasal drip that lingers
___	Post traumatic stress disorder (PTSD)
____	Problems finding the words you want

TABLE 30.4 The Oxidative Stress Index questionnaire.—cont'd

____ Rapid hair loss
____ Rapid heartbeat
____ Scaly skin
____ Seizures
____ Shortness of breath
____ Shoulder pain that lingers
____ Sinus pain
____ Skin mole growth
____ Skin rashes
____ Sleep less than 7 hours per night
____ Sleep more than 9 hours a night
____ Slow to heal from cuts, bruises or other injuries
____ Slurred speech
____ Smaller field of vision
____ Sore throat that doesn't heal
____ Stressed out most or all of the time
____ Stuffy nose
____ Swollen eye lids
____ Tics (involuntary movements)
____ Tingling in the hands or feet
____ Tire easily
____ Tired most of the time
____ Tooth pain
____ Tremors
____ Twitching
____ Unusual vaginal bleeding or discharge
____ Urination difficulty
____ Urination pain
____ Varicose veins
____ Vomiting
____ Wake up more than 3 times per night
____ Weakness
____ Weight gain
____ Weight loss (rapid)
____ Wheezing
____ Wrinkling or loss of tone in skin
____ Yawning frequently

Total Symptoms Checks ____

DIAGNOSTIC DATA

The following address results obtained from test doctors ordered done as part of annual examinations. Check all that apply to you.

(Continued)

TABLE 30.4 The Oxidative Stress Index questionnaire.—cont'd

____ High or low blood sugar
____ High or low BUN (blood urea nitrogen)
____ High C-Reactive protein
____ High or low calcium
____ High or low carbon dioxide (bicarbonate)
____ High or low chloride
____ High cholesterol
____ High or low creatinine
____ High glucose
____ High or low potassium
____ High PSA
____ High sedimentation rate
____ High or low sodium
____ High triglycerides
____ High white cell count
____ Low blood oxygen
____ Low potassium

TOTAL DIAGNOSTIC DATA CHECKS ______

PHARMACEUTICALS

Check each of the boxes that apply. If you regularly take five prescription drugs, for example, check **all** of the first 5 items, so that the total number of items checked equals the total number of prescriptions regularly taken.

____ 1 prescription
____ 2 prescriptions
____ 3 prescriptions
____ 4 prescriptions
____ 5 prescriptions
____ 6 prescriptions
____ 7 prescriptions
____ 8 prescriptions
___ 9 prescriptions
____ 10 or more prescriptions
____ Have a heart pacemaker

TOTAL PHARMACEUTICAL CHECKS ____

DIET

Check each item that applies to the foods that are part of the diet you regularly eat.

TABLE 30.4 The Oxidative Stress Index questionnaire.—cont'd

____ Artificial sweeteners for coffee or tea
____ Canned or frozen cooked foods regularly eaten (soups, pastas, meats)
____ Bread and pasta made primarily from white processed flour
____ Fast food frequently eaten
____ Fewer than 3 fruits or vegetables a day
____ Grilled, smoked or blackened meat, chicken or fish
____ Food high in fat (whole milk, cheeses, foods cooked with butter and animal fat)
____ Often eat processed foods (bacon, hot dogs, salami, sausages. deli meats
____ Eat red meat more than 2 times a week
____ Eat foods high in sugar (sweetened drinks and desserts)
____ Salty food

TOTAL DIET CHECKS ____

LIFE STYLE

These items refer to where you live, the type of work you do and chemicals you may be exposed to.

____ Are a farmer that regularly uses pesticides
____ Burn wood for heat or for cooking
____ Constantly use a cell phone
____ Drink chlorinated water
____ Drink more than one alcoholic drink per day
____ Exercise less than one half hour a week
____ Have mold in your home
____ Have new (less than 6 months old) carpet in your home
____ Have pets in your home that you are allergic to
____ Live down wind from a smoking industrial chimney
____ Live in a city with air quality alerts
____ Live or work close to a cell tower
____ Live or work near high voltage electrical transmission lines
____ Live near a heavily traveled highway or road
____ Live near a landfill
____ Live with a smoker
____ Regularly experience allergic reactions in your home
____ Regularly experience allergic reactions in your work place
____ Regularly use room or furniture deodorants
____ Regularly play contact sports
____ Regularly sleep with lights or television on
____ Regularly have outdoor lights shining into your bedroom while sleeping
____ Work as a toll booth collector
____ Work in very hot or very cold conditions regularly

(Continued)

TABLE 30.4 The Oxidative Stress Index questionnaire.—cont'd

____ Work in an adhesives or coatings manufacturing plant
____ Work in an agricultural chemical manufacturing plant
____ Work as an automobile, diesel or aircraft mechanic
____ Work with chemicals on the job regularly
____ Work in a dusty environment regularly
____ Work in a landfill
____ Work in a hair or nail salon
____ Work in metal refinery or mill
____ Work as a miner
____ Work in a noisy environment
____ Work in a paint, lacquer, stain or varnish manufacturing plant
____ Work as a painter
____ Work as a pilot or flight attendant
____ Work as a pesticide applicator
____ Work in a petroleum refinery
____ Work in a plastics manufacturing plant
____ Work in a plywood or particle board manufacturing plant
____ Work in a polluted environment (road paver, toll booth operator, for example)
____ Work in a water or sewage treatment plant
____ Work in wood treatment plant

TOTAL LIFESTYLE CHECKS ____

TOBACCO AND ALCOHOL USE

Tobacco use. Check **all** of the items that apply. For example, if you smoke one pack of cigarettes a day, check both of the first 2 items. If you smoke two packs a day, check all of the first 4 items.

____ Smoke 10 cigarettes or less daily (even just one)
____ Smoke a pack a day (20 cigarettes)
____ Smoke a pack and a half a day (30 cigarettes)
____ Smoke two packs a day (40 cigarettes)
____ Smoke more than two packs a day
____ Smoke 1-5 cigars a day
____ Smoke 6 or more cigars a day
____ Use smokeless tobacco

Alcohol consumption. Number of drinks consumed on a daily basis. Check the number that applies. The daily number of drinks is the OSI drink score.

____ 1 drink daily
____ 2 drinks daily
____ 3 drinks daily

TABLE 30.4 The Oxidative Stress Index questionnaire.—cont'd

____ 4 drinks daily
____ 5 or more drinks daily

TOTAL TOBACCO AND ALCOHOL CHECKS ____

TOTAL NUMBER OF CHECKS ____

TOTAL PHYSICAL DATA SCORE ____

OSI SCORE - TOTAL CHECKS ____

30.6 Additional applications

The Oxidative Stress Index (OSI), a measure of the likelihood of disease onset in people who are clinically ill as well as in those without prevalent disease, is presented. The OSI has also been modified for other health related and research applications. These include predicting which disease-causing parameters are more contributory to the onset of specific diseases (Alzheimer's disease in Chapter 23 as an example), how it may be applied as a public health survey instrument to evaluate disease onset as a function of residence proximity to points of environmental pollutant releases (as discussed in Chapter 33), as a tool to predict adverse drug reactions in those using polypharmacy and as a preliminary diagnostic tool in remote areas with limited available medical facilities. These OSI applications and others, as well as the use of a condensed version of the OSI, are discussed in the ensuing chapters.

The OSI cannot predict which disease one might be afflicted with, as all disease is associated with elevated oxidative stress. The OSI can, however, be a valuable tool that not only shows that disease onset is likely, but can also shed light on what one can do to lower OS and the likelihood of disease onset.

References

Albertsen, L.W., Schmidt, A.J., Grey, C., Jakson, R., Sorensen, H.T., Schmidt, M., 2020. The Danish comorbidity index for acure myocardial infarction (DANCAMI): development, validation and comparison with existing comorbidity indices. Clin. Epidemiol. 12, 1299–1311.

Charlson, M.E., Pompei, P., Ales, K.L., 1987. A new method of classifying prognostic comorbidity in longitudinal studies: development and validation. J. Chron. Dis. 40 (5), 373–383.

EPA, 2020. Air Quality Index (AQI). https://airnow.gov/index/cfm?action=aquabasics.aqi. (Accessed 4 June 2021).

Jia, C.E., Zhang, H.P., Lv, Y., Liang, R., Jiang, Y.G., Powell, H., et al., 2013. The asthma control test and asthma control questionnaire for assessing asthma control: systematic review and meta-analysis. J. Allergy Clin. Immunol. 131 (3), 695–703.

Liang, W., Liang, H., Ou, L., Chen, B., Chen, A., Li, C., et al., 2020. Development and validation of a clinical illness in hospitalized patients with COVID-19. JAMA Intern. Med. https://doi.org/10.1001/jamainternmed.2020. (Accessed 3 June 2021).

NIH, US Dept. of Health & Human Services. National Institutes of Health, 2021. Body Mass Index Tables. https://www.nhlbi.nih.gov/health/educational/lose_wt/BMI_tb.htm. (Accessed 5 June 2021).

Pahwa, R., Goyal, A., Bansal, P., Jialal, I., 2020. Chronic Inflammation. NIH govt books. https://www.ncbi.nlm.nih.gov/books/NBK493173/. (Accessed 30 June 2021).

Sabbagh, M.N., Malek-Ahmadi, R., Belden, C.M., Connor, D.J., Pearson, C., Jacobson, R., et al., 2010. Alzheimers Dis. 22 (3), 1015–1021.

Zeliger, H.I., 2016. Predicting disease onset in clinically healthy people. Interdiscipl. Toxicol. 9 (2), 39–54.

Zeliger, H.I., 2017. Oxidative stress index predicts disease onset. J. Med. Res. Pract. 6 (3), 86–92.

CHAPTER

31

OSI condensed questionnaire

31.1 Introduction

As discussed in Chapter 30, the Oxidative Stress Index (OSI) can be used to predict the likelihood of noncommunicative chronic disease onset, as well as for other applications. A condensed form of the OSI questionnaire, discussed here, enables meaningful OSI data to be obtained from answers to only eight questions, rather than from the 350+ items contained in the detailed questionnaire. In this chapter, the condensed OSI version is shown as being applicable for assessing general well-being and other applications (Zeliger, 2020a,b).

The detailed OSI presented in Chapter 30 is based upon responses to a detailed questionnaire that address all contributory items to a person's OS level. In addition to being a valuable predictor of disease onset likelihood, it is also an indicator of lifestyle changes that can be made to lower OS and thereby help prevent disease. The length of the detailed OSI questionnaire, however, does not readily lend itself to regular use in clinical settings. Presented here is the condensed form of the OSI questionnaire, containing only eight parts, which can be readily used as a screening device to measure total oxidative stress level. Analysis has shown that OSI values obtained from the condensed OSI form produce scores that are nearly identical to those obtained from the original detailed OSI form (Zeliger, 2020a,b). Accordingly, the condensed form can serve as a preliminary screening device that measures total OS and flags potential problems. Those with elevated OSIs in the condensed form can then use the detailed OSI form shown in Chapter 30 to help identify specific causes of elevated OS which can lead to clinically relevant follow up and preventive measures.

The eight components of the condensed OSI form are shown in Table 31.1. A discussion of each follows.

TABLE 31.1 Eight components of the condensed Oxidative Stress Index Form.

Age
Weight
Preexisting chronic conditions
Medications regularly taken

(Continued)

https://doi.org/10.1016/B978-0-323-91890-9.00017-9

TABLE 31.1 Eight components of the condensed Oxidative Stress Index Form.—cont'd

Genetics
Education level
Place of residence
Psychological stress

31.2 Age

Aging is accompanied by an increase in OS, generally starting around age 40 and continuing throughout life (Epel et al., 2004; Hou et al., 2015). Accordingly, one OSI point is scored for each decade starting at age 40. For one aged 40–49, 1 OSI point is added to his or her total, two points are added for those aged 50–59, etc (Zeliger, 2017).

31.3 Weight

Being overweight is associated with elevated OS and the onset of numerous diseases and excessive weight is a worldwide health issue that continues to grow. In the United States, more than 70% of adults are overweight, more than half of whom are obese (CDC, 2018). Diseases associated with excess weight include, but are not limited to; hypertension, dyslipidemia, type 2 diabetes, coronary heart disease, stroke, gallbladder disease, osteoarthritis, sleep apnea, breathing difficulties, anxiety, depression and several cancers (endometrial, breast, colon, kidney, liver and gallbladder (CDC, 2015)). Gender, height, weight and frame (thin, average, large) information enables one to determine if the person is of normal weight, is heavy or obese. Weight impact is addressed in the condensed OSI via body mass index (BMI) (Zeliger, 2017). Four weight categories are scored as follows.

BMI range	Classification	OSI score
19–24	Normal	0
25–29	Overweight	1
30–39	Obese	2
40 or more	Extremely obese	4

31.4 Preexisting chronic conditions

The number of prevalent diseases at the time of OSI measurement is a critical indicator of OS. It is well established that disease increases total OS and hence the OSI (Zeliger, 2016,

2017). All diseases have symptoms associated with them and these are individually addressed in the detailed OSI. The OSI condensed form does not probe individual symptoms. Rather, it has been found that assigning five points for each prevalent disease adequately incorporates elevations in OS associated with the prevalent diseases. Respondents are asked to list all diseases they currently have been diagnosed with. Multiplying the number of prevalent diseases by five yields the disease number in the condensed OSI.

31.5 Medications regularly taken

Essentially all pharmaceutical medications raise OS (Zeliger, 2017) and all have adverse drug reactions associated with them (Zeliger, 2019b). The detailed OSI form addresses these individually. In the condensed form, one OSI point is assigned for each medication taken.

31.6 Genetics

Genetics is well known to be a factor in most noncommunicable chronic diseases (Dato et al., 2013; Jiang et al., 2013; Guillaumet-Adkins et al., 2017). Indeed, many diseases, including Alzheimer's disease, Parkinson's disease and cancers, just to name and few, "run in families." Recently, epigenetics, as well, has been shown to lead to heritable diseases (Cencioni et al., 2013; Guillaumet-Adkins et al., 2017). Though all noncommunicable diseases are more prevalent in those whose ancestors have suffered from those diseases, parental disease is most closely associated with the likelihood of disease onset in an individual (Awdeh et al., 2006). Both genetic traits and epigenetic effects raise OS (Cencioni et al., 2013; Dato et al., 2013; Jiang et al., 2013; Guillaumet-Adkins et al., 2017). In the condensed OSI form, respondents are asked to check which of their prevalent diseases were also prevalent in their parents and each genetic link is assigned a value of one OSI point.

31.7 Education level

Socioeconomic status (SES) is well established as an indicator of detrimental lifestyle that raise OS (Mielck et al., 2014). These lifestyle choices include diets high in fats, sugar, salt or processed foods, tobacco, alcohol and recreational drug abuse, radiation exposure and the need to reside or work in a toxic environment Zeliger (2016). SES is also associated with an increased likelihood of having undiagnosed diseases (Bein et al., 2012; Mielck et al., 2014; Shaw et al., 2016).

The detailed OSI lists numerous items that address these points. In the condensed form, all of these are factored into SES as indicated by highest educational level achieved. Educational achievement has been shown to be a valid indicator of SES, with lower SES individuals more likely to lead unhealthy, OS raising life styles and to have undiagnosed diseases (Yin et al., 2017). In the condensed OSI form, five education levels are identified: some high school, high school graduate, some college, college graduate and graduate degree. These are assigned 5,4,3,2 and 1 OSI point values respectively.

SES can also be obtained from annual income information, but asking income information can be considered being nosy and discourage some people from completing the questionnaire. Hence, income is not used in the condensed OSI questionnaire.

31.8 Place of residence

In the 21st century, 90% of the world's people, regardless of SES, are exposed to air pollution (Combes and Franchineau, 2019; World Health Organization, 2019). All air pollutants are toxic, raise OS in a dose response relationship, impact the OSI, and are responsible for the onset of numerous diseases including respiratory diseases, cardiovascular diseases, several cancers and Alzheimer's disease (Kelly 2003; Riggs et al., 2020; Combes and Franchineau, 2019; Zhou et al., 2019; Xia et al., 2019; Kilian and Kitazawa, 2018).

A method to quantify air pollution impact on disease onset, termed the Air Quality Toxicity Index (AQTI) has been reported (Zeliger, 2020). This method, based upon the dose response relationship between toxic exposure and OS elevation (Zeliger, 2016 and the numerous references contained therein), enables the classification of air quality for individual locations to be calculated and reported as cumulative annual values.

In the United States, the Environmental Protection Agency (EPA) regularly measures air quality in multiple locations and reports the data on-line daily as the Air Quality Index (AQI) (EPA, 2019). Worldwide, other nations similarly report air quality data (Fino et al., 2021) and the World Air Quality Project also reports air quality data for hundreds of cities (World Air Quality Index Project, 2019). On an annual basis, these indices identify the number of days in which the air quality in a given locale is classified as either good, moderate, unhealthy for sensitive groups, unhealthy, very unhealthy or hazardous. Hazard numbers ranging from 1–6 have been assigned to the six air quality classifications, as follows (Zeliger, 2020).

Air quality classification	Hazard number
Good	1
Moderate	2
Unhealthy for sensitive groups	3
Unhealthy	4
Very unhealthy	5
Hazardous	6

The air pollution impact on OSI is obtained by multiplying the number of days per year for each of the six EPA classifications at the residence of the responder by the hazard number for

TABLE 31.2 EPA Air Quality Index classifications.

EPA AQI	EPA classification	Hazard number
0 to 50	Good	1
51 to 100	Moderate	2
101 to 150	Unhealthy for sensitive groups	3
151 to 200	Unhealthy	4
201 to 300	Very unhealthy	5
301 to 500	Hazardous	6

that classification and adding these up to yield the AQTI total as shown in Table 31.3. Also shown in Table 31.3. are the OSI air pollution severities (OSI—AP). These are assigned number values from 0 to 6, and entered into the OSI condensed form. It should also be noted that EPA publishes daily and annual air quality data for cities and counties in all U.S. states (EPA, 2019). The following exemplifies the Air Quality Toxicity Index (AQTI) calculation.

EPA (AQI) classifications for air quality are shown in Table 31.2 (EPA, 2019). Other air quality indices may be similarly classified. Hazard numbers are assigned here to each classification ranging from 1-6, with 1 for good air quality through 6 for hazardous air quality, as also shown in Table 31.2. The relevant hazard number for a given area is the number enetered into the OSI.

EPA also reports the number of days per year that fall into each air quality classification. The AQTI is calculated by multiplying the hazard number (HN) by the number of days in a given year that air quality corresponds to each number and then totaling the numbers to generate the AQTI for that location, as shown by the following equation.

$$\begin{aligned} \text{AQTI} = {} & (1) \times (\text{Days at HN } 1) + (2) \times (\text{Days at HN } 2) + (3) \times (\text{Days at HN } 3) + \\ & (4) \times (\text{Days at HN } 4) + (5) \times (\text{Days at HN } 5) + (6) \times (\text{Days at HN } 6) \end{aligned}$$

As examples, Table 31.3 shows the calculation of AQTI numbers that were thus generated for three American cities in the year 2018: Houston, Texas (HT); Los Angeles, California (LA); and Portland, Maine (PM).

AQTI values can be used to determine an area's overall annual air quality classification, as shown in Table 31.4.

As can be seen from the data for the three cities in Tables 31.3 and 31.4, Portland, Maine's 2018 air quality overall was good. Houston's 2018 air quality overall was unhealthy for sensitive groups and the air in Los Angeles 2018 overall was unhealthy 2018.

TABLE 31.3 AQTI calculation for three American cities.

	Multiplier Number	Days of each city at each level			Multiplier times days at each level		
		HT	LA	PM	HT	LA	PM
Good	1	166	35	311	166	35	311
Moderate	2	164	220	2	328	440	104
Unhealthy for sensitive groups	3	26	90	3	78	270	6
Unhealthy	4	7	19	0	28	76	0
Very unhealthy	5	2	1	0	10	5	0
Hazardous	6	0	0	0	0	0	0
			AQTI		650	826	421

HT, Houston Texas; *LA*, Los Angeles; California; *PM*, Portland Maine.

TABLE 31.4 Annual air quality classification as shown by AQTI.

Air quality classification	AQTI
Good	365–450
Moderate	451–600
Unhealthy for sensitive groups	601–750
Unhealthy	751–900
Very unhealthy	901–1050
Hazardous	1051 or higher

31.9 Chronic psychological stress

Psychological stress, anxiety and depression are associated with most diseases and are more pronounced as illnesses progress (Sahle et al., 2020). Psychological stress, anxiety and depression raise OS via the release of hormones that elevate OS. Chronic activation of this stress response system results in disease and triggers numerous health problems (Mayo Clinic, 2019). These include:

- Anxiety
- Depression
- Memory and concentration issues
- Digestive problems

TABLE 31.5 OSI condensed form point assignments.

Age	1 for each decade of age starting at age 40.
Personal	0,1,2, or 3, depending upon weight status
Preexisting conditions	5 per prevalent chronic disease
Medications	1 per prescription or overthe-counter medication regularly taken
Genetics	1 per each item checked
Education	5,4,3,2 or 1, depending upon highest education level achieved
Residence	1,2,3,4,5 or 6 depending upon air pollution level at residence
Stress	0,1,2 or 3, depending upon chronic psychological status

- Headaches
- Heart disease
- Sleep problems
- Weight gain

Respondents are asked to check if they often feel stressed, anxious or depressed. One point is assigned for each positive response.

The relevance of the OSI to predicting disease onset probability as well as for other applications uses has been previously established (Zeliger, 2017, 2019a,b,c,d). When the OSI condensed form values are used as presented here an excellent correlation between standard form and condensed form OSI values obtains, with condensed form values showing less than a 10% variation from standard form values (Zeliger, 2020a).

31.10 Condensed OSI form

As per the discussion above, the point values assigned to each of the parameters contained in the OSI condensed questionnaire are shown in Table 31.5.

The total OSI score is related to the likelihood of further disease onset as shown in Table 31.6 (Zeliger, 2017).

The OSI condensed questionnaire is shown in Table 31.7.

TABLE 31.6 OSI score and likelihood of disease onset.

OSI level	Disease onset likelihood
0–15	Indicative of good health
16–30	Disease onset predicted
31–45	Disease onset probable
46 or higher	Disease imminent

TABLE 31.7 OSI condensed questionnaire.

1. Age:	Age score 0. 1, 2, 3, 4, 5,6	________
2. Personal:	female _____ male ______ height _____ weight ______ BMI 0, 1, 2, or 3	________
3. Diseases:	All current diseases and diagnosed conditions Multiple number by 5	________
4. Medications:	Number of medications regularly taken ______	________
5. Family history:	Number of the above diseases parents had/have Mother ________ Father ________ Enter total number for both	________
6. Education:	Check highest level achieved some high school, enter 5 high school graduate, enter 4 some college, enter 3 college graduate, enter 2 graduate school, enter 1	________
7. Residence:	Air quality in City and State of residence City ________ State________ Good, enter 1 Moderate, enter 2 Unhealthy for sensitive groups, enter 3 Unhealthy, enter 4 Very unhealthy, enter 5 Hazardous, enter 6	________
8. Stress:	Check all that apply and enter total Do you often feel: Stressed _____ Anxious _____ Depressed _____	________
	Total	________

31.11 Condensed OSI applications

Though the condensed OSI score can predict the likelihood of disease onset, it cannot predict which disease(s) are likely to strike. Its applications include the following:

1. A part of routine medical examinations to alert clinicians to potential illness in seemingly healthy people.
2. An indicator of the need to have a person fill out the detailed OSI form which can alert clinicians to specific potential problems.
3. A screening aid in areas with limited medical personnel and facilities.
4. Aid in identifying lifestyle changes that will lower OS and likelihood of disease onset.
5. Taking of public health surveys to identify disease clusters arising from chronic exposures to toxic chemicals.
6. Help in making community medical need projections including estimating clinical staffing and resource needs.
7. Researching sequences of disease onset in those with multiple diseases.
8. Predicting numbers of people likely to ail with noncommunicative diseases in different geographic and demographic areas.
9. Predicting severity of new infectious and chronic noninfectious disease onset.

31.12 Conclusions

A questionnaire consisting of only eight parameters that are representative of all sources of oxidative stress can be used to determine the OSI. The eight parameters used to calculate the condensed form OSI all raise OS levels in dose response relationships (Zeliger, 2016). The following chapters address additional OSI applications.

References

Awdeh, Z.L., Yunis, E.J., Auden, M.J., Fici, C., Pugliese, A., Larsen, C.E., Alper, C.A., 2006. A genetic explanation for the rising incidence of type 1 diabetes, a polygenic disease. J. Autoimmun. 27 (3), 174–181.

Bein, T., Hackner, K., Zou, T., Schultes, S., Bosch, T., Schlitt, H.J., et al., 2012. Socioeconomic status, severity of disease and level of family members' care in adult surgical intensive care patients: the prospective ECSSTASI study. Intensive Care Med. 38, 612–619.

Centers for Disease Control and Prevention, 2015. Healthy weight. In: The Health Effects of Overweight and Obesity. https://cdc.gov/healthyweight/effects/index.html. (Accessed 25 January 2020).

CDC., 2018. Centers for Disease Control. National Center for Health Statistics. https://www.cdc.gov/nchs/hus/contents2018.htm#Table_021.

Cencioni, C., Spalotta, F., Martelli, F., Valente, S., Mai, A., Zeiher, A.M., Gaetano, C., 2013. Oxidative stress and epigenetic regulation in aging and are-related diseases. Int. J. Mol. Sci. 14, 17643–17663.

Combes, A., Franchineau, G., 2019. Fine particle environmental pollution and cardiovascular diseases. Metabolism 2019. https://doi.org/10.1016/j.metabol.2019.07.008. (Accessed 27 January 2020).

Dato, S., Crocco, P., D'Aquila, P., de Rango, F., Belllizi, D., Rose, G., Passarino, G., 2013. Exploring the role of genetic variability and lifestyle in oxidative stress response for healthy aging and longevity. Int. J. Mol. Sci. 14, 16443–16472.

EPA., 2019. Air Quality Index (AQI). https://airnow.gov/index.cfm?action=aquibasics.aqi. (Accessed 2 January 2020).

Epel, E.S., Blackburn, E.H., Dhabhar, F.S., Adler, N.E., Morrow, J.D., Cawthorn, R.M., 2014. Accelerated telomere shortening in response to life stress. Proc. Natl. Acad. Sci. USA 101 (49), 17312–17315.

Fino, A., Vichi, F., Leonardi, C., Mukhopadhyay, K., 2021. An overview of experiences made and tools used to inform the public on ambient air quality. Atmosphere 2021. https://doi.org/10.3390/atmos12111524. (Accessed 3 December 2021).

Guillaumet-Adkins, A., Yanez, Y., Preis, M.D., Palancia-Ballester, C., Sandoval, J., 2017. Ox Med Cell Longevity Article No. ID9175806. https://doi.org/10.1155/2017/9175806. (Accessed 23 January 2020).

Hou, L., Joyce, B.T., Gao, T., Liu, L., Zheng, Y., Penedo, F.J., et al., 2015. Blood telomere length attrition and cancer development in the normative aging study cohort. EBioMedicine 132 (6), 591–596.

Jiang, T., Yu, J.T., Tian, Y., Tan, L., 2013. Epidemiology and etiology of Alzheimer's disease. From genetic to non-genetic factors. Curr. Alzheimer Res. 9, 852–867.

Kelly, F.J., 2003. Oxidative stress: its role in air pollution and adverse health effects. Occup. Environ. Med. 60, 612–616.

Kilian, J., Kitazawa, M., 2018. The emerging risk of exposure to air pollution on cognitive decline and Alzheimer's disease - evidence from epidemiological and animal studies. Biomed. J. 41 (3), 141–162.

Mayo Clinic, 2019. Healthy Lifestyle Stress Management. https://www.mayoclinic.org/heaalthy-lifestyles-management/in-depth/stress/art-20046037.

Mielck, A., Vogelman, M., Reiner, L., 2014. Health-related quality of life and socioeconomic status: inequalities among adults with a chronic disease. BioMed Central 12 (58). http://www.hqlo.com/content/12/1/58. (Accessed 21 January 2020).

Riggs, D.W., Zafar, N., Krishnasamy, S., Yeager, R., Rai, S.N., Bhatnagar, A., O'Toole, T.E., 2020. Exposure to airborne fine particulate matter is associated with impaired endothelial function and biomarkers of oxidative stress and inflammation. Environ. Res. 2020. https://doi.org/10.1016/j.envres.2019.108890. (Accessed 30 December 2019).

Sahle, B.W., Chen, W., Melaku, Y.A., Akombi, B.J., Rawal, L.B., Renzaho, A.M.N., 2020. Association of psychosocial factors with risk of chronic diseases: a nationwide longitudinal study. Am. J. Prev. Med. 58 (2), e39–e50. https://doi.org/10.1016/j.amepre.2019.09.007. (Accessed 27 January 2020).

Shaw, K.M., Theis, K.A., Self-Brown, S., Roblin, D.W., Barker, L., 2016. Chronic Disease Disparities by County Economic Status and Metropolitan Classification, Behavioral Risk Factor Surveillance System, 2013. Centers for Disease Control. Preventing Chronic Disease. https://www.cdc.giv/pcd/issues/2016/16_0088.htm. (Accessed 21 January 2020).

World Air Quality Index Project, 2019. https://aqicn.com. (Accessed 27 January 2020).

Xia, S.Y., Huang, D.S., Jia, H., Zhao, Y., Lin, N., Mao, M.Q., et al., 2019. Relationship between atmospheric pollutants and risk of death caused by cardiovascular and respiratory diseases and malignant tumors in Shenyang, China from 2013 to 2016: an ecological research. Chin. Med. J. (Engl) 132 (19), 2269–2277.

Yin, H., Wu, Q., Cui, Y., Hao, Y., Liu, C., Li, Y., et al., 2017. Socioeconomic status and prevalence of chronic non-communicable diseases in Chinese women: a structural equation modelling approach. BMJ Open 2017 (7), e014402. https://doi.org/10.1136/bmjopen-2016-014402. (Accessed 22 January 2020).

Zeliger, H.I., 2016. Predicting disease onset in clinically healthy people. Interdiscipl. Toxicol. 9 (2), 39–54.

Zeliger, H.I., 2017. Oxidative stress index predicts disease onset. J. Med. Res. Prac. 6 (3), 86–92.

Zeliger, H.I., 2019. Predicting Alzheimer's disease onset. Eur. J. Med. Health Sci. 1 (1). https://doi.org/10.24018/ejmed.2019.1.1.16. (Accessed 28 January 2020).

Zeliger, H.I., 2019a. Oxidative stress index as a public health survey instrument. Eur. J. Med. Health Sci. 1 (2). https://doi.org/10.24018/ejmed.2019.1.2.28. (Accessed 28 January 2020).

Zeliger, H.I., 2019b. Predicting and reducing adverse drug reactions. Euro. J. Med. Health Sci. 1 (4). https://doi.org/10.24018/ejmed.2019.1.4.91 (Accessed 2 January 2020).

Zeliger, H.I., 2019c. Oxidative stress index: disease onset prediction and prevention. EC Pharm. Toxicol. 7.9 (2019), 1022–1036.

Zeliger, H.I., 2020a. Air quality toxicity index (AQTI): quantifying air pollution impact on disease onset. Eur. J. Med. Health Sci. 2 (1). https://doi.org/10.24018/ejmed.2020.2.1.143. (Accessed 28 January 2020).

Zeliger, H.I., 2020b. Oxidative stress index (OSI) condensed questionnaire. Eur. J. Med. Health Sci. https://doi.org/10.24018/ejmed.2020.2.1.163. (Accessed 30 June 2021).

Zhou, H., Wang, T., Zhou, F., Liu, Y., Zhao, W., Wang, X., et al., 2019. Ambient air pollution and daily hospital admissions for respiratory disease in children in Guiyang, China. Front. Pediatr. 2019. https://doi.org/10.3389/fped.2019.00400. (Accessed 30 January 2020).

CHAPTER

32

OSI and Alzheimer's disease

32.1 Introduction

People with chronic diseases and those who are disease free but living unhealthy lifestyles are known to be candidates for numerous noncommunicative diseases, as well as more frequent and more severe bouts with infectious diseases. The Oxidative Stress Index (OSI) has been shown as a method for assigning the probability of disease onset to all people, those clinically ill as well as those without prevalent disease (Zeliger, 2017).

The OSI, as originally formatted, does not predict which disease will more likely develop, only that further disease is predicted with increased OSI scores. It has been recently shown, however, that the OSI may also be used to demonstrate which parameters are more contributory to the onset of a particular disease when measured at the time of onset of that disease and applied to a study of Alzheimer's disease (AD) (Zeliger, 2019).

A modified version of the OSI can be used to predict the likelihood of AD onset in clinically healthy individuals, as well as to identify which known AD inducing parameters are most contributory to the likelihood of its onset (Zeliger, 2019).

32.2 Alzheimer's disease and oxidative stress

Oxidative stress is a crucial mechanistic component of AD onset (Liu et al., 2015; Huang et al., 2015; Kamat et al., 2016; Tonnies and Trushina, 2017) and has been definitively associated with brain neuroinflammation and blood-brain barrier dysfunction that is both a cause and consequence of AD generating hyper-permeability that leads to the absorption of, as well as generation of, OS inducing species in the brain (Zhang and Jiang, 2015; Erickson and Banks, 2013; Najjar et al., 2013).

As has also been discussed in earlier chapters of this book, it is also known that disease onset likelihood is related to total oxidative stress arising from numerous sources (Zeliger, 2016 and the references contained therein).

Oxidative Stress
https://doi.org/10.1016/B978-0-323-91890-9.00016-7

32.3 Alzheimer's disease and dose response relationship (DRR)

Total Oxidative Stress has been shown to be related to disease onset, whether coming from single sources or from combinations of multiple sources in a dose response relationship (DRR) (Zeliger, 2016). A single source example of this effect as it applies to AD is chronic exposure to radon (Lehrer et al., 2017). Multiple source examples are lifetime cigarette smoking (Mons et al., 2013; Durazzo et al., 2014) and simultaneous exposure to heavy metals or polynuclear aromatic hydrocarbons (Deng et al., 2019). DRRs for increased OS include exposures to trichloroethylene, perchloroethylene, air pollution, tobacco smoking, metals (including arsenic, lead, cadmium and mercury), polynuclear aromatic hydrocarbons, persistent organic pollutants (Kuang et al., 2013; Zeliger, 2016; Grova et al., 2019) and ultraviolet radiation (Agarwal et al., 1987).

32.4 Parameters known to increase likelihood of AD onset

AD is characterized by deposition of amyloid-beta plaques, hyperphosphorylated tau protein and neurofibrillary tangles (Kamat et al., 2016). It is well known that elevated oxidative stress is associated with increased likelihood of AD onset (Christen, 2000; Perry et al., 2002; Huang et al., 2016, Durazzo et al., 2014, Durazzo et al., 2014a). Several parameters have been shown to increase the likelihood of AD onset. These parameters, the symptoms associated with each and representative references for these are listed in Table 32.1. Oxidative stress has also been shown to be a crucial mechanistic component of AD onset (Liu et al., 2015; Huang et al., 2015; Kamat et al., 2016; Tonnies and Trushina, 2017).

As can be seen from Table 32.1, several of the symptoms associated with AD onset are associated with multiple parameters. Accordingly, symptom severity can increase with increasing symptom sources as can their contributions to total oxidative stress. The OSI takes this into account by including both parameters and symptoms in calculating total scores. It should also be noted that long-term exposures to persistent organic pollutants (POPs), though producing few or no discernible symptoms from chronic exposures to low concentrations, have also been associated with increased prevalence of AD (Grova et al., 2019).

32.5 Late onset AD

Late onset diseases, an exemplar being AD, manifest themselves only after long term (years) exposure to causative agents. Even early onset of disease (before age 65) follows many years of exposure to OS causing stimuli, and in the case of persistent organic pollutants, storage in the body's adipose tissue for up to decades (Cencioni et al., 2013; Zeliger, 2013; Zeliger and Lipinski, 2015; Grova et al., 2019).

TABLE 32.1 Parameters shown to increase the likelihood of AD onset and representative references for each.

Parameter	Symptoms	References
Air pollution	Eye irritation Nose irritation Throat irritation Coughing Chest tightness Dyspnea	Block and Calderon-Garciduenas (2009), Moulton and Yang (2012), Carey et al. (2018), Kilian and Kitazawa (2018)
Pesticides	Headache Nausea Dizziness Sweating Salivation Tearing Mucous production	Yan et al. (2016), University of Missouri Publications (2002)
Heavy metals	Headaches Constipation Weakness Fatigue Joint and muscle ache	Adlard and Bush (2018), Lee et al., 2018, NORD (2020)
Persistent organic pollutants	Skin irritation	Grova et al. (2019)
Radiation	Nausea Vomiting Diarrhea Headache Fever Dizziness Disorientation Weakness Fatigue Hair loss Infections Low blood pressure	Tang (2018), Lehrer et al. (2017), Litin (2021)
Smoking	Bad breath Cold hands, feet Frequent lung Infections Frequent flu Frequent common Colds Frequent bronchitis Frequent pneumonia Hypertension Taste and smell loss Easily fatigued Dyspnea Ongoing cough	Ott et al. (1998), Durazzo et al. (2014a), Healthgrades (2020)

(*Continued*)

TABLE 32.1 Parameters shown to increase the likelihood of AD onset and representative references for each.—cont'd

Parameter	Symptoms	References
Brain injury	Cognitive loss Memory loss Anxiety Depression Sleep difficulty	Turner et al. (2016), Yuan and Wang (2018), Breunig et al. (2013), Mayo Clinic (2022)
Chronic inflammation	Anxiety Body pain, arthralgia, myalgia Brain fog Depression Sedimentation rate elevation Excessive mucous production Frequent infections Gastrointestinal issues Bloating Abdominal pain Constipation Diarrhea Weight loss High C-Reactive protein High white cell count Low energy	Dunn et al. (2005), Misiak et al. (2012), Trempe and Lewis (2018), Pahwa et al. (2020)
Frequent infectious diseases	Viral, Fungal Bacterial Parasitic	Honjo et al. (2009), D'Aiuto et al. (2010), Karim et al. (2014), Sochocka et al. (2017), Honjo et al. (2009), D'Aiuto et al. (2010), Eimer et al. (2018), Readhead et al. (2018)
Unhealthy diet	Fatigue Brittle, dry hair Ridged, spoon shaped nails Diarrhea Apathy Irritability Lack of appetite Slow healing Premature aging	Luchsinger et al. (2002), Kirkpatick (2019)
Chronic Traumatic Encephalopathy	Difficulty thinking Memory loss Depression Anxiety Emotional instability Impaired judgment	Yanjun et al. (2017), Ramos-Cejudo et al. (2018), American Brain Foundation (2020)

TABLE 32.1 Parameters shown to increase the likelihood of AD onset and representative references for each.—cont'd

Parameter	Symptoms	References
Psychological stress (chronic)	Rapid heart rate Hypertension Fatigue Feeling overwhelmed Sleeping difficulty Poor problem solving Social withdrawal Feeling sad Feeling frustrated Inability to rest Difficulty controlling emotions	Machado et al. (2018), Justice (2018), Lindberg (2019)
Obesity	Excess body fat Dyspnea Increased perspiration Snoring Sleeping difficulty Fatigue Back and joint pain Depression	Profenno et al. (2010), Alford et al. (2018), Ma, et al. (2019)
Blood brain barrier breakdown	ADHD Autism (ASD) Brain fog Chronic fatigue Sudden onset headaches Memory loss Mood disorders Anxiety Depression Peripheral neuropathy Seizures	Lockhead et al. (2010), Nation et al. (2019), Mindd (2021)
Aging	Above age 40	Dato et al. (2013), Cencioni et al. (2013), Guillaumet-Adkins et al. (2017)
Genetics	Family AD history	Jiang et al. (2013), Dato et al. (2013), Guillaumet-Adkins et al. (2017), Cencioni et al. (2013), National Institute of Aging (2015)

(Continued)

TABLE 32.1 Parameters shown to increase the likelihood of AD onset and representative references for each.—cont'd

Parameter	Symptoms	References
AD comorbidities	Heart disease Glaucoma Hypertension Behavioral deficits Sleep disorders Cognitive deficits Bladder control loss Bowel control loss Obesity Type 2 diabetes Osteoporosis Sleep apnea	Bannon (2011); Bunn et al. (2014), Bauer et al. (2014), Poblador-Plou et al. (2014), Barbagallo and Dominguez (2014), Zeliger (2014); Naderali et al. (2009), Fafara et al. (2014), Karki et al. (2017), Alford et al. (2018), Doraiswamy et al. (2002)

32.6 The OSI and AD

The oxidative stress index can be applied to the study of Alzheimer's disease in three ways.

1. Incorporating the items in Table 29.1 to predict the likelihood of AD onset.
2. Using the data to identify which AD-causing parameters are the primary ones associated with disease onset.
3. To determine if there are ranges of OSI scores that correspond to AD onset odds.

32.6.1 Predicting AD onset

The OSI has been proposed as applicable to determine the likelihood of AD onset. One's oxidative stress onset potential can be determined from the fraction of AD associated parameters in an AD modified version of the OSI condensed questionnaire that are known to be associated with AD onset. The AD modified OSI condensed questionnaire is shown in Table 32.2. A supplemental AD associated parameters questionnaire, derived from Table 32.1, is listed in Table 32.3. The respondent is asked to complete both questionnaires so that likelihood of AD onset can be determined.

The AD likelihood ratio is defined as the ADSUPP divided by $\sum$ OSI. Subsequent follow up of statistically significant numbers of individuals can be used to create the relationship between ADSUPP/$\sum$ OSI ratio ranges and AD likelihood onset. Studies in this regard are underway as this book goes to press.

TABLE 32.2 AD modified version of the OSI condensed form.

1. Age:	Age score 0. 1, 2, 3, 4, 5,6	________
2. Personal:	female _____ male ______	
	height _____ weight ______	
	BMI 0, 1, 2, or 3	________
3. Diseases:	All current diseases and diagnosed conditions	________
	Multiple number by 5	
4. Medications:	Number of medications regularly taken ______	________
5. Family history:	Number of your current diseases parents had/have	________
	Mother ________Father ________	
	Enter total number for both	
6. Education:	Check highest level achieved	________
	some high school, enter 5	
	high school graduate, enter 4	
	some college, enter 3	
	college graduate, enter 2	
	graduate school, enter 1	
7. Residence:	Air quality in City and State of residence	________
	City _________ State_________	
	Good, enter 1	
	Moderate, enter 2	
	Unhealthy for sensitive groups, enter 3	
	Unhealthy, enter 4	
	Very unhealthy, enter 5	
	Hazardous, enter 6	
8. Stress:	Check all that apply and enter total	________
	Do you often feel: Stressed _____	
	Anxious _____	
	Depressed _____	
		OSI TOTAL ($\sum$ OSI)________

TABLE 32.3 Supplemental AD associated parameters questionnaire.

Family history
Check all family members who have/had Alzheimer's disease
____ Maternal grandmother
____ Paternal grandmother
____ Maternal grandfather
____ Paternal grandfather
____ Mother
____ Father
____ Sister (one check for each if more than one
____ Brother (one check for each if more than one)
Total family history ____
Symptoms
____ abdominal pain
____ anxiety
____ apathy
____ appetite loss
____ back pain
____ bad breath
____ bladder control loss
____ bloating
____ body pain, arthralgia, myalgia
____ bowel control loss
____ brain fog
____ brittle hair
____ chest tightness
____ chronic fatigue
____ cognitive loss
____ cold hands or feet
____ constipation
____ coughing
____ depression
____ diarrhea

TABLE 32.3 Supplemental AD associated parameters questionnaire.—cont'd

____ difficulty thinking
____ difficulty controlling emotions
____ disorientation
____ dizziness
____ dry hair
____ dyspnea
____ emotional instability
____ eye irritation
____ eye tearing
____ fatigue
____ feeling frustrated
____ feeling overwhelmed
____ feeling sad
____ fever
____ frequent body infections
____ frequent bronchitis
____ frequent common colds
____ frequent flu
____ frequent lung infections
____ frequent pneumonia
____ hair loss
____ have excess body fat
____ headache
____ impaired judgment
____ inability to rest
____ increased perspiration
____ irritability
____ joint paint
____ low blood pressure
____ low energy
____ nose irritation

(*Continued*)

TABLE 32.3 Supplemental AD associated parameters questionnaire.—cont'd

____ peripheral neuropathy

____ perspiration

____ poor problem-solving ability

____ premature aging

____ memory loss

____ mucous production

____ nausea

____ ridged spoon shaped nails

____ salivation

____ seizures

____ sleeping difficulty

____ slow healing

____ smell loss

____ snore a lot

____ socially withdrawn

____ sudden severe headache onset

____ taste loss

____ throat irritation

____ unable to rest

____ weakness

____ weight loss

Total symptoms ____

Tobacco use

Check one line for every 10 cigarettes smoked per day

____ ____ ____ ____ ____

____ use smokeless tobacco

Total tobacco use ____

Diagnoses

____ behavioral deficits

____ chronic periodontitis

____ cognitive deficits

____ elevated sedimentation rate

____ glaucoma

____ high blood pressure

TABLE 32.3 Supplemental AD associated parameters questionnaire.—cont'd

____ high C-Reactive Protein (CRP)

____ high white cell count

____ mood disorders

____ obesity

____ osteoporosis

____ sleep apnea

____ sleep disorders

____ type 2 diabetes

Total diagnosis ____

AD Supplemental total (ADSUPP) ____

32.6.2 Primary AD-causing parameters

The 17 parameters in Table 32.1 are known to be causative of AD onset. The AD modified OSI can be supplemented with an AD parameter questionnaire, shown in Table 32.4, at the time of first AD diagnosis to determine which of these parameters are the primary ones for an individual. The responses to it can be applied to predicting which of the AD individual causative parameters are the dominant ones in that individual. Collecting such information from statistically significant numbers of individuals is projected to be used to predict which AD causing parameters present the greatest threat to AD onset. Research in this area is also underway as this book goes to press.

32.6.3 AD onset odds

OSI data collected at the time of first AD diagnosis from statistically significant numbers of individuals can be used to determine the relationship between total OS and the likelihood of disease onset, i.e., to establish ranges of disease onset as a function of oxidative stress. This area is currently being studied.

32.7 AD prevention

Further evidence for the relationship of AD with OS comes from a consideration of lifestyle changes known to lower AD incidence. It is estimated that as many as half of AD cases are attributable to the following eight modifiable factors, all of which lower oxidative stress (Barnes and Yaffe, 2011). These include:

- Acting to prevent and manage type 2 diabetes
- Acting to prevent or treat hypertension

TABLE 32.4 Supplemental AD parameter questionnaire.

Check all items that apply.
____ AD comorbidities
____ aging – above age 40
____ air pollution – in residential area
____ blood brain barrier breakdown -from symptoms in Table 32.1
____ brain injury – from single incident of blow to the head
____ chronic inflammation – from symptoms in Table 32.1
____ chronic traumatic encephalopathy – if regularly play contact sports
____ frequent infectious diseases
____ genetics – if family history of AD
____ heavy metals – if work causes heavy metal exposure
____ obesity – if above healthy weight
____ persistent organic pollutants – if exposed to PCBs, dioxins or furans
____ pesticides – if a pesticide applicator
____ radiation – if regularly exposed to ionizing or microwave radiation
____ psychological stress – if anxious or depressed
____ smoking – smoke cigarettes or cigars
____ unhealthy diet – regularly eat high fat, high sugar, high salt or prepared foods

– Not smoking
– Maintaining physical activity
– Maintaining cognitive activity
– Treating depression, if present
– Preventing or reversing obesity
– Maintaining a healthy diet

A healthy diet is one that is low caloric, high in fruits and vegetables, whole grains, and nutritional antioxidants with and low in carbohydrates, saturated fats, red and processed meats. There are several versions of such a diet, the Mediterranean diet being the one most often followed. This diet includes the consumption of large quantities of extra virgin olive oil and moderate quantities of red wine (Dato et al., 2013). Other examples include the Okinawan diet, (followed by residents of Okinawa, Japan), the Dash diet (developed to stop hypertension) and the Portfolio diet (aimed at reducing cholesterol levels) (Wilcox et al., 2014).

It has also been reported that AD prevalence can be significantly reduced by properly treating the following conditions, which undermine immune system vitality, lead to chronic inflammation, and are known to elevate OS (Trempe and Lewis, 2018; Zeliger, 2017).

- chronic migraines
- mood disorders
- Eye diseases
- Metabolic syndromes
- chronic viral diseases
- hormonal diseases
- autoimmune diseases

Each of the AD causative factors discussed above raise OS, while all preventative factors lower OS. It is, therefore, reasonably hypothesized that avoiding other OS elevating factors such as chronic infections which stress the immune system and result in chronic inflammation, as well as chronic environmental exposures to toxic chemicals and radiation, which also elevate OS can delay onset of AD by years. Regular OSI monitoring can been used to demonstrate oxidative stress status.

32.8 Questionnaire use

Questionnaires, such as the Charlson Comorbidity Index, are routinely used to solicit patient background information on prevalent diseases, symptoms and likelihood of further disease onset (Charlson et al., 1987). The Alzheimer's Questionnaire is another example (Sabbagh et al., 2010). These and numerous others have established the validity of using patient questionnaires to gather essential health and diagnostic information. The OSI is another example of such an application.

32.9 Strengths and limitations

Strengths include:

1. The OSI is noninvasive, yet predicts OS levels associated with AD onset and offers insights into critical OS levels at which AD onset is likely. The OSI also can demonstrate which known AD causing parameters are most contributory to AD onset.
2. Individual items in the OSI are presented in alphabetical order and may differ widely from one item to the next. This is deliberate so that the person completing the OSI is required to consider each item alone, rather than as a part of a series of related questions, which could cause the responder to just check all the items in a single set.

Limitations include:

1. All parameters in the OSI carry equal weight. Clearly, some parameters are more detrimental to health to others. That said, more serious diseases will generally produce greater number of symptoms and require more medications than less severe ones.

2. Though preliminary studies strongly suggest that the relationships between the OSI and AD are valid, studies with statistically significant numbers of cases to prove these relationships remain to be completed.

32.10 Conclusions

From the results described above, it is clear that there is no single cause of AD, but that AD onset can be triggered by numerous parameters, all of which are associated with elevated OS.

It is shown that increased levels of OS are ultimately responsible for increased AD incidence, that all parameters which contribute to OS elevation increase the likelihood of AD onset and that total OS, measured via the OSI questionnaire can shed light on AD onset as well as determine which parameters are primary contributors to AD onset.

The results reported here suggest that the OSI can be used to similarly study other diseases.

References

Adlard, P.A., Bush, A.I., 2018. Metals and Alzheimer's disease: how far have we come in the clinic? J. Alzheim. Dis. 62, 1369–1379.

Agarwal, S., Ghosh, A., Chatterjee, S.N., 1987. Spontaneous release of malondialdehyde from ultra violet light exposed liposomal membranes. Z Naturforsch 42c 585–588.

Alford, S., Patel, D., Perakakis, N., Manzoros, C.S., 2018. Obesity as a risk factor for Alzheimer's disease: weighing the evidence. Obes. Rev. 19 (2), 269–280.

American Brain Foundation, 2020. What to Know About Chronic Traumatic Encephalopathy (CTE). https://www.ameicanbrainfoundation.org/what-to-know-about-cte/?gclid=EAIaIQobChMI9Yz861IDK8QIVbXFvBB3ohw1GEAAYAiAAEgJS2PD_BwE. (Accessed 4 July 2021).

Bannon, M., 2011. Co-morbidities and Alzheimer's disease: the need for vigilance. Q. J. Med. 104, 911–912.

Barbagallo, M., Dominguez, L.J., 2014. Type 2 diabetes and Alzheimer's disease. World J. Diabetes 5 (6), 889–893.

Barnes, D.E., Yaffe, K., 2011. The projected impact of risk factor reduction on Alzheimer's disease prevalence. Lancet Neurol. 10 (9), 819–828.

Bauer, K., Schwartzkopf, L., Graessel, E., Holle, R., 2014. A claims data-based comparison of comorbidity in individuals with and without dementia. BMC Geriatr. 14 (10). www.biomedcentral.com/1471-2318/14/10. (Accessed 12 January 2019).

Block, M.L., Caldersom-Garciduenas, L., 2009. Air pollution: mechanisms of neuroinflammation & CNS disease. Tends Neurosci. 32 (9), 506–516. www.biomedcentral.com/1741-7015/12/192. (Accessed 11 January 2019).

Breunig, J.J., Guillot-Sestier, M.V., Town, T., 2013. Brain injury, neuroinflammation and Alzheimer's disease. Aging Neurosci. 5. Article 5.

Bunn, F., Burn, A.M., Goodman, C., Rait, G., Norton, S., Robinson, L., et al., 2014. Comorbidity and dementia: a scoping review of the literature. BMC Med. 12, 192.

Carey, I.M., Anderson, H.R., Atkinson, R.W., Beevers, S.D., Cook, D.G., Strachan, D.P., et al., 2018. Are noise and air pollution related to the incidence of dementia? A cohort study in London, England. BMJ Open 8 (9), e022404. https://doi.org/10.1136/bmjopen-2018-022404. (Accessed 10 January 2019).

Cencioni, C., Spallotta, F., Martelli, F., Valente, S., Mai, A., Zeiher, A.M., Gaetano, C., 2013. Oxidative stress and epigenetic regulation in aging and age-related diseases. Int. J. Mol. Sci. 14, 17643–17663.

Charlson, M.E., Pompei, P., Ales, K.L., 1987. A new method of classifying prognostic comorbidity in longitudinal studies: development and validation. J. Chron. Dis. 40 (5), 373–383.

Christen, Y., 2000. Oxidative stress and Alzheimer's disease. Am. J. Clin. Nutr. 71 (Suppl. 1), 621S–629S.

D'Aiuto, F., NIbali, L., Parkar, M., Patel, S.J., Donos, N., 2010. Oxidative stress, systemic inflammation, and severe periodontitis. J. Dent. Res. 89 (11), 1241–1246.

Dato, S., Crocco, P., D'Aquila, P., de Rango, F., Bellizzi, D., Rose, G., Passarino, G., 2013. Exploring the role of genetic variability and lifestyle in oxidative stress response for healthy aging and longevity. Int. J. Mol. Sci. 14, 16443–16472.

Deng, Q., DAi, X., Feng, W., Huang, S., Yuan, Y., Xiao, Y., et al., 2019. Co-exposure to metals and polycyclic aromatic hydrocarbons, microRNA expression, and early health damage in coke oven workers. Environ. Int. 122, 369–380. https://doi.org/10.1016/j.envint.2018.11.056. (Accessed 8 January 2019).

Doraiswamy, P.M., Leon, J., Cummings, J.L., Martin, D., Neumann, P.J., 2002. Prevalence and impact of medical comorbidity in Alzheimer's disease. J. Gerontol. 57A (3), M173–M177.

Dunn, N.D.M., Mullee, M., Perry, V.H., Holmes, C., 2005. Association between dementia and infectious disease: evidence from a case-control study. Alzheimer Dis. Assoc. Disord. 19 (2), 91–94.

Durazzo, T.C., Mattsson, N., Weiner, M.W., 2014. Smoking and increased Alzheimer's disease risk: a review of potential mechanisms. Alzheimers Dement. (3 Suppl), S122–S145.

Durazzo, T.C., Mattsson, N., Weiner, M.W., Korecka, M., Trojanowski, J.Q., Shaw, L.M., 2014a. History of cigarette smoking in cognitively-normal elders is associated with elevated cerebrospinal fluid biomarkers of oxidative stress. Drug Alcohol Depend. 142, 262–268.

Eimer, W.A., Kumar, V., Navalpur Shanmugam, N.K., Mitchell, T., Washicosky, K.J., Gyorgy, B., et al., 2018. Alzheimer's disease-associated beta-amyloid is rapidly seeded by Herpesviridae to protect against brain infection. Neuron 100 (6), 1527–1532.

Erickson, M.A., Banks, W.A., 2013. Blood-brain barrier dysfunction as a cause and consequence of Alzheimer's disease. J. Cerebr. Blood Flow Metabol. 33, 1500–1513.

Fafara, A., Ciesielska, N., Damiza, A., Chatys, Z., Bentryn, D., Gajos, A., et al., 2014. Comorbidities disorders and Alzheimer's disease. J. Health Sci. 4 (6), 57–70.

Grova, N., Schroeder, H., Olivier, J.L., Turner, J.D., 2019. Epigenic and neurological implications with early life exposure to persistent organic pollutants. Int. J. Genomics. https://doi.org/10.1155/2019/2085496. (Accessed 5 July 2021).

Guillaumet-Adkins, A., Yanez, Y., Preis-Diaz, M.D., Calabria, I., Palancia-Ballester, C., Sandoval, J., 2017. Epigenetics and oxidative stress in aging. Ox. Med. Cell Longevity. https://doi.org/10.1155/2017/9175806. Article No. ID9175806.

Heathgrades, 2020. Smoking. https://www.healthgrades.com/right-care/quitting-smoking/smoking. (Accessed 4 July 2021).

Honjo, K., van Reekum, R., Verhoeff, N.P., 2009. Alzheimer's disease and infection: do infectious agents contribute to progression of Alzheimer's disease? Alzheimers Dement. 5 (4), 348–360.

Huang, T.T., Leu, D., Zou, Y., 2015. Oxidative stress and redox regulation on hippocampal-dependent cognitive functions. Arch. Biochem. Biophys. 576, 2–7.

Huang, W.J., Zhang, X., Chen, W.W., 2016. Role of oxidative stress in Alzheimer's disease (review). Biomed. Rep. 4, 519–522.

Jiang, T., Yu, J.T., Tian, Y., Tan, L., 2013. Epidemiology and etiology of Alzheimer's disease: from genetic to non-genetic factors. Curr. Alzheimer Res. 9, 852–867.

Justice, N.J., 2018. The relationship between stress and Alzheimer's disease. Neurobiol. Stress 8, 127–133.

Kamat, P.K., Kalani, A., Rai, S., Swarnkar, S., Tota, S., Nath, C., Tyagi, N., 2016. Mechanism of oxidative stress and synapse dysfunction in the pathogenesis of Alzheimer's disease: understanding the therapeutics strategies. Mol. Neurobiol. 53 (1), 648–661.

Karim, S., Mizra, K.M.A., Abuzenadah, A.M., Azher, E.I., Al-Qahtani, M.H., Sohrab, S.S., 2014. An association of virus infection with type 2 diabetes and Alzheimer's disease. CNS Neurol. Disord. Drug Targets 13 (3), 429–439.

Karki, R., Kodamullil, A.T., Hoffmann-Apitius, M., 2017. Comorbidity analysis between Alzheimer's disease and type 2 diabetes mellitus (T2DM) based on shared pathways and the role of T2DM drugs. J. Alzheim. Dis. 60, 721–731.

Kilian, J., Kitazawa, M., 2018. The emerging risk of exposure to air pollution on cognitive decline and Alzheimer's disease - evidence from epidemiological and animal studies. Biomed. J. 41 (3), 141–162.

Kirkpatrick, K., 2019. Sneaky signs of an unhealthy diet. Cleveland Clinic – Nutrition. https://clevelandclinic.org. (Accessed 4 July 2021).

Kuang, D., Zhang, W., Deng, Q., Zhang, X., Huang, K., Guan, L., et al., 2013. Dose-response relationships of polycyclic aromatic hydrocarbons exposure and oxidative damage to DNA and lipid in coke oven workers. Environ. Sci. Technol. 47 (13), 7446–7456.

Lee, H.J., Park, M.K., Seo, Y.R., 2018. Pathogenic mechanisms of heavy metal induced-Alzheimer's disease. Toxicol. Environ. Sci. 10 (1), 1–10.

Lehrer, S., Rheinstein, P.H., Rosenzweig, K.E., 2017. Association of radon background and total background ionizing radiation with Alzheimer's disease deaths in U.S. states. J. Alzheimers Dis. 59 (2), 737–741.

Lindberg, S., 2019. Psychological stress. Healthline. https://www.healthline.com/health/psychological-stress#signs. (Accessed 4 July 2021).

Litin, S.C., 2021. Radiation Sickness, fifth ed. Mayo Clinic Family Health Book.

Liu, Z., Li, T., Wei, N., Zhao, Z., Liang, H., Ji, X., et al., 2015. The ambiguous relationship of oxidative stress, tau hyperphosphorylation, and autophagy dysfunction in Alzheimer's disease. Oxid. Med. Cellular Longevity Article ID 352723. https://doi.org/10.1155/2015/352723. (Accessed 1 February 2019).

Lockhead, J.J., McCaffrey, G., Quigley, C.E., Finch, J., DeMarco, K.M., Nametz, N., Davis, T.P., 2010. Oxidative stress increases blood-brain barrier permeability and induces alterations in occludin during hypoxia-reoxygenation. J. Cerebr. Blood Flow Metabol. 30, 1625–1636.

Luchsinger, J.A., Tang, M.X., Shea, S., Mayeux, R., 2002. Caloric intake and the risk of Alzheimer disease. Ach. Neur. 59 (8), 1258–1263.

Ma, et al., 2019. Metabolically healthy obesity reduces the risk of Alzheimer's disease in elders. A longitudinal study. Aging 11 (23), 10939–10951.

Machado, A., Herrera, A.J., de Pablos, R.M., Espisona-Oliva, A.M., Sarmeinto, M., Ayala, A., et al., 2018. Chronic stress as risk factor for Alzheimer's disease. Rev. Neurosci. 25 (6), 785–804.

Mayo Clinic. 2022. Alzheimer's disease. www.mayoclinic.org/alzheimer'sdisease. Accessed August 25, 2022.

Mindd Foundation, 2021. What causes leaky brain? Repairing the blood brain barrier. https://mindd.org/leaky-brain. (Accessed 4 July 2021).

Misiak, B., Leszek, J., Kiejna, A., 2012. Metabolic syndrome, mild cognitive impairment and Alzheimer's disease—the emerging role of systemic low-grade inflammation and adiposity. Brain Res. Bull. 89 (3–4), 144–149.

Mons, U., Schottker, B., Muller, H., Kliegel, M., Brenner, H., 2013. History of lifetime smoking, smoking cessation and cognitive function in the elderly population. Eur. J. Epidemiol. 10, 823–831.

Moulton, P.V., Yang, W., 2012. Air pollution, oxidative stress, and Alzheimer's disease. J. Environ. Pub. Health. https://doi.org/10.1155/2012/47251. Article ID 472751. (Accessed 5 January 2019).

Naderali, E.K., Ratcliffe, S.H., Dale, M.C., 2009. Obesity and Alzheimer's disease: a link between body weight and cognitive function in old age. Am. J. Alzheimer's Dis. Other Dementias 24 (6), 445–449.

Najjar, S., Pearlman, D.M., Devinsy, O., Najjar, A., Zagzag, D., 2013. Neurovascular unit dysfunction with blood-brain barrier hyperpermeability contributes to major depressive disorder: a review of clinical and experimental evidence. J. Neuroinflammation 10, 142. http://www.jneuroinflammation.com/content/10/1/142. (Accessed 1 March 2019).

Nation, D.A., Sweeney, M.D., Montagne, A.M., Abhay, P., Sagare, A.P., D'Orazio, L.M., et al., 2019. Blood-brain barrier breakdown is an early biomarker of human cognitive dysfunction. Nature Magazine. https://www.nature.com/articles/s41591-018-0297-y. (Accessed 31 January 2019).

National Institute of Aging, 2015. United States National Institute of Health. https://www.nia.nih.gov/health/alzheimers-disease-genetics-fact-sheet. (Accessed 6 April 2019).

NORD, 2020. Heavy Metal Poisoning. https://rarediseases.org/rare-diseases/heavy-metal-poisoning/. (Accessed 4 July 2021).

Ott, A., Slooter, A.J., Hofman, A., van Harskamp, E., Witteman, J.C., Van Broeckhoven, C., et al., 1998. Smoking and risk of dementia and Alzheimer's disease in a population-based cohort study: the Rotterdam Study. Lancet 351 (9119), 1840–1843.

Pahwa, R., Goyal, A., Bansal, P., Jialal, I., 2020. Chronic inflammation. NIH govt books, 2020. https://www.ncbi.nlm.nih.gov/books/NBK493173/. (Accessed 30 June 2021).

Perry, G., Cash, A.D., Smith, M.A., 2002. Alzheimer's disease and oxidative stress. J. Biomed. Biotechnol. 2 (3), 12–123.

Poblador-Plou, B., Calderon-Larranage, A., Marta-Moreno, J., Hancco-Saavedra, J., Sicras-Mainar, A., Solijak, M., Prados-Torres, A., 2014. Comorbidity of dementia: a cross-sectional study of primary care older patients. BMC Psychiatr. 14 (84). www.biomedcentral.com/1471-244X/14/84. (Accessed 11 January 2019).

Profenno, L.A., Porsteinsson, A.P., Faraone, S.V., 2010. Meta-analysis of Alzheimer's disease risk with obesity, diabetes and related disorders. Biol. Psychol. 67 (6), 505–512.

Ramos-Cejudo, J., Wisniewski, T., Marmar, C., Zetterberg, H., Biennow, K., de Leon, M.J., Fossati, S., 2018. Traumatic brain injury and Alzheimer's disease: the cerebrovascular link. EBioMedicine 21–30. https://doi.org/10.1016/j.ebiom.2018.01.021. (Accessed 14 January 2019).

Redhead, B., Haure-Mirande, F.C.C., Richards, M.A., Shannon, P., Haroutunian, V., et al., 2018. Multiscale analysis of independent Alzheimer's cohorts find disruption of molecular, genetic, and clinical networks by human Herpesvisus. Neuron 99, 64–82.

Sabbagh, M.N., Malek-Ahmadi, R., Belden, C.M., Connor, D.J., Pearson, C., Jacobson, R., et al., 2010. Alzheimers Dis. 22 (3), 1015–1021.

Sochocka, M., Zwolinska, K., Leszek, J., 2017. The infectious etiology of Alzheimer's disease. Curr. Neuropharmacol. 15, 9967–1009.

Tang, F.R., 2018. Radiation and Alzheimer's disease. J. Alzheimer's Dis. Park. 8 (1). https://doi.org/10.4172/2161-0460.1000418. (Accessed 8 January 2019).

Tonnies, E., Trushina, E., 2017. Oxidative stress, synaptic dysfunction, and Alzheimer's disease. J. Alzheim. Dis. 57, 1105–1121.

Trempe, C.L., Lewis, T.J., 2018. Its never too early of too late-to end the epidemic of Alzheimer's by preventing or reversing causation from pre-birth to death. Front. Aging Neurosci. https://doi.org/10.3389/fragi.2018.00205. (Accessed 6 March 2019).

Turner, R.C., Lucke-Wald, B.P., Robson, M.J., Lee, M.L., Bailes, J.E., 2016. Alzheimer's disease and chronic traumatic encephalopathy. Baing Inj. 30 (11), 1279–1292.

University of Missouri Publications, 2002. Pesticide Poisoning Symptoms and First Aid. https://extension.missouri.edu/publications/g1915. (Accessed 4 July 2021).

Willcox, D.C., Scapagnini, G., Willcox, B.J., 2014. Healthy aging diets other than the Mediterranean: a focus on the Okinawan diet. Mech. Ageing Dev. 136–137, 148–162.

Yan, D., Zhang, Y., Liu, L., Yan, H., 2016. Pesticide exposure and risk of Alzheimer's disease: a systematic review and meta analysis. Sci. Rep. 6, 32222. https://doi.org/10.1038/srep32222. (Accessed 5 January 2019).

Yanjun, L., Yongming, L., Xiaotao, L., Zhang, S., Zhao, J., Zhu, X., 2017. Head injury as a risk factor for dementia and Alzheimer's disease: a systematic review and meta-analysis of 32 observational studies. PLoS One 12 (1), e0169650. https://doi.org/10.1371/journal.pone.0169650. (Accessed 13 January 2019).

Yuan, S.H., Wang, S.G., 2018. Alzheimer's dementia due to Suspected CTE from Subconcussive head impact. Case Rep. Neurol. Med. Article ID 7890269 https//doi.org/10.1155/2018/7890269. (Accessed 3 January 2019).

Zeliger, H.I., 2013. Exposure to lipophilic chemicals as a cause of neurological impairments, neurdevelopmental disorders and neurodegenerative diseases. Interdiscip. Toxicol. 6 (3), 103–110.

Zeliger, H.I., 2014. Co-morbidities of environmental diseases: a common cause. Interdisip. Toxicol. 7 (3), 101–106.

Zeliger, H.I., 2016. Predicting disease onset in clinically healthy people. Interdiscipl. Toxicol. 9 (2), 15–21.

Zeliger, H.I., 2017. Oxidative stress index predicts disease onset. J. Med. Res. Prac. 6 (3), 86–92.

Zeliger, H.I., 2019. Predicting Alzheimer's disease onset. Euro J. Med. Health Sci. 1 (1). https://doi.org/10.24018/ejned, 2019.1.1.16. (Accessed 8 July 2021).

Zeliger, H.I., Lipinski, B., 2015. Physiochemical basis of human degenerative disease. Interdiscipl. Toxicol. 8 (1), 39–54.

Zhang, F., Jiang, L., 2015. Neuroinflammation in Alzheimer's disease. Neuropsychiatric Dis. Treat. 11, 243–256.

CHAPTER

33

OSI public health surveys

33.1 Introduction

As discussed in Chapter 30, the Oxidative Stress Index (OSI) has been presented as a noninvasive, questionnaire-based diagnostic method for assigning the probability of disease onset to all people, those clinically ill as well as those without prevalent disease (Zeliger, 2017). Environmental exposures to toxic chemicals and other OS raising parameters elevate oxidative stress (OS) in a dose response relationship (Zeliger, 2016) and correspondingly produce higher OSI scores which are indicative of increased likelihood of disease onset (Zeliger, 2017).

This chapter addresses how the OSI can be modified to carry out public health surveys to determine increased risks for disease onset to individuals who are chronically exposed to environmental pollutants such as air pollutants emanating from chemical production and use, petroleum storage and transfer sites, power plant and other industrial stack emissions, leaking toxic landfills, mining operations, heavy metal recycling plants and other OS elevating factors such as electromagnetic radiation.

The rationale for using patient questionnaires is basd upon a long history of using these to illicit backgroud diagnostic information. Questionnaires, such as the Charlson Comorbidity Index (Charlson et al., 1987) and the Alzheimer's Questionnaire are routinely used to solicit patient background information on prevalent diseases, symptoms and the likelihood of further disease onset. Indeed, the Alzheimer (Sabbagh et al., 2010).

Using a modified OSI, members of communities impacted by chronic environmental spills and releases, as well as others regularly exposed to OS raising parameters, can be surveyed to determine how proximity to release points and durations of exposure can predict the likelihood of disease onset (Zeliger, 2019a,b). The OSI applications discussed here are based upon literature reviews of published studies on the causes of OS and OS-induced disease, methods of measuring OS, empirical and mechanistic associations between OS and disease onset (Zeliger, 2016 and the numerous references contained therein). The condensed form of the OSI questionnaire that is used to determine OS status contains questions that address all OS raising factors, including genetic factors, environmental exposures, prevalent diseases and lifestyle (Zeliger, 2020). Supplemental modifications to the OSI enable it to be used as

Oxidative Stress
https://doi.org/10.1016/B978-0-323-91890-9.00021-0

a public health survey instrument several specific applications. Three examples of such applications follow. These include:

1. Measuring OS impact as function of distance from a toxic chemical release point.
2. Assessing disease onset from direct chemical exposures as a function of time.
3. Assessing questionable toxic impact in situations where chronic sensual discomfort is prevalent.

The condensed form of the OSI suitable for such public health evaluations is shown in Table 33.1. Supplemental parts that are added for each are listed in Table 33.2.

33.2 Toxicity as a function of emission distance

It is well known that chronic exposures to chemical pollutants are a cause of numerous environmental diseases. For example, living proximate to heavily traveled highways results in a greater likelihood of respiratory disease onset and that this likelihood declines in a linear fashion as place of residence is distanced from such highways, where airborne concentrations of vehicle exhausts also decline linearly (Huynh et al., 2010). Accordingly, the closer one resides to a highway with heavy traffic, the greater the exposure to exhaust pollutants and the greater the oxidative stress impact on one's body. Thus, using the OSI, it becomes possible to numerically assign probabilities of disease onset related to such exposure as a function of distance from a disease-causing emission source. Adding the supplemental distancing questions to the OSI shown in Table 33.2 to those in the OSI questionnaire shown in Table 33.1 and surveying individuals at varying distances from a pollutant emission source up to distances where these emissions provide zero impact, i.e., where levels of the pollutant equal ambient background levels, enables one to assess the pollution (and hence the oxidative stress) impact of such emissions. It also enables the establishment of safe residential distances from the particular emission source (Zeliger, 2019a). The OSI questionnaire which has been modified to accomplish this is shown in Table 33.1.

The combined OSI total ($\sum$ OSI) from the OSI questionnaire plus the TPS score, relative to the OSI score alone, provides data that can be used to assess the impact of residence proximate to toxic emissions. Thus, the ratio of TPS divided by $\sum$ OSI shows the fraction of OS that is ascribable to residence proximate to a toxic release site and allows for the creation of a scale of values directly related to proximity to pollution releasing sites. Such data can also be used to establish safe residential distances from toxic emissions.

Often, environmental exposures to chemicals produce sensual discomfort, most notably odor. For example, those living proximate to venting petroleum storage tanks or chemical manufacturing plants often experience unpleasant odors associated with vapors venting from these tanks. Residents in such communities, at times, report increased incidences of cancers and other diseases following long-time residence despite air quality tests which have demonstrated low airborne concentrations of these vapors. This is consistent with the fact that odor thresholds are lower than air quality standards for most organic chemicals (Amoore and Hautala, 1983). Exposures to mixtures of chemicals, however, often produce adverse health effects at concentrations much lower than the effects caused by exposures to single

TABLE 33.1 Condensed OSI form.

1. Age:	Age score 0. 1, 2, 3, 4, 5, 6	________
	1 for each decade, starting with age 40	
2. Personal:	Female _____ male ______	
	Height _____ weight ______	
	BMI 0, 1, 2, or 3	________
3. Diseases:	All current diseases and diagnosed conditions	________
	Multiple number by 5	
4. Medications:	Number of medications regularly taken ______	________
5. Family history:	Number of your current diseases parents had/have	________
	Mother ________Father ________	
	Enter total number for both	
6. Education:	Check highest level achieved	________
	Some high school, enter 5	
	High school graduate, enter 4	
	Some college, enter 3	
	College graduate, enter 2	
	Graduate school, enter 1	
7. Residence:	Air quality in city and state of residence	________
	City ________ state________	
	Good, enter 1	
	Moderate, enter 2	
	Unhealthy for sensitive groups, enter 3	
	Unhealthy, enter 4	
	Very unhealthy, enter 5	
	Hazardous, enter 6	
8. Stress:	Check all that apply and enter total	________
	Do you often feel: Stressed _____	
	Anxious _____	
	Depressed _____	
	OSI TOTAL ________	

TABLE 33.2 OSI questionnaire supplement to assess oxidative stress as a function of residential distance and time of exposure to toxic chemical emission points.

Proximity to emission source	Score
0.1 mile or less	5
0.2–0.5 mile	4
0.6–1.0	3
1.1–2.0 miles	2
2.0–3.0 miles	1
Greater than 3.0 miles	0
____ proximity score	
Total proximity score (TPS)	______

chemicals (Zeliger, 2003, 2011). The $\sum$ OSI/TPS method just described is applicable for demonstrating if the health of those experiencing chronic chemical odors is impacted by the chemicals they are smelling. OSI as a function of distance from a release point can thus be used to demonstrate adverse health effects due to low levels of released toxic chemicals.

33.3 Disease impact as a function of exposure time

Disease onset from exposure to OS raising parameters, such as toxic chemicals, is well known to increase with time of exposure. The OSI can be used to predict the likelihood of disease onset as a function of time of exposure, by comparing OSI scores as functions of exposure times. Two examples of such of such exposure are polychlorinated biphenyl (PCB) exposure in schools and coal combustion residuals (CCA) exposure at CCA generating and use sites.

33.3.1 PCB exposure in schools

PCBs were used extensively in paints, caulks and fluorescent light ballasts in schools and other buildings constructed prior to 1977, when their production was discontinued (Herrick et al., 2004; Herrick et al., 2016). With time, deterioration and leakage has resulted in the presence of PCBs in quantities far exceeding safe limits (Thomas et al., 2012). As a result, it is estimated that as many as 26,000 schools in the United States could be contaminated with airborne PCBs and that approximately 14,000,000 students, teachers and staff are exposed to PCBs on a daily basis (DesRoches, 2016; Herrick et al., 2016).

PCBs are highly toxic, attack multiple systems in the body and cause numerous diseases. Health effects attributable to PCBs are listed in Table 33.3 (ATSDR, 2000).

TABLE 33.3 Health effects attributable to PCBs.

Respiratory
Cardiovascular
Gastrointestinal
Hematological
Musculoskeletal
Hepatic
Renal
Endocrine
Dermal
Ocular
Body weight
Immunological
Lymphoreticular
Neurological
Reproductive
Developmental
Genotoxic
Cancer

Exposures to PCBs in contaminated schools occur on a daily basis throughout the school year. In addition to chronic exposures to the same chemicals, PCBs accumulate in body fat and are retained in the body for as long as decades (ATSDR, 2000), thus providing for increased toxicity, oxidative stress, detrimental health and the OSI increase as a function of time. Accordingly, measuring OSI in statistically significant numbers of individuals over time can provide insight into onset times of the various diseases caused by PCBs. This can be accomplished by adding the supplement shown in Table 33.4 to the OSI questionnaire shown in Table 33.1.

Data collected via Table 33.4 for statistically significant numbers of individuals can provide insight into onset times for specific PCB-related diseases. The fraction of a person's total OS can be determined from the PCBSS divided by PCBSS + $\sum$ OSI. The greater the fraction, the greater the significance of PCB exposure to disease onset. Thus collected data can be used to establish critical ratio ranges for the danger of exposures to PCBs.

PCBs have been found in numerous buildings other than schools (Herrick et al., 2004). A list of these is shown in Table 33.5.

TABLE 33.4 OSI supplement for PCBs in school impact.

____ Number of years in school (score 1 point for each year)	
____ Number of diseases diagnosed with since starting at the school (score 1 point for each disease)	
____ Total PCB supplemental score − PCBSS (years + diseases scores)	
____ For each disease just listed, diagnosed, how many years into time at the school	
Did this occur	
Disease	Years of time at the school
____________________	____
____________________	____
____________________	____
____________________	____
____________________	____

TABLE 33.5 Buildings found to be contaminated with PCBs.

Government and business office
Elderly housing
University dormitory
Houses of worship
Hospitals
Museums
Hotels
Police stations

What has just been described for PCBs can similarly be applied to the dangers of exposures to other OS raising parameters, such as those found in "sick building syndrome," for example (Zeliger, 2011) to the onset of other diseases.

33.3.2 Fly ash toxicity

Coal combustion residuals (CCR), also known as coal ash, is produced as a by-product of coal combustion. CCR, which includes fly ash, bottom ash and flue gas desulfurization solids, is largely composed of small ($PM_{2.5}$ or less) particles that penetrate deep into the lungs when inhaled, are absorbed through skin pores and cell membranes and readily carried by blood and lymph to all body organs (Fisher et al., 1978; Mcelroy et al., 1982; Hagemeyer et al., 2019). The most common health effects associated with chronic exposure to CCR are listed in Table 33.6 (Gottlieb et al., 2010). In a work site investigated by this writer, more than half of those employed at a facility where CCR was generated and transported on-site for land fill were regularly exposed to high levels of CCR and diagnosed with one or more of the health effects listed in Table 33.6.

CCR is a complex mixture of crystalline silica, metals, polynuclear aromatic hydrocarbons and radionuclides (Hagemeyer et al., 2019).

A. Crystalline Silica

Crystalline silica is a major component of fly ash (Hicks and Yager, 2006). Its inhalation is well known to be a cause of lung cancer (IARC, 2018). Silica is also a cause of noncarcinogenic respiratory diseases including COPD and emphysema, as well as noncancerous lung and lymphatic tumors (Bagchi, 1992).

B. Metals

Depending upon specific location, CCR contains multiple different mixtures of the following metals. These, their target organs and diseases identified with them are listed in Table 33.7 (CDC, 2007).

C. Radionuclides

Coal contains radioactive isotopes of uranium, thorium and radon, as well as daughter radioisotopes of radium, bismuth, lead and potassium which are produced when uranium and thorium decay (USGS, 1997). Power generation combustion residues contain about 8.5%–13% of the original weight of the coal and almost all of the radioactivity except that from radon, which is gassed off. As a result, fly ash emits about 3–10 times as much radiation as coal does and thus poses a considerably greater danger to those exposed to it than coal does (Hvistendahl, 2007).

TABLE 33.6 Health effects commonly associated with chronic exposure to CCR.

Cancer: In multiple organs, nodules on throat, larynx and thyroid
Respiratory system: Asthma, emphysema, COPD
Skin: Rashes, scarring, discoloration, itching, psoriasis
Cardiovascular system: Heart failure, myocardial infarction,
Tachycardia, Raynaud's disease
Gastrointestinal system: Stomach, colon products
Immune system: Psoriatic arthritis
Neurological system: Headaches, memory loss, Parkinson's disease

TABLE 33.7 Metals contained in CCR, their target organs and diseases identified with them.

Antimony	Target organs are the eyes, skin, respiratory system and cardiovascular system. Causes irritation, cough, nausea, vomiting and diarrhea.
Arsenic	Target organs are the liver, kidneys, lungs, lymphatic system and prostate gland. Cancer sites include the lungs, lymphatic system, prostate and skin. Increased risks for developing lung cancer from arsenic exposures ensue whether arsenic is inhaled or ingested.
Beryllium	Target organs are the lungs and urinary tract. Cancer sites include the lungs, and urinary tract. Causes COPD.
Cadmium	Target organs are the respiratory system prostate and blood. Cancer sites include the lungs and prostate.
Chromium	Target organs, particularly for the +6 oxidation state (found at gavin), include the eyes, skin and respiratory system. Effects include lung fibrosis. Cancer site from inhalation is the lungs.
Cobalt	Target organs are the skin and respiratory system. Causes cough, dyspnea, wheezing, asthma and dermatitis.
Copper	Target organs are the eyes, skin, respiratory system, liver and Kidneys. Causes irritation of the eyes and nose and damage to the Liver and kidneys.
Manganese	Target organs are the respiratory system, central nervous system, blood and kidneys. Causes mental confusion, lower-back pain, pneumonitis and kidney damage.
Mercury	Target organs are the eyes, skin, respiratory system, central nervous system and kidneys. Causes eye and skin irritation, dyspnea, headache and memory difficulties.
Nickel	Target organs are the nasal cavities, lungs and skin. Causes lung cancer, sensitization dermatitis, allergic asthma and pneumonitis.
Selenium	Target organs are the eyes, skin, respiratory system, liver, kidneys, blood, spleen. Causes eye, skin, nose and throat irritation, dyspnea, bronchitis, dermatitis, eye and skin burns.
Thallium	Target organs are the eyes, respiratory system, centra; nervous system, liver, kidneys, gastrointestinal tract and hair. Causes nausea, diarrhea, vomiting, tremors, pulmonary edema, liver and kidney damage, convulsions and psychosis.
Tin	Target organs are the eyes, skin and respiratory system. Causes irritation of these organs.
Vanadium	Target organs are the eyes, skin and respiratory system. Causes irritation of eyes, skin and throat, cough, rales, wheezing bronchitis and dyspnea.
Zinc	Target organs are those in the respiratory system. Causes dry throat, cough, chest tightness, dyspnea, rales and reduced pulmonary function.

D. Polynuclear Aromatic Hydrocarbons (PAHs)

Numerous PAHs are found in CCR, including known or suspected carcinogens that include anthracene, benzo[a]anthracene, benzo[a]pyrene, chrysene, phenanthrene and pyrene (Srivastava et al., 1985; ATSDR, 2011).

As is evident from the above, each of the four different components in CCR are highly toxic to humans. As discussed previously in this book, there are several mechanisms by which such toxicity occurs, all of which result in elevated oxidative stress levels, leading to multiple diseases and numerous cancers

1. Insoluble particles, such as silica, act as irritants when attached to lung and other tissues, causing inflammation (Ryu et al., 2014; OSHA, 2020).
2. Insoluble metal containing particles react with body liquids to release soluble metal compounds (Jan et al., 2015).
3. Other toxins adsorb on particles and are carried by the particles into the blood stream and ultimately to sites where they are toxic (Luanpitpong et al., 2014).
4. The probability of mixture effects due to the presence of the lipophilic PAHs which facilitate the absorption of the hydrophilic metals (Zeliger and Lipinski, 2015).

CCR exposure can be used, via the following protocols, as a model to demonstrate how the OSI can be applied to predict disease onset in likelihood in employees regularly exposed to oxidation stress-raising toxic chemicals.

1. The onset of the diseases and symptoms listed in Table 33.6 in employees thus exposed can be compared to the prevalence of these in the general population of similar ages, data for which exists in state health department statistics.
2. Oxidative stress levels in the exposed individuals can be compared to those of unexposed cohorts using the condensed form of the OSI and comparisons of OSI scores made.
3. A supplement to the condensed OSI can be used to predict the time to onset for all of the diseases and symptoms in Table 33.6. The condensed OSI shown in Table 33.1, used with the supplement shown in Table 33.8, can be administered at time of first.

TABLE 33.8 OSI condensed for supplement for predicting time of onset for diseases associated with CCR exposure.

____ Age when first employed ____ Years of employment

List all diseases and medical conditions that have been diagnosed since first employed at CCR facility and years of employment when each was diagnosed.

Disease	Years at the CCR plant when diagnosed
____________________	____
____________________	____
____________________	____
____________________	____
____________________	____
____________________	____

employment and annually to employees exposed to CCA to determine if their oxidative stress levels have been elevated. OSI can then be monitored as function of exposure time to CCR.

It should be noted that many of the diseases that are caused by exposure to CCR are also attributable to other exposures. In the example of CCR exposed employees cited above, a preponderance of those with respiratory disease also smoked tobacco, also a well-defined cause of emphysema, COPD, asthma and lung cancer. Exposure to CCR and smoking, two OS elevating parameters, are additive, and would be expected to produce enhanced and earlier onset of these diseases. This effect was indeed observed in this group, compared to nonCCR exposed cohorts and demonstrated the dose response relationship between total exposure and elevated oxidative stress.

33.4 Summary

As all factors known to raise OS have been incorporated into the modified OSI questionnaire shown in Table 33.1, it now becomes possible to determine a person's total OS status that incorporates chronic exposure to a pollutant source as a function of distance from release and/or time of exposure. The OSI is noninvasive, yet predicts OS levels and offers insights into which parameters are the most contributory to disease onset.

The OSI can be used to measure disease onset likelihood when used as a public health survey instrument to survey effects as a function of distance from an emission source, as well as the effects of exposure to constant levels of OS raising parameters.

Strengths of these applications include the noninvasive nature of the OSI, the ability to factor out other items via cohort testing, the ability to determine what the impacts are on individuals and what deleterious health effects a community chronically exposed to OS raising parameters can anticipate and take steps to prevent.

Limitations include the fact that respondents are asked to estimate distances from emission sources to their residences and/or to report disease onset times. With a small number of individual respondents, these can lead to errors. With statistically significant numbers of respondents, however, such limitations are minimized and reliable survey results regarding the health impacts of toxic environmental contaminant exposure can be determined.

References

Amoore, J.E., Hautala, E., 1983. Odor as an aid to chemical safety: odor thresholds compared with threshold limit values or volatilities for 214 industrial chemicals in air and water dilution. J. Appl. Toxicol. 3 (6), 272–290.

ATSDR, Agency for Toxic Substances and Disease Registry (ATSDR). 2000, 2000. Toxicological Profile for Polychlorinated Biphenyls (PCBs). Atlanta, GA: U.S. Department of Health and Human Services. Public Health Service. (Accessed 13 July 2021).

ATSDR, 2011. Polycyclic Aromatic Hydrocarbons (PAHs). What Health Effects Are Associated with PAH Exposure? (Accessed 14 July 2021).

Bagchi, N., 1992. What makes silica toxic? Br. J. Ind. Med. 49 (3), 163–166.

CDC, 2007. NIOSH Pocket Guide to Chemical Hazards. Dept of Health and Human Services. Centers for Disease Control and Prevention. DHHS (NIOSH) Publication No. 2005-149. (Accessed 14 July 2021).

Charlson, M.E., Pompei, P., Ales, K.L., 1987. A new method of classifying prognostic comorbidity in longitudinal studies: development and validation. J. Chron. Dis. 40 (5), 373–383.

DesRoches, D., 2016. 14 Million US Kids Could Be Exposed to Toxic PCBs at School. WNPR. https://revealnews.org/blog/14-million-kids-could-be-exposed-to-toxic-pcbs-at-school. (Accessed 13 July 2021).
Fisher, G.L., Prentice, B.A., Silberman, D., Ondov, J.M., Bierman, A.H., Ragaini, R.C., McFarland, A.R., 1978. Physical and morphological studies of size-classified coal fly ash. Environ. Sci. Technol. 12 (4), 447–451.
Gottlieb, B., Gilbert, S.G., Evans, L.G., 2010. Physicians for Social Responsibility. Coal Ash, the Toxic Threat to Our Health and Environment. www.psr.org. (Accessed 14 July 2021).
Hagemeyer, A.N., Sears, C.G., Zierold, K.M., 2019. Respiratory health in adults residing near a coal-burning power plant with coal ash storage facilities. A cross-sectional epidemiological study. Int. J. Environ. Res. Publ. Health 16, 3642, 2019.
Herrick, R.F., McClean, M.D., Meeker, J.D., Baxter, L.K., Waymouth, G.A., 2004. An unrecognized source of PCB contamination in schools and other buildings. Environ. Health Perspect. 112 (10), 1051–1053.
Herrick, R.F., Stewart, J.H., Allen, J.G., 2016. Review of PCBs in US schools: a brief history, estimate of the number of impacted schools, and an approach for evaluating indoor air samples. Environ. Sci. Pollut. Res. Int. 23 (3), 1975–1985.
Hicks, J., Yager, J., 2006. Airborne crystalline silica concentrations at coal-fired power plants associated with coal fly ash. J. Occup. Environ. Hyg. 3 (8), 448–455.
Huynh, P., Salam, M.Y., Morphew, T., Kwong, K.Y.C., Scott, L., 2010. Residential proximity to freeways is associated with uncontrolled asthma in inner-city Hispanic children and adolescents. J. Allergy 2010, 7. https://doi.org/10.1155/2010/157249. (Accessed 6 June 2019). Article ID 157249.
Hvistendahl, M., 2007. Coal ash is more radioactive than nuclear waste. Sci. Am.. http://www.scientificamerican.com/articlecfm?id=coal-ash-is-more-radioactive-than-nuclear-waste. Accessed August 26, 2022.
IARC, 2018. Silica Dust, Crystalline, in the Form of Quartz or Cristobalite, pp. 355–405.
Jan, A.T., Azam, M., Siddiqui, K., Ali, A., Choi, I., Haq, Q.M.R., 2015. Heavy metals and human health: mechanistic insight into toxicity and counter defense system of antioxidants. Int. J. Mol. Sci. 16 (12), 29592–29630.
Luanpitpong, S., Wang, L., Rojanasakul, Y., 2014. The effects of carbon nanotubes on lung and cellular behaviors. Nanomedicine (Lond) 9 (6), 895–912.
Mcelroy, M.W., Carr, R.C., Ensor, D., Markowski, G.R., 1982. Size distribution of fine particles from coal combustion. Science 215 (4528), 13–19.
OSHA, 2020. Control Silica Dust. Health Effects. https://www.osha.gov/silica-crystalline/health-effects. (Accessed 21 July 2021).
Ryu, H.J., Seong, N.W., So, B.J., Kim, J.H., Hong, J.S., Park, M.K., et al., 2014. Int. J. Nanomed. 9 (Suppl. 2), 127–136.
Sabbagh, M.N., Malek-Ahmadi, R., Belden, C.M., Connor, D.J., Pearson, C., Jacobson, R., et al., 2010. Alzheimers Dis. 22 (3), 1015–1021.
Srivastava, V.K., Srivastava, P.K., Misra, U.K., 1985. Polycyclic aromatic hydrocarbons of coal fly ash: analysis by liquid-gas chromatography using nematic liquid crystals. J. Toxicol. Environ. Health 15 (2), 333–337.
Thomas, K., Xue, J., Williams, R., Jones, P., Whitaker, D., 2012. Polychlorinated Biphenyls (PCBs) in School Buildings: Sources, Environmental Levels and Exposures. United States Environmental Protection Agency. Office of Research and Development. National exposure research laboratory. EPA/600R-12/051. September 30, 2012. (Accessed 13 July 2021).
USGS, 1997. Fact Sheet FS-163-97. Radioactive Elements in Coal and Fly Ash: Abundance, Forms, and Environmental Significance. https://pubs.usgs.gov/fs/1997/fs163-97/FS-163-97html. (Accessed 14 July 2021).
Zeliger, H.I., 2003. Toxic effects of chemical mixtures. Arch. Environ. Health 58 (1), 23–29.
Zeliger, H.I., 2011. Human Toxicology of Chemical Mixtures, second ed. Elsevier, London.
Zeliger, H.I., 2016. Predicting disease onset in clinically healthy people. Interdiscip. Toxicol. 9 (2), 15–21.
Zeliger, H.I., 2017. Oxidative stress index predicts disease onset. J. Med. Res. Prac. 6 (3), 86–92.
Zeliger, H.I., 2019a. Oxidative stress index: disease prediction and prevention. EC Pjarm Toxicol. 7 (9), 1022–1036.
Zeliger, H.I., 2019b. Oxidative stress index as a public health survey instrument. Euro. J. Med. Health Sci. 1 (2). https://doi.org/10.2418/ejmed.2019.1.2.28. (Accessed 8 July 2021).
Zeliger, H.I., 2020. Oxidative stress index (OSI) condensed questionnaire. Eur. J. Med. Health Sci. https://doi.org/10.24018/ejmed.2020.2.1.163. (Accessed 13 July 2021).
Zeliger, H.I., Lipinski, B., 2015. Physiochemical basis of human degenerative disease. Interdiscip. Toxicol. 8 (1), 39–54.

CHAPTER

34

Predicting COVID-19 severity

34.1 Introduction

As discussed in the preceding chapters, chronic OS is associated with the onset of numerous non-communicable diseases (Zeliger, 2016). OS is also associated with the onset and severity of infectious viral, bacterial, fungal and parasite-carried diseases via immune system suppression (Hughes, 1999; Akaike, 2001; Splettstoesser and Schuff-Warner, 2002; Xu et al., 2015). People with non-communicable diseases subsequently have a high incidence of other non-communicable diseases, as well as more frequent and severe bouts with infectious diseases (Andresen et al., 2006; Zeliger et al., 2012; Zeliger, 2014). As examples, viral infections have been shown to be co-morbid with Alzheimer's disease (Honjo et al., 2009) and co-morbidity has been established between tuberculosis or malaria and type 2 diabetes (Marais et al., 2013).

The severity of viral infections has been found to increase with the prevalence of pre-existing conditions, all of which are associated with elevated oxidative stress (Schwartz, 1996; Peterhans, 1997; Beck et al., 2000). COVID-19 is no exception to this finding (Liang et al., 2020; Del Valle et al., 2020; Delgado-Roche and Mesta, 2020; Derouchie, 2020; Kalem et al., 2021; Aukac et al., 2021) and accordingly, the Oxidative Stress Index has been proposed and demonstrated to be predictive of the of severity of COVID-19 in those infected with it (Zeliger and Kahaner, 2020; 2020a).

34.2 Oxidative stress and COVID-19 severity

Elevated oxidative stress has been shown to be directly related to the likelihood of falling ill with non-communicable disease, as well as the onset, frequency and severity of infectious viral, bacterial, fungal and parasite-carried diseases. These enhanced effects have been shown to prevail in an OS induced dose response relationship no matter what the cause(s) of OS elevation. OS-raising causes can include age, gender, environmental exposures toxic chemicals and radiation, diet, smoking, alcohol consumption and other lifestyle choices, pre-existing disease, disease symptoms, medications regularly taken and chronic psychological stress as well as combinations of these (Zeliger, 2016).

Oxidative Stress
https://doi.org/10.1016/B978-0-323-91890-9.00013-1

Primate modeling demonstrates the probability of a dose-dependent relationship between inhalation of SARS-CoV-2 and COVID-19 onset (Dabisch et al., 2021). This finding is consistent with similar relationships in the onsets of other oxidative stress mediated diseases.

Oxidative stress-elevating factors shown to increase COVID-19 severity are listed in Table 34.1 (Foldi et al., 2020; Ssentongo et al., 2020; Yang et al., 2020; Fang et al., 2020; Zheng et al., 2020; Ciarello et al., 2020; Ellington et al., 2020; Bourdrel et al., 2021; Emami et al., 2020; Zhang et al., 2020; Zhou et al., 2020; Marin et al., 2021; Gazzaz, 2021; Brosseau et al., 2021). These factors fall into three categories that include physical status, environmental exposures and pre-existing disease.

TABLE 34.1 Oxidative stress-elevating factors shown to increase COVID-19 severity.

Physical status
Age
Male gender
Pregnancy
Environmental exposures
Smoking
Air pollution
Pre-existing disease
Diabetes
Hypertension
Cancer
Cardiovascular disease
Asthma
Chronic obstructive pulmonary disease
Chronic liver disease
Hepatitis B
HIV
Autoimmune disease
Thyroid disease
Autoimmune disease
Obesity
Abnormal lipid metabolism
Digestive disease
Endocrine disease

As discussed in previous chapters of this book, elevated oxidative stress has been shown to be an indicator of the disease onset likelihood. The relevance of the OS levels in predicting COVID-19 severity is based on the following:

1. All parameters listed in Table 34.1 are known to increase oxidative stress.
2. COVID-19 disease has been shown to be of greater severity and resulting in increased numbers of deaths in people who are impacted by each of the OS raising parameters in Table 34.1 (Emami et al., 2020; Zhang et al., 2020; Zhou et al., 2020; Marin et al., 2021).
2. Men have higher OS levels than women and have been shown to be more susceptible to COVID-19 than women (Brunelli et al., 2014; Kander et al., 2017; Brosseau et al., 2021).
3. Oxidative stress is elevated during pregnancy and pregnancy has been identified as a risk factor for severe COVID-19 (Chiarello et al., 2020; Brosseau et al., 2021).
4. As discussed in Chapter 4, air pollution (AP) is a major contributor to oxidative stress. AP is believed to increase COVID-19 severity by decreasing immune system response, thus facilitating both viral penetration and viral replication (Bourdrel et al., 2021).
5. Viral diseases are more prevalent and more severe in those who smoke (Gualano et al., 2008; Zeliger, 2016, 2019). Smoking tobacco elevates OS and smoking increases COVID-19 severity (Zeliger, 2016, 2019; Fang et al., 2020).
6. Obesity, which raises OS, is a risk factor for increased viral disease and has been shown to increase COVID-19 severity (Honce and Schultz-Cherry, 2019; Derouiche, 2020; Delgado-Roche and Mesta, 2020; Martin et al., 2021; Aukac et al., 2021).
7. Lower socioeconomic status, shown to be associated with oxidative stress (Zeliger, 2020), is also associated with increased incidence of COVID-19 (Mena et al., 2021).
8. Lifestyles that reduce OS, such as following the Mediterranean diet, have been proposed to protect against COVID-19 severity (Hopkins, 2020).
9. In the United States, young residents of the deep south and mid-south states have been found to be at much higher risk of death from COVID-19 than their similarly aged co-horts in other states. This effect has been attributed to a higher prevalence of heart disease, lung disease, diabetes, obesity and tobacco smoking (Keiser Family Foundation, 2021).

34.3 Oxidative stress index and COVID-19 severity

The Oxidative Stress Index (OSI), a non-invasive questionnaire based numerical value, has been shown to be applicable in predicting the severity of COVID-19 after its onset or if, indeed, disease severity is a function of total oxidative stress, as it the case with other viral infections (Zeliger and Kahaner, 2020, 2020a).

The condensed form of the OSI (Zeliger, 2020), shown in Table 34.2, readily lends itself to rapid production of a numerical OSI value. This can be accomplished via a paper questionnaire or via an electronic app that can be accessed from computers or smart phones to instantly produce numerical OSI values which can be applied on both micro and macro scales.

TABLE 34.2 Oxidative Stress Index condensed form.

1. Age:	Age score 0. 1, 2, 3, 4, 5, 6	________	
	1 for each decade, starting with age 40		
2. Personal:	Female _____	Male _____	
	Height _____	Weight _____	
	BMI 0, 1, 2, or 3		________
3. Diseases:	All current diseases and diagnosed conditions		________
	Multiple number by 5		
4. Medications:	Number of medications regularly taken ________		________
5. Family history:	Number of your current diseases parents had/have		________
	Mother ________	Father ________	
	Enter total number for both		
6. Education:	Check highest level achieved		________
	Some high school, enter 5 High school graduate, enter 4 Some college, enter 3 College graduate, enter 2 Graduate school, enter 1		
7. Residence:	Air quality in city and state of residence		________
	City ________	State ________	
	Good, enter		1
	Moderate, enter		2
	Unhealthy for sensitive groups, enter		3
	Unhealthy, enter		4
	Very unhealthy, enter		5
	Hazardous, enter		6
8. Stress:	Check all that apply and enter total		________
	Do you often feel: Stressed		_____
		Anxious _____	
		Depressed _____	
			OSI TOTAL ________

On a micro scale, using the condensed form of the OSI, individuals can be pre-screened to determine if their pre-existing conditions suggest increased likelihood of COVID-19 onset and disease severity.

On a macro scale, results from OSI screening can potentially be used to predict the number of COVID-19 cases that can be anticipated in a given locale and thus help health care professionals predict the amount of care that will be required and allocate resources appropriately.

COVID-19 severity has been classified as asymptomatic, mild, moderate.

34.4 Validation of OSI use in predicting COVID-19 severity

Evidence for using the OSI to predict severity of COVID-19 has been produced from the Clinical Risk Score Calculation (CRSC) data collected by Liang et al. (2020). The CRSC is based upon clinical data ascertained at the time of hospital admission for patients in 575 hospitals in China following the COVID-19 outbreak.

Though Liang et al. do not specifically address oxidative stress (OS) or OSI, all the parameters shown by them to increase COVID-19 severity are known to elevate OS and raise OSI scores. The clinically supported results reported in the Liang paper precisely dovetail and support the use of the OSI to predict COVID-19 severity.

34.4.1 Clinical risk score calculation (CRSC)

The recently developed clinical risk score calculation (CRSC) (Liang, 2020) shows that one's pre-existing conditions at the time of presentation with COVID-19 are indicative of ultimate disease severity in a mathematically related matter (Liang et al., 2020). The CRSC is based upon age, lifestyle, pre-existing conditions and clinical data ascertained at the time of hospital admission for patients in 575 hospitals in China following the COVID-19 outbreak and shows that such data can be used to predict the severity of COVID-19. Variables included in the CRSC include demographic, co-morbidities, symptoms and laboratory data, all of which are known to elevate oxidative stress and are included in the OSI. These are shown in Table 34.3.

TABLE 34.3 CRSC variables known to elevate OS.

Demographic
Age
Smoking status
Co-morbidities
COPD
Diabetes
Hypertension

(*Continued*)

Cardiovascular disease
Cerebrovascular disease
Hepatitis B
Cancer
Chronic kidney disease
Immunodeficiency
Symptoms
Fever
Congestion
Headache
Dry cough
Productive cough
Sore throat
Fatigue
Dyspnea
Myalgia/arthralgia
Chills
Skin rash

Laboratory testing was carried out in support of CSRS clinical evaluations. The data collected was adapted into a calculation using the following 10 parameters to predict probability of severe disease onset.

X-ray abnormality	**Yes or No**
Age	Years
Hemoptysis	Yes or No
Dyspnea	Yes or No
Co-morbidities[a]	Number
Unconsciousness	Yes or No
Cancer	Yes or No
Neutrophil/lymphocytes	0–80
Lactate dehydrogenate	0–1500 U/L
Direct bilirubin	0–24 μmol/L

[a]Co-morbidities included in the CRSC calculation included COPD, hypertension, diabetes, heart disease, chronic kidney disease, Hepatitis B and immunodeficiency.

34.4.2 Comparison of the OSI and the CRSC

Both the OSI and CRSC demonstrate that a patient's health and lifestyle data can be used to predict COVID-19 severity and that predictions regarding severity can be mathematically obtained. Though the CRSC does not relate relevant disease severity factors to oxidative stress, all the factors that enter into the CRSC calculation (age, detrimental lifestyle choices pre-existing diseases and symptoms associated with these) and clinical data have elevated oxidative stress associated with them. The CRSC and the OSI both demonstrate that higher numerical scores are mathematically related to disease severity.

In lieu of clinical evaluations, the OSI uses eight parameters, all of which elevate OS and have been shown to be incorporated into the clinical parameters relied upon by the CRSC (Zeliger, 2017). In addition, the OSI includes environmental, lifestyle and socioeconomic parameters not included in the CRSC.

The OSI parameters include:

Age
Gender, height and weight
Pre-existing disease(s)
Medications regularly taken
Genetics
Socioeconomic status as indicated by education level
Residence as an indicator of pollution exposure
Chronic psychological stress.

By incorporating all the above, the OSI not only predicts the probability of COVID-19 severity in coronavirus-exposed individuals, but also illustrates why some people who are exposed to the virus develop more severe symptoms than their cohorts.

The following scale of OSI values is proposed as indicative of the severity of COVID-19 in exposed individuals.

0–15	Asymptomatic to mild
16–30	Mild to moderate
31–45	Moderate to severe
46 or greater	Severe to critical

34.5 Covid-19 as a cause of oxidative stress

COVID-19 is both a cause and a consequence of oxidative stress. It releases a pro-inflammatory cytokine storm giving rise to a massive increase in oxidative stress that both exacerbates the disease and gives rise to long-term, post-acute effects in the organs of multiple body systems (Alwazeer et al., 2021; Farshidfar et al., 2021; Graciano-Machuca et al., 2021;

TABLE 34.4 Body systems impacted long-term following acute COVID-19.

Cardiovascular
Pulmonary
Gastrointestinal
Endocrine
Neurological
Hematologic
Dermatologic
Urinary
Multisystem inflammatory syndrome in children (MIS-C)

Nalbandian et al., 2021; Xie et al., 2022; Shahbaz et al., 2022). The list of systems impacted long-term following acute COVID-10 is shown in Table 34.4.

As this book goes to press (2022) the full impact of long-term COVID-19 remains to be determined. As discussed in Part II of this book, diseases in all body systems are oxidative stress-mediated. Hence, as a massive OS producer, the list of multiple long-term effects of COVID-19 is anticipated to grow.

34.6 Summary

Elevated oxidative stress is associated with severity of COVID-19. The Oxidative Stress Index, a non-invasive, self-administered questionnaire, provides a basis for measuring oxidative stress and predicting an ill individual's propensity for disease severity. The OSI can also be used to predict the likelihood of COVID-19 onset prior to the manifestation of disease symptoms.

References

Akaike, T., 2001. Role of free radicals in viral pathogenesis and mutation. Rev. Med. Virol. 11 (2), 87–101.

Alwazeer, D., Liu, F.F.C., Wu, X.Y., LeBaron, T.W., 2021. Combating oxidative stress and inflammation in COVID-19 by molecular hydrogen therapy: mechanisms and perspectives. Oxid. Med. Cell. Longev. https://doi.org/10.1155/2021/5513868. Accessed March 6, 2022.

Andresen, H.M., Regueira, H.T., Leighton, F., 2006. Oxidative stress in critically ill patients. Rev. Med. Chile 134 (5), 649–656.

Aukak, K., Ozsurekci, Y., Yayla, B.C.C., Gyrlevik, S.L., Oygar, P.D., Bolu, N.B., et al., 2021. Oxidant and antioxidant balance in patients with COVID-19. Pediatr. Pulmonol. https://doi.org/10.1002/ppul.25549. Accessed July 22, 2021.

Beck, M.A., Handy, J., Levander, O.A., 2000. The role of oxidative stress in viral infections. Ann. NY Acad. Sci. 917, 906–912.

Bourdrel, T., Annesi-Maesano, I., Alahmad, B., Maesano, C.N., Bind, M.A., 2021. The impact of outdoor air pollution on COVID-19: a review of evidence from in vitro, animal, and human studies. Eur. Respir. Rev. https://doi.org/10.1183/16000617.0242-2020. Accessed March 5, 2022.

Brosseau, L.M., Escandon, K., Ulrich, A.K., Rasmussen, A.L., Roy, C.J., Bix, G.J., et al., 2021. Severe acute respiratory syndrome coronavirus 2 (SARS-Co V-2) dose, infection and disease outcomes for coronavirus disease 2019 (COVID-19): a review. Clin. Infect. Dis. ciab903. https://doi.org/10.1093/cid/ciab903. Accessed March 4, 2022.

Brunelli, E., Domanico, F., La Russa, D., Pellegrino, D., 2014. Sex differences in oxidative stress biomarkers. Curr. Drug Targets 15 (8), 811–815.

Ciarello, D.I., Abad, C., Rojas, D., Toledo, F., Vazquez, C.M., Mate, A., et al., 2020. Oxidative stress: normal versus preeclampsia. BBA – molecular basis of disease 165354. https://doi.org/10.1016/j.bbadis.2018.12.005. Accessed March 4, 2022.

Dabisch, P.A., Biryukov, J., Beck, K., Boydston, J.A., Sanjak, J.S., Herzog, A., et al., 2021. Seroconversion and fever are dose-dependent in a nonhuman primate model of inhalational COVID-19. PLoS Pathog. 17 (8). https://doi.org/10.1371/journal.ppat.1009865. Accessed March 6, 2022.

Delgado-Roche, L., Mesta, F., 2020. Oxidative stress as a key player in severe acute respiratory syndrome coronavirus (SARS-CoV) infection. Arch. Med. Res. 51, 384–387.

Del Valle, D., Kim-Schulze, S., Huang, H.H., Beckmann, N.D., Nienberg, S., Wang, B., et al., 2020. An inflammatory cytokine signature predicts COVID-19 severity and survival. Nat. Med. 26, 1636–1643.

Derouiche, S., 2020. Oxidative stress associated with SARS-Cov-2 (COVID-19) increases the severity of the lung disease – a systematic review. J. Infect. Dis. Epidimiol. 6 (3). https://doi.org/10.23937/2474-3658/151021. Accessed July 23, 2021.

Ellington, S., Strid, P., Tong, V.T., Woodworth, K., Galang, R.R., Zambrano, L.D., et al., 2020. Characteristics of Women of Reproductive Age with Laboratory-Confirmed SARS-CoV-2 Infection by Pregnancy Status – United States, January 22–June 7, 2020. Morbid Mortal Weekly Report, June 26, 2020. MMWR Weekly/Vol. 69/No. 25.

Emami, A., Javanmardi, F., Pirbonyeh, N., Akbari, A., 2020. Arch. Acad. Emerg. Med. 8 (1), e35. Accessed March 31, 2020.

Fang, X., Li, S., Yu, H., Wang, P., Zhang, Y., Chen, Z., et al., 2020. Epidemiological, comorbidity factors with severity and prognosis of COVID-19: a systematic review and meta-analysis. Aging (Albany NY) 12 (13), 12493–12503.

Farshidfar, F., Koleini, N., Ardehali, H., 2021. Cardiovascular complications of COVID-19. JCI Insight 6 (13), e14980. htpps://doi.org/10.1172/jciinsight.14980. Accessed March 6, 2022.

Foldi, M., Farkas, N., Kiss, S., Zadori, N., Vancsa, S., Szako, L., et al., 2020. Obesity is a risk factor for developing a critical condition in COVID-19 patients: a systematic review and meta-analysis. Obes. Rev. https://doi.org/10.11111/obr.13095. Accessed March 4, 2022.

Gazzaz, Z.J., 2021. Diabetes and COVID-19. Open Life Sci. 16, 297–302.

Graciano-Machuca, O., Villegas-Rivera, G., Lopez-Perez, I., Macias-Barrigan, J., Sifuentes-Franco, S., 2021. Multisystem inflammatory syndrome in children (MIS-C) following SARS-CoV-2 infection: role of oxidative stress. Front. Immunol. https://doi.org/10.3389/fimmu.2021.723654. Accessed March 6, 2022.

Gualano, R.C., Hansen, M.J., Vlahos, R., Jones, J.E., Park-Jones, R.A., Deliyannis, G., et al., 2008. Cigarette smoke worsens lung inflammation and impairs resolution of influenza infection in mice. Respir. Res. 9, 53. https://doi.org/10.1186/1465-9921-9-53. Accessed February 28, 2020.

Honce, R., Schultz-Cherry, S., 2019. Impact of obesity on Influenza A virus parthenogenesis, immune response and evolution. Front. Immunol. 10 (10), 1071. https://doi.org/10.3389/fimmu.2019.01071. Accessed April 4, 2020.

Honjo, K., von Reeku, R., Verhoeff, N.P., 2009. Alzheimer's disease and infection: do infectious agents contribute to progression of Alzheimer's disease? Alzheimer's Dementia 5 (4), 348–360.

Hopkins, 2020. Take your diet to the Mediterranian. https://www.hopkinsmedicine.org/healthlibrary/conditions/adult/cardiovascular_diseases/atherosclerosis. Accessed April 4, 2020.

Hughes, D.A., 1999. Effects of dietary antioxidants on the immune function of middle-aged adults. Proc. Nutr. Soc. 58, 79–84. https://doi.org/10.3390/v10080392. Accessed March 14, 2020.

Kalem, A.K., Kayaaslan, B., Neselioglu, S., Eser, F., Hasanoglu, I., Aypak, A., et al., 2021. A useful and sensitive marker in the prediction of COVID-19 and disease severity: Thiol. Free Rad Biol Med 166, 11–17.

Kander, M.C., Cui, Y., Liu, Z., 2017. Gender difference in oxidative stress: a new look at the mechanisms for cardiovascular diseases. J. Cell Mol. Med. 21 (5), 1024–1032.

Keiser Family Foundation, 2021. State COVID-19 Data and Policy Actions. KFF. https://www.kff.org/coronavirus-covid-19/issue-brief/state-covid-19-data-and-policy-actions/. Accessed July 22, 2021.

Liang, W., Liang, H., Ou, L., Chen, B., Chen, A., Caichen, L., et al., 2020. Development and validation of a clinical risk score to predict the occurrence of critical illness in hospitalized patients with COVID-19. JAMA Intern. Med. https://doi.org/10.1001/jamainternmed.2020.2033.

Marais, B.J., Lonroth, K., Lawn, S.D., Migliori, G.B., Mwaba, K., Glaziou, P., et al., 2013. Tuberculosis comorbidity with communicable and non-communicable diseases: integrating health services and control efforts. Lancet Infect. Dis. 13 (5), 436–448.

Marin, B.G., Aghagoli, G., Lavine, K., Yang, L., Siff, E.J., Chiang, S.S., et al., 2021. Predictors of COVID-19 severity: a literature review. Rev. Med. Virol. 31 (1), 1–10.

Mena, G.E., Martinez, P.P., Mahmud, A.S., Marquet, P.A., Buckee, C.O., Santillana, M., 2021. Socioeconomic status determines COVID-19 incidence and related mortality in Santiago, Chile. Science 373, 924. https://doi.org/10.1126/science.abg5298. Accessed July 23, 2021.

Nalbandian, A., Sehgal, K., Gupta, A., Madhavan, M.V., McGroder, C., Stevens, J.S., et al., 2021. Post-acute COVID-19 syndrome. Nat. Med. 27, 601–615.

Peterhans, E., 1997. Oxidants and antioxidants in viral diseases: disease mechanisms and metabolic regulation. J. Nutr. 127 (5 Suppl. l), 9625–9655.

Schwartz, K.B., 1996. Oxidative stress during viral infections: a review. Free Radic. Biol. Med. 21 (5), 641–649.

Shahbaz, F.F., Martins, R.S., Umair, A., Ukrani, R.D., Jabeen, K., Sohail, M.R., Khan, E., 2022. A review of coronaviruses associated with Kawasaki disease: possible implications for pathogenesis associated with COVID-19. Clin. Med. Insights Pediatr. 12, 1–12. https://doi.org/10.1177/11795565221075319. Accessed March 6, 2022.

Splettstoesser, W.D., Schuff-Warner, P., 2002. Oxidative stress in phagocytes-the enemy within. Microsc. Res. Tech. 57 (6), 441–455.

Ssentongo, P., Seentongo, A.E., Heilbrunn, E.S., Ba, D.M., Chinchilli, V.M., 2020. Association of cardiovascular disease and 10 other pre-existing comorbidities with COVID-19 mortality: a systematic review and meta-analysis. PLoS One. https://doi.org/10.1371/journal.pone.0238215. Accessed March 4, 2022.

Xie, J., Liu, L., Hongyan, L., Che, J.H., Xie, W., 2022. Ink melanin from Sepiapharaonis ameliorates colitis in mic via reducing oxidative stress, and protecting the intestinal mucosal barrier. Food Res. Int. 2022. https://doi.org/10.1016/j/foodres.2021.110888.

Yang, J., Zheng, Y., Gou, X., Pu, K., Chen, Z., Guo, Q., et al., 2020. Prevalence of comorbidities and its effects in patients infected with SARS-CoV-2: a systematic review and meta-analysis. Int. J. Infect. Dis. 94, 91–95.

Zeliger, H.I., Pan, Y., Rea, W.J., 2012. Predicting co-morbidities in chemically sensitive individuals from exhaled breath analysis. Interdiscipl. Toxicol. 5 (3), 123–126.

Zeliger, H.I., 2014. Co-morbidities of environmental diseases: a common cause. Interdisip. Toxicol. 7 (3), 117–122.

Zeliger, H.I., 2016. Predicting disease onset in clinically healthy people. Interdisip. Toxicol. 9 (2), 39–54.

Zeliger, H.I., 2017. Oxidative Stress Index predicts disease onset. J. Med. Res. Pract. 6 (3), 86–92. https://doi.org/10.20936/jmrp/17/03/003. Accessed April 1, 2020.

Zeliger, H.I., 2019. Oxidative stress index: disease prediction and pevention. EC Pharmacol. Toxicol. 7 (9), 1022–1036.

Zeliger, H.I., 2020. Oxidative Stress Index (OSI) condensed questionnaire. http://doi.org/10.24018/ejmed.2020.2.1.163. Accessed March 30, 2020.

Zeliger, H.I., Kahaner, H., 2020. Can the oxidative stress index predict the severity of COVID-9? Eur. J. Med. Health Sci. https://doi.org/10.24018/ejmed.2020.2.2.233. Accessed July 22, 2021.

Zeliger, H.I., Kahaner, H., 2020a. Oxidative stress index predicts the severity of COVID-19. Eur. J. Med. Health Sci. https://doi.org/10.24018/ejmed.2020.2.4.369. Accessed July 22, 2021.

Zhang, J.J., Dong, X., Cao, Y.Y., Yuan, Y.D., Yang, Y.B., Yan, Y.Q., et al., 2020. Clinical characteristics of 140 patients infected with SARS-CoV-2 in Wuhan, China. Allergy. https://doi.org/10.1111/all.14238. Accessed April 4, 2020.

Zheng, Z., Peng, F., Xu, B., Zhao, J., Liu, H., Peng, J., et al., 2020. Risk factors of critical & mortal COVID-19 cases: a systematic literature review and meta-analysis. J. Infect. https://doi.org/10.1016/j.jinf.2020.04.021. Accessed March 4, 2022.

Zhou, F., Yu, T., Du, R., Guohui, F., Liu, Y., Liu, Z., et al., 2020. Clinical course and risk factors for mortality of adult inpatients with COVID-19 in Wuhan, China: a retrospective cohort study. Lancet. https://doi.org/10.1016/S0140-6736(20)30566-3. Accessed April 2, 2020.

PART IV

Prevention

CHAPTER

35

Disease prevention: oxidative stress control, antioxidants, and social factors

35.1 Introduction

The World Health Organization estimates that almost one quarter of all disease globally and 33% of disease in children under the age of five is caused by environmental exposures that can be averted (WHO, 2021). Disease prevention is dependent upon two interrelated factors, preventing overproduction of oxidative stress and maintaining essential levels of antioxidants.

35.2 Oxidative stress overproduction control

As discussed in the chapters of Part I of this book, and listed in Table 35.1, there are many causes of overproduction of oxidative stress, which is additive from multiple sources in a dose response relationship and responsible for the onset of almost all diseases.

Though all oxidative stress cannot be eliminated, lifestyle accommodations can, to a great extent, reduce OS to safe levels. These, labeled micro actions, can be accomplished by a person individually up to a point. A significant amount of OS reduction, however, is dependent upon societal action. These, labeled macro actions, range from local government intervention to national and international action and cooperation. Examples of macro actions are banning smoking tobacco public gathering places, outlawing the use of chemicals like PCBs and multinational treaties to reduce global warming.

Items 1–11 in Table 35.1, as discussed in Part I of this book, are related solely to chemically induced elevation of oxidative stress. Oxidative stress due to these parameters can be lowered by eliminating or reducing exposures to the chemicals involved. OS due to radiation exposure can similarly be lowered by limiting or eliminating, where possible, exposure to electromagnetic radiation. Items 12–13 have more complex oxidative stress involvement and are addressed individually in the following sections.

Oxidative Stress
https://doi.org/10.1016/B978-0-323-91890-9.00034-9

TABLE 35.1 Causes of oxidative stress overproduction.

1. Residence
2. Employment
3. Air quality
4. Water quality
5. Clothing worn
6. Chemicals in the home
7. Building materials and furnishings
8. Alcohol consumption
9. Tobacco use
10. Recreational drugs
11. Pharmaceuticals regularly taken
12. Radiation
13. Diet
14. Treating prevalent illness
15. Psychological stress
16. Global warming

35.3 Antioxidants

Antioxidants (AOs) are the first line of defense against excessive oxidative stress. AOs have two functions, neutralization of free radicals and protection of cells against the toxic effects of free radicals. Several body mechanisms act to carry out these functions via endogenous and exogenous AIs.

35.3.1 Antioxidant function

AOs function via free radical chain termination or prevention of free radical formation. Free radical chemistry proceeds by chain reactions, such as that in lipid peroxidation, in which radicals are repeatedly produced, with the reaction chain terminated when the last-produced uncoupled electron is paired. AIs serve as chain termination species (Young and Woodside, 2001). Examples of chain breaking AOs include vitamin C, vitamin E and carotenoids.

Free radical prevention is accomplished via two mechanisms. The first is suppression of chain reaction initiation via scavenging initiating free radicals by superoxide dismutase, catalase and glutathione, as examples. The second is the stabilization of free radicals by transition metals, examples being iron and copper (Pham-Huy et al., 2008).

35.3.2 Antioxidant categories

AOs fall into two categories; endogenous (enzymatic) and (exogenous) non-enzymatic (Birben et al., 2012). Endogenous AOs are the body's innate defenders against the deleterious effects of oxidative stress. Exogenous AOs, derived from dietary components are essential supplements for the endogenous species. All AOs are essential, with each having a different antioxidant function (Pham-Huy et al., 2008; Rizzo et al., 2010; Birben et al., 2012; Liguori et al., 2018); Primary examples of each are listed in Table 35.2.

35.4 Antioxidants and disease prevention

Numerous studies have investigated the roles of specific AOs and their effects on risk of disease onset (Donaldson, 2004; Tan et al., 2018, and references contained therein). Illustrative examples are shown in Table 35.3.

35.5 Diet

The Mediterranean diet (MEDD) is widely acclaimed as a healthy way to eat. Adherence to this diet provides antioxidants to combat oxidative stress and reduces uptake of oxidative stress-raising foods. Natural antioxidants contained in MEDD foods that have been demonstrated to lower oxidative stress (Buffo et al., 2011). Classes of these are listed in Table 35.4.

Other components of the MEDD include:

1. Limiting red meats and processed foods in favor of fresh fish and seafood.
2. Flavoring foods with herbs and spices in place of salt.
3. Replacing animal fats such as butter and lard with unsaturated vegetable oils, particularly extra virgin olive oil.

Numerous studies have attributed health benefits of the Mediterranean Diet (Harvard Public Health, 2021; Dutta, 2021). These are listed in Table 35.5.

TABLE 35.2 Endogenous (enzymatic) and exogenous (dietary) antioxidants.

Endogenous	Exogenous
Superoxide dismutase (SOD)	Vitamin A
Catalase	Vitamin C
Glutathione peroxidase	Vitamin E
Glutathione transferase	β-carotene (carotenoid)
Thioredoxin	Lycopene
Peroxiredoxin	Flavonoids (polyphenols)
Melatonin	Omega-3 and omega-6 fatty acids
	Selenium
	Glutathione

TABLE 35.3 Examples of disease prevalence impacts lowering effects of specific antioxidants.

Antioxidant	Diseases	Impacts
Glutathione	Cardiovascular	Lowers risk
	Type 2 diabetes	Protects against diabetic neuropathy
	Parkinson's	Reduces free radicals associated with neurological complications
Polyphenols	Cardiovascular	Reduce risk of death
	Type 2 diabetes	Improves glucose control and insulin sensitivity
	Cancer	Lower risk of colorectal cancer
Carotenoids	Alzheimer's	Lowers risk of onset
	Cardiovascular	Reduces risk of coronary artery disease and stroke
	Hypertension	Lowers blood pressure
	Macular degeneration	Lowers risk
	Osteoporosis	Lowers bone fracture risk
Zinc	Type 2 diabetes	Improves insulin sensitivity
Vitamin C	Cardiovascular	Lowers risk

TABLE 35.4 Natural antioxidants containing foods in the MEDD.

Fruits
Vegetables
Legumes
Whole grains
Nuts
Seeds
Red wine

MEDD recommended oxidative stress-lowering foods to eat and those to avoid or limit are listed in Table 35.6 (Dai et al., 2008; Detopoulou et al., 2021).

The foods listed in Table 35.4 all contain antioxidant phytochemicals. The different types of phytochemical-containing compounds foods in which they are found are listed in Table 35.7 (Zeliger, 2019; Sweeney, 2021).

TABLE 35.5 Health benefits attributed to the MEDD.

1. Slowed down aging and prolonging of life
2. Weight loss promotion
3. Lowered incidence of obesity
4. Improved immune system health
5. Improved psychological well-being, reduced anxiety and depression
6. Reduced risk of heart attack and heart disease
7. Reduced risk of stroke
8. Reduced risk of Alzheimer's disease
9. Reduced symptoms of multiple sclerosis
10. Lowered risk of type 2 diabetes by as much as 60%
11. Reduced symptoms of type 2 diabetes
12. Reduced incidence of breast cancer
13. Reduced incidence of colorectal cancer
14. Reduced incidence of lung cancer
15. Lowering of high blood pressure
16. Lowering of elevated cholesterol
17. Lowered incidence of ADHD in children

TABLE 35.6 MEDD recommended foods to eat and those to avoid.

Eat	Avoid or limit
Whole grains	Sugar, high fructose corn syrup
Freshly prepared food	Processed food
Fish and seafood	Fatty red meat
Herbs and spices	Salt
Extra virgin olive oil	Animal fat (butter, lard)
Canola oil	Coconut oil
Fresh vegetables	Trans fat
Fresh fruit	Saturated fat
Tree nuts	Artificial flavors
Seeds	Artificial colors
Legumes	Sulfites and other preservatives
Red wine[a]	MSG

[a]*Red, purple or blue grapes can be substituted for red wine by those limiting alcohol intake to achieve the same benefit.*

The large number of antioxidants and the different foods in which they are found point out the need to eat a varied diet thus ensuring an adequate supply of varied antioxidants, all of which provide unique benefits to different organs and body systems.

TABLE 35.7 Phytochemical-containing compounds and foods in which they are found.

1. Allium sulfur compounds – leeks, onions and garlic
2. Anthocyanins – eggplant grapes and berries.
3. Beta carotene – pumpkins, mangoes, apricots, carrots, spinach and parsley
4. Catechins – red wine and tea
5. Copper – seafood, lean meat, milk, nuts and legumes
6. Cryptoxanthins – red peppers, pumpkins and mangoes
7. Flavonoids – tea, red wine, citrus fruits, onions and apples
8. Indoles – broccoli, cabbage and cauliflower
9. Lignans – sesame seeds, bran, whole grains and many vegetables
10. Lutein – corn, spinach and other leafy greens
11. Lycopene – tomatoes, pink grapefruit and watermelon
12. Manganese – seafood, lean red meat, milk and nuts
13. Polyphenols – thyme and oregano
14. Vitamin C – oranges, berries, kiwi, mangoes, broccoli, spinach and peppers
15. Vitamin E – vegetable oils, nuts, avocados, seeds and whole grains
16. Zinc – seafood, lean meat, milk and nuts

Macro actions also can address diet. In industrialized societies, few people can grow food that it is not tainted by pesticides and other man-made chemicals. It is assumed that commercially produced food is safe to eat and that existing government regulations guaranty the safety of purchased food. In the United States and elsewhere in the industrialized world, however, the sale of chemically contaminated food proceeds without even warning the consumer of the dangers. Government can take macro actions to protect its citizens (Zeliger, 2011). Regulatory steps to ensure the safety of foods are listed in Table 35.8.

TABLE 35.8 Regulatory steps that can ensure food safety.

1. Requiring that foods contain labels that identify the presence of:
   Pesticides
   Hormones
   Antibiotics
   Genetically modified ingredients
   All added chemicals.
2. Banning the incorporation of known toxic chemicals into food.
3. Requiring thorough testing of new food sources, such as genetically modified foods before they are approved for use.
4. Promoting healthy diets and discouraging production and consumption of processed foods and foods that are high in sugar, saturated fats, salt and chemical additives. It has been estimated that 30%–50% of cancers can be prevented modification of life styles, of which an unhealthy diet is the primary culprit (Khan et al., 2010; WHO, 2021a).
5. Banning the use of phthalates and bisphenol A in food packaging, dishes and eating utensils.

This discussion would not be complete without addressing the production of healthy food rather than empty calories. Historically, farmers grew diversified crops and raised small numbers of animals. Today, farmers grow enormous quantities of single crops and raise huge numbers of cattle, hogs and fowl and are paid by how many calories they produce, rather than by how much nutrition is contained in the food. Genetically modified corn, fat cows and high fat content milk are more valuable to today's farmers than producing foods with good nutritional value. Corn is an excellent example of this phenomenon. Currently in the United States, 90 million acres are devoted to growing it. Modern farming advancements have led to corn fields that yield more than 170 bushels per acre, compared with 20 bushels per acre 100 years ago (Wisconsin State Farmer, 2021). This translates into the production of an astounding 15 million calories per acre, much of which is converted into the high fructose corn syrup (extensively used in fast food and sweetened drinks) that is detrimental to oxidative stress levels.

It must also be noted that eating healthy is much more expensive than eating high calorie processed foods containing unhealthy quantities of fats, sugar, salt and chemicals. In the United States, as elsewhere in the industrialized world, corn, wheat and soy farmers are subsidized while those growing fruits and vegetables and raising organically fed cows, pigs and poultry receive very little or no government support (FAO, 2018). This results in much higher prices for high quality food and largely precludes economically disadvantaged people from access to healthy food.

35.6 Treating prevalent illness

All illnesses and their associated symptoms elevate oxidative stress (Zeliger, 2016). Accordingly, curing where possible, or treatment of uncurable disease to reduce symptoms can lower OS levels and thereby prevent or delay the onset of co-morbidities.

On a micro level, the individual can seek medical care and counseling for disease treatment. On a macro level, government can make health care available and provide health education for all its citizens.

35.7 Psychological stress

As discussed in Chapter 13, mental health is just as important as physical health, as imbalances in both elevate oxidative stress. In many ways, stress caused by worry, fear, anxiety, depression or post-traumatic stress disorder (PTSD) is more difficult to treat than infectious or environmental illness, yet, emotional stress is a major contributor to overall oxidative stress.

On a micro level, the individual can seek professional care for anxiety, depression and other psychiatric diseases, and adjust lifestyle to reduce stress. Such adjustment can include physical as well as mental and emotional activities, examples of which are listed in Table 35.9.

On a macro level, society must recognize that a healthy state of mind is as essential as a healthy physical state. The following can be instituted:

TABLE 35.9 Stress reducing steps.

Walking
Biking
Swimming
Fishing
Playing golf
Dancing
Socialize with a friend or neighbor
Reading books
Playing cards, or board games
Working on jigsaw puzzles
Knitting
Sewing
Quilting
Gardening

1. Psychological and psychiatric care made available to all and covered by all health insurance plans.
2. Stress hot lines created and manned by people qualified to offer consultation to those in need of it.
3. Recognizing that a "mental health" sick day off from work or school can be just as important as staying home with the flu.
4. Recognize and support the use of alternative stress therapies.

35.7.1 Alternative stress therapies

Mental health professionals can provide stress-reducing therapy to lower oxidative stress. Western medicine, however, cannot always provide all the stress-reduction answers. For many, alternative therapies have been shown to be applicable.

35.7.1.1 Sweat lodge participation

Sweat Lodges have long been used by Native Americans to cleanse the body of toxins and restore good health. There are many variations on how Sweat Lodges are operated. For example, fasting is required prior to entry in some and length of stay in a sweat lodge can vary as well. All, however, are akin to sauna in that heat is used to induce sweating and thereby eliminate body toxins (Livingston, 2010).

35.7.1.2 *Environmental chamber use*

In this form of therapy, patients live in chambers or rooms that are devoid of pollutants for several days. They breathe air, eat food and drink water free of toxic chemicals and allow their bodies to naturally metabolize and expel the toxins they hold. This type of therapy has been shown to work very well for patients with severe environmental allergies who generally report that their symptoms are greatly reduced following treatment (Rea, 1996).

35.7.1.3 *Chinese herbal medicine*

Chinese herbal medicine has been used to promote good health for centuries. Some herbs are specified for treating specific single diseases, while others are used to treat multiple diseases. Recent research has reported on herbal use to reduce inflammation and lower oxidative stress (Yang et al., 2019).

35.7.1.4 *Accupuncture*

Acupuncture is a primary component of Chinese medicine that is used extensively to treat a myriad of diseases. Acupuncture has been shown to relieve inflammation, reduce oxidative stress and eliminate pain. Recent studies have shown that it reduces memory impairment and animal models suggest that laser acupuncture may one day be a treatment for autism (Su et al., 2020).

35.7.1.5 *Yoga*

Yoga, a part of traditional Indian culture, has been shown in several studies to lower blood pressure and even prevent its onset. Doing Yoga reduces the synthesis of a body protein associated with elevated stress. These effects are consistent with other studies that have definitively shown that yoga lowers oxidative stress (Patil et al., 2014).

35.7.1.6 *Meditation*

Those who practice meditation find that it lowers body stress and brings them an inner peace. Recent studies have shown that meditation also slows down the aging of body cells, a finding that has been ascribed to the fact that regular meditation lowers oxidative stress. It has been found that meditation increases the quality of life for patients with ALS (amyotrophic lateral sclerosis) (Martinelli et al., 2011).

35.7.1.7 *Aromatherapy*

Aromatherapy is the practice of using the aroma of sensually pleasing natural oils that are extracted from plants to enhance psychological and physical well-being. The oils may be inhaled or applied to the skin either gently or by message. Users of aromatherapy report that it reduces tension and anxiety and generally induces a state of relaxation. These effects are consistent with reductions in oxidative stress (Aponso et al., 2020).

35.7.1.8 *Hypnosis*

Recently, hypnosis has been suggested as a new tool to reduce oxidative stress and combat disease. Such treatment is still in its early stages and does not have the long histories of other alternative treatments. It is presented here as an example of the ongoing development of new methods to treat oxidative stress and reduce disease prevalence (Cozzolino and Celia, 2021).

35.7.1.9 Pets

Most households in the United States have at least one pet. Pets provide comfort, reduce stress and promote exercising (Jennings, 1997). The Centers for Disease Control lists the following health benefits of having a pet, all of which result in reduced oxidative stress (CDC, 2021).

- Decreased blood pressure.
- Lower cholesterol levels.
- Reduced triglyceride levels.

35.8 Exercise

The value of regular exercise cannot be overstated. Numerous studies have shown that regular moderate aerobic exercise lowers blood pressure, reduces the risk of heart disease, stroke, respiratory diseases and helps in treating type 2 diabetes. It has been shown that aerobic exercise lowers oxidative stress in the brain, can preserve brain function as people age and helps prevent the onset of dementias, including Alzheimer's Disease (Briones and Touyz, 2009).

On a micro level, the individual can almost always find a way to regularly exercise. Oxidative stress-reducing exercises not limited to intense workouts. Less strenuous exercises such as walking, dancing, easy swimming, biking, jogging and working with light weights are effective as well. Exercises can be done in the home by walking, or climbing up and down stairs. Workouts can even include walking in place. One suggestion, for example, is that when watching television, stand up and walk around or walk in place during the commercials. Doing so for a 1-h program provides up to a 20-minute workout. No matter which form exercise takes, doing it for at least 150 min per week provides an essential step in reducing oxidative stress and this total need not come from a single form of exercise, but can be made up of several different exercise types (Laskowski, 2021; Pinterest, 2021).

Though society can't exercise for a person, on a macro level it can take action to promote and facilitate exercising. Such steps include.

1. Fostering the building of communities that promote running, jogging, walking, swimming, biking and other forms of exercise.
2. Making roads wide enough to accommodate runners, joggers, walkers and bikers as well as cars and trucks.
3. Funding community centers that provide exercise opportunities not otherwise available to many individuals, such as swimming pools, tracks for running and walking, tennis and basketball courts.
4. Sponsoring competitive events such as 5K running and walking races that encourage people to seriously exercise.

On a micro level the individual can embrace any available alternative therapy to lower oxidative stress, but should first seek medical help to rule out other possible OS raising causes and ascertain if a proposed alternative therapy presents a health risk.

On a macro level, society, and to some extent the medical community, should keep an open mind to alternative treatments. Medical insurance plans should cover costs for alternative therapies when these have demonstrated their ability to lower stress.

35.9 Treating illness

As previously discussed, ailing with any disease, be it one that is curable or a chronic disease that is not, results in an increase in oxidative stress which can trigger the onset of additional disease.

On a micro level, the individual can aggressively treat illness so that the oxidative stress it produces is reduced. Almost equally as important in treating disease is the need to lower oxidative stress from sources not associated with an illness, as the onset of co-morbidity is due to total oxidative stress in the body irrespective of what the sources are (Zeliger, 2016). Accordingly, when ill, continuing to follow a healthy diet, avoiding toxic chemical exposures and avoiding, to the extent possible, the other causes of oxidative stress is essential. The faster the initial curable disease is cured or the symptoms of chronic illnesses are alleviated, the less likely it is that additional disease will ensue.

People with chronic illnesses whose symptoms wax and wane such as asthma, hypertension, chronic fatigue syndrome, multiple chemical sensitivity and irritable bowel syndrome, can take steps to reduce the symptoms when they flare up, as these symptoms are large contributors to elevated oxidative stress.

On a macro level, society must consider health care a right, not a privilege and make medical treatment universally available. Society must also provide education to all, making them aware of the need to treat all disease as soon as possible and how to do so.

As all disease is oxidative stress-mediated, all steps taken to treat an existing disease and reduce its symptoms will also reduce the likelihood of additional disease onset.

35.10 Inequalities

In a free and open society, there will always be inequality among its citizens. Those of exceptional intellect, talent, economic standing and social status will have greater opportunities and rewards than those not so fortunate. After disease, poverty is the greatest barrier to partaking in what life has to offer. Poor people are last to get new innovations, are also often denied healthy nutrition and medical care and are often forced to live and work in contaminated environments, all of which result in increased oxidative stress and disease. Studies have shown that socioeconomically disadvantaged people have greater numbers of diseases and earlier deaths than those in the social and economic mainstream and the elite do. As individuals and as a society, we must decide what is a right and what is a privilege. Are clean air and water a right? If so, polluters are denying us our rights. Is available health care and access to healthy nutrition and living a right? If so, it should be provided to all (Binkley, 2020).

On a micro scale, the individual, regretfully, can have only a limited effect on inequality. Though there are numerous examples of individuals who have pulled themselves up by their boot straps and succeeded despite the circumstances into which they were born, for far too many these examples are only the exceptions that prove the rule.

On a macro level, providing equal opportunity for all is rightly the function of society as a whole and its representative government. Steps that society can take to alleviate the health effects of inequality are:

1. Adopt universal health care to provide disease prevention and treatment to all at its earliest stages.
2. Provide education essential to teach all about the causes of disease and the steps that can be taken to prevent and alleviate it.
3. Work to prevent the spread of communicable disease for all. It is in our personal best interest as well in our humanitarian interest that our neighbors stay healthy, as infectious diseases can affect all.
4. Protect the public from exposures to toxic chemicals, radiation and other preventable causes of disease by ensuring a healthy environment for all.
5. Ensure a healthy and nutritious food supply for all.

35.11 Summary

Disease prevention is dependent upon numerous factors, all of which, however, center on control of oxidative stress levels. These include an appreciation of the multiple disease-causative mechanisms, addressing the role of antioxidants and application of preventive measures an individual can take and those measures only society as a whole can take.

References

Aponso, M., Patti, A., Bennett, L.E., 2020. Dose-related effects of inhaled essential oils on behavioural measures of anxiety and depression and biomarkers of oxidative stress. J. Ethnopharmacol. https://doi.org/10.1016/j.jep.2019.112469. Accessed December 6, 2021.

Binkley, C.E., 2020. A right or a privilege? How to practically and ethically reconcile two opposing views of health care. Markula Center for Applied Ethics. http://www.scu.edu/ethics-spotlight/post-election-reconciliation/a-right-or-aprivilege-how-to-practically-and-ethically-reconcile-two-opposing-views-of-health-care/. Accessed December 7, 2021.

Birben, E., Sahiner, U.M., Sackesen, C., Erzurum, S., Kalayci, O., 2012. Oxidative stress and antioxidant defense. WAO J. 9–19.

Briones, A.M., Touyz, R.M., 2009. Moderate exercise decreases inflammation and oxidative stress in hypertension. But what are the mechanisms? J. Hypertens. 54, 1206–1208.

Buffo, M., Lamuela-Raventos, R., Salas-Salvado, J., 2011. Mediterranean diet and oxidation: nuts and olive oil as important sources of fat and antioxidants. Curr. Top. Med. Chem. 11 (14), 1797–1810.

CDC, 2021. Centers for Disease Control and Prevention. Keeping Pets and People Healthy. https://cdc/gov/healthypets/keeping-pets-and-people-healthy/how.html. Accessed December 6, 2021.

Cozzolino, M., Celia, G., 2021. The psychological genomics paradigm of hypnosis and mind-body integrated psychotherapy: experimental evidence. Am. J. Clin. Hypn. 64 (2), 123–138.

Dai, J., Jones, D.P., Goldberg, J., Ziegler, T.R., Bostick, R.M., Wilson, P.W., et al., 2008. Association between adherence to the Mediterranean diet and oxidative stress. Am. J. Clin. Nutr. 88 (5), 1364–1370.

Detopoulou, P., Demopoulos, C.A., Antonopoulou, S., 2021. Micronutrients, phytochemicals and Mediterranean diet: a potential protective role against COVID-19 through modulation of PAF actions and metabolism. Nutrients. https://doi.org/10.3390/nu13020462. Accessed December 6, 2021.

Donaldson, M.S., 2004. Nutrition and cancer: a review of the evidence for an anti-cancer diet. Nutr. J. 3, 19–25.

Dutta, S.S., 2021. What Are the Health Benefits of the Mediterranean Diet? https://www.news-medical.net/health/What-are-the-Health-Benefits-of-the-Mediterranean-diet.aspx. Accessed December 6, 2021.

FAO, 2018. The Future of Food and Agriculture – Alternative Pathways to 2050. Rome. 224 pp. Licence: CC BY-NC-SA 3.0 IGO.

Harvard Public Health, 2021. Harvard T.H. Chan school of public health. Diet review: Mediterranean diet. https://www.hsph.harvard.edu/nutritionsource/healthy-weight/diet-reviews/mediterraneand-diet/. Accessed December 6, 2021.

Jennings, L.B., 1997. Potential benefits of per ownership in health promotion. J. Holist. Nurs. 15 (4), 358–372.

Khan, N., Afaq, F., Mukhtar, H., 2010. Lifestyle as a risk factor for cancer: evidence form human studies. Cancer Lett. 293 (2), 133–143.

Laskowski, E.R., 2021. How much should the average adult exercise every day? Mayo Clinic 2021. https://mayoclinic.org/exercise/faq-20057916. Accessed December 7, 2021.

Liguori, I., Russo, G., Curcio, F., Bulli, G., Aran, L., Della-Morte, D., et al., 2018. Clin. Interv. Aging 13, 757–772.

Livingston, R., 2010. Medical risks and benefits of the sweat lodge. J. Alternative Compl. Med. 16 (6), 617–619.

Martarelli, D., Cocchioni, M., Scuri, S., Pompei, P., 2011. Diaphragmatic breathing reduces exercize-induced oxidative stress. Evidenced-Based Comp. Alt. Med. https://doi.org/10.1093/ecam/nep169. Accessed December 6, 2021.

Patil, S.G., Dhanakshirur, G.B., Aithala, M.R., Naregal, G., Das, K.K., 2014. Effect of yoga on oxidative stress in elderly with grade-1 hypertension: a randomized controlled study. J. Clin. Diagn. Res. 8 (7), BC04–BC08.

Pham-Huy, L.A., He, H., Pham-Huy, C., 2008. Free radicals, antioxidants in disease and health. Int. J. Biomed. Sci. 4 (2), 89–96.

Pinterest, 2021. Exercise during Commercials. https://www.pinterest.com/elmogirl7/exercises-during-commercials/. Accessed December 7, 2021.

Rea, W.J., 1996. Chemical Sensitivity. Lewis Publishers, Boca Raton, Florida.

Rizzo, A.M., Berselli, P., Zava, S., Montorfano, M., Corsetto, P., Berra, B., 2010. Endogenous antioxidants and radical scavengers. Adv. Exp. Med. Biol. 698, 52–67.

Su, X.T., Wang, L., Ma, S.M., Cao, Y., Yang, N.N., Lin, L.L., et al., 2020. Mechanisms of acupuncture in the regulation of oxidative stress in treating ischemic stroke. Oxid. Med. Cell. Longev. https://doi.org/10.1155/2020/7875396. Accessed December 6, 2021.

Sweeney, E., 2021. 110 foods you can eat on the Mediterranean diet – from hummus to beets to ...octopus? Parade. https://parade.com/983137/ericasweeney/mediterranean-diet-food-list. Accessed December 6, 2021.

Tan, B.L., Norhaizan, M.E., Liew, W.P.P., Rahman, H.S., 2018. Antioxidant and oxidative stress: a mutual interplay in age-related diseases. Front. Pharmacol. https://doi.org/10.3389/fphar.2018.01162. Accessed December 10, 2021.

WHO, 2021. Environmental Health Impacts. About the Environmental Burden of Disease. https://www/who.int/activities/environmental-health-impacts. Accessed December 6, 2021.

WHO, 2021a. Cancer. https://who.int/news-room/facts-in-pictures/detail/cancer. Accessed December 7, 2021.

Wisconsin State Farmer, 2021. USDA reports mixed results for corn and soybean yields. https://www.wisfarmer.com/story/news/2021/08/17/usda-reports-mixed-results-cor-and-soybean-yields/8170324002/. Accessed December 6, 2021.

Yang, X., He, T., Han, S., Zhang, X., Sun, Y., Xing, Y., Shang, H., 2019. The role of traditional Chinese medicine in the regulation of oxidative stress in treating coronary heart disease. Oxid. Med. Cell. Longev. 2019. https://doi.org/10.1155/2019/3231424. Accessed December 6, 2021.

Young, I., Woodside, J., 2001. Antioxidants in health and disease. J. Clin. Pathol. 54 (3), 176–186.

Zeliger, H.I., 2011. Human toxicology of chemical mixtures. Elsevier, London.

Zeliger, H.I., 2016. Predicting disease onset in clinically healthy people. Indertiscip. Toxicol. 9 (2), 39–54.

Zeliger, H.I., 2019. A Pound of Prevention for a Healthier Life. Universal Publishers, Boca Raton, Florida.

CHAPTER

36

Global warming, oxidative stress and disease

36.1 Introduction

The World Health Organization has labeled global warming (also called climate change) as the world's greatest threat to health (WHO, 2021). Micro actions individuals can take to prevent global warming-caused elevation of oxidative stress and disease are:

1. Avoiding exposure to extreme heat and cold where possible.
2. Avoiding severe physical exertion on very hot, very cold and high air pollution days.

Society cannot protect the individual from elevated oxidative stress caused by global warming after the fact. Rather, society can only help by taking the following macro actions reduce global warming, thereby preventing oxidative stress elevation:

1. Stop burning carbon-based fuels, whose combustion products induce the greenhouse effect responsible for ocean warming and rising sea levels.
2. Convert to green energy (solar, wind, fuel cell and geothermal) sources for pollution-free production of electricity.
3. Take steps to insure an ample supply of healthy drinking water worldwide.
4. Act to eliminate malnutrition and promote healthy diets.

It is conservatively estimated that the Earth will warm by 2°C (3.6°F) or more by the year 2100. Such warming will cause the oceans to rise, the Arctic to melt, induce mass migrations, cause mass extinctions of species, reduce fresh water availability, reduce crop production and cause numerous fierce storms and wildfires.

Should this forecast come to past, it will be a major cause of disease increase. Such an effect, for example, has been demonstrated by a study linking global warming to increases in diabetes. This study concludes that global warming leads to interruptions in the supply of fresh foods and results in the consumption of high calorie, less essential nutrient-containing carbohydrates in place of healthy fruits and vegetables leading to obesity, a known risk of diabetes (Barnes, 2011).

Oxidative Stress
https://doi.org/10.1016/B978-0-323-91890-9.00030-1

As the Earth continues to warm, profound changes will be experienced (Lindsey, 2020). In many ways the world in 2050 will bear little resemblance to one that was present at the start of the 21st century. Climate scientists predict that many cities situated on the oceans' shores will disappear and hundreds of millions of people living on coastlines around the world will be forced to migrate away from flooded homes, fields and businesses. Drinking water supplies will dwindle as sea water infiltrates fresh water supplies and wars will be fought over water. Farmland will be lost and food supplies will become scarce in some parts of the world. Resultant decreases in nutrition and increases in stress will raise oxidative stress levels, with resulting onset of numerous diseases.

It is essential to recognize that contributions to global warming need to be considered on a worldwide basis. It makes no difference whether coal is burned in West Virginia or Beijing, China to produce electricity. The warming of the world will be impacted identically by either. It is also necessary to recognize that there is nowhere on Earth that one can hide from the impacts of global warming. Even those living solitary lives north of the Arctic Circle will feel the impacts of a warmer Earth, as warmer temperatures alter migration patterns of animals necessary for food and the later onset of winter and earlier start of the spring thaw make essential travel across frozen rivers more perilous (Macdonald et al., 2005).

Recent studies have demonstrated that the amount of airborne pollution that travels around the world is much greater than previously believed (WHO, 2016). Oxidative stress inducing air pollutants generated in Asia cross the Pacific Ocean and airborne particulates generated by dust storms in Africa cross the Atlantic Ocean to the United States. The World Health Organization has predicted that cancer risks due to these pollutants are four times greater and rising than previously estimated (Weiderpass, 2016). Many in the industrialized world believe that their wealth will shield them from the effects of global warming. Alas, we all breathe the same air when we step outdoors.

The ideas just presented are frightening, yet they will serve as prophesies if global warming continues as currently forecast. It is hoped that actions can be taken to prevent the onset of these events. One consequence of global warming is already upon us. Global warming is making us sicker and the rate of illness continues to increase as the world gets warmer. A joint study undertaken by The Lancet and University College - London concluded: "Climate change is the biggest health threat of the 21st century" (Costello, 2009) It is the almost unanimous belief among scientists the world over that this is not an understatement. How global warming leads to increased disease levels is addressed here.

36.2 Global warming, oxidative stress

Increases in oxidative stress due to global warming are due to the following parameters:

Population
Migration
Energy use
Urbanization
Industrialization and air pollution
Atmospheric and oceanic energy

Air pollution
Water pollution
Increased pesticide use
Disease increase
Mold growth
Absorption of toxins
Contaminated food
Food quality decline
Absorption of toxins
Increased stress
Species extinction

36.2.1 Population

The population of the world currently (2022) is close to eight billion people and is projected to be 9.8 billion by 2050, an increase of 30% (UN, 2021). The world will have to provide almost one third more food, drinkable water, shelter and energy, than it does today. Supplying all these additional resources will result in a corresponding increase in pollution and other sources of oxidative stress.

36.2.2 Migration

Forced migration due to disappearance of land and land left barren will result in hundreds of millions of people living in squalor for extended periods of time. Such living inevitably leads to increases in vector-borne disease, psychological stress and malnutrition, all major contributors to oxidative stress.

36.2.3 Energy use

As time marches on the world becomes more industrialized and worldwide per capita use of energy continues to rise. When coupled with growing population, energy consumption is rising at an astronomical rate. The U.S. Energy Information Administration predicts that 56% more energy will be used in 2040 than was used worldwide in 2013 (EIA, 2013). This will correspond to an enormous growth in pollution exposure and its related oxidative stress increase.

36.2.4 Urbanization

The continued decline of agrarian societies, increase in industrialization and mass migration of populations will result in more and more people relocating to cities. Cities use more energy, produce more pollution and serve as centers of pollution concentration to a much greater extent than rural areas (UN Habitat, 2014).

36.2.5 Industrialization and air pollution

Air pollution caused by industrialization and associated energy production are by far the greatest contributors to pollution by toxic chemicals. As air pollution increases, it is accompanied by a corresponding increase in the generation of oxidative stress causing chemicals, which in turn, lead to increases in disease. The recent experience in China illustrates this point. China has gone from a largely agrarian society in 1950 to the world's leader in industrialization today (2022). During this time span, coal burning in China has increased from a few million tons in 1950 to greater than 4 billion tons each year, most of it used to produce electricity (He et al., 2020). Correspondingly, air pollution levels in 70 Chinese cities have led to warnings being issued to its citizens to avoid breathing outdoor air as much as possible. Air pollution levels in parts of China are at times as much as 100 times higher than those in the United States, and asthma rates in China have soared as a result (Jin et al., 2017).

36.2.6 Atmospheric and oceanic energy

36.2.6.1 Extreme weather

Global warming is increasing the amount energy getting trapped on earth, creating an imbalance that increases temperatures of the atmosphere and the oceans. This, in turn, leads to increases in extreme weather events; very high air temperatures, severe storms (tornadoes, hurricanes), floods, droughts, heat waves and cold waves (Buis, 2020).

Increased global warming has led to record high temperatures being recorded year after year. The human body is designed to live in environments with temperatures in the range of 4–35°C (Pitts, 1985) In India, for example, summer temperatures are now commonly as high as 52°C (125°F). It is feared that South Asia may be too hot for humans to live in by the year 2100 (Aljazeera, 2021).

Heightened storm intensities trigger infrastructure destruction which results in the release of toxic chemicals to the air and water, as, for example, occurred with Hurricane Harvey in Houston, Texas in the summer of 2017 (NOAA, 2017). Destructive storms can also lead to interruptions in food and drinking water supplies. These effects cause oxidative stress and disease to increase via air pollution, water pollution and by increasing the spread of infectious diseases. It is estimated that extreme weather will kill 150,000 people annually in Europe by the beginning of the next century (BBC, 2017).

Wildfires, which are naturally ignited by lightning, are essential for the regeneration of forests, as overgrown forests do not permit enough sunlight penetration to ground level to start new tree seedling growth (California Fire, 2021). Following fires, saplings quickly take root and the forests regenerate. There can be too much of a good thing, however. Global warming produces more violent storms that result in more frequent and intense lightning strikes which ignite increased numbers of wildfires that impart much greater than necessary forest loss. Global warming is also responsible for severe droughts that dry out shrubs and trees, making them easily ignitable and leading to forest fires. Global warming is also the cause of stronger winds that help spread wildfires which consume greater numbers of trees and shrubs than would be burned by natural fires, thus leading to the release of much greater quantities of carbon dioxide and toxic smoke. As previously discussed in Chapter 26, wood smoke contains many of the same oxidative stress inducing chemicals found in diesel fumes and cigarette smoke.

36.2.7 Air pollution

As discussed in previous chapters, air pollution is associated with increased oxidative stress and the onset of numerous diseases. Global warming increases air pollution in several ways.

1. Causing increased combustion of carbonaceous fuel via wildfires.
2. Increasing evaporation of volatile and semi-volatile organic compounds.
3. Acceleration of atmospheric chemical reactions, giving rise to secondary pollutants that are more toxic than the ones from which they are derived. An example of this is the formation of dangerous levels of ozone from the reaction of chemicals like benzene with oxides of nitrogen which are produced from the burning of diesel fuel (Jia and Xu, 2014).
4. Producing stronger winds that blow pollutants to places far removed from their sources.
5. Raising respiratory irritating pollen levels caused by increases in atmospheric carbon dioxide.
6. Increasing humidity leading to increases elevated levels of mold growth and resultant formation of mycotoxins.

36.2.8 Water pollution

Many of the chemicals released when fuels are burned settle to the ground or are carried down by rain. Once on the ground, these are washed into rivers, streams, lakes and estuaries thus contaminating the water we drink.

Another source of water pollution comes from rising sea levels which cause brackish waters containing large quantities of organic matter, to infiltrate drinking water sources. Such contamination dictates that increased quantities of chlorine be used to disinfect water so it is safe to drink. As discussed in Chapter 5, chlorination of drinking water produces disinfection bye products that elevate oxidative stress and are associated with numerous health effects including spontaneous abortion.

As the temperature of water containing toxic chemicals increases, the rates with which these chemicals react also increases. Thus, for example, perchloroethylene decomposes at an accelerated rate to produce increased quantities of other carcinogens as water temperatures rise (Parsons et al., 1984). Solubility of petroleum chemicals, pesticides and heavy metals also increase as water temperatures rise, further increasing the toxicity of the water.

Much of the municipal drinking water in the United States is contaminated with pharmaceuticals that pass through our bodies, or are discarded by being flushed. These contaminants pass unchanged through waste water treatment systems and make their way into drinking water. Philadelphia, Pennsylvania's drinking water, for example, has been found to contain 56 different pharmaceuticals, including antibiotics, anticonvulsants and sex hormones. Other water supplies have been found to contain opioids, including Oxycodone (Meza, 2008). All these pharmaceuticals elevate oxidative stress.

Polluted water can also lead to the following oxidative stress raising effects:

- Illness caused by eating fish and shellfish living in contaminated waters.
- Illness caused by swimming and bathing in chemically polluted waters.

– Increased absorption of bacteria including E. coli and Salmonella, viruses, algae and parasites which flourish in waters rich in fertilizer runoff (Low et al., 2013).

36.2.9 Increased pesticide use

Global warming results in greater use of pesticides for several reasons. These include:

1. Increases in numbers of insects, crop diseases and weed growth.
2. Increased evaporation of applied pesticides.
3. Increased runoff after application.
4. Increased loss due to wind.
5. Decreased effectiveness.

36.2.9.1 Increased numbers of insects, crop diseases and weeds

Global warming results in the migration of insect species to environments in which they could not previously survive. An example is the northern migration of the Mountain Pine Beetle, which attacks trees, in California. A warmer and wetter environment also increases the populations of insects already living in an area. As a result of such effects, larger quantities of pesticides are required to control insect populations (CPR, 2021).

Crop diseases, which are caused by parasites, fungi, bacteria or viruses have also been shown to migrate as a result of global warming an example being the Alfalfa Stem Nematode, which attacks alfalfa crops and results in the need for application of greater quantities of pesticides to combat these invaders (Roos et al., 2011).

Weed migration, too, results from global warming, an example being the Yellow Star Thistle, a plant that whose range has been dramatically increased in Northern California and Nevada. Warmer temperatures also result in the accelerated growth of native weeds, requiring increased use of herbicides (Patterson et al., 1999).

36.2.9.2 Increased evaporation of applied pesticides

Most pesticides evaporate slowly at ordinary temperatures, but as temperatures rise, evaporation rates increase, raising airborne pesticide levels of and leaving less to attack pests, thereby requiring the application of additional pesticide quantities (SLU, 2021).

36.2.9.3 Increased runoff after application

Global warming brings with it more frequent storms and, in many areas, greatly increased rainfall. These effects, in turn, produce greater runoff that carries applied pesticides away from fields to which they were applied. This has the dual effect of contaminating waterways and requiring additional pesticide application to protect crops (SLU, 2021).

36.2.9.4 Increased loss due to wind

Pesticides are applied to fields either by broadcasting along the ground or by air spray from crop dusters. Both methods produce airborne droplets and particles that are easily blown great distances from their application points. Global warming produces winds that are stronger than usual breezes, resulting in applied pesticides missing their targets, contaminating adjacent areas and requiring that additional amounts of insecticides and herbicides be applied (SLU, 2021).

36.2.9.5 *Decreased effectiveness*

Warmer temperatures decrease the effectiveness of pesticides. It has been shown, for example, that as temperatures increase, the efficacy of Permethrin, a pesticide used to control mosquito populations, decreases dramatically (Whiten and Patterson, 2016).

All pesticides have several things in common. They are toxic to people and the environment, increase oxidative stress and their use raises the incidence of disease.

36.2.10 Disease increase

Global warming is responsible for disease increase in four ways. These are:

1. Increased pollen counts
2. Spread of infectious diseases
3. Antibiotic resistance
4. Mold growth

36.2.10.1 *Increased pollen counts*

Warmer temperatures in areas with ample water supply promote the growth of pollen producing plants. Higher temperatures extend the growth ranges, the lengths of growing seasons of plants and generate stronger winds that spread the pollen produced over wider areas. Higher pollen counts are associated with elevated oxidative stress leading to respiratory and immunological disease (Bacsi et al., 2005).

36.2.10.2 *Spread of infectious diseases*

Temperature increases are conducive to increasing insect populations and to their geographic spread, leading to larger numbers of disease carrying insects such as mosquitoes and ticks and their spread to ever widening areas. Global warming has been shown to be responsible for the spread of mosquito borne diseases including Malaria, West Nile Virus and Zika Virus and tick-borne diseases such as Lyme disease and Rocky Mountain Spotted Fever. Lyme disease, for example, has now spread to almost 50% of the counties in the United States (CDC, 2020).

Higher temperatures also lead to increases in house fly populations. House flies carry hundreds of bacteria, many of which cause typhoid, cholera and dysentery in people. When a house fly lands on food, it deposits these bacteria on it. The greater the number of flies, the more likely the disease-spreading potential (NEHA, 2021).

36.2.10.3 *Antibiotic resistance*

Recent research has shown that global warming is a factor in antibiotic resistance in bacteria such as E. coli and others (Rodreguez-Verdugo et al., 2020).

36.2.10.4 *Mold growth*

Mold growth is accelerated by warmth and moisture. As global warming contributes to both in some areas, it is responsible for increased levels of mold and for the production of mycotoxins which contaminate grains, nuts and other foods and are associated with liver disease, kidney disease, allergies and various cancers (Science Direct Topics, 2021).

36.2.11 Contaminated food

Global warming-caused climbing temperatures, rising humidity levels, increased numbers of heavy downpours, storms and floods cause increases in populations of pathogens such as E. coli and Salmonella that are responsible for food-borne illnesses to proliferate (Miraglia, 2009). Heavy rains and floods spread viruses and fecal bacteria in the fields where food is grown, resulting in food contamination and resultant disease.

Food spoilage is also accelerated by warmer temperatures. Higher temperature caused-increases in the numbers of pest insects and parasites in growing fields results in the use of ever-increasing quantities of disease-causing pesticides which are absorbed by growing crops and delivered to people eating these crops.

36.2.12 Food quality decline

Sixty percent of the world's food supply comes from rice, wheat and corn and in many Asian countries, three quarters of the calories eaten by people come from rice. Recent studies in China and Japan have demonstrated how global warming reduces the nutritional quality of rice by lowering protein, iron and zinc levels by 5–20% and B-Vitamin levels by 10–45%. This decline in the quality of rice is estimated to affect 600 million people. The losses in nutritional quality are associated with increased atmospheric levels of carbon dioxide. Carbon dioxide causes rice plants to grow larger and produce more carbohydrates, but does not affect the amount of rice's other nutrients. Thus, eating the same quantity of carbohydrate rich rice gives a person more calories, but fewer essential nutrients (Gazella, 2019).

Heat causes food to deteriorate in quality and rot faster. The higher the temperatures, the faster the spoilage occurs. Losses of food occur due to storms and drought during the growing season, harvesting, packing, transporting, distributing to stores, spoilage while on store shelves and spoilage in homes prior to being eaten. It is estimated that 40% of the food grown in the United States spoils and is not eaten (USDA, 2021). It should also be noted that spoiled foods become nesting grounds for disease bearing insects, bacteria, viruses and parasites.

36.2.13 Absorption of toxins

Heat raises the concentrations toxic chemicals in air and water. Higher temperatures are also responsible for increased absorption of these toxins through the skin, lungs and digestive system as we breathe, eat and drink, necessitating lifestyle changes to reduce the chances being harmed by such chemicals. For example, strenuous exercise and its associated heavy breathing on hot polluted days results in the inhalation of larger than normal quantities of toxins and people are regularly advised by health officials to curtail heavy exertions on days with poor air quality to protect themselves from chemical overexposures. In the extreme, as in many Chinese cities that are affected by massive air pollution, face masks are required to filter out at least some of the chemical pollutants. As the globe warms, chemical contamination continues to grow to where it is entirely possible that face masks and other protective devices will be required most days to limit toxic absorption that increases oxidative stress and leads to disease onset (Akomeah et al., 2004).

36.2.14 Increased stress

Living with increased levels of disease, floods, droughts, storms, extreme heat or cold, wars and mass migration dramatically affects peoples' psyches. Stress, depression and anxiety associated with living through such traumas, as discussed in Chapter 13, greatly increases oxidative stress and leads to the onset of additional disease. This additional psychological burden also contributes to drug and alcohol abuse, which further elevates oxidative stress.

36.2.15 Species extinction

Biodiversity, the presence of diverse populations of animals and plants, is essential for existence and survival of life on Earth. Global warming, however, has resulted in the extinction of numerous animals and plants and threatens an even greater extinction as the world continues to warm. It is beyond the scope of this book to examine this subject in detail. Following, however, are examples of the losses of different species and the impacts of these losses on disease proliferation and indeed on the overall health of our planet. These include:

- Animal extinction
- Insect extinction
- Parasite extinction
- Fish migration
- Coral reef destruction
- Tree population loss

36.2.15.1 Animal extinction

The African cheetah is the world's fastest animal. It can accelerate faster than any current sports car, going from zero to 60 miles per hour in 3 seconds. Sadly, the sperm counts of these giant cats have declined to levels that are one tenth those of house cats. Cheetahs historically have preyed upon Thompson's gazelles, whose meat is high in protein and other nutrients essential for producing high sperm counts in these big cats. Global warming has brought droughts to gazelle habitats that has decimated the natural grasses that they feed on. This, in turn, has resulted in a huge decline in the Thompson's gazelle population, causing the cheetahs to feed on other prey that is deficient in the protein necessary for maintaining their sperm counts at high enough levels to successfully reproduce. Though there are still wild cheetahs in Africa, their numbers have declined by 80% and they now face extinction (National Geographic, 2020). Cheetahs are but one example of animals essential for a healthy environment facing extinction.

36.2.15.2 Insect extinction

Insects are the most populous and diverse animals in the world. Though they often intrude on our living space, sometimes stinging us, destroying crops, denuding trees and decimating entire planted fields, are essential for human survival. Eighty percent of wild plants depend upon insects for pollination and they are the food for 60% of the world's birds and numerous fish. Insects also serve as the Earth's janitors, by burying animal dung, for example. If insects disappear, the world's ecosystems will collapse. Yet, that is exactly what is happening right now (Halsch et al., 2021). Consider the following examples:

1. Honeybee numbers in North America have dropped by almost 60% in the last 70 years (Duran, 2017).
2. British moth populations are declining by 30% every 10 years (Fox et al., 2021).
3. The weight of flying insects captured in German nature preserves has decreased by an average of more than 75% per year in the past quarter century (Daley, 2017).

36.2.15.3 *Parasite extinction*

Parasites, organisms that earn a living by feeding on other living species, include harmful leeches, ticks, bedbugs and tapeworms. All parasites, however, are not dangerous to humans. About half of the world's 7.7 million known living species are parasites and most serve very valuable functions in the balance of nature (Panko, 2017). Parasites control the populations of many species of plants and animals that could otherwise overwhelm nature's balance. These include providing nutrients for plants, warding off crop-destroying insects such as aphids and combating autoimmune diseases that include multiple sclerosis (MS), asthma and irritable bowel syndrome (Preston and Johnson, 2010).

Global warming is causing the populations of parasites to sharply decline. It is estimated that as many as a third of the world's parasites will be extinct by the year 2070, with the total impact of such an extinction anticipated to be dire.

36.2.15.4 *Fish migration*

Fish are essential sources of protein for much the world's population. Cold water living ocean fish, such as Atlantic cod, have historically been a major food source for North American and European people. Though cod populations have precipitously declined due to over fishing, rising ocean water temperatures in the northern Atlantic Ocean are also impacting their numbers. Cod, as well as lobsters and other fish dependent upon cold water, are moving north and being replaced by warmer water loving fish, such as ocean Black Bass and some crab species, that are now finding the northern waters to be suitable habitats. Such migrations are not always beneficial. Green crab populations are currently soaring in Maine coastal waters, as these crabs are feeding upon native clams and drastically reducing clam numbers. The rapidly rising temperatures in the Gulf of Maine are resulting in a declining lobster population that is expected to drop by 40–60% over the next 30 years as the lobsters continue migrate north (Goldfarb, 2017).

36.2.15.5 *Coral reef destruction*

Rising global temperatures are increasing the amount of dissolved carbon dioxide in the world's oceans. This, in turn, increases the acidity of ocean water, a condition which leads to coral reef death. Coral reefs are also being destroyed by chemicals in sunscreens that leech nutrients out of the coral. The loss of coral disrupts the growth of fish and other marine life dependent upon it for survival, thus further reducing essential biodiversity (NOAA, 2021).

36.2.15.6 *Tree population loss*

Higher temperatures and greater rainfall in northern forests are changing tree profiles for these forests. The following example demonstrates this phenomenon:

In northern New York, Vermont, New Hampshire and Maine, maple trees are being replaced by beech trees. Unlike maples, beeches are susceptible to beech bark disease which causes them to die young. Overall, this is producing a reforestation that is changing the habitat of animals that are dependent upon the forests for food and shelter (Stanke et al., 2021; Manomet, 2018).

36.3 Stress

The effects of global warming discussed above are major contributors to increased stress levels in people (Weir, 2016). As also discussed in chapter 13, psychological stress is known to elevate oxidative stress and lead to the onset of numerous diseases.

36.4 Human extinction?

The average temperature in world has risen dramatically in the last 75 years and continues to rise at an accelerated pace. Scientists are in almost universal agreement that the warming of our planet is due to human activity and, if not reversed, will have dire consequences for life on Earth. As discussed above, impacts include dramatic increases in disease and the rendering of huge parts of the world uninhabitable. Are we approaching mankind's last generation?

People account for only 0.01% of life on Earth, yet have managed to destroy more than 80% of the planet's wild animals and half of its plants since the advent of civilization (National Park Service, 2021). Almost the entire population of wild animals has been replaced with domestic livestock. Forests have been converted to farmlands requiring massive quantities of polluting synthetic fertilizers and pesticides that lead to further loss of biodiversity.

Extinction due to global warming, however, may not limited to plants and animals. As discussed above, we are now experiencing ever increasing profound impacts on human life that may threaten our survival. These effects on humans include:

- Lower life expectancy
- Loss of life due to extreme weather
- Decline in fertility

36.4.1 Lower life expectancy

As discussed in the earlier chapters, the incidences of life-shortening infectious as well as non-communicable diseases has dramatically risen since the advent of the Industrial Revolution and has been further accelerated by the global warming that was ultimately triggered. In recent years, there has been a drop of life expectancy in the industrialized world. In the United States, for example, life expectancy declined by a year and a half in 2020 (CDC, 2021; PRB, 2020). Similar declines continue to occur in industrialized societies worldwide.

36.4.2 Loss of life due to extreme weather

Extreme heat and storms triggered by global warming can kill large numbers of people. Hurricane Maria, which devastated Puerto Rico in 2017, for example, claimed almost 3000 lives. It is estimated that extreme weather will kill 152,000 lives per year in Europe by the year 2100 (Forzien et al., 2017). Ever increasing heat waves and extreme cold snaps always claim human lives. As these events increase in frequency and ferocity, however, the number of deaths will only accelerate.

36.4.3 Decline in fertility

No animal or plant species can survive without reproducing itself. In the industrialized world, people are not producing enough offspring to maintain populations and birth rates. In 2017, U.S. birth rates dropped to a 30 year low and far below the population replacement rate (PRB, 2020; Smith, 2021). Why is this so? Partial reasons are: economic, with fewer people feeling that they can support more children; fear of global warming impacts; malnutrition; and ongoing wars. By far the primary reason for this phenomenon, however, is the alarming decline in human fertility, primarily in men.

Sperm counts of men living in the industrialized world have dropped by more than half in the past generation and based on a review of more than 2500 studies, it is clear that sperm counts of men living in North America, Europe, Australia and New Zealand declined by 50–60%. between 1973 and 2011. This decline is attributed to environmental pollution by endocrine disrupting chemicals (EDCs) which act like the female hormone estrogen in males (Levine et al., 2017). These include the following "Dirty Dozen" persistent organic pollutants, heavy metals and other organic compounds, whose spread is facilitated by global warming. These and their exposure sources are listed in Table 36.1 (Stockholm Convention Secretariat, 2019):

The EDCs just listed are but a few of numerous others (UN, 2018; HKSAR, 2012; Edlists, 2021). All EDCs increase oxidative stress and all are associated with the onset of numerous other diseases. Global warming accelerates the spread of EDCs around the world and from pole to pole due to increased evaporation, stronger winds, ocean currents, more frequent storms and increased solubilities in warmer water.

In 1976, a chemical explosion in Seveso, Italy released a giant cloud of dioxin that impacted many people. Twenty-two years later, boys aged one to nine when they were exposed to the dioxin cloud had sperm counts that were 43% lower than those who were not so exposed (Barrett, 2011).

Sperm counts in men living in the developing world at this time remain higher than those of men living in the industrialized world. This is changing with time. China was, until recently, a largely agrarian society. As it industrialized, however, sperm counts of its men have dropped precipitously. In 1986, sperm counts in more than half of sperm donors in the Chinese province of Hunan had semen that was acceptable for use. Fifteen years later, less than one in five sperm donors had acceptable semen (Levine et al., 2017; Liu et al., 2020).

Declining fertility, though primarily due to decreased sperm counts in men, is also caused by impacts of the reproductive tracts of women. For example, BPA exposure has been shown to reduce the viability of eggs and prevent embryo implantation (Huo et al., 2015). A study in

TABLE 36.1 Endocrine disrupting chemicals affecting sperm counts and their sources.

Chemical	Exposure sources
Dirty dozen	
Aldrin	Pesticide
Chlordane	Termite control
DDT	Insect control
Dieldrin	Termite and textile pest control
Endrin	Crop spray
Heptachlor	Soil insect control
Hexachlorobenzene	Fungi control on food crops
Mirex	Fire ant control
Toxaphene	Insect control
Polychlorinated biphenyls (PCBs)	Heat exchange fluids
Polychlorinated dibenzo-p-dioxins	Combustion by products
Polychlorinated dibenzofurans	Combustion by products
Heavy metals	
Lead	Leaded paint, old water pipes
Arsenic	Some ground water, treated wood, pesticides
Mercury	Fish, contaminated drinking water
Other organic compounds	
Bisphenol A (BPA)	Food can liners and plastics
Phthalates	Plastics, food packaging, cosmetics
Atrazine	Weed killer used on corn, sugarcane, lawns and golf courses
Perfluorinated chemicals (PFCs)	Non-stick cookware
Polybrominated diphenyl ethers (PBDEs)	Flame retardants in childrens' clothing and upholstery
Organophosphate pesticides	Insecticides on fruits and vegetables
Glycol ethers	Cleaners, cosmetics, theatrical smoke, electronic cigarettes (vape)

Australia has found that women who eat more fast foods high in oxidative stress producing sugar, salt, processed meat and synthetic chemical additives and less fresh fruit and vegetables take longer to get pregnant than women who consume healthier diets (Pultarova, 2018).

Are we nearing the birth of the last generation of people? Will our great grandchildren be the last humans?

References

F. Akomeah, T. Nazir, G.P. Martin, M.B. Brown, Effect of heat on percutaneous absorption and skin retention of three model penetrants, Eur. J. Pharmaceut. Sci. 21 (2–3) (2004) 335–345.

Aljazeera, Millions in India's northern states sizzle in severe heatwave. https://www.aljazeera.com/news/2021/7/2/india-severe-heatwave-northern-states-delhi, July 7, 2021. Accessed December 7, 2021.

A.S. Barnes, The epidemic of obesity and diabetes, Tex. Heart Inst. J. 38 (2) (2011) 142–144.

J.R. Barrett, Window for dioxin damage: sperm quality in men born after the Seveso disaster, Environ. Health Perspect. 119 (5) (2011) A219.

A. Basci, N. Dharajiya, B.K. Choudhury, S. Sur, I. Boldogh, Effect of pollen-mediated oxidative stress on immediate hypersensitivity reactions and late-phase inflammation in allergic conjunctivitis, J. Allergy Clin. Immunol. 116 (4) (2005) 836–843.

BBC, Extreme weather could kill up to 152,000 a year in Europe by 2100. https://www.bbc.com/news/world-europe-40835663, August 5, 2017. Accessed December 7, 2021.

A. Buis, How climate change may be impacting storms over Earth's tropical oceans, Ask NASA Climate, https://climate.nasa.gov, March 12, 2020.

California Fire, Benefits of fire. www.fire.ca.gov, 2021. Accessed December 7, 2021.

CDC, Lyme disease transmission. https://www.cdc.gov/lyme/transmission, 2020. Accessed December 7, 2021.

CDC, Life expectancy in the U.S. declined a year and half in 2020. https://www.cdc.gov/pressroom/nchs_press_releases, 2021. Accessed December 7, 2021.

A. Costello, M. Abbas, A. Allen, S. Ball, S. Bell, R. Bellamy, et al., UCL-Lancet commission on managing the health effects of climate change. https://www.ucl.ac.uk/global-health/research/a-z/lancet-commission-climate-change, 2009. Accessed December 7, 2021.

CPR, Pesticides and climate change. https://www.pesticidereform-change, 2021. Accessed December 7, 2021.

J. Daley, Over three quarters of flying insects disappear from German nature preserve, Smithsonian, https://www.smithsonianmag.com/smart-news/german-preserves-see-76-percent-decline-flying-insects-180965328/, 2017. Accessed December 7, 2021.

L. Duran, The buzz on climate change: its bad for bees, Conservation, https://www.conservation.org/blog/the-buzz-on-climate-change-its-bad-for-bees, 2017. Accessed December 7, 2021.

Edlists, Substances identified as endocrine disruptors at EU level. https://edlists.org/the-ed-lists/list-i-substances-identified-as-endocrine-disruptors-by-the-eu, 2021. Accessed December 7, 2021.

EIA, Today in energy. EIA projects world energy consumption will increase 56% by 2040. www.eia.gov, 2013. Accessed December 7, 2021.

G. Forzien, A. Cescatti, F. Batista, L. Feyen, Increasing risk over time of weather-related hazards to European population: a data-driven prognostic study, Lancet Planet. Health 1 (2017) e200–e208.

R. Fox, E.B. Dennis, C.A. Harrower, D. Blumgart, J.R. Bell, P. Cook, et al., The State of Britain's Larger Moths 2021, Butterfly Conservation, Rothamsted Research and UK Centre for Ecology & Hydrology, Wareham, Dorset, UK, 2021.

K.A. Gazella, Climate change and food quality, Nat. Med. J. 11 (8) (2019). https://www.naturalmedicinejournal.com/journal/2019-08/climate-change-and-food-quality. Accessed December 7, 2021.

B. Goldfarb, Yale Environment 360. Feeling the heat: how fish are migrating form warming waters. https://e360.yale.edu/features/feeling-the-heat-warming-oceans-drive-fish-into-cooler-waters, 2017. Accessed December 7, 2021.

C.A. Halsch, A.M. Shapiro, J.A. Fordyce, C.C. Nice, J.H. Thorne, D.P. Wastjen, M.L. Forister, Insects and recent climate change, Proc. Natl. Acad. Sci. USA (2021), https://doi.org/10.1073/pnas.2002543117. Accessed December 7, 2021.

G. He, J. Lin, Y. Zhang, W. Zhang, G. Larangeira, C. Zhang, et al., Enabling a rapid and just transition away from coal in China, One Earth 3 (2) (2020) 187–194.

HKSAR, Endocrine disrupting chemicals in food. Centre for food safety, food and environmental hygiene dept., the government of the Hong Kong special administrative region, Risk Assessment Report No, 48, https://www.cfs.gov.hk/programme_rafs/files, 2012. Accessed December 7, 2021.

X. Huo, D. Chen, Y. He, W. Zhu, W. Zhou, J. Zhang, Bisphenol-A and female fertility a possible role of gene-environment interactions, Int. J. Environ. Health 12 (9) (2015) 11101–11116.

L. Jia, Y.F. Xu, Studies of ozone formation potentials for benzene and ethylbenzene using a smog chamber and model simulation, Hyan Jing KE Xue. Abstract only, https://pubmed.ncbi.nlm.nih.gov/24812939/, 2014. Accessed December 7, 2021.

L. Jin, X. Luo, P. Fu, X. Li, Airborne particulate matter pollution in China; a chemical mixture perspective from sources to impacts, Natl. Sci. Rev. 4 (4) (2017) 593–610.

H. Levine, N. Jorgensen, A. Martino-Andrade, J. Mendiola, D. Weksler-Derri, I. Mindlis, et al., Temporal trends I sperm count: a systematic review and meta-regression analysis, Hum. Reprod. Update 23 (6) (2017) 646–659.

J. Liu, Y. Dai, Y. Li, E. Yuan, Q. Wang, Y. Guan, A longitudinal study of semen quality among Chinese sperm donor candidates during the past 11 years, Sci. Rep. (2020), https://doi.org/10.1038/s41598-020-67707. Accessed December 7, 2021.

R. Lindsey, Climate change: global sea level, U.S. National Oceanographic and Atmospheric Administration, www.noaa.gov, 2020. Accessed December 7, 2021.

S.X.Z. Low, Z.Q. Aw, B.Z.L. Loo, K.C. Lee, J.S.H. Oon, C.H. Lee, M.H.T. Ling, Viability of Escherichia coli ATCC 8739 in nutrient broth, Luria-Bertani broth and brain heart infusion over 11 weeks, Electron. Phys. 5 (1) (2013) 576–581.

R.W. Macdonald, T. Harner, J. Fyfe, Recent climate change in the Arctic and its impact on contaminant pathways and interpretation of temporal tend data, Sci. Total Environ. 342 (1–3) (2005) 5–86.

Manomet, The end of Maple syrup? Climate change threatens Maine's iconic Maple sugar industry. https://www.manomet.com, 2018. Accessed December 7, 2021.

J. Meza, AP investigation finds traces of medicine in Philadelphia water, The Daily Pennsylvanian. Accessed December 7, 2021, April 8, 2008.

M. Miraglia, H.J.P. Marvin, G.A. Kleter, P. Battilani, C. Brera, F. Coni, et al., Climate change and food safety: an emerging issue with special focus on Europe, Food Chem. Toxicol. 47 (5) (2009) 1009–1021.

National Geographic, Cheetahs: on the brink of extinction, again. www.nationalgeographic.org/article/cheetahs-brink-extinction-again/, June 4, 2020. Accessed December 7, 2021.

National Park Service, Plants and climate change, 2021. https://www.nps.gov/articles/000/plants-climateimpact.htm#-text=Rising temperatures lead to more.them be less productive. Accessed August 31, 2022.

National Research Council, Climate Change: Evidence and Causes: Update 2020, The National Academic Press, Washington, DC, 2020, https://doi.org/10.17226/25733. Accessed December 7, 2021.

NEHA, Musca domestica. Transmitted diseases, Vectors, https://www.neha.org/vector/common-housefly, 2021. Accessed December 7, 2021.

NOAA, Hurricane Harvey & its impacts on southeast Texas (August 25–29, 2017). https://www.weather.gov/hgx/hurricaneharvey, 2017.

NOAA, How does climate change affect coral reefs?. https://oceanservice.noaa.gov/facts/coralreef-climate, 2021. Accessed December 7, 2021.

B. Panko, The world's parasites are going extinct. Here's why that's a bad thing, Smithsonian Magazine, https://smithsomianmag.com/science-nature/parasites-are-going-extinct-heres-why-thats-a-bad-thing-180964808/, September, 2017. Accessed December 7, 2021.

F. Parsons, P.R. Wood, J. DeMarco, Transformations of tetrachloroethylene and trichloroethylene in microcosms and ground water, J. AWWA 76 (1984) 56–59.

D.T. Patterson, J.K. Westbrook, R.J.V. Joyce, P.D. Lingren, J. Rogasik, Weeds, insects and disease, Clim. Change 43 (1999) 711–727.

J.A. Pitts, The Human Factor, U.S. National Aeronautics and Space Administration publication no. 85-215-26, 1985.

PRB, Why is the U.S., birth rate declining?, Population Research Bureau; 75(1), https://www.prb.org/resources/why-is-the-u-s-birth-rate-declining/, 2020. Accessed December 7, 2021.

D.L. Preston, P. Johnson, Ecological consequences of parasitism, Nat. Educ. Knowl. 3 (10) (2010) 47.

T. Pultarova, Eating fast food may affect a woman's fertility, CBS News, https://www.cbsnews.com/news/eating-fast-food-may-affect-womens-fertility-study-finds/, May 4, 2018. Accessed December 7, 2021.

A. Rodriguez-Verdugo, N. Lozano-Huntelman, M. Cruz-Lova, V. Savage, P. Yeh, Compounding effects of climate warming and antibiotic resistance, Science (2020), https://doi.org/10.1016/j.sci.2020.101024. Accessed December 7, 2021.

J. Roos, R. Hopkins, A. Kvarnheden, C. Dixelius, The impact of global warming on plant diseases and insect vectors in Sweden, Eur. J. Plant Pathol. 129 (1) (2011) 9–19.

Science Direct Topics, Mycotoxins. https://www.sciencedirect.com/topics/agricultural-and-biological-sciences/mycotoxins, 2021. Accessed December 7, 2021.

SLU, Climate change and pesticides. https://www.slu.se/en/Collaborative-Centres-and-Projects/SLU-centre-for-pesticides-in-the-environment/information-about-pesticides-in-the-environment/, 2021. Accessed December 7, 2021.

R.P. Smith, Male fertility is declining – studies show that environmental toxins could be a reason, The Conversation, https://theconversation.com/male-fertility-is-declining-studies-show-that-environmental-toxins-could-be-a-reason-163795, 2021.

H. Stanke, A.O. Finley, G.M. Domke, A.S. Weed, D.W. MacFarlane, Over half of western United States' most abundant tree species in decline, Nat. Commun. (2021), https://doi.org/10.1038/s41467-020-20678-z. Accessed December 7, 2021.

Stockholm Convention, The 12 initial POPs under the Stockholm Convention. http://chm.pops.int/TheConvention/The POPs/The12InitialPOPs/tabid/296/Default.aspx, 2019. Accessed December 7, 2021.

UN, Persistent organic pollutants (POPs) and pesticides. https://www.unep.org/cep/pesistent-organic-pollutants-pops-and-peticides, 1995. Accessed December 7, 2021.

UN Habitat, Urban energy. www.unhabitat.org/topic/energy, 2014. Accessed December 8, 2021.

UN, UN list of identified endocrine disrupting chemicals. https://www.chemsafetypro.com/Topics/Restriction/UN_list_identifed_endocrine_disrupting_chemicals-EDCs.html, 2018. Accessed December 7, 2021.

UN, World population projected to reach 9.8 billion in 2050 and 11.2 billion in 2100. https://www.un.org/en/desa/world-population-projected-reach-98-billion-2050-and112-billion-2100, 2021. Accessed December 7, 2021.

USDA, Food Waste FAQs, U.S. Department of Agriculture, Washington, DC, 2021. https://www.usda.gov/foodwaste/faqs. Accessed December 7, 2021.

E. Weiderpass, Air Pollution as a Major Risk Factor for Cancer, World Health Organization. International Agency for Research on Cancer, Lyon, France, 2016.

K. Weir, Climate change is threatening mental health, J. Am. Psychol. Assn. (2016). https://www.apa.org/monitor/2016/07-08/climate-change. Accessed December 7, 2021.

S.R. Whiten, R.K.D. Peterson, The influence of ambient temperature on the susceptibility of Aedes aegypti (Diptera: Culicidae) to the pyrethroid insecticide permethrin, J. Med. Entomol. 53 (1) (2016) 139–143.

WHO, Ambient Air Pollution: A Global Assessment of Exposure and Burden of Disease, World Health Organization, 2016. ISBN 9789241511353.

WHO, Climate change. https://www.who.int/health-topics/climate-change#tab=tab_1, 2021. Accessed December 7, 2021.

Index

Note: 'Page numbers followed by *f* indicate figures and *t* indicate tables.'

A

Acetaminophen (APAP), 72
Acquired Immune Deficiency Sydrome (AIDS), 264
Acupuncture, 451
Acute inflammation, 101
Acute lymphoblastic leukemia (ALL), 273
Acute myeloid leukemia (AML), 273
Additivity, 18
Adenosine triphosphate (ATP), 211
Adequate sleep, 137
Adipocytes, 242
Adipose tissue (AT), 239–244
  and persistent organic pollutants, 242–243
    as lipotoxins, 244
    as obesogens, 244
    release into body, 243–244
Adverse drug reactions (ADRs), 143–146, 376
  effects, 145–146
Aerosols, 84
Age, 319
  age-related disease, 168–170
    cancer, 170
    cardiovascular disease, 168–169
    immune system impact, 169
    neurodegenerative disease, 169
    psychological stress, 169
    type 2 diabetes, 169
  OSI condensed questionnaire, 392
Aging, 103–106, 159, 307–308, 311
  age-related disease and oxidative stress, 168–170
  clinical conditions associated with, 167
  hallmarks of, 170–173, 170t, 229–234, 230t
  lowering age of disease onset, 168
Agricultural chemicals, 53. *See also* Water polluting chemicals
  agricultural pollutants in U. S. drinking water, 54t–56t
Air pollutants, 419
Air pollution (AP), 102, 336, 433, 460–461
  ambient air pollutants, 37–44
  confined air pollutants, 44
  and oxidative stress, 44–45
    toxic chemicals in air, 38t–44t
Air Quality Index (AQI), 394
Air Quality Toxicity Index (AQTI), 394–395
Airborne allergens, 336
Airway inflammation, 339
Alcohol, 377
  alcohol metabolism, 73–74
  consumption, 82
  effect
    on nutrients and pharmaceuticals, 72
    on toxicities of chemicals, 72–73
  metabolism, 73–74, 347
  and oxidative stress, 74
Alcoholic liver disease (ALD), 347
Allergic sensitizations, 336
Allowable chemical impurities in food, 115–116
Altered cellular communication (ACC), 173
Alzheimer's disease (AD), 31–32, 106, 157, 159, 168, 215, 291, 363, 401
  co-morbidity data, 294
  and DRR, 402
  hallmarks of, 291
  late onset, 402
  likelihood ratio, 406
  mechanisms of Alzheimer's disease onset, 294–295
  onset odds, 411
  onset prediction, 406
    AD modified version of OSI condensed form, 407t
    supplemental AD associated parameters questionnaire, 408t–411t
  OSI and, 406–411
  and oxidative stress, 401
  parameters known to increase likelihood of AD onset, 402
  prevention, 411–413
  questionnaire use, 413
  risk factors for, 292–293, 292t–293t
  strengths and limitations, 413–414

Alzheimer's Questionnaire, 413, 419
Ambient air pollutants, 37–44
Ambient temperature extremes, 137
American Conference of Governmental Industrial Hygienists (AGCIH), 16
Amino acid sensing mechanisms, 232
Amyotrophic lateral sclerosis (ALS), 106, 162, 451
Anabolic agents, 112
Androgen receptors (ARs), 269–270
Animal extinction, 465
Animal fur and dander, 33–34
Animal ingestion, 112
Antagonism, 18
Anti-citrullinated protein antibodies (ACPAs), 322
Anti-cyclic citrallinated peptides (antiCCP), 317
Antibiotics, 112, 337
Antimicrobial agents, 115
Antioxidants (AO), 220, 444–445
  categories, 444–445
    endogenous, 445t
  and disease prevention, 445
    disease prevalence impacts lowering effects of specific antioxidants, 446t
  function, 444
Aromatherapy, 451
Artificial colors, 116–119
Asbestos, 30–31
Aspartame, 120
Asthma, 453, 466
  biomarkers of, 334–335
  cross sensitization, 339
  mechanisms of asthma and oxidative stress, 339–340
  phenotypes of, 329–334
  risk factors for, 335–338
    air pollution, 336
    antibiotics, 337
    atopy and allergic sensitizations, 336
    diet, 337
    epigenetics, 336
    ethnicity, 335
    family history, 335
    gender, 335
    genetics, 336
    microbial respiratory infections, 337
    obesity, 337
    occupational exposure, 337–338
    tobacco smoke exposure, 336
  symptom triggers, 329
Atherosclerosis (ATS), 285
  hallmark of, 285
  mechanism of onset, 286–287
  oxidative stress and, 287
  progression, 285
  risk factors, 285, 288t
Atmospheric energy, 460
Atopy, 336
ATR-X syndrome, 161
Attention deficit hyperactivity disorder (ADHD), 116, 203
Autism spectrum disorders, 177
Autoimmune inflammatory disease, 317
Azo dyes, 8

B

Bacterial agents, 102
Bacterial infections, 337
Benzene, 94
  toxicity, 72
1,4-benzenediol (BD), 94
1,2,4-benzenetriol (BTL), 94
Biking, 452
Biological effects, 87–88
Biomarkers, 184, 354
Bipolar disorder, 157
Bisphenol-A (BPA), 304
Bleaching agents, 114
Blood eosinophil count, 335
Body mass index (BMI), 239, 320, 364, 373, 392
Bone loss, 323
Breast cancer, 266–268
Bronchial hyper-responsiveness, 329
Butylated hydroxyl anisole (BHA), 115, 204
Butylated hydroxyl toluene (BHT), 115, 204

C

C-reactive protein (CRP), 318
Caffeine, 72
Calorie restriction, 232
Cancer(s), 71, 77–78, 102–103, 159, 168, 170, 244
  cancer-causing chemicals, 8
  cancer-related inflammation, 254–255
  initiation promotion and progression, 256–258
  metastasis, 255
  progression, 255
    breast cancer, 266–268
    colorectal cancer, 270–271
    exogenous carcinogens, 258–260
    food and, 260
    hallmarks, 252–256
    initiation promotion and progression, 256–258
    intristic risks, 251
    kidney cancer, 276–277
    leukemia, 273–275
    lung cancer, 262–266
    mechanisms associated with specific cancers, 262–277

melanoma, 271–273
metals, metalloids and, 260
metastasis, 261–262
non-Hodgkin lymphoma, 275
non-instristic risks, 251–252
prostate cancer, 269–270
Carbon dioxide, 464
Cardiac antioxidant defense system, 83
Cardiotoxicity, 89
Cardiovascular diseases (CVD), 21, 45, 159, 168–169, 188, 285, 363
Cardiovascular systems, 154
Cartilage damage, 323
β-cell
dysfunction, 308
functionAbsorption, 3, 307
of toxins, 464
Cells, 211
Cellular senescence (CS), 172, 233
Centers for Disease Control (CDC), 137
Centers for Disease Control and Prevention, 239–240
Central nervous system (CNS), 72
Chamber process, 7
Charlson Comorbidity Index, 413, 419
Chelating agents, 114
Chemicals
characteristics, 10–11
chemically caused disease and oxidative stress, 26
diseases caused by, 24–25
exposure as cause of T2D, 301–307
lipophilic chemicals to cause T2D, 304–307
type 2 diabetes clusters, 303–304
in food packaging, 121–127
historical perspective, 7–10
hydrophilic organic compounds, 11–12
lipophile, 21
lipophilic organic chemicals, 11
metals, 13
mixtures, 7, 15–16, 21, 93, 162
additivity, 18
antagonism, 18
potentiation, 18–19
synergism, 19
nonmetallic inorganic chemicals, 14
reactions in groundwater, 56–63
sequential absorption, 21–24
solvents and pollutants, 348–349
toxicity, 241–244
adipose tissue and persistent organic pollutants, 242–243
POPs as lipotoxins, 244
POPs as obesogens, 244
POPs release into body, 243–244
traditional toxicology, 16–17
unanticipated effects of mixtures, 19–20
in water, 63
Chinese herbal medicine, 451
Chloralkali process, 8
Chlorination, 64–65
Chlortetracycline, 91
Cholesterol, 287
Chromasomal instability (CIN), 271
Chromatin remodeling, 161
factors, 231
Chronic chemical exposure, 301
Chronic diseases, 144, 153, 229
Chronic fatigue syndrome, 363, 453
Chronic inflammation, 102–106, 167
aging, 103–106
cancer, 102–103
chronic non-infectious diseases, 102
chronic trauma, 106
non-healing wounds, 106
persistent infection, 102
Chronic inflammatory skin disease, 45
Chronic kidney disease (CKD), 353
causes of, 353
hallmarks of, 353–354
and oxidative stress, mechanisms of, 354–356
risk factors for, 354
Chronic lymphocytic leukemia (CLL), 273
Chronic metabolic Disease, 299
Chronic myeloid leukemia (CML), 273
Chronic non-infectious diseases, 102
Chronic obstructive pulmonary disease (COPD), 83, 188, 229
Chronic OS, 431
Chronic psychological stress, 154, 396–397
Chronic sleep deprivation, 139
Chronic stress, 154–157
Chronic trauma, 106
Chronic traumatic encephalopathy (CTE), 106
Chronological aging, 167
Cigarettes, 77
Circadian cycle interruption, 139
Circadian rhythms, 139
Cirrhosis, 345
Clear cell renal cell carcinoma (ccRCC), 276
Climate change, 457–458
Clinical Risk Score Calculation (CRSC), 435–436
comparison of OSI and, 437
variables known to elevate OS, 435t–436t
Clusters, 24, 303–304
Coal ash. *See* Coal combustion residuals (CCR)
Coal combustion residuals (CCR), 422, 424–425
Cocaine abuse, 73

Coffin-Lowry syndrome, 161
Cognitive biomarkers, 291
Cognitive impairment, 291
Colorectal cancer (CRC), 270–271
Comorbidity/comorbidities, 240–241, 359–362, 431
  representative list of obesity co-morbidities, 241t
Confined air pollutants, 44
  sources, 45t
Contaminated food, 464
Coral reef destruction, 466
Corn, 449
Coronary heart disease (CHD), 83
Cortisol, 223–224
COVID-19, 45, 431
  as cause of oxidative stress, 437–438
  oxidative stress, 431–433
    index, 433–435
  validation of OSI use in predicting COVID-19 severity, 435–437
    comparison of OSI and CRSC, 437
    CRSC, 435–436
CpG island methylator phenotype (CIMP), 271
Crop diseases, increased numbers of, 462
Cross sensitization, 339
Crystalline silica (CS), 30, 425
Cytochrome P450 2E1 (CYP2E1), 72–74
Cytochrome P450, 349
Cytokine factors, 30–31

D

Dairy farmers, 83
Dancing, 452
Dementia pugilistica. *See* Chronic traumatic encephalopathy
Dense nonaqueous phase liquid (DNAPL), 66–67
Depression, 203
Deregulated nutrient sensing, 171–172
Detoxification, 112
Detrimental lifestyle, 393
Diabetes, 169, 203, 244, 308
  aging, 311
  epigenetics, 311
  obesity rates, 311
  projection, 311
1,1-dichloro ethylene (1,1-DCE), 56–58
Diesel exhaust particles (DEPs), 215
Diet, 263, 321, 337, 377, 445–449
  health benefits attributed to MEDD, 447t
  MEDD recommended foods to eat and those to avoid, 447t
  natural antioxidants containing foods in MEDD, 446t
  phytochemical-containing compounds and foods, 448t
  regulatory steps ensure food safety, 448t
Dietary choices, 128–131
  animal protein, 130
  carbohydrates, 129–130
  fats, 130–131
  nitrates, nitrites and nitrosamines, 130
  salt, 131
Disease(s), 101, 154–157, 159, 195, 375, 457
  associated with oxidative stress, 184–187, 185t–187t
  comorbidities, 359–362
    diseases associated with oxidative stress, 184–187, 185t–187t
    multimorbidity, 188–190
    oxidative stress, common thread, 362–364
    systems, organs, oxidative stress and disease, 184
  factors elevate, 196t
  impact as function of exposure time, 422–428
    fly ash toxicity, 424–428
    PCB exposure in schools, 422–424
  increase, 463
    antibiotic resistance, 463
    increased pollen counts, 463
    mold growth, 463
    spread of infectious diseases, 463
  onset
    lowering age of, 168
    prediction, 206, 371
  and oxidative stress, 195
  prevention
    antioxidants, 444–445
    causes of oxidative stress overproduction, 444t
    diet, 445–449
    exercise, 452–453
    inequalities, 453–454
    oxidative stress overproduction control, 443
    psychological stress, 449–452
    strategies, 203–205
    treating illness, 453
    treating prevalent illness, 449
  rate synergism, 78–83
    lung cancer, 78–83
    noncarcinogenic synergism, 83
Disinfection by-products (DBPs), 64–65
Distal metastasis, 256
DNA
  damage, 30, 87, 231
  fragmentation, 90
  free radical reactions with, 217
  lesions, 169
  methylation, 161, 231, 268
  mutations, 163
Dose response relationships (DRRs), 16, 402
Doxycycline, 91

Drinking water, 64, 206
Drugs, 347–348
Dust, 33–34
  inhalation, 320

E

e-cigarettes. *See* Electronic cigarettes
Electromagnetic radiation (EMR), 87
  electromagnetic spectrum, 87–88
  ionizing radiation, 88–89
    and chemical mixtures, 89
  nonionizing radiation, 92–95
  radiation and oxidative stress, 95–96
  ultraviolet radiation, 89–90
    and toxic chemical mixtures, 90–92
  visible light radiation, 95
Electromagnetic spectrum, 87–88
Electronic cigarettes, 84–85
  chemicals contained in electronic cigarettes, 84t
Elevated oxidative stress, 177
Empirical data analysis, 377
Emulsification, 114
Endocrine disrupting chemicals (EDCs), 468
Endogenous antioxidants, 444–445
Endogenous chemical species, 150
Endogenous oxidative stress, 340
Endogenous steroids, 112
Endothelial dysfunction, 286
Environmental chamber use, 451
Environmental chemicals, 345
Environmental disease, 177
Environmental magnetic fields, 94
Environmental pollutants, 419
Environmental Protection Agency (EPA), 37, 394
  National Air Toxics assessment, 37
Environmental tobacco smoke (ETS). *See* Passive smoking; Tobacco smoking
Enzymatic systems, 212
Enzymes, 245
Epidemiological studies, 337
Epigenetics, 161–162, 311, 336
  alterations, 171, 231–232
  chemical environmental and factors in epigenetic effects, 162
  oxidative stress in genetic and, 163–164
Esthetic and storage additives, 121
Ethanol, 90
Ethnicity, 320, 335
Ethylene diaminete-traacetic acid (EDTA), 66, 114
Everything Added to Food in United States (EUFUS), 116
Excess ROS production, 355
Excipients in pharmaceuticals, 146
  pharmaceutical excipients and adverse effects, 149t–150t
  representative list of excipients FDA, 147t–148t
Exercise, 452–453
Exogenous antioxidants, 444–445
Exogenous carcinogens, 258–260
  mixtures, 258–259
  resonance stabilization, 260
  sequential absorption, 259–260
  total lipophilic load, 259
Exogenous chemicals, 220, 348–349
  species, 150
Exogenous lipophilic chemicals, 252
Exogenous oxidative stress, 340
Exogenous steroids, 112
Exposure to secondhand smoke. *See* Passive smoking
Extracellular amyloid-β plaques, 291, 294
Extracellular matrix (ECM), 323
Extreme weather, 460
  loss of life due to, 468
Extremely low frequency waves (ELF), 92

F

Fasting glucose level (FGL), 299
Fatty acids (FAs), 219
Fatty foods, 243
Federal Food, Drug and Cosmetic Act of 1938 (FD&C), 115
Fenton reaction, 215–217
Fertility, 65
  decline in, 468–469
Fertilizers, 111
Fibers
  dust, 33–34
  oxidative stress, 29–33
  sources and content, particle and, 29
    airborne and water carried particles and fibers, 31t
Fibroblast-like synoviocytes (FLSs), 322–323
Fibromyalgia syndrome (FM), 120
Fish migration, 466
Flavor enhancers, 120–121
Fly ash, 33
  toxicity, 424–428
    health effects commonly associated with chronic exposure to CCR, 425t
    metals contained in CCR, 426t
    OSI condensed for supplement for predicting time of onset for diseases, 427t
Food, 260
  allowable chemical impurities in, 115–116
  animal ingestion, 112
  artificial colors, 116–119
  chain, 65, 111

Food (*Continued*)
chemicals
in food packaging, 121–127
impurities in, 115
and oxidative stress, 128
preservatives in, 128
dietary choices and oxidative stress, 128–131
esthetic and storage additives, 121
flavor enhancers, 120–121
irradiated, 127–128
mercury in, 113–114
persistent organic pollutants, 112–113
preparation, 114–115
quality decline, 464
uptake from soil and plant surfaces, 111
volatile organic compounds in, 121
Food and Drug Administration (FDA), 78, 115, 146
Fractional exhaled nitric oxide (FeNO), 334–335
Free radicals, 3
antioxidants, 220
diseases associated with, 220
Fenton reaction, 215–217
hypothalamus-pituitary-adrenal axis, 223
immune system and, 220–223
prevention, 444
reactions
with DNA, 217
with lipids, 219
with proteins, 217
signaling, 220
stability, 212–215

G

Gender, 319, 335
Genetics, 159, 263, 320, 336
chemical environmental and factors in epigenetic effects, 162
factors, 354
manipulations, 233
OSI, 375
condensed questionnaire, 393
oxidative stress in genetic and epigenetic effects, 163–164
Genomic instability, 170–171, 229–231
Glioblastoma cells (GMB), 89
Global warming, 457–467
absorption of toxins, 464
air pollution, 461
atmospheric and oceanic energy, 460
contaminated food, 464
disease increase, 463
energy use, 459
food quality decline, 464
increased pesticide use, 462–463
increased stress, 465
industrialization and air pollution, 460
migration, 459
population, 459
species extinction, 465–467
urbanization, 459
water pollution, 461–462
Glutathione (GHS), 74
Glutcocorticoid hormone, 223–224
Gut microbiotica, 321

H

Haber process, 8
Hallmarks, 254–256
additional cancer hallmarks, 256
of aging, 170–173, 229–234
altered cellular communication, 173
altered intercellular communication, 234
cellular senescence, 172, 233
deregulated nutrient sensing, 171–172, 232
epigenetic alterations, 171, 231–232
genomic instability, 170–171, 229–231
loss of proteostasis, 171, 232
mitochondrial dysfunction, 172, 232–233
stem cell exhaustion, 172, 233–234
telomere attrition, 171
telomere shortening, 231
of Alzheimer's disease, 291
biomarkers, 292t
of atherosclerosis, 285
of cancer, 252–254
emerging hallmarks, 253–254
enabling characteristics, 254
cancer hallmark regulation, 256
cancer-related inflammation, 254–255
of chronic kidney disease, 353–354
systemic hallmarks, 255–256
distal metastasis, 256
global inflammation, 255
immunity inhibition, 255
metabolic changes to cachexia, 255
primary tumor-metastasis network, 255
propensisty to thrombosis, 255
Haloacetic acids (HAAs), 64–65
Haloacetonitriles (HANs), 64–65
Hazard number (HN), 395
Heavy metals, 31–32, 66, 73
Heavy rains and floods, 464
Hepatitis C, 350–351
Hepatitis C virus (HCV), 350
Herpes Zoster virus, 363
Heterocyclic aromatic amines (HAAs), 260

Heterogenous disease, 335
High density lipoproteins (HDL), 299, 356
Histone modification, 161
Human culture, 111
Human extinction, 467–469
decline in fertility, 468–469
loss of life due to extreme weather, 468
lower life expectancy, 467
Huntington's disease, 162
Hydrogen peroxide (H_2O_2), 74
Hydrophile mixtures, 21
Hydrophilic organic compounds, 11–12
Hydroxyl radical, 215
4-hydroxynonelal (4-HNE), 295
Hyperglycemia, 299
Hypertension, 168–169, 353, 363, 453
Hypnosis, 451–452
Hypothalamus-pituitary-adrenal axis (HPA axis), 223

I
Illness, 453
Immediately dangerous to life or health (IDLH), 17
Immune systems, 67, 106, 154, 184, 254, 350
and free radicals, 220–223
autoimmune disease, 223
endogenous immune system free radicals, 223
exogenous immune compromising agents, 222
function, 103
impact, 169
Immunity inhibition, 255
Immunoglobin E (IgE), 335
Impaired antioxidant defense system, 356
Inactive ingredients, 143
Indoor air polluted with chemicals, 37
Industrial chemicals, 53
industrial pollutants in U. S. drinking water, 59t–63t
Industrialization pollution, 460
Inequalities, 453–454
Infectious diseases, 187
spread of, 463
Inflammation, 287, 308
acute, 101
chronic, 102–106
and oxidative stress, 106–107
Inflammatory cells, 254
Inflammatory theory, 323
Inhalation, 340
Inorganic mercury, 113
Insects
extinction, 465–466
increased numbers of, 462
Insulin resistance, 305
Intercellular communication, 234
Interleukin-13 (IL-13), 89
Interleukin-17 (IL-17), 335
Interleukin-4 (IL-4), 334
International Agency for Research on Cancer (IARC), 264
Ionizing radiation (IR), 88–89. *See also* Nonionizing radiation
and chemical mixtures, 89
Irritable bowel syndrome, 453, 466

J
Jogging, 452
Juberg-Marsidi syndrome, 161

K
Kidneys, 353
cancer, 276–277
Kupffer cells, 349

L
Laboratory testing, 436
Leukemia, 273–275
Lifestyle
OSI, 377
condensed questionnaire, 391
Lipids, 229
free radical reactions with, 219
lipid-laden foam cells, 285
peroxidation, 219
Lipolytic processes, 242
Lipophiles, 21, 301–303
mixtures, 21
Lipophilic carcinogenic chemicals, 258
Lipophilic chemicals, 360
to cause T2D, 304–307
pharmaceuticals, 306–307
plastic exudates, 304–305
polluted air, 305
POPs, 304
tobacco smoke, 306
Lipophilic organic chemicals, 11
K_{ow} values for homologous series of n-alcohols, 11t
Liver cirrhosis, 345
liver fibroproliferative diseases, 345–351
Liver damage, 350
Liver disease, 351
Liver fibroproliferative diseases, 345–351
alcoholic liver disease, 347
chemical solvents and pollutants, 348–349
drugs, 347–348
hepatitis C, 350–351
nonalcoholic fatty liver disease, 349–350
oxidative stress and liver disease, 351

Liver fibrosis, 345
Liver proliferative diseases, 347
Low density inflammation, 362
Low density lipoproteins (LDL), 355
Low molecular weight hydrocarbons (LMWHCs), 11, 359
Low-density lipoprotein (LDL), 287, 299
Lower life expectancy, 467
Lung cancer, 78–83, 262–266
  causes of, 263–266
    biomass, wood and coal smoke, 264
    genetics, 263
    infections, 264
    obesity, 264
    outdoor air pollution, 264–266
    tobacco smoking, 263
  occupational chemical exposure, 266
  radon, 266
Lyme disease, 463
Lymphocytes, 242

M

Macrophage, 287
Malondialdehyde (MDA), 184, 295
  as indicator of oxidative stress, 195–199
  measurement, 371
Marine species, 33
Matrix metalloproteinase-1 (MMP-1), 95
Maturation stage of rheumatoid arthritis onset, 322
Medications, OSI condensed questionnaire, 393
Meditation, 451
Mediterranean diet (MEDD), 337, 445
Melanocytes, 272
Melanoma, 271–273
  vitamin A, 272
  vitamin C, 272
  vitamin D, 272
  vitamin E, 273
  vitamin K, 273
Mercury, 113–114
Mesothelioma, 263
Metabolic changes to cachexia, 255
Metabolic diseases, 159
Metabolic syndrome, 203, 363
Metalloids, 260
Metals, 13, 260, 425
Metastasis, 253, 261–262
2-methoxyethanol (2-ME), 93
Methyl ethyl ketone (MEK), 116
Methyl isobutyl ketone (MIBK), 17
Methyl mercury (MeHg), 113
Methylparaben (MP), 92
Microbeads, 33
Microbial respiratory infections, 337
Microsatellite instability (MSI), 271
Milligrams per cubic meter of air (MPCM), 17
Minamata disease, 114
Minimum observed effect level (MOEL), 16
Minimum-Observed-Adverse-Effect-Level (MOAEL), 19
Mitochondria, 163–164, 211–212
Mitochondrial damage, 74
Mitochondrial dysfunction, 172, 232–233, 287, 339, 349
Mitochondrial substrate load, 245
Mitomycin C (MMC), 93
Mixtures, 258–259
Modified OSI, 419–420
Mold growth, 463
Monosodium glutamate (MSG), 115, 120
Mood biomarkers, 291
Mucin, 83
Mucosal inflammation, 321
Multi-morbid diseases, 177
Multimorbidity, 188–190, 202–203
Multiple chemical sensitivity, 453
Multiple sclerosis (MS), 466
Musculoskeletal diseases, 159, 363
Mutagenesis, 93

N

N,N-dimethyl-m-toluamide (DEET), 73
N-methyl-D-aspartate (NMDA), 120
N-methyl-N′-nitro-N-nitrosoguanidine (MNNG), 94
N-nitrosamines (NNSAs), 263
National Cancer Institute, 249
National Institute of Occupational Safety and Health (NIOSH), 16
Neurodegenerative diseases, 159, 169, 363
Neurofibrillary tangles, 295
Neuroimmunology, 156–157
Neuroinflammation, 295
Neurological disease, 157
Neurotoxic Aβ oligomer peptides, 295
Nicotine amide dinucleotide phosphate (NAD(P)H), 355
Nitric oxide (NO), 335
  synthase, 212
4-nitroquinoline-1-oxide (4NQO), 93
Nitrous oxide (NO), 286
No observed adverse effect level (NOAEL), 16, 19
No observed effect concentration (NOEC), 16
No observed effect level (NOEL), 16, 18
Non communicative disease (NCD), 359
Non-coding RNAs (ncRNAs), 268
Non-healing wounds, 106

Non-Hodgkin lymphoma (NHL), 275
Nonalcoholic fatty liver disease (NAFLD), 349–350
Noncarcinogenic synergism, 83
Noninfectious diseases, 203
Nonionizing radiation, 92–95
  ELF radiation and chemical mixtures, 94–95
  RF radiation and chemical mixtures, 93
Nonmetallic inorganic chemicals, 14
Nutrients, 253
  effect of alcohol on, 72
  sensing, 232

O

Obesity, 205, 308, 320, 337, 363
  adipose tissue and chemical toxicity, 241–244
  biomarkers of, 240
  causes of, 240
  comorbidities, 240–241
  and oxidative stress, 244–245
    enzymes, 245
    factors, 245
    mitochondrial, 245
  psychological impact of, 245
  rates, 311
  statistics, 239–240
Occupational asthma (OA), 337
Occupational Safety and Health Administration (OSHA), 16
Oceanic energy, 460
  extreme weather, 460
Octanol, 10–11
Oil disease, 303
Onset of disease, 161
Organic chemicals, 10
Organic compounds, 45
Organic mercury pollution, 114
Organic synthesis, 7
Organo-chlorine pesticides (OCs), 11, 301
Organs, 184
∑OSI/TPS method, 420–422
Osteoarthritis (OA), 317
Oxalic acid, 7
Oxidation, 217
Oxidative stress (OS), 3–4, 29–33, 47, 74, 84, 101, 106–107, 150, 154, 163, 168–170, 184, 195, 272, 287, 294, 308, 339–340, 345, 351, 354, 362–364, 401, 419, 431–433, 435, 457–467
  absorption of toxins, 464
  additivity, 202
  air pollution, 44–45, 461
    toxic chemicals in air, 38t–44t
  ambient air pollutants, 37–44
  asbestos, 30–31
  atmospheric and oceanic energy, 460
  biomarker serum MDA, 371
  cancer, 170
  cardiovascular disease, 168–169
  confined air pollutants, 44
  contaminated food, 464
  covid-19 as cause of, 437–438
    body systems impacted long-term following acute COVID-19, 438t
  crystalline silica, 30
  and disease, 184
  disease increase, 463
  energy use, 459
  exogenous and endogenous causes, 4t
  factors elevate oxidative stress, 196t
  fly ash, 33
  food quality decline, 464
  in genetic and epigenetic effects, 163–164
  heavy metals, 31–32
  immune system
    impact, 169
    responses to inflammation, 107t
  increased pesticide use, 462–463
  increased stress, 465
  industrialization and air pollution, 460
  levels, 449
  limits to oxidative stress-mediated disease prevention, 205–206
  malondialdehyde as indicator of, 195–199
  measurement, 3–4
  mechanisms of asthma and, 339–340
    endogenous oxidative stress, 340
    exogenous oxidative stress, 340
  mechanisms of chronic kidney disease and, 354–356
    excess ROS production, 355
    impaired antioxidant defense system, 356
  migration, 459
  neurodegenerative disease, 169
  OS-caused disease, 4
  OS-elevating agent, 202
  overproduction control, 443
  oxidative stress-elevating factors shown to increase COVID-19 severity, 432t
  oxidative stress-mediated chronic disease, 239
  plastics, 33
  from polluted water and soil, 67
  population, 459
  psychological stress, 157, 169
  radionuclides, 32
  species extinction, 465–467
  total OS, 188
  type 2 diabetes, 169
  urbanization, 459

Oxidative stress (OS) (*Continued*)
water pollution, 461–462
Oxidative Stress Index (OSI), 5, 371, 389, 391, 401, 419, 433
and AD, 406–411
AD onset odds, 411
predicting AD onset, 406
primary AD-causing parameters, 411
additional applications, 389
age, 392
basis, 373
chronic psychological stress, 396–397
comparison of CRSC and, 437
condensed form, 434t
condensed OSI applications, 399
condensed OSI form, 397
OSI condensed form point assignments, 397t
OSI condensed questionnaire, 398t
OSI score and likelihood of disease onset, 397t
and COVID-19 severity, 433–435
education level, 393–394
eight components of condensed OSI Form, 391t–392t
genetics, 393
medications, 393
parameters raise oxidative stress and lead to likelihood of disease onset, 372t
place of residence, 394–395
annual air quality classification as shown by AQTI, 396t
AQTI calculation for three American cities, 396t
EPA Air Quality Index classifications, 395t
preexisting chronic conditions, 392–393
public health surveys
disease impact as function of exposure time, 422–428
toxicity as function of emission distance, 420–422
questionnaire, 373–377
diagnostic data, 376
diet, 377
genetics, 375
lifestyle, 377
pharmaceuticals regularly taken, 376
physical data, 373–375, 375t
prevalent diseases and conditions, 375
symptoms, 375, 376t
tobacco and alcohol use, 377
score and disease, 377
oxidative stress index questionnaire, 378t–389t
validation of OSI use in predicting COVID-19 severity, 435–437
weight, 392
Oxidative Stress Index air pollution severities (OSI–AP), 394–395
Oxidatively modified LDL particles (OxLDL), 287
Oxygen, 211, 253

P

P-aminobenzoic acid (PABA), 91
Parasite extinction, 466
Parental disease, 159
Parkinson's disease, 31–32, 106, 157, 159, 171, 215, 363
Part per billion (PPB), 17
Particles
dust, 33–34
and fiber sources and content, 29
airborne and water carried particles and fibers, 31t
oxidative stress, 29–33
Parts per million (PPM), 17
Passive smoking, 77
Pathogenesis, 263
Perchloroethylene (PCE), 56–58
Perfluorinated substances (PFAS), 67
Perfluorooctane sulfonate (PFOS), 127
Perfluorooctanoic acid (PFOA), 127
Peridontal disease, 321
Periodontitis (PD), 321
Periostin, 335
Peripheral nervous system, 113
Permissible exposure level (PEL), 16
Persistent infection, 102
Persistent organic pollutants (POPs), 11, 111–113, 205, 241–242, 258, 301, 304, 359, 402
as lipotoxins, 244
mixture, 304
as obesogens, 244
release into body, 243–244
Pesticides, 53, 303, 462–463
decreased effectiveness, 463
increased evaporation of, 462
increased loss due to wind, 462
increased numbers of insects, crop diseases and weeds, 462
increased runoff after application, 462
Petrolatum, 92
Pets, 452
Phagocytes, 187
Pharmaceutical drugs (PDs), 144
Pharmaceuticals, 306–307, 376
adverse drug reactions, 144–146
contributing factors to elevation of ADR prevalence, 144t
effect of alcohol on, 72
excipients in, 146

and oxidative stress, 150
use of, 143–144
Phenotype, 329–334
connections, 362
factors, 354
Photosensitizers, 91
Phototoxicity, 90–91
Phthalates, 305
Physical disease, 153
Physiological function, 103
Physiological processes, 167
Phytochemical-containing compounds foods, 446
Plant absorption of soil toxins and bioacculation, 65–66
Plastic exudates, 304–305
Plastics, 33
Polluted air, 305
Polybrominated biphenyls (PBBs), 301
Polybrominated diphenyl ethers (PBDEs), 11, 113
Polychlorinated biphenyls (PCBs), 8, 11, 301, 422
exposure in schools, 422–424
buildings found to be contaminated with PCB, 424t
health effects attributable to PCB, 423t
OSI supplement for PCBs in school impact, 424t
Polychlorinated dibenzo-p-dioxins (PCDDs), 113
Polychlorinated dibenzofurans (PCDFs), 113
Polyethylene, 8
Polyethylene terephthalate (PET), 8, 127
Polymerization aids, 127
Polynuclear aromatic hydrocarbons (PAHs), 11, 66, 82, 258, 426
Polypharmacy, 146
Polyvinylchloride (PVC), 8, 304–305
Post-traumatic stress disorder (PTSD), 449
Potentiation, 18–19
Prairie grass, 66
Preexisting chronic conditions, OSI condensed questionnaire, 392–393
Preexisting disease, 359
Pretreatment, 72
Primary AD-causing parameters, 411
supplemental AD parameter questionnaire, 412t
Primary tumor-metastasis network, 255
Priority pollutants, 47–53
EPA 129 priority pollutants and their product sources, 49t–53t
Prostate cancer, 269–270
Protein kinase C (PKC), 308
Proteins, 229
damage, 87
free radical reactions with, 217
Proteostasis, 232
loss of, 171, 232
Psychiatric diseases, 157
Psychological impact of obesity, 245
Psychological stress, 101, 153, 169, 449–452
alternative stress therapies, 450–452
chronic stress and disease, 154–157
and oxidative stress, 157
reducing steps, 450t
Punch-drunk syndrome, 106

Q

Questionnaires, 413, 419
Quinoline Yellow (QY), 121

R

Race, 320
Radiation and oxidative stress, 95–96
Radioactive isotopes, 32
Radiofrequency waves (RF), 92
radiation, 93
waves, 92
Radionuclides, 32, 425
Rain water, 47
Reactive airways dysfunction syndrome (RADS), 20
Reactive nitrogen species (RNS), 3, 31–32, 130, 211, 229
Reactive oxygen species (ROS), 3, 31–32, 211, 229
effects of, 308
ROS-induced DNA damage, 170
Replicative DNA polymerases, 231
Residence, OSI condensed questionnaire, 394–395
Resonance stabilization, 260
Respiratory chain, 172
Respiratory diseases, 45, 159, 359–360
Respiratory viral infections, 321
Retinol, 72, 90
Rheumatoid arthritis (RA), 317
biomarkers of, 318
complications of, 318
mechanisms of rheumatoid arthritis onset, 322–323
onset
fulminant, 322–323
maturation, 322
mechanisms of, 322–323
targeting, 322
triggering, 322
risk factors for, 318–321
age, 319
diet, 321
dust inhalation, 320
gender, 319
genetics/familial history, 320
gut microbiotica and mucosal inflammation, 321

Rheumatoid arthritis (RA) (*Continued*)
history of live births, 319
obesity, 320
peridontal disease, 321
race/ethnicity, 320
smoking, 320
viral infections, 321
Risk factors, 285, 288t
for Alzheimer's disease, 292–293
for type 2 diabetes, 300–301

S

Semiquinone radicals, 215
Senescence-associated secretory phenotype (SASP), 172
Sensitization, 336
Sequential absorption, 21–24, 259–260
Serum malondialdehyde, 195, 199
Sewage treatment, 67
Short term exposure limit (STEL), 16
Sick building syndrome, 424
Skin cancer, 271
Sleep apnea, 138–139
Sleep deprivation
circadian cycle interruption, 139
diseases at elevated risk for onset in people with sleep apnea, 139t
and oxidative stress, 139–140
prevalent health conditions and lifestyle choices, 139
sleep apnea, 138–139
sleep deprivation and oxidative stress, 139–140
temperature extremes, 137
Sleep interruption, 137
Small airways, 329
Smoking, 287, 320
Socioeconomic status (SES), 393
Soil
formation, 65
and plant surfaces, 111
pollution, 65
chemical reactions in groundwater, 56–63
disinfection bye-products, 64–65
effects of mixtures, 66–67
oxidative stress from polluted water and soil, 67
plant absorption of soil toxins and bioacculation, 65–66
water polluting chemicals, 47–53
Species extinction, 465–467
animal extinction, 465
coral reef destruction, 466
fish migration, 466
insect extinction, 465–466
parasite extinction, 466
tree population loss, 466–467
Specific absorption rate (SAR), 93
Sputum eosinophils, 334
Stabilization, 224
Statistics, 239–240
Stem cell exhaustion, 172, 233–234
Stockholm Convention, 242–243
Stress, 467
therapies, 450–452
acupuncture, 451
aromatherapy, 451
Chinese herbal medicine, 451
environmental chamber use, 451
hypnosis, 451–452
meditation, 451
pets, 452
sweat lodge participation, 450
yoga, 451
triggers, 154
Styrofoam, 8
Superoxide dismutases (SOD), 350–351
Sweat lodge participation, 450
Swimming, 452
Synaptic impairment, 291
Synergism, 19, 87
Synergistic effects, 84, 89

T

Telomeres, 105
attrition, 171
shortening, 231
Temozolomide, 89
Temperature extremes, 137
CDC recommended hours of sleep by age, 138t
2,3,7,8-tetrachlorodibenzo-p-dioxin (TCDD), 21, 113, 303
Tetrahydrocannabinol (THC), 73
Threshold limit values (TLV), 16
Thrombosis, propensisty to, 255
Time weighted average (TWA), 16
Tobacco, 377
and cancer, 78
electronic cigarettes, 84–85
and oxidative stress, 84
smoke, 77–83, 259, 301, 306
and disease rate synergism, 78–83
exposure, 336
lung cancer, 78–83
noncarcinogenic synergism, 83
smoking, 263
toxicity, 77–78
Total lipophilic load, 244, 259
Total oxidative stress, 371

Total oxidative stress and disease
  disease onset prediction, 206
  disease prevention strategies, 203–205
  diseases and oxidative stress, 195
  limits to oxidative stress-mediated disease prevention, 205–206
  malondialdehyde as indicator of oxidative stress, 195–199
  multimorbidity, 202–203
  obesity, 205
  oxidative stress additivity, 202
Toxic chemicals, 162. *See also* Lipophilic chemicals
  release, 420
Toxicity, 77–78
  effect of alcohol on toxicities of chemicals, 72–73
  as function of emission distance, 420–422
    condensed OSI form, 421t
    OSI questionnaire supplement to assess oxidative stress, 422t
Toxicological data, 16–17
Toxins, absorption of, 464
Traditional toxicology, 16–17
  chemical impact, 16–17
  toxicological data, 16–17
Trans fatty acids (TFAs), 131
Trans-1,2-dichloroethylene (1,2-DCE), 56–58
Traumatic brain injury (TBI), 106
Tree population loss, 466–467
1,1,1-trichloroethane (TCA), 56–58
Trichloroethylene (TCE), 56–58, 66
Triglycerides, 242
Trihalomethanes (THMs), 64–65
Tumor necrosis factor (TNF), 349
Tylenol, 8
Type 2 diabetes (T2D), 169, 299, 359–360
  chemical exposure as cause of, 301–307
  clusters, 303–304
  complications, 308
  diabetes, 308
  effects of ROS, 308
  mechanisms of type 2 diabetes onset, 307–308
  and oxidative stress, 308
  risk factors for, 300–301, 302t
Type 2 inflammation (T2I), 334

U

U.S. Department of Agriculture (USDA), 115
Ultrafine particles, 29
Ultraviolet radiation (UV), 89–90
  radiation, 272
  and toxic chemical mixtures, 90–92
Urban runoff, 53
  water pollutants in U.S. drinking water, 57t–58t

V

Vapes. *See* Electronic cigarettes
Vicia faba, 93
Vinyl chloride (VC), 56–58
Viral agents, 102
Viral infections, 321, 337
Viruses, 321
Visible light radiation, 95
Vitamin A, 272
Vitamin C, 272
Vitamin D, 90, 272
Vitamin E, 273
Vitamin K, 273
Volatile organic compounds (VOCs), 66, 121
  in food, 121

W

Walking, 452
Wane, 453
Water partition coefficients, 10–11
Water polluting chemicals, 47–53
  agricultural chemicals, 53
  industrial chemicals, 53
  priority pollutants, 47–53
  urban runoff, 53
Water pollution, 461–462
  chemical reactions in groundwater, 56–63
  disinfection by-products, 64–65
  effects of mixtures, 66–67
  oxidative stress from polluted water and soil, 67
  plant absorption of soil toxins and bioacculation, 65–66
  water polluting chemicals, 47–53
Waterborne paints, 20
Wax, 453
Weeds, increased numbers of, 462
Weight, OSI condensed questionnaire, 392
White adipose tissue (WAT), 11, 205
White blood cell, 93
World Air Quality Project, 394
World Health Organization (WHO), 37, 239, 299, 350, 457–458

Y

Yoga, 451